GONG-SoHO24: A New Era of Seismology of the Sun and Solar-like Stars

Journal of Physics: Conference Series Volume 271

Aix-en-Provence, France
27 June – 2 July 2010

Editor:

Thierry Appourchaux

ISBN: 978-1-61782-657-3
ISSN: 1742-6588

Printed from e-media with permission by:

Curran Associates, Inc.
57 Morehouse Lane
Red Hook, NY 12571

Some format issues inherent in the e-media version may also appear in this print version.

Printed by Curran Associates, Inc. (2011)

For permission requests, please contact the Institute of Physics at the address below.

Institute of Physics
Dirac House, Temple Back
Bristol BS1 6BE UK

Phone: 44 1 17 929 7481
Fax: 44 1 17 920 0979

techtracking@iop.org

Additional copies of this publication are available from:

Curran Associates, Inc.
57 Morehouse Lane
Red Hook, NY 12571 USA
Phone: 845-758-0400
Fax: 845-758-2634
Email: curran@proceedings.com
Web: www.proceedings.com

TABLE OF CONTENTS

LOCAL HELIOSEISMOLOGY:

SOLAR DIAMETER, IRRADIANCE AND ACTIVITY:

SOLAR AND STELLAR MODELLING:

LOW DEGREE STELLAR SEISMOLOGY:

FIRST RESULTS FROM SPACE MISSIONS:

CONVECTION, DYNAMO AND FLOWS:

PROSPECTIVE:

FORWORD:

Thierry Appourchaux

Volume 271 (2011) of the *Journal of Physics: Conference Series* provides a record of the invited and contributed talks, and of the posters presented at the GONG2010–SoHO24 conference entitled 'A new era of seismology of the Sun and solar-like stars'. The conference was held from 27 June 2010 to 2 July 2010 in Aix-en-Provence, France. More than 120 scientists from all over the world attended the conference.

I would like to express my gratitude for the the financial support from the following organisations: Université Paris-Sud; the Centre National d'Etudes Spatiales (CNES); the Programme National des Relations Soleil–Terre (PNST) and the Programme National de Physique Stellaire (PNPS) (both programmes under the umbrella of the Institut National des Sciences de l'Univers, INSU); INSU of the Centre National de la Recherche Scientifique (CNRS); the SoHO project of the European Space Agency (ESA), and the Science Programme of ESA; the Global Oscillations Network Group (GONG); and finally the European Aeronautic Defence and Space Company (EADS).

The Scientific Organizing Committee comprised Thierry Appourchaux (chairman, Institut d'Astrophysique Spatiale, Orsay, France), Frank Hill (co-chairman, GONG / National Solar Observatory, Tucson, Arizona, United States), Annie Baglin (Observatoire de Paris-Meudon, France), William Chaplin (University of Birmingham, United Kingdom), Jørgen Christensen-Dalsgaard (Aarhus Universitet, Denmark), Thierry Corbard (Observatoire de la Côte d'Azur, Nice, France), Bernhard Fleck (European Space Agency), Laurent Gizon (Max-Planck-Institut für Sonnensystemforschung, Lindau, Germany), Travis Metcalfe (National Center for Atmospheric Research, Boulder, Colorado, United States), Michael Thompson (Sheffied University, United Kingdom; High Altitude Observatory, Boulder, Colorado, United States) and Jesper Schou (Stanford University, California, United States).

The Editorial Committee of these proceedings was composed of Thierry Appourchaux (chairman), Annie Baglin, William Chaplin, Jørgen Christensen-Dalsgaard, Laurent Gizon, Michael Thompson, Takashi Sekii (National Astronomical Observatory of Japan, Tokyo) and John Leibacher (IAS, Orsay, France; GONG / NSO, Tucson, Arizona, United States).

This volume consists of 86 articles organised in sections reflecting the scientific programme of the conference:

- 012083-012086 Prospective

All papers are freely accessible on the internet, in colour, at http://iopscience.iop.org/1742-6596/271/1, and an interactive picture of the conference is available in the attached PDF.

I am also grateful to the Local Organizing Committee for making this conference a success: Catherine Cougrand (secretary), Stéphane Caminade (web designer), Delphine Prival (administration) and Jean-Paul Rozet (logistics). I could not have done it without your help!

Last but not least, let me also thank the official photographer of the conference, Pierre Assus, for producing excellent photographs, including the group photo. Please feel free to send me an e-mail at Thierry.Appourchaux@ias.u-psud.fr if you would like copies of these pictures.

Thierry Appourchaux
Editor

Orsay, France
24 December 2010

IAS
orsay

cnrs
dépasser les frontières

UNIVERSITÉ
PARIS-SUD 11

esa
NASA
SOHO

Teide
Udaipur
Learmonth
Mauna Loa
Big Bear
Cerro Tololo
GONG

esa
EADS
astrium

NASA

Peer review statement:

All papers published in this volume of *Journal of Physics: Conference Series* have been peer reviewed through processes administered by the proceedings Editors. Reviews were conducted by expert referees to the professional and scientific standards expected of a proceedings journal published by IOP Publishing.

Solar and Heliospheric Observatory

Progress in sunspot helioseismology

A. C. Birch

NWRA, CoRA Division, 3380 Mitchell Lane, Boulder C0 80301 USA

E-mail: `aaronb@cora.nwra.com`

Abstract. Local helioseismology is a set of methods that are used to study wave propagation and infer physical conditions in the solar interior. Sunspots are a particularly challenging target for local helioseismology. In this review, I will show that some new methods (magnetoconvection simulations and numerical wave propagation simulations) lead to shallow sunspot models that are apparently inconsistent with traditional inferences from local helioseismology. In addition, I will show that inferences for the depth structure of moat flows are not in general agreement either.

1. Introduction

Local helioseismology is a collection of methods that are used to study the three dimensional structure and dynamics of the solar interior. All of these methods are based on the interpretation of measurements of solar oscillations. Recent reviews of local helioseismology are given by [1–5].

Sunspots are one of the most challenging targets for local helioseismology. Detailed reviews of this topic were given by [1] and [6]. Here, I will focus on very recent work on this topic. The two main questions about sunspots are the nature of their subsurface structure (i.e. the three-dimensional distribution of sound speed, magnetic field, density, etc.) and the associated subsurface flows. I will first discuss the question of structure (§2) and then review recent work on subsurface flows (§3). In section 4, I will discuss some potential avenues for further progress in the helioseismology of sunspots.

2. Sunspot Structure

As summarized by Figure 1, there is not general consensus regarding the subsurface structure of sunspots. As discussed by Crouch et al. [7], models can be categorized as "deep" models in which significant deviations from quiet-Sun structure extend below a few Mm and "shallow" models where the structure beneath the sunspot is very similar to that of the quiet Sun at depths below a few Mm.

Traditional time-distance helioseismology measurements have led to "deep" models that feature a near-surface layer (surface to roughly 3 Mm depth) with reduced wave speed and a deeper region (roughly 3 to to 15 Mm) of increased wave speed (e.g. [11; 12; 16–18] among many others). Ring-diagram inversions have led to models in which the wave speed is significantly different than the quiet Sun sound speed at depths well below a few Mm (e.g. [19; 20]). Comparisons of time-distance and ring-diagram inversions for the same Active Region using the same input data sets have shown that the two types of inversions can give quite different results [5; 6; 18; 21], with the caveat that the horizontal resolution of the two methods is quite different.

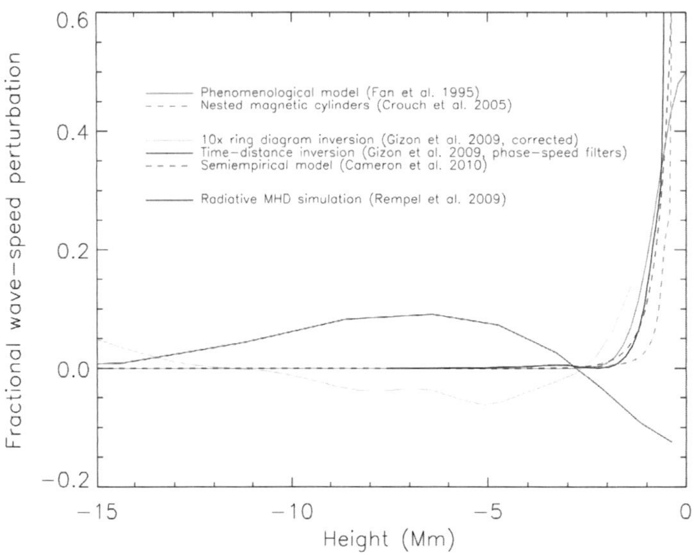

Figure 1. Fractional wave-speed perturbations for a variety of models for the structure of sunspot umbrae. For magnetic models, the wave-speed is taken to be the fast-mode speed. For non-magnetic models the sound speed is shown. The solid red curve shows the phenomenological model of Fan et al. [8] which was a sound-speed model designed to explain the Hankel analysis measurements of [9]. The dashed red curve shows the results of a non-linear inversion for a simplified magnetic model [10], also aimed at explaining the measurements of [9]. The solid green curve is the result of a ring diagram inversion, with the amplitude increased by a factor of ten, as the ring diagram inversion does not have the spatial resolution to isolate the umbra. The solid blue curve shows the result of a traditional time-distance inversion, and is qualitatively similar to what has been seen in other time-distance inversions (e.g. [11; 12]). The dashed blue curve is from the semi-empirical model of [13], other aspects of this model are shown in Figure 2. The black solid line shows the structure from the magnetoconvection simulation of [14]. This is Figure 16 from [5].

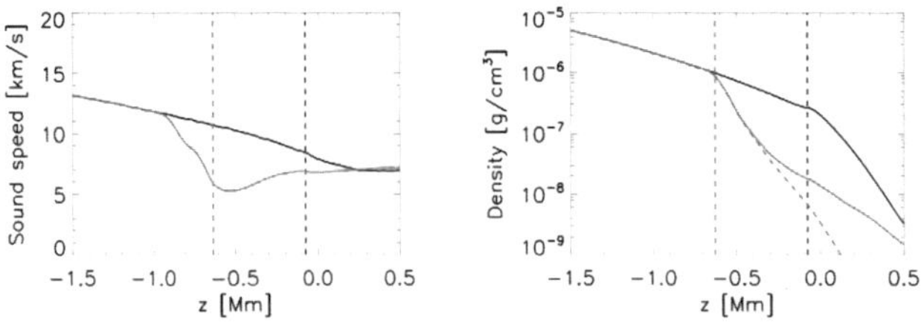

Figure 2. Models for the sound-speed (left) and density (right) for the quiet Sun (black) and umbra (blue) for empirical sunspot model of Cameron et al. [13]. The red dashed line in the right-hand panel shows the density from the umbral model of Maltby et al. [25]. In both panels the vertical black (red) dashed lines show optical depth of one in the quiet Sun (umbral) model. In this model, the umbral sound speed and density are indistinguishable from their values in the quiet Sun below a depth of about one Mm. Adapted from Figure 1 from [13].

In the "shallow" models, the near-surface wave speed is enhanced relative to its value in the quiet Sun. Models of this type have been inferred from Hankel analysis [8] and [10], from direct simulation of magnetoconvection in a sunspot-like feature [14], and from magnetic wave propagation simulations [13].

At this point, it is not clear what the sources of the disagreement are. Gizon et al. [5] suggested some possible issues, including among many others: 1) linear forward models (and inversions) may not be capable of capturing the large effect of sunspots on solar oscillations (in some models the perturbations to the wave speed are more than 100% of the quiet Sun sound speed) and 2) traditional forward models do not explicitly include magnetic fields (the effect of the magnetic field is included in a phenomenological sense in a "wave-speed" perturbation). In addition, it may not be sufficient to describe sunspots simply in terms of a local change in the sound speed, but rather it may be important to include an explicit model for the Wilson depression (this will be discussed in §2.2). In the following subsections, I will review recent work that may be helpful in resolving the apparent incompatibility of the "deep" and "shallow" models.

2.1. radiative transfer effects

Rajaguru et al. [22] presented observations of wave propagation in a sunspot made using different parts (corresponding to different formation heights) of a magnetically sensitive line (Fe I 6173 Å) and a magnetically insensitive line (Fe I 7090 Å). The line cores (wings) are thought to be formed at a height of about 270 km (20 km) above the continuum optical depth of unity ([22] and references therein). They showed that the sunspot supported upward propagating waves at frequencies below the acoustic cutoff frequency (this has been seen before, e.g., [23]). In addition, they showed that the observed travel-time shifts depend on the formation height at which the Doppler velocities are measured. For the case of ingoing waves, the choice of observation height can cause an effect of about 10 s in magnitude, while for outgoing waves the choice of observation height leads to an effect of 40 s (see Fig. 3). Especially for the case of outgoing waves, this can be a large fraction of the total signal (typically about a minute). In addition, they observed differences of up to about 15 s (at 3.5 mHz) between travel times measured in the magnetically sensitive and insensitive lines. This difference between the observations made in magnetically sensitive and insensitive lines is potentially extremely important as current methods of local helioseismology make no attempt at a formal accounting for the radiative transfer that connects plasma motions with the observed Dopplergrams.

2.2. progress in forward modeling

Forward modeling is the process of predicting the local helioseismic observations (e.g., time-distance travel-time shifts) that would be expected to result from a particular model of the solar interior (e.g., a model for the subsurface structure of a sunspot). In the past year there has been progress in not only computing forward models for sunspots, but of finding sunspot models that can largely explain the observed travel-time shifts (and the measured wave absorption as well).

Cameron et al. [13] presented a method for constructing a model sunspot based on standard models for umbrae, penumbrae, and quiet Sun together with a simple self-similar model for the magnetic field. In this model, the fast-mode speed is greatly enhanced due to the magnetic field near the surface and the sunspot looks much like the quiet Sun below a depth of 1 Mm (see Figs. 1 and 2).

Cameron et al. [13] used a series of wave propagation simulations to show that this sunspot model was able to explain, within the observational errors, observations of wave propagation (both phase shifts and amplitude changes) through a sunspot made using MDI data [24]. This work shows one example of a "shallow" sunspot model that appears to be consistent with helioseismic observations. It is important to note that this work does not use observations of solar

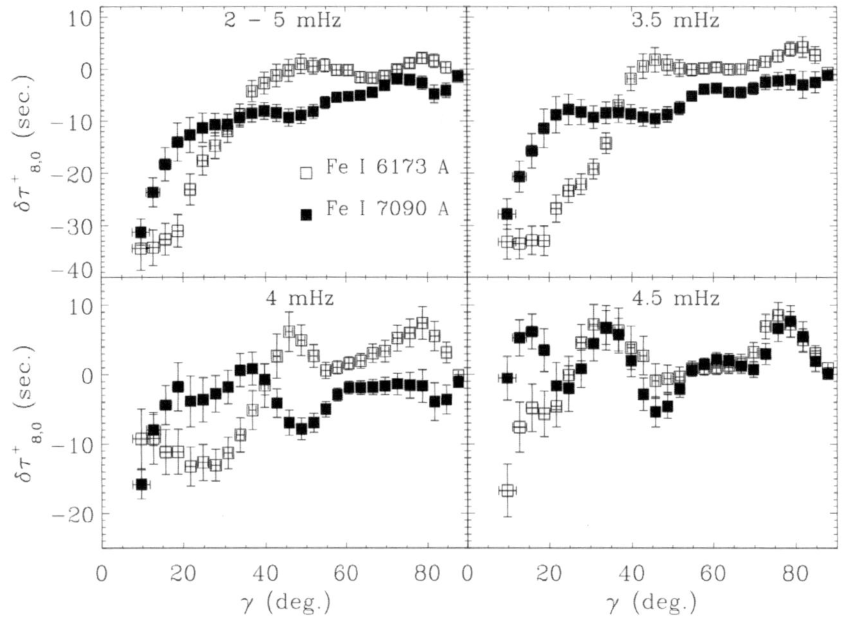

Figure 3. Changes in the outgoing travel times as functions of the inclination γ of the magnetic field. The change in the travel times is defined as $\delta\tau^+_{8,0} = \tau^+_8 - \tau^+_0$ where τ^+_8 is the outgoing travel time measured near the wings of the line (formation height of about 20 km above the photosphere) and $\delta\tau_0$ is the outgoing travel-time shift measured in the core of the line (formation height of about 270 km above the photosphere) at the travel distance of $\Delta = 16.95$ Mm. The open square symbols show travel-time shifts for a magnetically sensitive line (Fe I 6173 Å) and the filled square symbols are for a magnetically insensitive line (Fe I 7090 Å). This is a modified version of Figure 3a from [22].

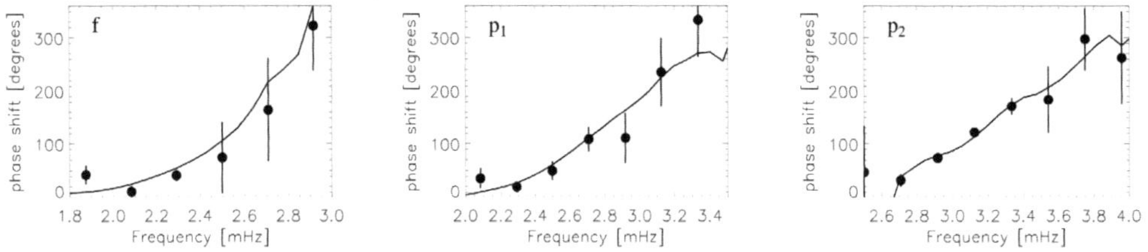

Figure 4. Phase shifts for wave propagating through the empirical sunspot model of Cameron et al. [13]. The panels show phase-shifts measured from MDI data (filled circles with error bars) and from the numerical wave propagation simulation (solid lines) for the f (left panel), p_1 (middle panel), and p_2 (right panel) modes. The simulation also reproduces the observed changes in the amplitudes of the waves that propagate through the sunspot. Adapted from Figure 9 from [13].

oscillations inside the penumbra or umbra of the sunspot, but rather is based on measurements of the travel-time shifts and amplitude changes in waves that have traveled through the sunspot. As a result, the potentially complicated radiative transfer in sunspots is not a factor in the observations (though may certainly play a role in the wave propagation).

Lindsey et al. [26] also showed an example of a "shallow" sunspot model. This work was

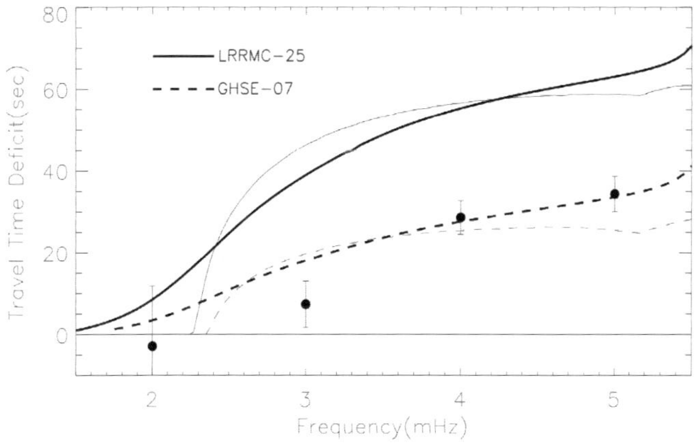

Figure 5. Comparison of measured and theoretical travel-time shifts for nearly vertically propagating waves incident on a sunspot umbra from below. The observations are shown as filled circles with error bars; the error bars represent the scatter in a sample of two sunspots observed for a few days with the MDI instrument. The heavy curves show the results of finite wavelength forward calculations for the umbral models LRRMC-25 (solid) and GHSE-07 (dashed). The thin curves show the results of ray theory calculations for the same umbral models. The LRRMS-25 model is based on a magnetoconvection simulation carried out using the method of [14]. The GHSE-07 is a modification of LRRMS-25 that was designed to provide a better fit to the observations. This is based on Fig. 11 from [26] and is reproduced by permission of the AAS.

based on a simple one-dimensional vertical wave propagation calculation aimed at understanding observations of almost vertically propagating waves interacting with sunspot umbrae. In the approximations that 1) the acoustic waves are propagating vertically, 2) the horizontal variations in the sunspot structure are not important, and 3) the magnetic field is vertical, the wave propagation calculation reduces to a simple one-dimensional non-magnetic problem. Lindsey et al. constructed a model umbra "GHSE-07" that produces travel-time shifts that mostly fit the observations. This model does not require an enhancement in the sound speed to explain the travel-time deficit. Rather, the travel times are reduced by the reduction in the path length due to the lower temperature of the umbra. This reduction in path length more than compensates for the reduced sound speed.

Braun et al. [27] compared travel-time shifts measured from MDI observations of sunspots with travel-time shifts measured from a magnetoconvection simulation. Figure 6 shows a comparison of mean (i.e. average of ingoing and outgoing) travel-time shifts averaged over the penumbrae for two different sunspots (one seen in the high-resolution MDI data and one in the MDI full-disk data) with mean travel-time shifts measured from the vertical velocity at fixed optical depth $\tau = 0.1$ from a magnetoconvection sunspot simulation [14]. The simulation reproduces the general trends seen in the observed travel-time shifts for the phase-speed filters centered at 14.9 km/s, 17.5 km/s, and 25.8 km/s. The simulation is not as close to the observations for the case of the smallest phase-speed filter; this is also the filter for which the largest difference between the two observed sunspots is seen.

Parchevsky et al. [28] showed numerical simulations of wave propagation through a model sunspot. The aim of the simulations was to understand the observed variation of travel-time shifts with the angle between the magnetic field vector and the line of sight [29–31]. The results suggested that the magnetic field does indeed introduce a dependence of the travel-time shifts on the line-of-sight direction. Rajaguru et al. [32] observed vertically propagating waves in a

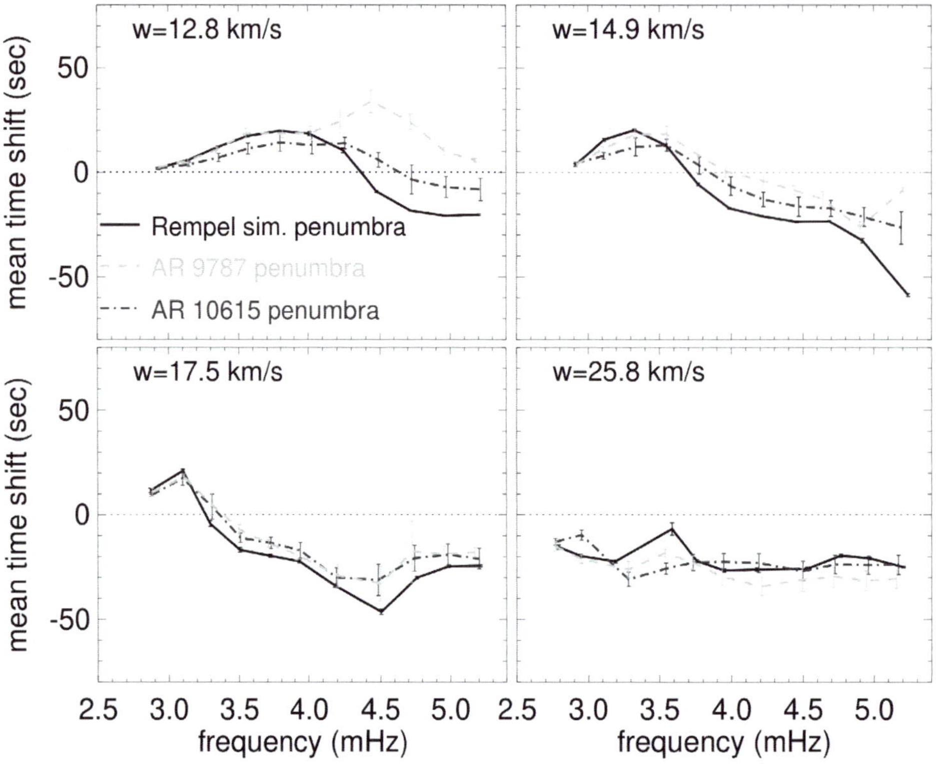

Figure 6. Travel-time shifts measured from full disk MDI data for AR9787 (dashed), from high-resolution MDI data for AR10615 (dot-dashed), and from the magnetoconvection sunspot simulation carried out by Rempel using the method of Rempel et al. [14]. Each panel is labeled with the central phase speed of the phase-speed filter used in the measurements. The simulation largely reproduces the observed frequency variations that are seen in the MDI observations, except for the case of frequencies above about 4 mHz in the filter with smallest phase speed. This is Figure 2 from [27].

sunspot penumbra and proposed that phase shifts resulting from these waves combined with radiative transfer effects can also cause apparent line-of-sight variations of travel-time shifts. Further work is needed to disentangle the radiative transfer issues from the wave propagation.

2.3. measurement issues

In principle, the details of a procedure for measuring travel-time shifts (i.e., data-analysis filters, definition of travel-time shift) can be accounted for in the forward modeling process (e.g. [33]). In practice, the situation in sunspots may be more complicated than can be easily accounted for using traditional ray-approximation forward models (e.g. [34]) or models based on perturbation theory around quiet Sun models (e.g. [35]). As a result, there are "systematic effects" (by which I mean effects that are not accounted for in a forward model) in the local helioseismology of sunspots. In this subsection I will describe a few investigations of potential systematic effects that have appeared in the literature very recently.

Travel-time shifts measured using time-distance helioseismology and helioseismic holography, in general, depend on the frequencies and phase speeds of the waves that contribute to the measurement (e.g. [36] among many others). As a result, data analysis filters (i.e. phase-speed or ridge filters) can have substantial effects on the measurements (e.g. [3; 6; 37]). In modeling these effects, the detailed shape of the power spectrum of the waves can play an important role (e.g. [6]). Zhao et al. [12] used intensity observations from the SOT instrument on the Hinode spacecraft [38] to test the impact of phase-speed filters on time-distance measurements. They found that the phase-speed filters caused up to a 40% change in observed travel-time shifts, with the largest changes occurring at the shortest travel distances. This result demonstrates the importance of accounting for the phase-speed filters in the forward modeling process. In addition, Zhao et al. carried out inversions for the subsurface structure of a sunspot and found a result that was qualitatively similar to that of [11] and the results shown in Figure 1, both of which were obtained from MDI Doppler velocity measurements.

Braun et al. [39] showed that the spectral content of helioseismic holography cross-correlations measured in sunspot umbrae and in the quiet Sun are very different; the cross-correlations in the umbra have less power at high frequency than those in the quiet Sun. This effect causes changes in the positions of the phase peaks in the cross covariance that can be confused with travel-time shifts depending on the details of the fitting procedure that is applied.

3. Subsurface flows associated with sunspots

The traditional approach to ring-diagram analysis has been to carry out one-dimensional (depth only) inversions while assuming that the subsurface flows and structure are horizontally homogeneous across the measurement patch. Featherstone et al. [15] show a generalization of this procedure to the case of three-dimensional inversions for subsurface flows. This inversion approach is different than that usually employed in time-distance helioseismology as it does not assume horizontal translation invariance of the noise covariance matrix. As a result, the inversion cannot be done using a MCD-type [40] decomposition into horizontal Fourier modes. The inversion is therefore expensive and iterative methods are employed to obtain approximate solutions.

Figure 7 shows one example inversion result from [15]. The horizontal resolution of the inversion decreases with depth (short wavelength modes have shallow lower turning points). Notice that the moat flow can be seen clearly at the depths of 0.2 and 4 Mm. It is less visible at 7 Mm and cannot be clearly seen at 11 Mm. There is no hint of a return (inward) moat flow. This result is apparently inconsistent with the shallow inflows seen using time-distance helioseismology [e.g. 41]. It is possible, however, that the subsurface structure of the moat flow varies strongly from one sunspot to another, or over the life cycle of sunspots. A direct comparison using the same observational data for the same sunspot at the same time period is needed.

Braun et al. [39] showed flows obtained from an inversion of travel-time differences measured using helioseismic holography of HMI data (see Fig. 2 from that paper). The near surface (0-3 Mm) moat flow was similar to that seen in Figure 7. A comparison between the moat flow inferred from time-distance, ring-diagrams, and helioseismic holography in the style of Figure 1 has not been done.

4. Discussion

There is not general agreement on the subsurface structure of sunspots. Some methods lead to "deep" sunspot models which feature substantial wave-speed variations (relative to the quiet Sun sound speed at the same depth) at depths below a few Mm. Other methods lead to "shallow" sunspot models in which the wave speed is essentially the same as the quiet Sun sound speed at depths below a few Mm. The sources of this disagreement are not known.

In the past year, however, there has been progress in many aspects of the local helioseismology of sunspots. It is now feasible to carry out simulations of wave propagation through model sunpots and make direct comparisons between these simulations and measurements [13]. Using

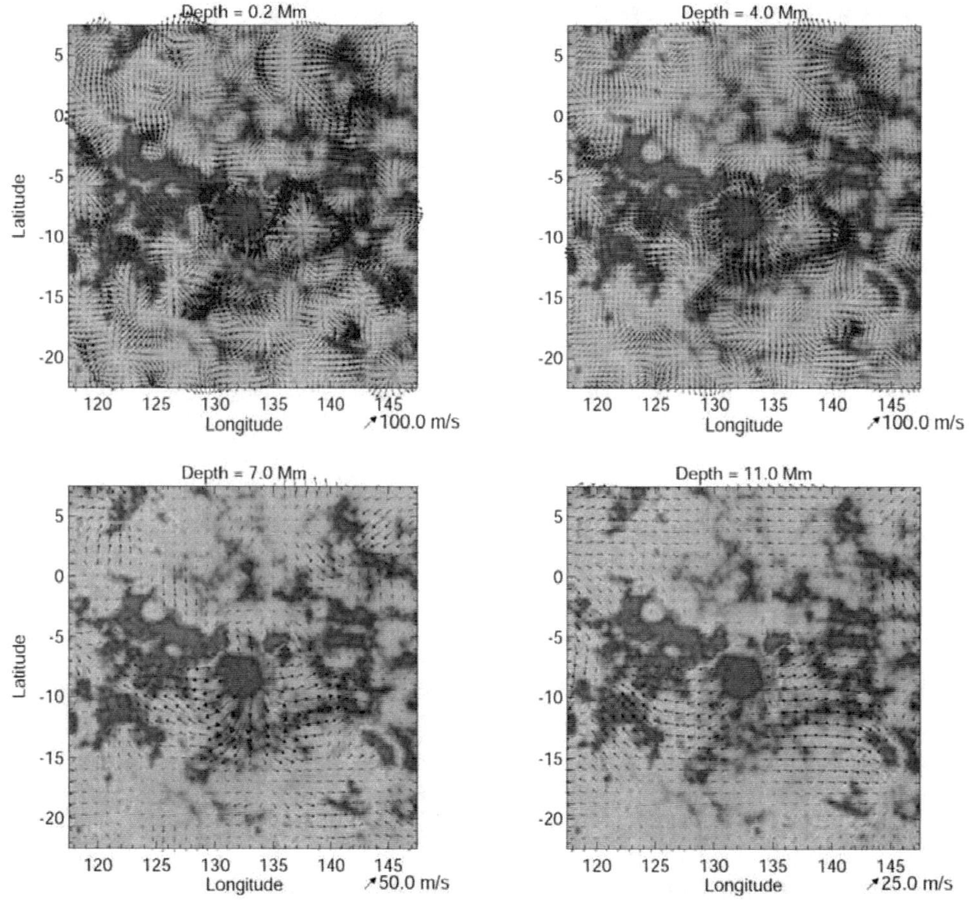

Figure 7. Three-dimensional ring diagram inversions for flows around a sunspot. From top left to bottom right, the panels show slices through the inferred flow field at depths 0.2, 4, 7, & 11 Mm. In all of the panels, the arrows show the horizontal flows and the color scale shows an MDI line-of-sight magnetogram. Notice that the spatial resolution is best near the surface and decreases with increasing depth. There is a clear moat flow surrounding the sunspot (dark green feature in the magnetogram at Longitude 132° and Latitude -8°) seen near the surface. The moat flow can be seen in the inversion at 7 Mm depth but is not obvious at 11 Mm depth. The supergranulation signal also decays with depth and is not clearly visible at or below 7 Mm. Adapted from [15].

this approach, it will be possible to determine if any particular sunspot model is compatible with observations. This is a crucial test of any proposed sunspot model. In addition, direct numerical simulations of magnetoconvection that include sunspot-like magnetic structures will provide synthetic data for testing methods of local helioseismology [14].

A direct comparison of the subsurface nature of moat flows inferred from time-distance methods, ring-diagrams, and helioseismic holography has not been done. This comparison is important and should be carried out.

Acknowledgments

ACB thanks N. Featherstone, C. Lindsey, D. Braun, R. Cameron, and S.P. Rajaguru for providing figures, acknowledges support from NASA contracts NNG07EI51C and NNH09CE41C, and thanks R. Cameron, A. Crouch, and D. Braun for helpful comments.

References

[1] Gizon L and Birch A C 2005 *Living Reviews in Solar Physics* **2** 6

[2] Chou D 2007 *New Solar Physics with Solar-B Mission* (*Astronomical Society of the Pacific Conference Series* vol 369) ed K Shibata, S Nagata, & T Sakurai p 313

[3] Thompson M J and Zharkov S 2008 Sol. Phys. **251** 225–240

[4] Schunker H 2010 *ArXiv e-prints* (*Preprint* 1005.2526)

[5] Gizon L, Birch A C and Spruit H C 2010 ARA&A **48** 289–338 (*Preprint* 1001.0930)

[6] Moradi H, Baldner C, Birch A C, Braun D C, Cameron R H, Duvall T L, Gizon L, Haber D, Hanasoge S M, Hindman B W, Jackiewicz J, Khomenko E, Komm R, Rajaguru P, Rempel M, Roth M, Schlichenmaier R, Schunker H, Spruit H C, Strassmeier K G, Thompson M J and Zharkov S 2010 Sol. Phys. **267** 1–62

[7] Crouch A D, Birch A C, Braun D C and Clack C T M 2010 *Proceedings IAU symposium No. 273* ed Choudhary D B and Strassmeier K G

[8] Fan Y, Braun D C and Chou D 1995 ApJ **451** 877

[9] Braun D C 1995 ApJ **451** 859

[10] Crouch A D, Cally P S, Charbonneau P, Braun D C and Desjardins M 2005 MNRAS **363** 1188–1204

[11] Kosovichev A G, Duvall Jr T L and Scherrer P H 2000 Sol. Phys. **192** 159–176

[12] Zhao J, Kosovichev A G and Sekii T 2010 ApJ **708** 304–313 (*Preprint* 0911.1161)

[13] Cameron R H, Gizon L, Schunker H and Pietarila A 2010 Sol. Phys. DOI 10.1007/s11207-010-9631-3

[14] Rempel M, Schüssler M and Knölker M 2009 ApJ **691** 640–649 (*Preprint* 0808.3294)

[15] Featherstone N A, Hindman B W and Thompson M J 2010 *These Proceedings*

[16] Kosovichev A G 1996 ApJ **461** L55

[17] Couvidat S, Birch A C and Kosovichev A G 2006 ApJ **640** 516–524

[18] Gizon L, Schunker H, Baldner C S, Basu S, Birch A C, Bogart R S, Braun D C, Cameron R, Duvall T L, Hanasoge S M, Jackiewicz J, Roth M, Stahn T, Thompson M J and Zharkov S 2009 Space Sci. Rev. **144** 249–273 (*Preprint* 1002.2369)

[19] Basu S, Antia H M and Bogart R S 2004 ApJ **610** 1157–1168

[20] Bogart R S, Basu S, Rabello-Soares M C and Antia H M 2008 Sol. Phys. **251** 439–451

[21] Gizon L, Schunker H, Baldner C S, Basu S, Birch A C, Bogart R S, Braun D C, Cameron R, Duvall T L, Hanasoge S M, Jackiewicz J, Roth M, Stahn T, Thompson M J and Zharkov S 2010 Space Sci. Rev. 99 DOI 10.1007/s11214-010-9688-1

[22] Rajaguru S P, Wachter R, Sankarasubramanian K and Couvidat S 2010 ApJ **721** L86–L91 (*Preprint* 1009.2350)

[23] Finsterle W, Jefferies S M, Cacciani A, Rapex P, Giebink C, Knox A and Dimartino V 2004 Sol. Phys. **220** 317–331

[24] Scherrer P H, Bogart R S, Bush R I, Hoeksema J T, Kosovichev A G, Schou J, Rosenberg W, Springer L, Tarbell T D, Title A, Wolfson C J, Zayer I and MDI Engineering Team 1995 Sol. Phys. **162** 129–188

[25] Maltby P, Avrett E H, Carlsson M, Kjeldseth-Moe O, Kurucz R L and Loeser R 1986 ApJ **306** 284–303

[26] Lindsey C, Cally P S and Rempel M 2010 ApJ **719** 1144–1156

[27] Braun D C, Birch A C, Crouch A D and Rempel M 2010 *IAU 273: Physics of Sun and Star Stops Conference Proceedings*

[28] Parchevsky K V and Kosovichev A G 2009 ApJ **694** 573–581 (*Preprint* 0806.2897)

[29] Schunker H, Braun D C, Cally P S and Lindsey C 2005 ApJ **621** L149–L152

[30] Zhao J and Kosovichev A G 2006 ApJ **643** 1317–1324 (*Preprint* arXiv:astro-ph/0602260)

[31] Schunker H, Braun D C, Lindsey C and Cally P S 2008 Sol. Phys. **251** 341–359 (*Preprint* 0801.4448)

[32] Rajaguru S P, Sankarasubramanian K, Wachter R and Scherrer P H 2007 ApJ **654** L175–L178 (*Preprint* arXiv:astro-ph/0611897)

[33] Gizon L and Birch A C 2002 ApJ **571** 966–986

[34] Kosovichev A G and Duvall Jr T L 1997 *SCORe'96 : Solar Convection and Oscillations and their Relationship* (*Astrophysics and Space Science Library* vol 225) ed F P Pijpers, J Christensen-Dalsgaard, & C S Rosenthal pp 241–260

[35] Birch A C, Kosovichev A G and Duvall Jr T L 2004 ApJ **608** 580–600

[36] Braun D C and Birch A C 2008 Sol. Phys. **251** 267–289 (*Preprint* 0802.2652)

[37] Birch A C, Braun D C, Hanasoge S M and Cameron R 2009 Sol. Phys. **254** 17–27

[38] Kosugi T, Matsuzaki K, Sakao T, Shimizu T, Sone Y, Tachikawa S, Hashimoto T, Minesugi K, Ohnishi A, Yamada T, Tsuneta S, Hara H, Ichimoto K, Suematsu Y, Shimojo M, Watanabe T, Shimada S, Davis J M, Hill L D, Owens J K, Title A M, Culhane J L, Harra L K, Doschek G A and Golub L 2007 Sol. Phys. **243** 3–17

[39] Braun D C, Birch A C, Crouch A D and Rempel M 2010 *These Proceedings*

[40] Jacobsen B H, Moller I, Jensen J M and Efferso F 1999 *Phys. Chem. Earth* **24** 215

[41] Zhao J, Kosovichev A G and Duvall Jr T L 2001 ApJ **557** 384–388

Ring-analysis flow measurements of sunspot outflows

Nicholas A Featherstone

JILA and Department of Astrophysical and Planetary Sciences, University of Colorado, Boulder, CO 80309-0440

Bradley W Hindman

JILA and Department of Astrophysical and Planetary Sciences, University of Colorado, Boulder, CO 80309-0440

Michael J Thompson

High Altitude Observatory, National Center for Atmospheric Research, Boulder, CO 80307-3000

E-mail: `feathern@solarz.colorado.edu`

Abstract. We present a local helioseismological analysis of the convective flows around sunspots. The flow deductions were obtained from MDI Dopplergrams using ring-analysis techniques to measure Doppler shifts induced in solar acoustic oscillations. A novel multi-scale 3-D inversion procedure was used to self-consistently combine all ring-analysis data taken from a mosaic of analysis tiles spanning the solar disk. The inversion is multi-scale because in addition to folding together information from tiles located at different positions on the solar surface, it is capable of incorporating tiles of different sizes, thus, enabling fine control of the horizontal resolution and the probing depth. The inversion is based on sensitivity kernels computed with the Born approximation. Our inversion results indicate that outflow observed at the surface surrounding sunspots persists to unexpectedly deep depths ($\sim$ 7 Mm). These outflows appear to have two components, being comprised of a superficial moat flow and a deeper outflow whose strength peaks around 5 Mm.

1. Introduction

Highly vigorous convection, driven by strong radiative cooling, occurs on many horizontal scales in the upper convection zone, ranging from granules to supergranules. Ring analysis, a form of local helioseismology, has been widely used to sample this region with data from the Michelson Doppler instrument (MDI) aboard the SOHO spacecraft, revealing much about the near-surface nature of the plasma flows. Ring-analysis studies have revealed the presence of large-scale flows, now called solar subsurface weather (SSW) that may be the combined signature of deep giant-cell convection and global-scale circulations [7]. As these evolving flows in the upper convection zone exhibit strong linkage with photospheric magnetism, and likely have a significant role in the advection and redistribution of magnetism, much attention has been devoted to the understanding the properties of flow in active regions and around magnetic structures [8, 9].

Ring analysis has also been used to study the subsurface thermodynamic structure of sunspots [2] and the surface outflows around sunspots [9, 11] The outflows around sunspots with depth, however, remain unexplored using this technique, but have been studied extensively using time-distance [13, 6, 14]. Curiously, these studies yield inconsistent results concerning the deep flows around sunspots. We outline a method for the 3-dimensional (3-D) inversion of ring-analysis results incorporating measurements made using multiple mode orders and multiple-analysis region sizes that tile the full solar disk. As a demonstration, we apply this technique to the inversion of flows beneath sunspots using MDI data from January of 2002 and compare our results to those found using time-distance.

2. Inversion Technique

Ring-analysis techniques assess subsurface flow velocities from the Doppler shifts that flows induce in the sun's acoustic oscillations or p modes. These shifts manifest in power spectra made from the surface Doppler signal obtained within a patch on the sun as shifts in frequency of the resonant oscillations. Flow measurements made in this way represent spatial averages over the flow beneath that patch of the sun. Measurements are accordingly carried out over analysis regions of different sizes to obtain reconstructions of the solar flow field with varying levels of horizontal resolution. Smaller measurement regions provide a more horizontally localized, near-surface average of the flow, while measurements from larger regions permit the analysis of longer-wavelength, more deeply penetrating modes, allowing one to assess the deeper subsurface flows with albeit lower horizontal resolution.

One challenge in such schemes has been the reconciliation of measurements resulting from ring analyses of multiple region sizes. Moreover, for a given analysis-region size, complementary measurements are produced using different radial mode orders with different spatial wavenumbers. These measurements represent unique averages over the solar subsurface flow field and flow maps obtained from different subsets of this data will necessarily disagree with one another. To that end, we have developed a 3-D inversion technique that allows us to simulataneously invert data from a number of different modes and region sizes to obtain a self-consistent map of the horizontal flow of the sun over a range of depths.

The aim of our inversion algorithm is to deduce the solar flow field over one large patch of the solar disk, $v(\boldsymbol{r})$, from the ring measurements, u, and their errors, σ. We accomplish this through a regularized-least-squares inversion (RLS)[4] that trades off between the solution's goodness of fit to the data and the smoothness of the solution. We wish to invert for the flow field over the entire solar disk at once, but owing to computational constraints, we can presently only invert regions 45° square in a reasonable amount of time. Instead, a full-disk solution must be constructed by tiling the disk with smaller inversion regions that overlap by an arbitrary amount and "stitching" these inversion regions together. We are thus faced with an additional constraint that we want the solution for some region of the solar disk, α, to join smoothly to the different overlapping regions β for which we also perform inversions. We accomplish this with an iterative inversion procedure, solving for the $v(\boldsymbol{r})$ that minimizes the functional

$$\sum_i \frac{1}{\sigma_i^2}\left(u_i^\alpha - \int_\odot K_i(\boldsymbol{r}) v_k^\alpha(\boldsymbol{r}) d\boldsymbol{r}\right)^2 + \lambda \int \hat{\mathcal{L}} v_k^\alpha(\boldsymbol{r}) d\boldsymbol{r} + \mathcal{R}_k^\alpha = M \tag{1}$$

at each iteration k. Unique measurements are indicated by the index i, and K_i is the corresponding flow sensitivity kernel. The smoothness trade-off parameter is denoted by λ and the smoothing function by $\hat{\mathcal{L}}$, which we take to be

$$\hat{\mathcal{L}} v(\boldsymbol{r}) = |\boldsymbol{\nabla} v(\boldsymbol{r})|. \tag{2}$$

Minimizing the functional M is the standard RLS approach with the exception of the $\mathcal{R}$ term. This additional term ensures the smoothness of the solution across regional boundaries, and we

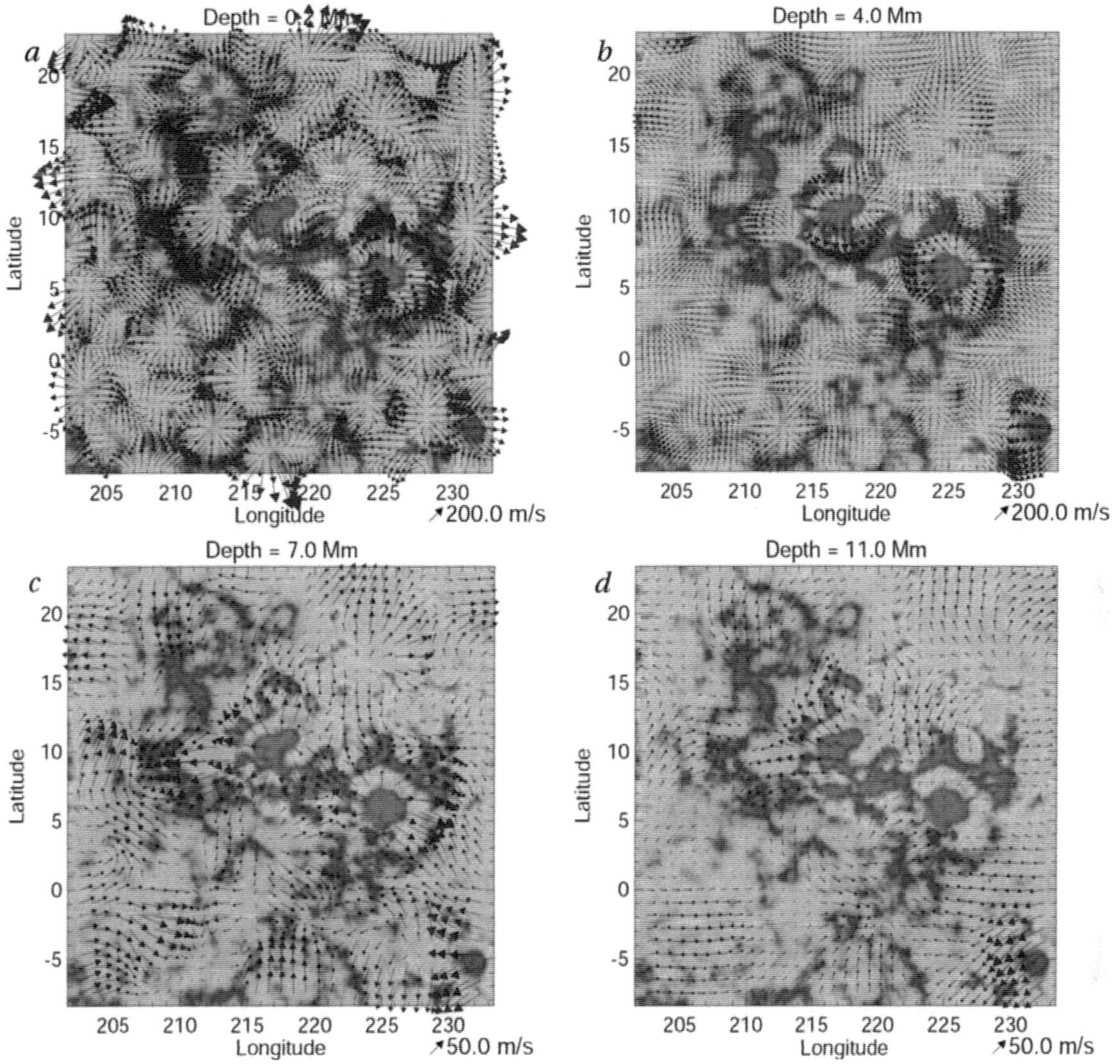

Figure 1. Flow field realized through 3-D inversion of ring-analysis measurements of MDI obtained in January 2002. Horizontal velocity vectors (blue) overlay the magnetogram for that day (positive field in red, negative in green). At a depth of 0.2 Mm (a), the cellular structure of supergranulation is apparent. Strong outflows are visible around the two sunspots at longitude 215° and 225° down to a depth of 7 Mm (b, c). Below 7 Mm (d), flows are decidely larger scale and exhibit no outflows from the sunspots.

choose to define it as

$$\mathcal{R}_k^\alpha = \Lambda \sum_{\beta \neq \alpha} \int (v_k^\alpha(\boldsymbol{r}) - v_{k-1}^\beta(\boldsymbol{r}))^2 \boldsymbol{dr}, \tag{3}$$

where Λ is an additional trade-off parameter. This term penalizes the mismatch between the solution at the current iteration and the overlapping solutions from the prior iteration. This procedure requires the inversion of two matrices: one including $\mathcal{R}$ and one without it (for the first iteration). Convergence is typically achieved in 3-5 iterations.

The results presented in these proceedings have been constructed from power spectra of MDI

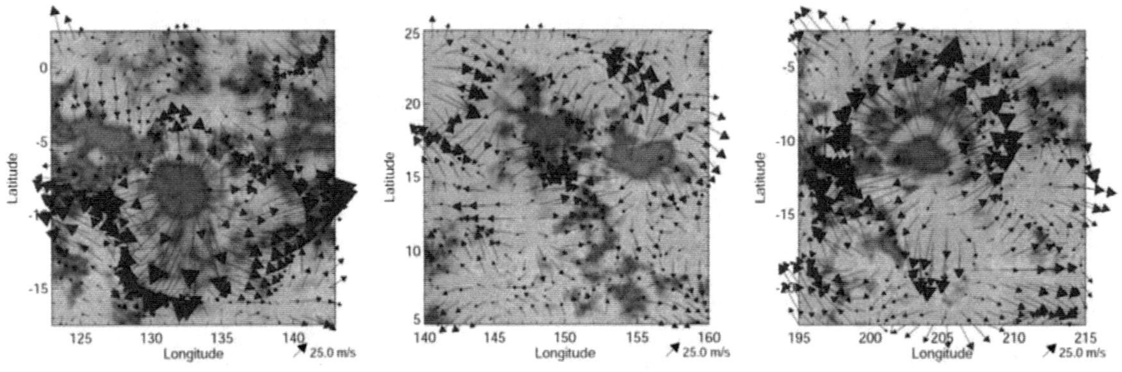

Figure 2. Representative sunspot outflows from CR 1985 at a depth of 6 Mm. Such outflows are typical of the flow field at this depth in the vicinity of sunspots for this rotation.

Dopplergrams using three different region sizes of 2°, 4°, and 16° in heliographic angle. These Dopplergrams have been differentially tracked in latitude at the Snodgrass rotation rate [12] and apodized to 1.875°, 3.75°, and 15° respectively before creating power spectra.

We have incorporated measurements from radial mode orders n from zero to six. Our inversion grid extends in depth from the surface to a depth of 25 Mm. We employ a variable horizontal resolution with depth, adopting 0.9375° at the surface, 1.875° at the mid-depths, and 7.5° near the bottom of the domain. Inversion regions for these analyses are 37.5° on a side, overlapping nearest-neighbor regions by 7.5°. Nine such regions were arranged in a 3x3 matrix to cover the solar disk. The sensitivity kernels employed in this analysis have been computed using solar model S [3] and using the Born approximation, thereby accounting for the effects of wave scattering. Further details concerning the sensitivity kernels may be found in [1].

3. Flows around Sunspots

As a demonstration of the flow fields obtained through our 3-D ring-analysis inversions, we have examined MDI data from Carrington Rotation (CR) 1985. The existence of several magnetic active regions during this rotation allows us to examine the nature of subsurface flows around sunspots. The resulting flow field for a small portion of the solar disk (30° on a side) from 17 January 2002, is shown in figure 1 at four depths. The cellular pattern of supergranulation is cleary visible at the near-surface depth (0.2 Mm), and regions of divergence and convergence formed from the intersection of these cells tend to correspond with the presence of magnetic features (colored underlay). As the depth increases, the size scale of the sampled flows increases and the speed of the flows falls.

One of the more interesting features in this map is the presence of a pair of outflows centered on the sunspots at longitudes 215° and 225° . These outflows, corresponding to the moat flow observed at the surface, persist from the surface down to a depth of about 7 Mm. By 11 Mm, outflows around these sunspots give way to larger-scale meandering motions that freely pass through regions of strong magnetism. The presence of deep subsurface outflows around sunspots has proven to be an ubiquitous feature of the sunspots in our inversions of MDI data from CR 1985. Other representative examples of sunspot outflow from this same rotation are shown in Figure 2 (here at a depth of 6 Mm). In all cases, an umistakable outflow signature is visible at depth, although the outflows around trailing spots are less organized.

To explore the depth dependence of these outflows in more detail, we plot the outflow as a function of depth for the sunspot in figure 1 located at longitude 225°. In figure 3a we show the radial outflow from the sunspot's center averaged in azimuth about the sunspot for

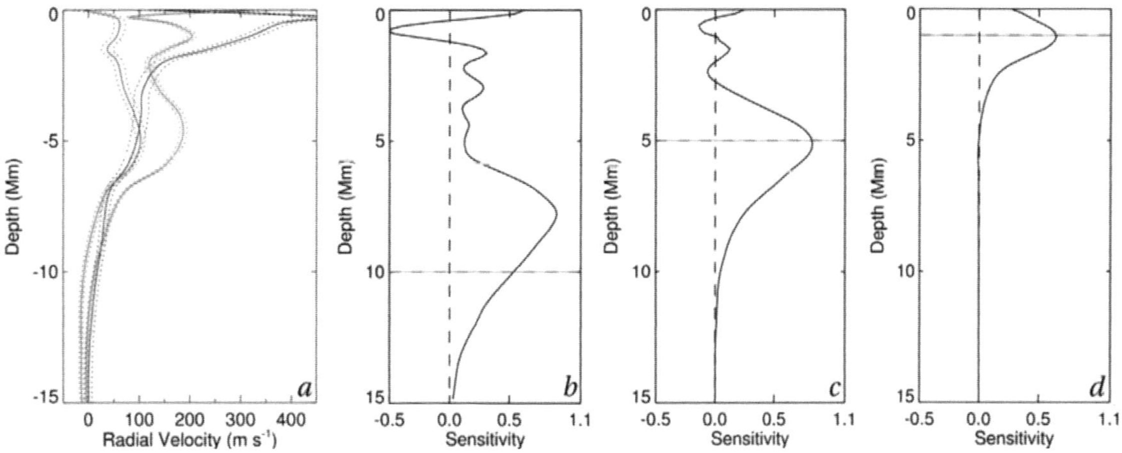

Figure 3. (*a*) Average outflow with depth around the sunspot from the sunspot in figure 1 located at 225° in longitude. Shown is the outflow (averaged in azimuth and over one day's worth of data) at a distance of 17 Mm (black), 23 Mm (red), and 28 Mm (blue) from the sunspot's center. Error bounds are denoted by dotted lines. Outflows are visible at all depths, but decay sharply below 7 Mm. Representative averaging kernels versus depth are shown in *b*-*d* with the horizontal line indicating the kernel's target depth. Averaging kernels are much narrower near the surface than at depth.

three different radial distances. Outflow is strongest near the surface and closest to the umbra-penumbra boundary. We note that the outflow is positive down to about 10 Mm, becoming weakly negative below this depth, and approaching zero beyond 15 Mm (where our averaging kernels become poor). The outflow structure transitions relatively sharply from strong outflow around 7 Mm to insignificant outflow at 10 Mm. Outflows exhibit two maxima; one near the surface and one near 5 Mm at the radii shown in Figure 3*a*. We may thus be seeing the combined signature of the surface moat flow and a distinct deeper outflow. All the sunspots we have examined in this Carrington rotation display this two component structure.

In Figure 3(*b*-*d*) we also show depth cuts through the accompanying averaging kernels resulting from this inversion. For target depths near the surface, these kernels are well localized, but broaden considerably as the target depth deepens. In some cases, narrow negative sidelobes do appear. These horizontal widths of our averaging kernels widen with depth as well, transitioning from a width of about 1° at the surface where the measurements from our 2° regions largely determine the solution, to about 8° at depth where the 16° regions govern the solution. The vertical width of the kernels is such that the true transition between strong and weak outflow (between 7 and 10 Mm) may in fact be much narrower than indicated by figure 3*a*.

4. Discussion

We have outlined a novel technique for the inversion of ring-analysis results that allows for the simultaneous inversion of measurements spanning the solar disk by tiling it with several smaller regions. This method allows for the self-consistent inversion of measurements made from a variety of analysis-region sizes, allowing us to incorporate both high- and low- resolution data into our inversion.

One of the more notable features we find is the ubiquitous outflow from sunspots. Zhao et al. [13, 14] have investigated the nature of outflow and inflows around sunspots using *p*-mode time-distance procedures. These studies suggest that the surface outflow observed in earlier f-mode time-distance studies [6] around sunspots is a surface effect, persisting to a depth of $\sim$ 1.5 Mm. Below this depth, and down to about 5 Mm, this trend reverses and an inflow is seen. This picture is, however, inconsistent with the findings of [6], who found outflows down to a depth of ~5 Mm using time-distance measurements incorporating both f-mode and p-mode data to study a sunspot in NOAA Region 9787.

Our ring analysis has found outflows that span a broad range of depths for essentially all observed sunspots, and seems consistent with the results of [6]. The lack of any outflow feature observed below a depth of about 9 Mm in the measurements of Zhao et al. [13] also agrees with our findings. The regularization used in our inversion might allow the imprinting of surface flows into the solution at depth. To minimize this effect, we repeated our inversion around the sunspots in figure 1 using only p-mode measurements. We found persistant outflows with depth even in the absence of the f-mode surface flow data.

The presence of two flow components, a superficial moat flow and a more deeply rooted outflow, is unexpected, particularly given that the time-distance studies of [6] have not reported such a subsurface structure (though they did indicate that outflows increased steadily from the surface to a depth of about 5 Mm). As the moat flow decays more sharply with radius than does the deeper outflow, these two flow structures may be driven by different physics that affect time-distance and ring-analysis techniques differently. We plan to pursue a more extensive study of the subsurface structure of sunspot flows in [5].

5. Acknowledgements

The authors would like to thank Juri Toomre, Deborah Haber, and Aaron Birch for many useful discussions and insights. Support for this work was provided through NASA grants NNG06GD97G, NNX07AH82G, NNX08AJ08G, NNX08AQ28G, and NNX09AB04G.

[1] Birch A C, Gizon L, Hindman B W and Haber D 2007 *ApJ* **662** 730
[2] Bogart R S, Basu S, Rabello-Soares M C, Antia H M 2008 *Solar Phys.* **251** 439
[3] Christensen-Dalsgaard J et al. 1996, *Science* **272** 1286
[4] Craig I J D and Brown J C 1986 *Inverse Problems in Astronomy* (Bristol, England: Adam Hilger Ltd)
[5] Featherstone N A, Hindman B W, Thompson M J and Toomre J 2010 *ApJ* in preparation
[6] Gizon L et al. 2009 *Space Sci. Rev.* **144** 249
[7] Haber D A, Hindman B W, Toomre J, Bogart S, Larsen R M and Hill F 2002 *ApJ* **570** 855
[8] Hindman B W, Haber D A and Toomre J 2006 *ApJ* **653** 725
[9] Hindman B W, Haber D A and Toomre J 2009 *ApJ* **698** 1749
[10] Komm R, Howe R, Hill F, Miesch M, Haber D and Hindman B 2007 *ApJ* **667** 571
[11] Moradi et al 2010 *Solar Phys. in press*
[12] Snodgrass H B 1984 *Solar Phys.* **94** 13
[13] Zhao J, Kosovichev A G, Duvall T L Jr. 2001 *ApJ* **557** 384
[14] Zhao J, Kosovichev A G and Sekii T 2010 *ApJ* **708** 304

Kinetic and magnetic helicities in solar active regions

Ram Ajor Maurya[1], **Ashok Ambastha and Vema Reddy**

Udaipur Solar Observatory, Physical Research Laboratory, Udaipur-313001, INDIA.

E-mail: [1]`ramajor@prl.res.in`

Abstract. We have studied the kinetic and magnetic helicities in sub-photospheric flows and photospheric magnetic fields, respectively, of a sample of 91 ARs of solar cycle 23. Hemispheric trend is investigated in the kinetic helicity of sub-photospheric flows averaged in the depth range of 2.5-12 Mms. Magnetic helicity parameters for the ARs are derived using photospheric vector magnetograms to examine their correlation with the corresponding kinetic helicities. We found no significant association between the two helicity parameters.

1. Introduction

Ordering of magnetic fields at various layers and scales in active regions (ARs) are observationally evident which may result from coherent patterns of surface flows [1], subphotospheric flows [2], and/or emergence of twisted sub-photospheric fluxes [3]. A photospheric manifestation of these patterns is the large-scale non-potential fields in flaring ARs. These non-potential fields are generally marked by twisted, or sheared, photospheric magnetic fields near flare sites [4].

In recent years, a topological property called magnetic helicity has been used to characterize the observed large-scale magnetic patterns which includes a subset of twisted or sheared fields, and also provides a measure of magnetic linking and kinking [5]. It is believed that the main source of the magnetic fields on the solar surface is a dynamo action [6] which sits somewhere either at the bottom of the convection zone or in a thin region called overshoot zone; a layer between the convection and radiation zones.

The necessary condition for the kinetic dynamo action is that the underlying velocity field ($\mathbf{u}$) lacks reflection symmetry, i.e., it has handedness [7, 8]. One of the natural measures of the lack of the reflection symmetry is the kinetic helicity H_k, representing the extent to a corkscrew-like motion. If a parcel of fluid is moving and rotating about an axis parallel to the direction of motion, it will have positive helicity. The existence of kinetic helicity has been observed in several physical and astrophysical systems, often a consequence of differential rotation. Therefore, kinetic helicity provides an important mechanism for the production of large scale hydrodynamic and magnetic structures.

The main source of magnetic helicity observed on the Sun is assumed to be the kinetic helicity of turbulent flows in the convection zone [9]. It twists the rising flux tube [2] and the observed photospheric magnetic fields. The other possibility of twists is the dynamo action [10]. Magnetic helicity of solar ARs shows a hemispheric trend; it is positive (negative) in the southern (northern) hemisphere. This phenomenon is called the "hemispheric sign rule (HSR)" of helicity. Photospheric vector magnetic fields of ARs reveal that on an average solar ARs have a small but statistically significant mean twist in magnetic helicity [2, 11] that is left-handed (right-handed) in the northern (southern) hemisphere. However, some theoretical [12] and observational [13] analyses show opposite hemispheric trends during the beginning of a cycle.

Kinetic helicity measurement in the solar convection zone may provide a direct inference of α-effect. Its measurement in the interior of solar ARs has become possible after the advent of local helioseismology, albeit only to a limited depth of a few Mms. Therefore, as to what extent the kinetic helicity measurements would explain the α-effect is somewhat uncertain. The hemispheric trend of kinetic helicity has been studied by several workers [14–19]. However, the relation between H_k of the underlying flows and H_m of external magnetic fields in ARs is, in general, still not well understood.

In this paper, we report on the statistics of the hemispheric trend of kinetic and magnetic helicities and their association. In Section 2, we describe the methodology to determine these helicity parameters. Section 3 describes the observational data and the method of analysis. Results and conclusions of our analysis are given in Sections 4 - 5.

2. The Basic Formalism

Magnetic helicity is a global quantity. But one cannot observe a full flux system volume because an AR is extended both below and above the visible surface. Therefore, information of the twist in magnetic fields is derived from the force-free parameter α, used as a measure of magnetic helicity. For a linear force-free field, it is given by (cf., [9]),

$$\alpha_z = (\nabla \times \mathbf{B})_z / B_z = \left(\frac{\partial B_y}{\partial x} - \frac{\partial B_x}{\partial y} \right) \frac{1}{B_z} \tag{1}$$

While calculating the values of α_z in an AR, we selected only those pixels where the values are above the accuracy of measurements. Further, to get the imbalance of twist (right/left handedness) in an AR, we took the average (α^z_{av}) over the entire region. Similarly, for calculating the kinetic helicity of an AR, we have considered only the vertical component of kinetic helicity density [20], which is given by

$$h^z_k = \left(\frac{\partial u_y}{\partial x} - \frac{\partial u_x}{\partial y} \right) u_z \tag{2}$$

3. The Observational Data and Analysis

We obtained the photospheric vector magnetograms from MSFC and Hinode/SOT, while, Doppler observations were taken from GONG for 91 ARs having good data sets. For selecting these ARs for our analysis, we used the following procedure: First, we selected ARs observed from July 2001 to August 2007 lying within the central longitude and latitude range of (-40, 40). Then we shortlisted the ARs depending upon their areas and the availability of required data.

The magnetic field components corresponding to the Hinode observations were derived from the Stokes' profiles using Unno-Rachkowsky [21, 22] inversion code provided in SolarSoft. We resolved the usual 180° ambiguity in the transverse components of all vector magnetograms using the acute angle method. Further, to avoid any projection effects, we transformed the maps to the disc center [23].

All three components (u_x, u_y and u_z) of the velocity vector ($\mathbf{u}$) are required to determine the parameter h^z_k of an AR with depth (see Eqn.2). We derived the horizontal components (u_x, u_y) using inversion [24] of p-mode parameters obtained from ring-diagrams of 128×128×1664 data cubes [25]. The vertical component (u_z) was derived from the divergence of horizontal components, assuming mass conservation [20, 26]. This gives the average flow in an AR with a horizontal spatial resolution of 16° while the vertical resolution varies from 0.1 to 0.2 Mm with depth in the depth range of 0.0 – 14.0 Mm.

Previous studies have revealed the presence of bipolar structure in sub-photospheric flows of several ARs [27–30]. This implies that there are two different types of twists in their underlying flows. To deal with this issue, we have computed vertical kinetic helicity densities $(h^z_{\mathrm{k1}})_{\mathrm{av}}$ and $(h^z_{\mathrm{k2}})_{\mathrm{av}}$ averaged over the two depth ranges corresponding to the bipolar flow structures, viz., 0.0-2.5 and 2.5-12 Mm, respectively.

4. Results and Discussions

Results of our analysis of the 91 selected ARs are shown in Figures 1 - 3.

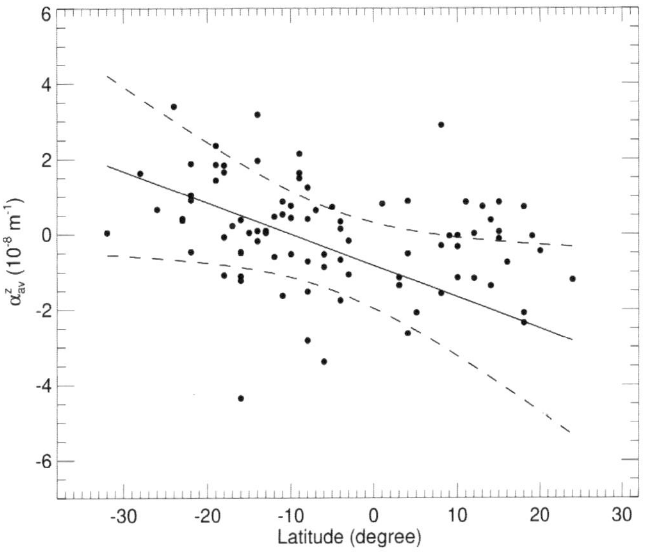

Figure 1. Latitudinal distribution of the force-free parameter α^z_{av}. The solid line represents a straight line fit through the data points while the dashed curves correspond to 95% confidence intervals.

4.1. Latitudinal Distribution of Magnetic Helicity

Figure 1 shows the latitudinal distribution of magnetic helicity parameter α^z_{av} obtained for our sample of ARs. We found 66% (63%) ARs located in the northern (southern) hemisphere with negative (positive) α^z_{av}, in agreement with previous reports [2, 11, 18]. The data points, however, show a large scatter in both the hemispheres raising the following questions: Is the hemispheric sign statistically significant? Do the α^z_{av} values have any relation with latitude? To address these questions, we fitted a straight line through the measured data points using a linear regression model (the solid line in Figure 1). Slope of the fitted line implies that the magnitude of α^z_{av} increases with latitude. However, this inference is not fully satisfactory because there is no known reason for a straight line to be the appropriate fit. But at the same time, one can not justify higher order polynomial fits. We calculated the average α^z_{av} to evaluate statistical significance of the hemispheric trend. These are found to be -1.39×10^{-8} $(+3.05 \times 10^{-9})$ m^{-1} for the northern (southern) hemispheres, confirming the hemispheric trend of the estimated α^z_{av} values for the ARs.

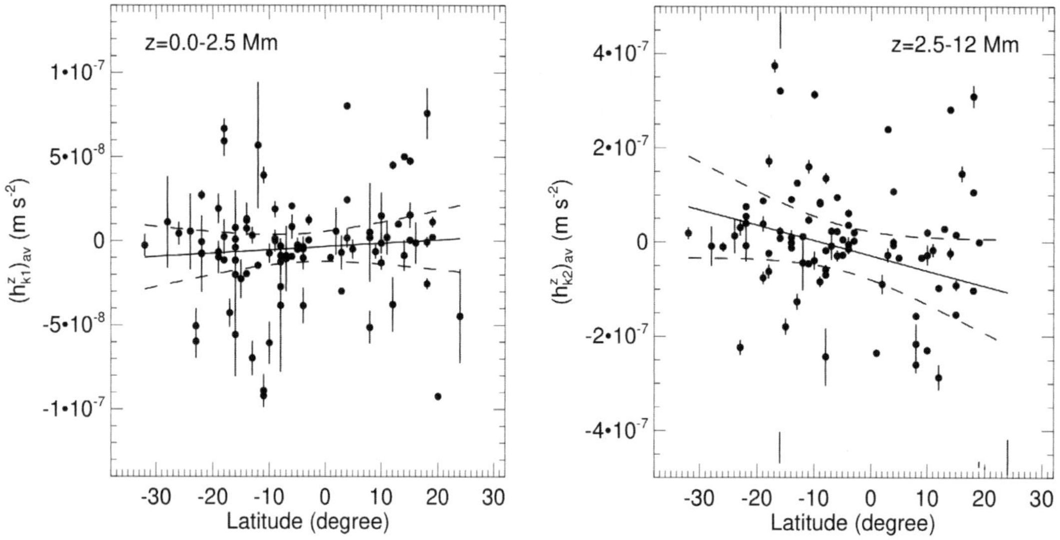

Figure 2. Latitudinal distribution of the vertical component of kinetic helicity density $(h^z_{\rm k})_{\rm av}$ averaged over the depth ranges 0-2.5 Mm (left column) and 2.5-12 Mm (right column). The solid lines represent straight line fits through the data points while the dashed curves correspond to 95% confidence ($\sim 2\sigma$) levels.

4.2. Latitudinal Distribution of Kinetic Helicity

Figure 2 shows the latitudinal distribution of the vertical component of kinetic helicity densities $(h^z_{\rm k1})_{\rm av}$ and $(h^z_{\rm k2})_{\rm av}$ averaged over depth ranges 0.0-2.5 Mm (left panel) and 2.5-12 Mm (right panel), respectively. There is no obvious hemispheric preference observed for $(h^z_{\rm k1})_{\rm av}$ as there are 47% (53%) ARs in the northern (southern) hemisphere having negative (positive) values. But $(h^z_{\rm k2})_{\rm av}$ shows a significant trend of HSR as 69% (56%) ARs in the northern (southern) hemisphere show negative (positive) helicity. The average value of $(h^z_{\rm k})_{\rm av}$ for northern (southern) hemisphere is found to be $+1.88 \times 10^{-9}$ (-7.62×10^{-9}) m s^{-2}. This confirms the opposite hemispheric trend for $(h^z_{\rm k1})_{\rm av}$. The average value of $(h^z_{\rm k2})_{\rm av}$ for northern (southern) hemisphere is -7.0×10^{-8} $(+1.7 \times 10^{-8})$ m s^{-2}, which further confirms the HSR for $(h^z_{\rm k2})_{\rm av}$.

4.3. Relation between the Kinetic and Magnetic Helicities

As mentioned in Section 1, magnetic field lines are rooted beneath the photosphere where they interact with the sub-photospheric flows. Therefore, one may expect the twists of magnetic and velocity fields to have some association. To examine this possibility, we plotted the force-free parameter $\alpha^z_{\rm av}$ against the sub-photosphere twist parameters, $(h^z_{\rm k})_{\rm av}$, as shown in Figure 3. Further, to get a quantitative measure of their relationship, we computed the Pearson correlation coefficients (r) among these parameters.

Figure 3 shows that there is no significant correlation between the force-free parameter $\alpha^z_{\rm av}$ and the twist of sub-photospheric flows. The correlation parameter suggests that $\alpha^z_{\rm av}$ is anti-correlated ($r \approx$-0.03 and -0.07) with $(\omega^z_2)_{\rm av}$ and, $(h^z_{\rm k1})_{\rm av}$ and mildly correlated ($r \approx$+0.11 and +0.23) with $(\omega^z_1)_{\rm av}$ and $(h^z_{\rm k2})_{\rm av}$.

5. Summary and Conclusions

From our study of the twists in photospheric magnetic and sub-photospheric velocity fields of 91 ARs selected in the period of July 2001-August 2007, we derive the following conclusions:

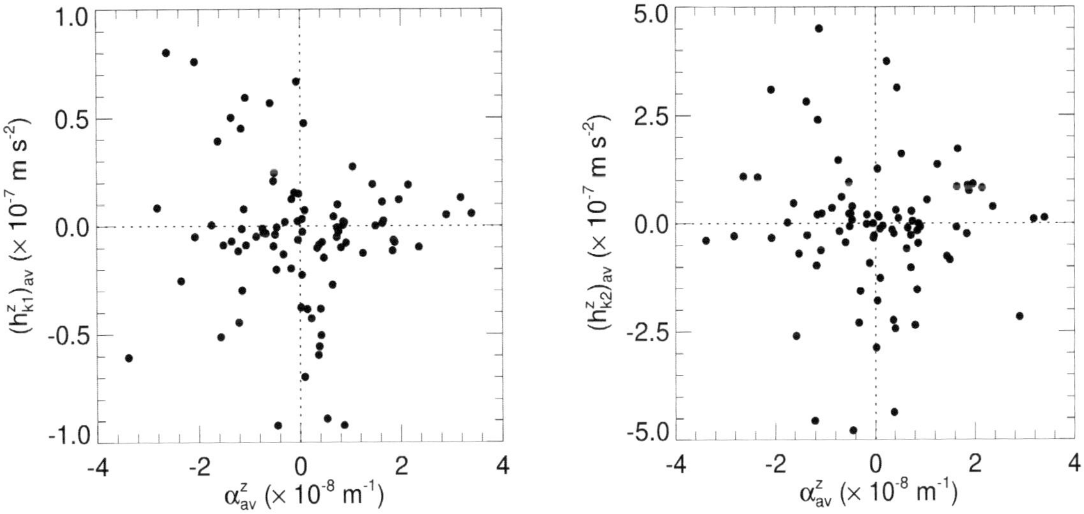

Figure 3. Association of magnetic helicity parameter α^z_{av} with the twist parameters of sub-photospheric flows for the sample of 91 ARs.

- Magnetic helicity parameter $\alpha^z_{\rm av}$ shows a significant hemispheric trend, in agreement with previous reports.
- No clear hemispheric trend is observed for average vertical kinetic helicity in the depth range 0.0-2.5 Mm while a strong hemispheric trend is discernible in the depth range 2.5-12 Mm.
- No unambiguous association is found between the twists of surface magnetic fields and sub-surface flows.

In summary, statistically significant and unambiguous association is not found between the topology of photospheric magnetic field and the sub-photospheric flows. Our analysis further supports the results from a recent study by Gao et al. [18]. It should be noted that there are some factors which may have affected these results. We have used only a single vector magnetogram map corresponding to an AR to derive the photospheric twist, while, the twist in sub-photospheric flow is derived using 1664 minutes' data cubes. Also, the magnetic and Doppler field observations were obtained from different instruments. The results are expected to improve with the recent availability of higher resolution observations from space borne instruments, such as, the Solar Dynamics Observatory (SDO).

Acknowledgments
This work utilizes data obtained by the GONG program operated by AURA, Inc. and managed by the National Solar Observatory under a cooperative agreement with the National Science Foundation, U.S.A. The vector magnetogram data were obtained from MSFC and Hinode/SOT. Hinode is a Japanese mission developed and launched by ISAS/JAXA, with NAOJ as domestic partner and NASA and STFC (UK) as international partners. It is operated by these agencies in co-operation with ESA and NSC (Norway). R. A. Maurya thanks the LOC for providing the financial support to attend the meeting that led to many fruitful discussions.

References

[1] van Ballegooijen A A 1999 *American Geophys. Union Geophys. Monograph Ser.* **111** 213–220

[2] Longcope D W, Fisher G H and Pevtsov A A 1998 *Astrophys. J.* **507** 417–432

[3] Longcope D W and Welsch B T 2000 *Astrophys. J.* **545** 1089–1100

[4] Hagyard M J, Teuber D, West E A and Smith J B 1984 *Solar Phys.* **91** 115–126

[5] Berger M A and Field G B 1984 *J. Fluid Mech.* **147** 133–148

[6] Parker E N 1955 *Astrophys. J.* **122** 293–314

[7] Pouquet A, Frisch U and Leorat J 1976 *J. Fluid Mech.* **77** 321–354

[8] Moffatt H K 1978 *Magnetic field generation in electrically conducting fluids* 1st ed (Cambridge University Press)

[9] Pevtsov A A, Canfield R C and Metcalf T R 1994 *Astrophys. J.* **425** L117–L119

[10] Choudhuri A R 2003 *Solar Phys.* **215** 31–55

[11] Pevtsov A A, Canfield R C and Metcalf T R 1995 *Astrophys. J.* **440** L109–L112

[12] Choudhuri A R, Chatterjee P and Nandy D 2004 *Astrophys. J.* **615** L57–L60

[13] Tiwari S K, Venkatakrishnan P and Sankarasubramanian K 2009 *Astrophys. J.* **702** L133–L137

[14] Duvall Jr T L and Gizon L 2000 *Solar Phys.* **192** 177–191

[15] Gizon L and Duvall Jr T L 2003 *GONG+ 2002. Local and Global Helioseismology: the Present and Future* ed. H. Sawaya-Lacoste (ESA SP-517, Noordwijk:ESA), 43–52

[16] Zhao J 2004 *Inference of solar subsurface flows by time-distance helioseismology* Ph.D. thesis Stanford University

[17] Komm R, Howe R, Hill F, Miesch M, Haber D and Hindman B 2007 *Astrophys. J.* **667** 571–584

[18] Gao Y, Zhang H and Zhao J 2009 *Mon. Not. Roy. Astron. Soc.* **394** L79–L83

[19] Gizon L, Birch A C and Spruit H C 2010 *Ann. Rev. Astron. Astrophys.* **48** 289–338

[20] Maurya R A 2010 *A Study of Oscillations in Solar Active Regions* Ph.D. thesis Udaipur Solar Observatory, Physical Research Laboratory, India

[21] Unno W and Kato S 1962 *Pub. Astron. Soc. Japan* **14** 417–430

[22] Rachkowsky D N 1967 *Izv. Krymsk. Astrofiz. Obs.* **37** 56

[23] Venkatakrishnan P and Gary G A 1989 *Solar Phys.* **120** 235–247

[24] Gough D 1985 *Solar Phys.* **100** 65–99

[25] Hill F, Bolding J, Toner C, Corbard T, Wampler S, Goodrich B, Goodrich J, Eliason P and Hanna K D 2003 *GONG+ 2002. Local and Global Helioseismology: the Present and Future* ed. H. Sawaya-Lacoste (ESA SP-517, Noordwijk:ESA), 295–298

[26] Komm R, Corbard T, Durney B R, González Hernández I, Hill F, Howe R and Toner C 2004 *Astrophys. J.* **605** 554–567

[27] Komm R, Howe R, Hill F, González Hernández I and Toner C 2005 *Astrophys. J.* **630** 1184–1193

[28] Maurya R A, Ambastha A and Tripathy S C 2009 *Astrophys. J.* **706** L235–L239

[29] Maurya R A and Ambastha A 2010 in *Astrophys. Space Sci. Proc.*, Magnetic Coupling between the Interior and the Atmosphere of the Sun, ed. S. S. Hasan & R. J. Rutten (Berlin: Springer), 516

[30] Maurya R A and Ambastha A 2010 *Astrophys. J.* **714** L196–L201

Stochastic Seismic Emission from Acoustic Glories in Solar Active Regions

Alina Donea, Marie Newington

School of Mathematical Sciences, Monash University, 3800, Victoria, Australia

E-mail: alina.donea@monash.edu

Abstract. Helioseismic images of active regions show enhanced seismic emission in 5 mHz oscillations in a halo surrounding the active region called the "acoustic glory". In this paper we analyse the high-frequency power excess surrounding two active regions that occurred during the "shy" ascending phase of the solar cycle 24, at the beginning of 2010. This study compares the acoustic properties of seismic emission from acoustic glories with that from the quiet Sun. The power distribution of quiet-Sun seismic emission far from solar activity is exponential, as for random Gaussian noise, and therefore not episodic. The magnitudes of the acoustic glories and their seismic structure allow us to make predictions of the seismic behaviour of active regions and compare the data with present theoretical models.

1. Introduction

The wealth of data from the Michelson Doppler Imager (Scherrer et al., 1995) onboard the Solar and Heliospheric Observatory (SOHO/MDI) satellite has provided us with images of solar active regions (AR) that scatter or absorb or convert sound waves in a very efficient way. These images revealed important properties of the acoustic regions, such as 'acoustic moats' and 'acoustic glories' (Braun et al. 1998, Lindsey and Braun 1998a, Braun and Lindsey 1999). Local helioseismology techniques such as acoustic holography have been used to map the seismic powers around active regions (Braun and Lindsey 2000). The aim of acoustic holography is to identify and estimate the strength of seismic sources, absorbers or scatterers of sound waves at the solar photosphere and also in the solar interior. The computational work consists in the reconstruction of the coherent acoustic source from the observed oscillations of the photosphere, reversed back in time, to their original location in the solar convection zone.

'Acoustic moats' extend from sunspots to areas far away from these highly magnetic regions. The absorption and scattering properties of acoustic moats may be explained by the existence of a convection-type plasma cell flowing outward at high speeds just beneath the photosphere (Braun et al. 1998). 'Acoustic glories' form a seismic halo around only complex active regions, with a high power at frequencies close to 5 mHz (Donea et al. 2000). Only large multipolar magnetic regions will present complex glories; some will display conspicuous point-like seismic emitters of higher frequency seismic power. In fact, a measurable enhancement in seismic emission actually exists around single monopolar sunspots (Lindsey and Braun 1999a), approximately 2.5% above that of the seismic emission from the surrounding quiet Sun, but this is subtle and diffuse. The strong output of high-frequency energy can therefore offer new insights into the physical processes occurring in and around active regions.

We have searched for active regions in the ascending phase of the solar cycle 24. The aim is to image the high frequency acoustic glories, the halo that shows sharply *enhanced* seismic emission, largely from small, point-like elements that tend to form thin, beady strings. We used the MDI/SOHO (www.soi.standford) database. The year 2010 in the solar cycle 24 had started with a few small active regions; then, the solar cycle had again a quiet time around March – April 2010. Interestingly, during June–August 2010, the Sun produced some "perfect" monopolar sunspots, such as the 11092. Two large enough ARs have drawn our attention. They had a full development on February 07, and February 14 and were observed intensively by the MDI instrument. In this paper, we have analysed the seismic enhancements in the 5 mHz acoustic glories of the AR of the year 2010, and compared the stochastic emission signal with previous work.

2. Acoustic glories in seismic power maps

We have analysed the active region, AR 8996 generated on May 18, 2000, at the maximum of the solar cycle 23 . In Fig 1a we mapped the 4.5–5.5 mHz band seismic power of AR 8996 and compared this with the AR 8179 (Fig 1b). One can see features such as strings of small-scale emitter, bright individual small seismic sources. Fig 1b shows a seismic emission map of AR 8179 in the 4.5–5.5 mHz band, made from MDI Doppler observations integrated over the 24 hr period beginning at March 15, 1998. The temporal character of the acoustic glory was largely analysed in Donea et al. (2000). We will use these images as reference work when compare acoustic glories of ARs in the rise phase of the cycle 24 with other ARs. As we emphasized in the Introduction we were interested in looking at two most recent ARs with good observational coverage by MDI/SOHO.

The acoustic glories appear as bright haloes surrounding the active regions. They are largely comprised of small, discrete seismic emitters that tend to cluster in strings in low-magnetic regions. The individual small-scale emitters comprising the strings are at the acoustic diffraction limit of the 5 mHz acoustic images attainable from the medium-resolution MDI images, ~3 Mm.

Fig 1 actually illustrates similarities in the acoustic glories of the two AR. Both ARs occurred during a solar maximum. Mostly remarkable is the conspicuous emitters in the acoustic glory surrounding the active regions and their preference for beading along the magnetic neutral line separating the two active regions (AR 8996 and 8998). The acoustic glories are generally sustained at a significantly greater level than from the quiet Sun. A further detailed analysis of AR 8996 will be presented in Donea and Newington (2011).

3. Results

The two active regions of the new cycle 24 were magnetically less complex than the ARs discussed in the previous section. The active region AR11045 was located at the position N24E01, on February 07 2010. MDI/SOHO continuously observed the active region from 10:19 – 15:26 UT. The active region AR11046 was located at N23W25 (similar latitude with AR 11045) and was observed by MDI from 16:21 – 23:02 UT. We remapped the solar full disk images onto Postels projections with the scale of the map 0.002 $R_{\odot}$ = 1.4 Mm per pixel over 256 pixels, so the extent of the maps are about 358.4 Mm in each direction.

Holographic images of seismic emission in the two active regions of year 2010 show high-frequency acoustic glories extending beyond the surface magnetic regions (left panel in Figures 2 and 3). However, AR 11045 and AR 11046 show only relative weak, diffuse acoustic glories that do not present intense point-like elements of enhanced seismic emission at 5 mHz, as shown for the ARs in the Figure 1. Some string-like features can be seen on the egression power map in Figure 2 (integrated over 5 h and 24 h, respectively).

Acoustic power halos at 5 mHz, are also seen in the right panels in Figures 2 and 3, as localized enhancement of the surface disturbance that registers the arrival of an underlying

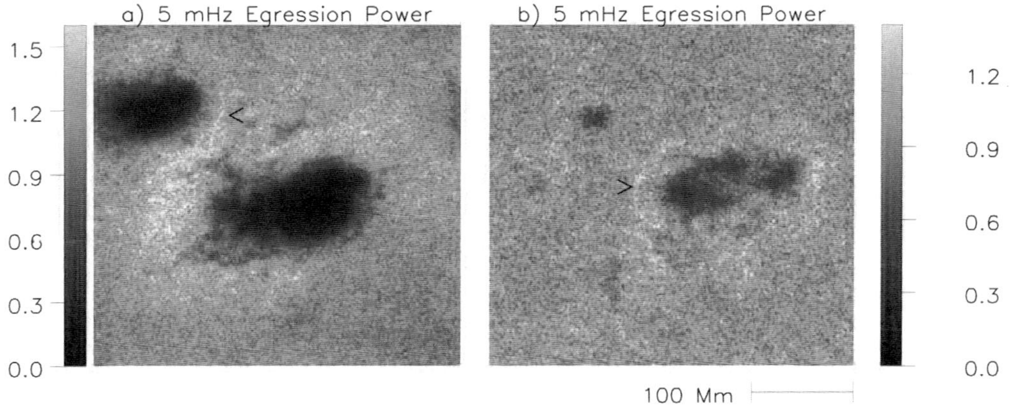

Figure 1. Seismic egression-power maps (a) of AR 8998 (May 18, 2000) in a 1 mHz band centered at 5 mHz, integrated over 24-hr interval (b) of AR 8179 (March 15, 1998). Reference arrow heads indicate the locations of the beading of small-scale elements of enhanced seismic emission which comprise the 5 mHz solar acoustic glories. The maps are normalised to the quiet sun areas.

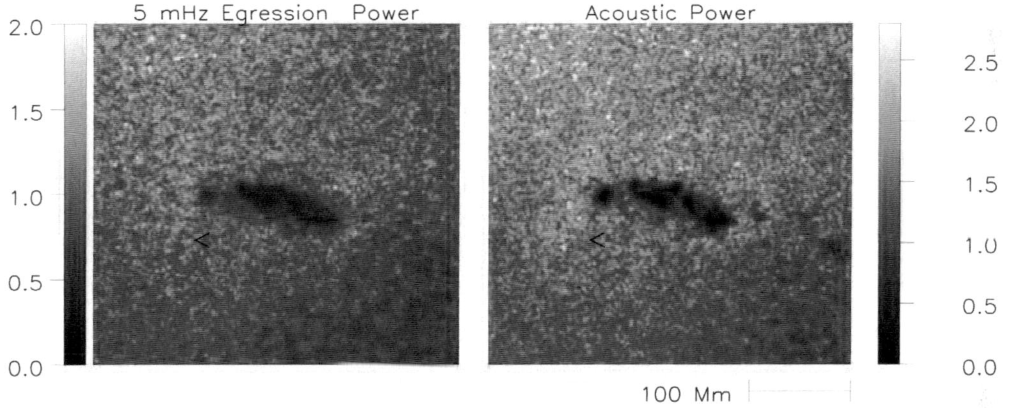

Figure 2. 5 mHz Egression power map (left) and the 5 mHz acoustic power map (right) of AR 11045 averaged over the full data set on February 07 2010. The coloured scales indicate egression power and acoustic power, normalized to unity for the quiet Sun. A halo of excess emission ('acoustic glory) is seen in the left map.

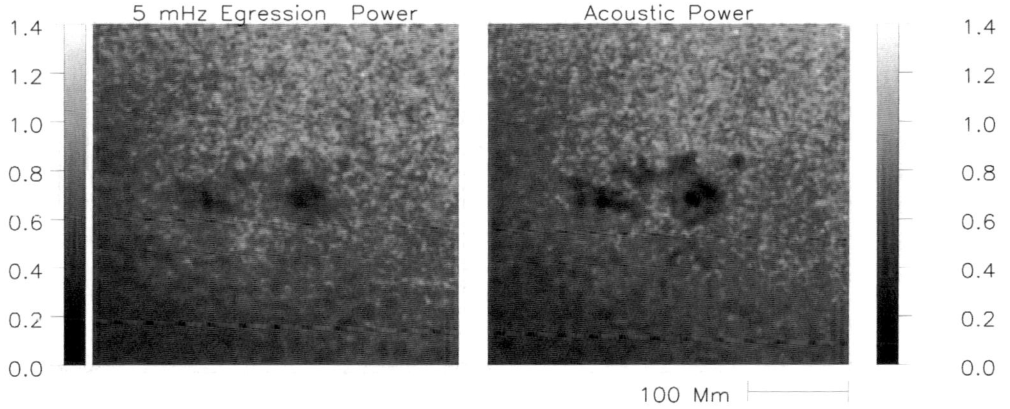

Figure 3. Same as Figure 2 for AR 11046 on February 14 2010.

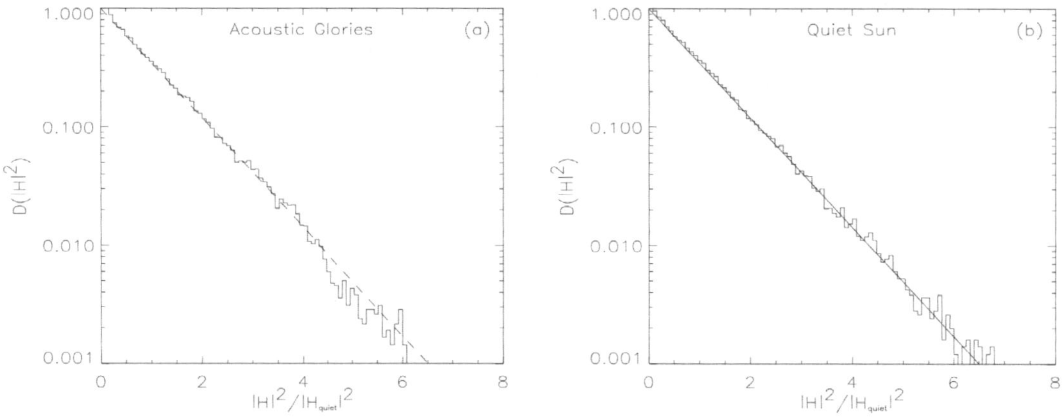

Figure 4. Histograms of 5-mHz egression power in (a) the acoustic glory (left frame) and in (b) the quiet Sun (right) for AR 11045 on February 07 2010.

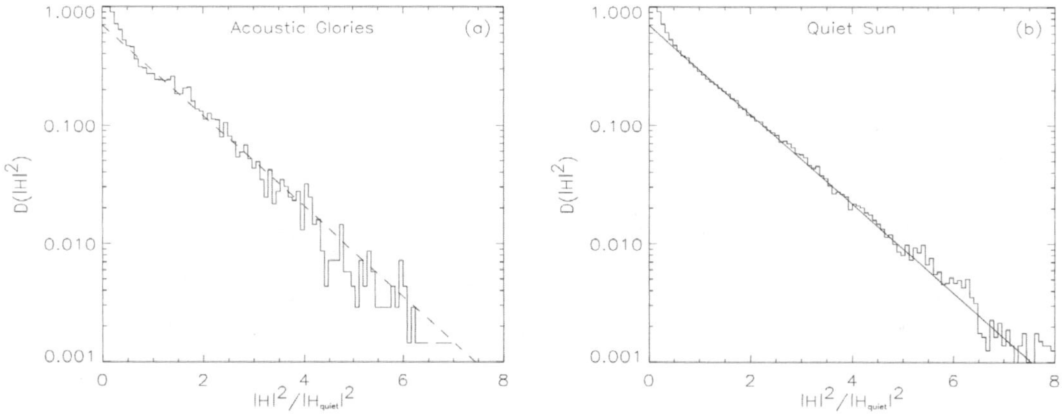

Figure 5. Distributions of 5-mHz egression power in (a) the acoustic glory (left frame) and in (b) the quiet Sun (right) for AR 11046 on February 14 2010.

wave (Schunker and Braun 2010).

The distribution of seismic power emanating from the most intense elements that comprise the acoustic glories is likewise exponential out to approximately $4|H|^2_{quiet}$. However, the behaviour of the distribution at a higher seismic power (the concave down shape starting at approximately $4|H|^2_{quiet}$) perhaps indicates a regime of saturation of the egression power in acoustic glories. This is still a puzzle. At this point, as Donea, Lindsey and Braun (2000) emphasized, we suggest that the sustained emission that emanates from acoustic glories involves a substantially different mechanism than that which operates in the quiet Sun.

The histograms of acoustic glories and the quiet sun are shown in the Figure 4 and 5 (we have analysed about 30 small-scale seismic elements for glories and 200 small-scale elements for the quiet sun). First, we should notice that the 5 mHz egression power glories represent the contribution from the acoustic radiation that propagates downwards from a focal point/ a source located nearby the active region, and later is refracted back into the photosphere. Secondly, the acoustic 5 mHz enhancement can also be attributed to the locally generated acoustic radiation from the subphotosphere (in the quiet Sun only the locally generated acoustic radiation is present). Both phenomena will contribute to the final emission in acoustic glories. The high seismic emission from the small-scale seismic sources in the acoustic glories could be also a result of all the above phenomena; it is hard to know which of these two contributions dominate.

If an egression power timeseries contains a component of acoustic emission that is *substantially episodic*, this should result in a significant departure of the egression-power distribution, D, from the nominal exponential profile.

The next step was to look at the the egression power time–series representing acoustic emitters in the quiet Sun. Control measurements from the quiet Sun were taken from the 5 mHz egression power maps. The results are shown in the Figures 4 and 5 (right plots). For random Gaussian noise, the distribution, $D(|H^2|)$ in egression power $|H|^2$ should be simply exponential:

$$D(|H^2|) \;=\; \exp\left(-|H|^2/H_{quiet}^2\right), \tag{1}$$

with H_{quiet} representing the mean power of the quiet Sun noise.

Distributions D of egression power $|H|^2$ from the quiet Sun far from magnetic regions show a profile that is accurately exponential. This is consistent with sustained, random Gaussian noise. In particular, for quiet sun, the histogram $D(|H|^2)$ shows some increase at the higher values of its argument, $|H|^2$, by the contamination of the outlying neighbourhood of the active region by occasional glory-like emission (right plots in Figures 4 and 5).

4. Conclusion

This work confirms that it is only around large-multipolar active regions that the acoustic glory is prominent, showing seismic emission averaging roughly 15% in excess of the mean quiet Sun. The glories are identified with seismic holography analysis sensitive to propagating waves.

High frequency small-scale seismic emitters are localised within acoustic glories with an episodic temporal behaviour. A small number of these small scale emitters is enough to contaminate the quiet sun region surrounding an AR and produce acoustic glories. The highest possible spatial resolution data from the Solar Dynamics Observatory (SDO) should be able to help us answering the question about "how many seismic emitters will generate a significantly seismic glory around an active region?". The new data will also improve the errors in the statistics, when dealing with 30 small-scale seismic sources, for examples.

The average spatial extension of the acoustic glories around active regions is 20 – 30 Mm. The most interesting structure of the acoustic glories, the bead alignmenet of emitters suggests that the emission sites lie preferably near low-magnetic areas, and sometimes follow the magnetic neutral line between two polarities of the host active region. This is an area with near horizontal magnetic field. This aspect will also be analysed with the new SDO data. Therefore, any theory intended to explain the acoustic glories and the beading of seismic emitters would probably refer to the interaction of acoustic waves with the magnetic field.

5. References

[1] Braun, D and Lindsey, C 1999 *ApJ* **513** L79

[2] Braun, D and Lindsey, C. 2000 *Solar Phys* **192** 307

[3] Braun, D, Lindsey, C, Fan, Y and Fagan, M 1998 *ApJ* **502** 968

[4] Donea, A, Lindsey, C and Braun, D 2000 *Solar Phys* **192** 321

[5] Donea A and Newington M 2011 it Solar Phys, in preparation

[6] Lindsey, C and Braun, D 1998a, *ApJ* **499** L99

[7] Scherrer, P.H., Bogart, R.S., Bush, R.I., Hoeksema, J.T., Kosovichev, A.G., Schou, J., Rosenberg, W., Springer, L., Tarbell, T.D., Title, A., Wolfson, C.J., Zayer, I., MDI Engineering Team: 1995 *Solar Phys.* **162** 129

[8] Schunker H and Braun D 2010 *Solar Phys.*, in press

Local helioseismology of sunspot regions: comparison of ring-diagram and time-distance results

A.G. Kosovichev[1], S. Basu[2], R. Bogart[1], T.L. Duvall, Jr[3], I. Gonzalez-Hernandez[4], D. Haber[5], T. Hartlep[1], R. Howe[4], R. Komm[4], S. Kholikov[4], K.V. Parchevsky[1], S. Tripathy[4], J. Zhao[1]

[1] Stanford University, Stanford, CA, USA; [2]Yale University, New Haven, CT 06520, USA; [3]Solar Physics Laboratory, Goddard Space Fight Center, NASA, Greenbelt, MD 20771, USA; [4]National Solar Observatory, Tucson, AZ 85719, USA; [5]JILA, University of Colorado, Boulder, CO 80309, USA

E-mail: `AKosovichev@solar.stanford.edu`

Abstract. Local helioseismology provides unique information about the subsurface structure and dynamics of sunspots and active regions. However, because of complexity of sunspot regions local helioseismology diagnostics require careful analysis of systematic uncertainties and physical interpretation of the inversion results. We present new results of comparison of the ring-diagram analysis and time-distance helioseismology for active region NOAA 9787, for which a previous comparison showed significant differences in the subsurface sound-speed structure, and discuss systematic uncertainties of the measurements and inversions. Our results show that both the ring-diagram and time-distance techniques give qualitatively similar results, revealing a characteristic two-layer seismic sound-speed structure consistent with the results for other active regions. However, a quantitative comparison of the inversion results is not straightforward. It must take into account differences in the sensitivity, spatial resolution and the averaging kernels. In particular, because of the acoustic power suppression, the contribution of the sunspot seismic structure to the ring-diagram signal can be substantially reduced. We show that taking into account this effect reduces the difference in the depth of transition between the negative and positive sound-speed variations inferred by these methods. Further detailed analysis of the sensitivity, resolution and averaging properties of the local helioseismology methods is necessary for consolidation of the inversion results. It seems to be important that both methods indicate that the seismic structure of sunspots is rather deep and extends to at least 20 Mm below the surface, putting constraints on theoretical models of sunspots.

1. Introduction: sunspot models and helioseismology

Sunspots and active regions are key elements of solar magnetism. Solar magnetic field is generated by a dynamo process in the interior. It appears on the surface in the form of highly concentrated magnetic structures, sunspots, often surrounded by plages representing large areas of diffuse magnetic field. The mechanism of formation and stability of sunspots is currently not understood. Observations of the solar surface provide evidence that sunspots are formed by merging together small-scale magnetic elements, which appear on the surface as emerging magnetic flux. After sunspots are formed, they may be stable for several weeks. Long-living sunspots are often accompanied by new emerging magnetic flux events. The decay of sunspots is associated with a rapid diffusive process in the form of enhanced moat flow and magnetic field submergence. It is quite clear that all these processes are controlled by subsurface dynamics, but how turbulent convection can lead to formation of such stable self-organized magnetic structures is a great puzzle.

Recent numerical simulations of magnetic structures on the Sun lead us to two basic theoretical ideas about the sunspot formation and stability. The first idea comes from the simulations of Rempel et al [1], in which the sunspot structure is supported by fixing a localized magnetic field concentration at the bottom boundary. These simulations show formation of mean converging downflows around a pore-like structure, which does not have a penumbra. When the penumbra-like magnetic structure is modeled (by setting up appropriate upper boundary conditions) then the simulations show that the subsurface dynamics is dominated by outflows, similar to the Evershed effect. These flows are driven by the overturning convection and Lorentz force near the surface, but the diverging flows extend through the whole depth of the computational domain [2]. A possible explanation for the deep diverging upflows is that the penumbra partially block the turbulent convective heat flux, and it needs to be transported by upflows (Rempel, private communication). In both cases, once the bottom boundary condition is released the structure disappears on the time scale of the convective turn-over time. Thus, for long-term stability in this model, it is necessary that the sunspots are formed by strong-field flux tubes anchored in the deep interior, where the convective turn-over time is large, and therefore, the lifetime of sunspots corresponds to this turn-over time.

Another important idea comes from the simulations of spontaneous formation of magnetic structures from an initially uniform magnetic field, recently obtained by Kitiashvili et al [3]. In this model, a stable pore-like structure (without penumbra) is formed by merging small-scale magnetic elements (flux tubes) initially concentrated by vortexes (whirlpool-like structures) in the intergranular lanes. The magnetic field strength in this self-organized magnetic structure is about 1.5 kG at the surface and 6 kG in the interior. The boundary conditions in these simulations only keep the mean magnetic field strength constant in the simulation domain, and no artificial boundary conditions are used to maintain the structure. It is intrinsically self-maintained. The principal mechanism of the structure formation and stability is associated with strong converging downdrafts. The sunspot penumbra has not been simulated for the self-organized structure model, but a separate computation of magnetoconvection in highly inclined strong magnetic fields shows that the Evershed flows are likely to be confined in the top 1Mm deep layer [4; 5]. The magnetic pore-like structure has an internal cluster-type structurization. The simulations indicate that for formation of a large sunspot-like structure it is necessary to increase the depth of the simulation domain.

These two radiative MHD simulations provide nice examples of the classical Parker's sunspot dilemma: monolithic vs cluster model [6]. The Rempel's model belongs to the monolithic type while Kitiashvili's simulations give a demonstration of a cluster-type model. Observations of sunspots on the solar surface cannot resolve this dilemma. It seems that the process of sunspot formation and the flow dynamics dominated by converging flows is similar to the cluster model until penumbra is developed [7; 8]. Both MHD models strongly suggest that the sunspot structure and flows extend into the deep interior because of the long-term stability requirement. But so far, the models have been calculated only for relatively shallow computational domains (6-8 Mm deep) due to the current computing power limitations.

In addition to the MHD sunspot models, several magnetostatic models were proposed (e.g. [9; 10]). These models do not include flows and are calculated using the hydrostatic pressure balance for prescribed parametric distributions of magnetic field and density or temperature. The main goal of these models is to provide background models for linear wave simulations, used for testing local helioseismology inferences. These models can be quite shallow in terms of the relative seismic perturbations.

The surface structure and dynamics of sunspots have been studied in detail by high-resolution imaging and spectro-polarimetric measurements. Studying the interior properties can be done only by local helioseismology methods, and this is a very challenging task. Helioseismic inversions have to rely on our understanding of the oscillation physics in a very complicated environment of the highly turbulent magnetized plasma. In addition, they represent a classical ill-posed problem, which does not have a unique solution. Also, for computational efficiency the relationship between the observational data and interior properties is linearized. In the case of strong perturbations, when more accuracy is required, an iteration scheme can be applied. The advantage of this approach is that it allows to investigate subsurface structures and dynamics without a priori model constraints (usually, only with smoothness constraints), and also to explore the sensitivity and resolution of helioseismology techniques. The accuracy of these techniques must be investigated by numerical simulations. Currently, there are several local helioseismology methods, including the ring-diagram analysis, acoustic holography and time-distance helioseismology. The current status of the local helioseismic diagnostics of sunspots was recently reviewed by Kosovichev [11].

Our goal is to investigate and compare the inversion results for subsurface sound-speed variations of AR 9787, obtained by two different local helioseismology techniques, the ring-diagram analysis and time-distance helioseismology. This sunspot was selected for comparing techniques at a HELAS workshop [12]. The work presented here was initiated at recent meetings of the LoHCo (Local Helioseismology Comparison) team to understand why the two techniques produced different results in the case of active region 9787. In this paper, we point out that the quantitative comparison of the ring-diagram and time-distance inversion results is not straightforward. It must take into account differences in the sensitivity and spatial resolution of these methods. In particular, the nonuniform distribution of the oscillation power in active region and differences in the averaging kernels of the inversions may have significant effects. Despite the differences, we find that both techniques indicate that sunspots and active regions are associated with sound-speed perturbations in the deep layers of the upper convection zone. These results are not consistent with "shallow" models of sunspots.

2. Local helioseismology of AR 9787

2.1. Structure and evolution of active region

Active region NOAA 9787 was observed by SOHO/MDI on January, 20-28, 2002. It had a complex magnetic structure, Beta-Gamma, according to the NOAA classification. Its location was at 6 degrees south latitude and 130 degrees Carrington longitude. The leading sunspot had initially a round magnetic structure, but then it became significantly distorted (Fig. 1a), probably because of new emerging magnetic flux. The total magnetic flux during the interval of rotation between two symmetrical, relative to the central meridian, locations ($\pm 30^\circ$) increased by more than 20%, from 7.9×10^{21} Mx to 9.7×10^{21} Mx. The leading sunspot was surrounded by a strong plage region with magnetic field reaching ~ 750 G. The plage region produced significant seismic travel-time anomalies comparable with the anomalies of the sunspot (Fig. 1b).

The travel-time anomaly in the sunspot region exhibits the sign reversal: the travel time perturbations relative to a quiet Sun region are positive for the short travel distances, in the range of 0.54-1.02 and 1.02-1.38 degrees, and negative for longer distances. In the plage region, the travel-time anomalies are negative for all distances. This fact has important implications for comparison of the ring-diagram and time-distance results.

2.2. Ring-diagram analysis

The ring-diagram analysis was carried out by using two fitting methods, described by Haber et al [14] and Basu et al [15]. These two techniques give similar results. We plan to discuss their comparison in a separate paper. Here we present the results obtained by the technique [15]. The same technique was used for the analysis presented in [12]. The results of this paper were

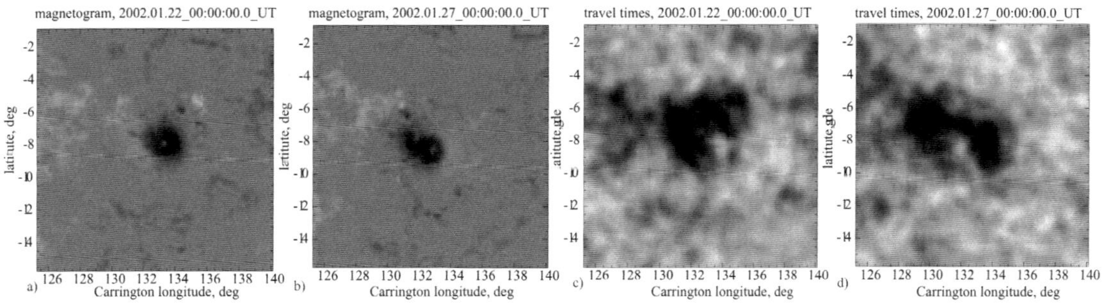

Figure 1. SOHO/MDI magnetograms of a 15-degree area around the leading sunspot of AR 9787, at two location approximately (a) 30 degrees West, and (b) 30 degrees East of the central meridian. The magnetograms are remapped into the heliographic coordinates using the Postel's projection with the resolution of 0.12 deg/pixel. The gray scale corresponds to the magnetic field strength from -1500 G (black) to 1500 G (white). c-d) SOHO/MDI mean travel time of acoustic waves for the distance range of 1.38–1.86 degrees of the 15-degree areas of AR 9787 as in panels a) and b). The gray scale corresponds to the travel times from 30.15 min (black) to 30.6 min (white). The travel times are calculated with the resolution of 0.24 deg/pixel and shown after 2x2-pixel smoothing.

substantially different from the previous ring-diagram studies of other active regions [15; 16]. During our investigation, we found that the results originally published in [12] were obtained for AR 9829 observed in February 2002 during the following Carrington rotation. The active region 9829 had the Carrington coordinates close to the coordinates of AR 9787. Presumably, AR 9829 was a remnant of the decaying AR 9787. When this region appeared on the East limb on February 16, 2002 (and was given the new NOAA number), the sunspot area was reduced by a factor of 10, from 400 to 40 millionths of the solar hemisphere. Two days later, by February 18, sunspots in this region completely disappeared. Only a plage area without sunspots remained during that Carrington rotation. Thus, the original ring-diagram results in [12] were mostly for the plage region without sunspots. This partly explains the difference from the previous ring-diagram results.

We have carried out inversions of the frequency difference measured by the ring-fitting technique [15] for a 15-degree square region containing AR 9787, and a similar quiet-Sun region without sunspots. We used two independent inversion codes. Both codes are based on the variational principle for the frequency difference [17], but employ different inversion methods: Optimally Localized Averaging (OLA) [18] and Subtractive Optimally Localized Averaging (SOLA) [15]. In Figure 2a, we show the inversion results of the frequency difference for the relative squared sound-speed perturbations obtained by these methods: OLA (dash-dots with errorbars), and SOLA (solid curve). For comparison we plot also the results published in [12] (dashed curve). The main difference is that the results [12] showed a strong near-surface positive variation and very small variations in the deeper interior, while the new results show mostly negative near-surface variations and significant positive variations in the deep layers. The results obtained by both, the OLA and SOLA inversion techniques, show a characteristic two-layer sound-speed structure. However, there are significant quantitative differences. The SOLA results show smaller variations, and the region of transition from the negative to positive variations appears shifted by 3-4 Mm down in depth. The reasons for these differences have to be investigated. They may be attributed to differences in the regularization (smoothness) parameters, which control the spatial resolution of the inversion results, also to differences in

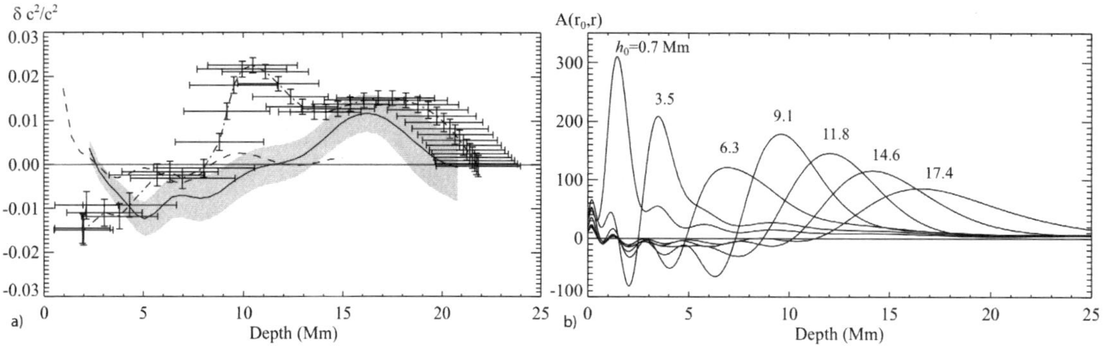

Figure 2. a) Comparison of the ring-diagram inversion results obtained for AR 9787 by two inversion technique: SOLA (solid curve with gray area indicating error estimates), and OLA (dash-dots with error bars), and the results of Gizon et al. [12] (dashed curve). The horizontal errorbars show an effective spatial resolution - spread of the averaging kernels, as defined by the inversion theory [18]. b) A sample of localized averaging kernels for the OLA ring-diagram inversion. The values of h_0 are the corresponding target depths in Mm.

the localization and shape of the averaging kernels, and other effects.

In general, it is important to realize that the inversion results do not give estimates precisely at a given depth, but represent a convolution with the averaging kernels. A sample of the averaging kernels obtained by the OLA method is illustrated in Fig. 2b. These kernels have small negative sidelobes and also are asymmetric. They show a good localization in depth between 1.5 Mm and 15 Mm. However, we were not able to obtain the good localization in the shallow and deeper layers. The averaging kernels of the SOLA method are shown in [15]. They have a good localization in the same range of depth, and more regular shape, but appear somewhat broader in the near surface layers.

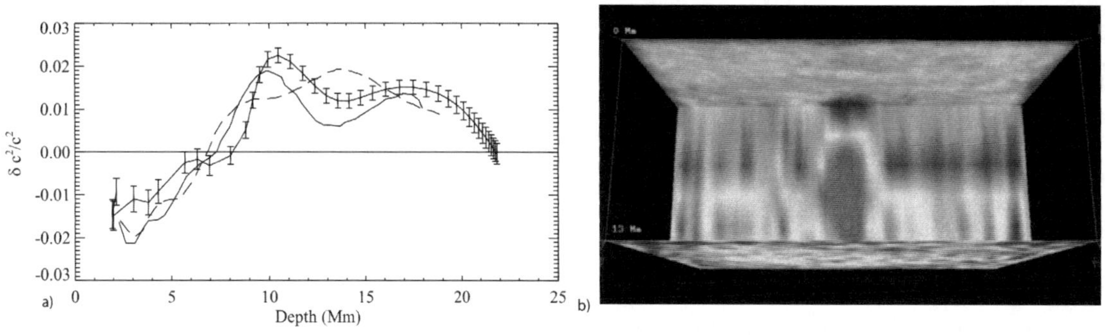

Figure 3. a) Comparison of the ring-diagram inversion results for AR 9787 obtained in this paper by the OLA inversion method (solid curve with error bars) with the inversion results of Bogart et al [16] two other active regions, AR 9906 (dashed curve) and AR 10793 (dot-dashed curve). b) A vertical cut through the 3D wave-speed structure of AR 9787 obtained by the time-distance inversion technique. Red color shows positive variations; blue color shows negative variations.

Our results for AR 9787 are in good agreement with the previous ring-diagram analysis of other active regions. A comparison with two active regions studied in [16] is shown in Fig. 3a.

According to the statistical study of Baldner et al [19], in most active regions the sound-speed variation obtained by the ring-diagram analysis has a negative variation near the surface and a positive variation in the deeper layer. Thus, the helioseismic structure of AR 9787 is not much different from other active regions.

2.3. Comparison with results of time-distance helioseismology

It has been shown by Basu et al [15] and Bogart et al [16] that this typical two-layer structure is qualitatively consistent with the time-distance inversion results [13; 20–24]. However, the quantitative comparison is not straightforward because the ring-diagram and time-distance methods have quite different spatial and temporal resolutions. The time-distance method attempts to resolve structures close to the half-wavelength resolution limit (2-3 Mm), on the time scale of 8 hours, while the ring-diagram technique has a typical spatial resolution of about 180 Mm (15 heliographic degrees) and the time scale of 24 hours or longer. Thus, the regions analyzed by the ring-diagram method are much larger than typical sunspots (for illustration see Fig. 1). Most of the oscillation signal in these areas comes not from sunspots, in which the oscillation power is suppressed, but from surrounding plage regions. The plage regions may have significant helioseismic effects (Fig. 1c-d), and thus they can make significant contribution to the mean seismic sound-speed profiles. These facts must be taken into account for quantitative comparisons between the ring-diagram and time-distance inversion results. In addition, as we have pointed out, the localization and spread of the averaging kernels also contribute to the inferred sound-speed profiles. Therefore, a direct point-to-point comparison of the sound-speed profiles obtained by different inversion techniques without taking into account differences in the sensitivity and averaging kernels, as presented for instance in [12], should be interpreted carefully.

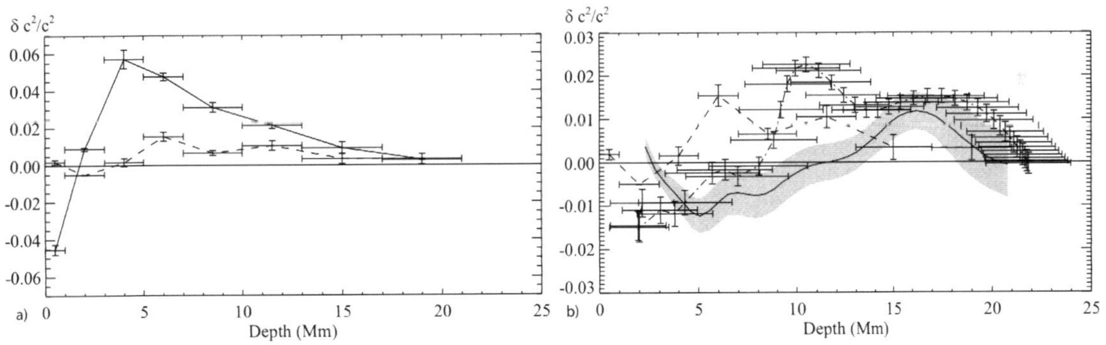

Figure 4. a) Relative variations of the seismic sound-speed structure with depth for AR 9787 obtained from the time-distance inversions: solid curve shows the results averaged over the sunspot; dotted curve shows the results of averaging for the 15-degree ring-diagram region, weighted with the local oscillation power. The horizontal bars show the width of the averaging layers in the time-distance inversion procedure [25]. The error estimates are obtained from RMS variations calculated for three different 8-hour periods. b) Comparison of the seismic sound-speed perturbations for AR 9787 obtained by the ring-diagram inversion techniques: SOLA (solid curve) and OLA (dash-dots) and by the power-weighted averaging of the time-distance inversion results (dashed curve).

To demonstrate the importance of these differences we carried out analysis of AR 9787 by the time-distance helioseismology method, using the codes developed for the SDO/HMI data analysis pipeline [25]. The time-distance inversion results are obtained in the ray-path approximation, and are shown in Fig. 3b. They reveal the typical two-layer seismic structure. In Figure 4a, we plot the depth dependence of the sound-speed variations averaged over the sunspot area (solid curve) and the variations averaged over the 15-degree ring-diagram analysis area (dashed curve). In the latter case, the sound-speed variations were averaged with weights proportional to the relative acoustic power distribution in this area. In this case, the sound-speed variation is much weaker, and also has a different depth dependence. In particular, the transition region from the negative to positive variations and the maximum of the positive variation are shifted to deeper layers.

In Figure 4b, we compare the ring-diagram inversion result obtained by the SOLA and OLA inversion methods and the power-weighted average of the time-distance inversion results. These results have similar amplitude variations and both show the two-layer structures. However, the transition between the negative and positive variations occurs at different depths: ~ 4 Mm for the time-distance result, and $\sim 5 - 8$ Mm for the ring-diagram inversions. While formally this difference is within the resolution of the averaging kernels, the sound-speed structure obtained by the ring-diagram technique appears systematically more spread with depth than the structure obtained by the weighted averaging of the time-distance results.

It will be interesting to investigate this difference in more detail. We plan to continue this investigation for other active regions and also by using numerical wave simulations [26; 27] for various models of the subsurface sound-speed and magnetic field structures.

3. Conclusions

Our investigation of the subsurface seismic structure of AR 9787 shows that the inversion results obtained by two different methods of local helioseismology, the ring-diagram analysis and time-distance helioseismology, are consistent with most of the previous results for other active regions, revealing the characteristic two-layer structure with a negative variation of the sound speed in a shallow subsurface layer and a positive variation in the deeper interior. However, there are significant quantitative differences between the inversion results obtained by the different techniques and different inversion methods. In particular, the seismic structure of the active region inferred by the ring-diagram method appears more spread with depth than the structure obtained from the time-distance technique.

In this paper, we point out that the quantitative comparison of the inversion results is not straightforward because of the substantially different spatial resolutions of the helioseismology methods. The quantitative comparison must take into account differences in the sensitivity and resolution. In particular, because of the acoustic power suppression the contribution of the sunspot seismic structure to the ring-diagram signal can be substantially reduced. We show that taking into account this effect reduces the difference in the depth of the sound-speed transition region. In this analysis, we assumed that the time-distance inversion results can be averaged over the ring-diagram analysis area, with weights proportional to the acoustic power distribution. Intuitively, this certainly appears to be a sensible thing to do. However, this assumption needs to be confirmed by numerical simulations. Further detailed analysis of the sensitivity, resolution and averaging properties is necessary for consolidation of the ring-diagram and time-distance inversion results.

Our results obtained by the two local helioseismology methods indicate that the seismic structure of sunspots is probably rather deep, and extends to at least 20 Mm below the surface. If confirmed by further studies this conclusion has important implications for development of theoretical models of sunspots.

Acknowledgement. We thank C. Balder for providing the ring-diagram fitting data and inversion results obtained by the SOLA technique, and helpful discussions.

References

[1] Rempel M, Schüssler M and Knölker M 2009 ApJ **691** 640–649

[2] Rempel M D 2010 *American Astronomical Society Meeting Abstracts* (*American Astronomical Society Meeting Abstracts* vol 216) p 211.05

[3] Kitiashvili I N, Kosovichev A G, Wray A A and Mansour N N 2010 ApJ **719** 307–312

[4] Kitiashvili I N, Bellot Rubio L R, Kosovichev A G, Mansour N N, Sainz Dalda A and Wray A A 2010 ApJ **716** L181–L184

[5] Kitiashvili I N, Kosovichev A G, Wray A A and Mansour N N 2009 ApJ **700** L178–L181

[6] Parker E N 1979 ApJ **230** 905–923

[7] Vargas Domínguez S, de Vicente A, Bonet J A and Martínez Pillet V 2010 A&A **516** A91

[8] Vargas Domínguez S, Rouppe van der Voort L, Bonet J A, Martínez Pillet V, Van Noort M and Katsukawa Y 2008 ApJ **679** 900–909

[9] Cameron R H, Gizon L, Schunker H and Pietarila A 2010 Sol. Phys. 167

[10] Khomenko E and Collados M 2008 ApJ **689** 1379–1387

[11] Kosovichev A 2010 Sol. Phys. submitted (*Arxiv e-prints* 1010.4927)

[12] Gizon L, Schunker H, Baldner C S, Basu S, Birch A C, Bogart R S, Braun D C, Cameron R, Duvall T L, Hanasoge S M, Jackiewicz J, Roth M, Stahn T, Thompson M J and Zharkov S 2009 Space Sci. Rev. **144** 249–273

[13] Kosovichev A G, Duvall Jr T L and Scherrer P H 2000 Sol. Phys. **192** 159–176

[14] Haber D A, Hindman B W, Toomre J, Bogart R S, Thompson M J and Hill F 2000 Sol. Phys. **192** 335–350

[15] Basu S, Antia H M and Bogart R S 2004 ApJ **610** 1157–1168

[16] Bogart R S, Basu S, Rabello-Soares M C and Antia H M 2008 Sol. Phys. **251** 439–451

[17] Gough D O and Kosovichev A G 1988 *Seismology of the Sun and Sun-Like Stars* (*ESA Special Publication* vol 286) ed E J Rolfe pp 195–201

[18] Kosovichev A G 1999 *Journal of Computational and Applied Mathematics* **109** 1–39

[19] Baldner C S, Bogart R S, Basu S and Antia H M 2009 *Astronomical Society of the Pacific Conference Series* (*Astronomical Society of the Pacific Conference Series* vol 416) ed M Dikpati, T Arentoft, I González Hernández, C Lindsey, & F Hill p 119

[20] Jensen J M, Duvall Jr T L, Jacobsen B H and Christensen-Dalsgaard J 2001 ApJ **553** L193–L196

[21] Jensen J M, Duvall Jr T L and Jacobsen B H 2003 *GONG+ 2002. Local and Global Helioseismology: the Present and Future* (*ESA Special Publication* vol 517) ed H Sawaya-Lacoste pp 315–318

[22] Couvidat S, Birch A C and Kosovichev A G 2006 ApJ **640** 516–524

[23] Couvidat S, Birch A C, Rajaguru S P and Kosovichev A G 2006 *Solar Activity and its Magnetic Origin* (*IAU Symposium* vol 233) ed V Bothmer & A A Hady pp 75–76

[24] Zhao J, Kosovichev A G and Sekii T 2010 ApJ **708** 304–313

[25] Zhao J, Couvidat S, Bogart R, Parchevsky K, Birch A, Duvall Jr T, Kosovichev A and Beck J 2010 *Solar Phys.* in press

[26] Hartlep T, Zhao J, Mansour N N and Kosovichev A G 2008 ApJ **689** 1373–1378

[27] Parchevsky K V and Kosovichev A G 2009 *Astronomical Society of the Pacific Conference Series* (*Astronomical Society of the Pacific Conference Series* vol 416) ed M Dikpati, T Arentoft, I González Hernández, C Lindsey, & F Hill p 61

The thermal structure of sunspots from ring diagram analysis

Charles S. Baldner[1], Richard S. Bogart[2], and Sarbani Basu[1]

[1] Astronomy Department, Yale University, P.O.Box 208101, New Haven, CT, 06530-8101, USA
[2] Hansen Experimental Physics Laboratory, Stanford University, Palo Alto, CA, 94305, USA

E-mail: `charles.baldner@yale.edu`

Abstract. We present a large sample of 264 active regions from solar cycle 23, analysed using ring diagrams. The frequencies of these rings are inverted to determine the thermal structure (sound speed and adiabatic index) of these regions as a function of radius. The large sample allows us to describe in a statistically significant way how the thermal properties of the outer layers of the Sun change in the presence of magnetic fields.

1. Introduction

Understanding the subsurface structure of sunspots is one of the current major avenues of research in solar physics. Helioseismology provides a unique opportunity to determine the structure of sunspots empirically, but this determination involves substantial difficulties. In this work, we present the analysis of a large number of ring diagrams of active regions taken from the Michelson Doppler Imager (MDI) instrument on board the Solar and Heliospheric Observatory (SOHO).

Ring diagrams are three-dimensional power spectra of localised areas on the solar surface [1]. The presence of strong surface magnetic fields in active regions is known to change the mode parameters of ring diagrams [2, 3, 4, 5]. Inversions of ring diagram frequencies for structure have been performed on small numbers of rings [6, 7] to determine the changes in sound speed and adiabatic index. In these works, sound speed and adiabatic index were found to be enhanced in the layers between approximately $0.975R_{\odot}$ and $0.985R_{\odot}$, and depressed in the shallower layers between $0.99R_{\odot}$ to $0.998R_{\odot}$.

Previous studies have used small numbers of active regions — typically less than 20. In this work, we present structure inversions of a much larger sample of active regions. The mode parameters of this sample were described in [5]. In section 2 we describe the sample and the inversion techniques, in section 3 we present the results, and in section 4 we compare the results to previous studies and discuss further work.

2. Method

In the present work, we have a sample of 264 active regions. This sample was previously described in [5]. The ring diagrams are constructed from Michelson Doppler Imager (MDI) full disk Doppler-grams. The region is a $16^{\circ} \times 16^{\circ}$ patch tracked across the disk for 8192 minutes, projected to a square grid using Postel's projection, and appodized to a uniform circular

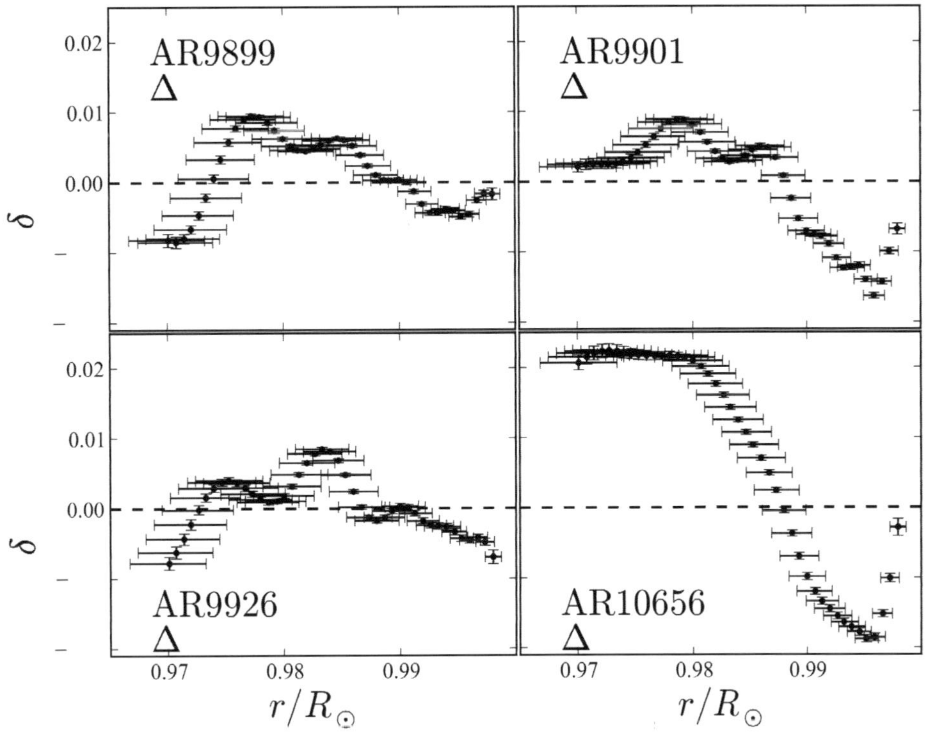

Figure 1. Inversions for squared adiabatic sound speed c^2, shown for four representative regions.

aperture. The ring diagram itself is the three dimensional power spectrum of the tracked velocity data [8].

Active regions are selected from the NOAA active region catalog. Construction of ring diagrams is limited by the availability of suitable data, which is typically only guaranteed during MDI dynamics campaigns, which run two to three months of every year. We require data coverage of at least 85%. The quality of the rings suffers greatly for lower data coverage rates. For every active region in the sample, we also construct one or two comparison ring diagrams of nearby quiet sun regions. These regions are tracked at the same solar latitude to minimise systematic effects in the frequency differences due to geometry.

To characterise each ring we use a measure of the total unsigned line-of-sight magnetic flux. This measure is the Magnetic Activity Index (MAI), defined in [6]. In this study, we invert the differences of the mode frequencies between the active region and the quiet comparison region, thus each measurement is characterised by the difference in MAI, Δ MAI. The sense of the inversion and the Δ MAI is active minus quiet.

The power spectra are fit using a 13-parameter function defined in [8]. The characteristics of the mode parameters were described in [5]. The frequency differences can then be inverted for structural quantities. In this work, we use Subtractively Optimised Local Averages (SOLA) as described in [9]. The inversion minimises the difference between the inversion averaging kernel and a target kernel. In our case, the form of the target kernel is fixed with a variable parameter controlling the width. We also remove a smoothly varying function of frequency from

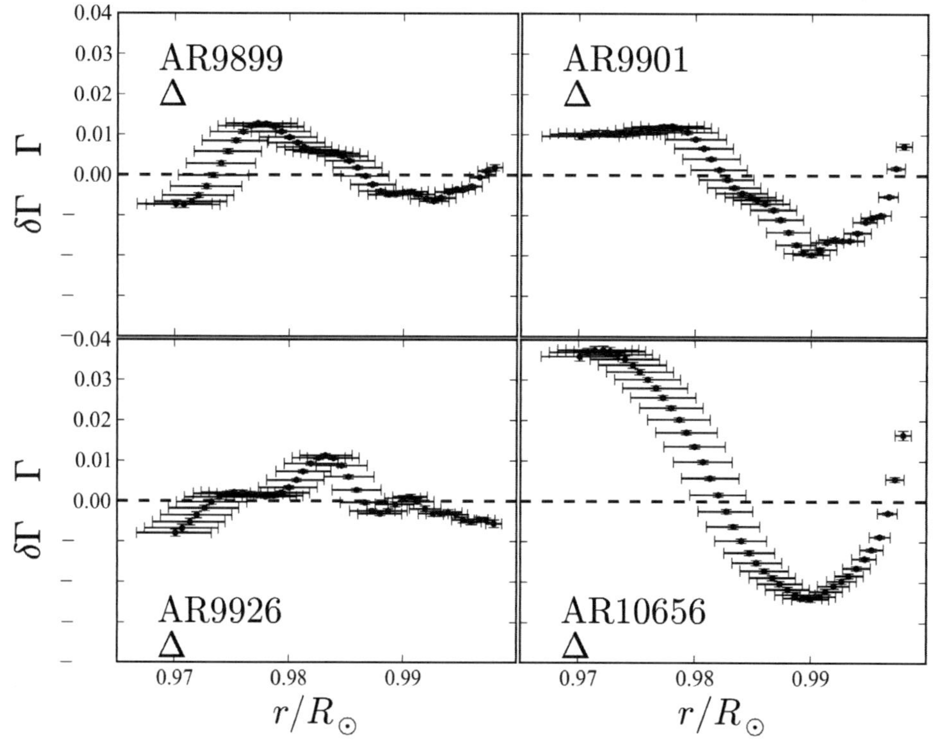

Figure 2. Inversions for adiabatic index Γ_1, shown for four representative regions.

the frequency differences. This function is commonly referred to as the surface term, which we construct using basis splines. In our inversions, then, we have four free parameters: the width of the target kernel, the error suppression parameter, the cross-term suppression parameter, and the number of knots in the surface term. The cross-term arrises from the need for two thermodynamic quantities to fully describe the changes we invert for. In this work, both the sound speed and adiabatic index cross-terms are density. These are the standard four free parameters in SOLA inversions of helioseismic data (e.g. [6]).

The principal effort in performing an inversion for structure from helioseismic data is the difficulty in choosing the appropriate inversion parameters. A thorough exploration of the parameter space was performed on a small subset of the rings, including comparisons to rings published in earlier works. It was found that good inversions could be obtained for all the rings from a small set of combinations of parameter values. The differences in inverted quantities and averaging kernels between inversions with these different parameter sets are easily seen, and so the appropriate value for the inversion parameters could quickly be chosen for the remaining rings.

3. Results

Inversions for the difference in squared sound speed c^2 and adiabatic index Γ_1 were performed for all regions in the sample. Figure 1 shows example sound speed inversions for four different rings with a range of active region strengths as a function of depth. Inversion parameters were chosen

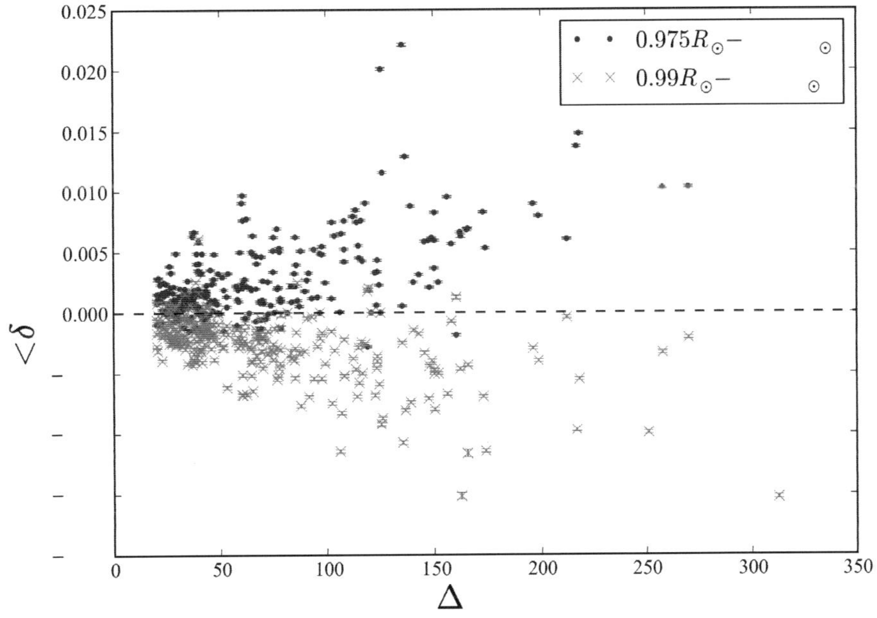

Figure 3. Averages of inversions for c^2 over two depth ranges are shown, plotted as a function of Δ MAI. Blue points are averages of inverted sound speed between $0.975R_\odot$ and $0.985R_\odot$; red crosses are averages of inverted sound speed between $0.99R_\odot$ and $0.998R_\odot$.

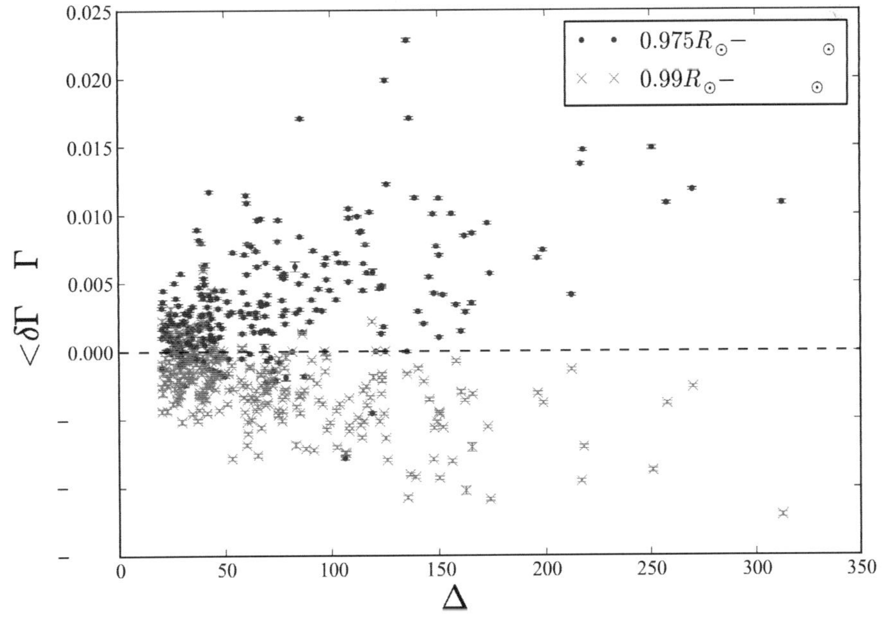

Figure 4. Averages of inversions for Γ_1 over two depth ranges are shown, plotted as a function of Δ MAI. Blue points are averages of inverted adiabatic index between $0.975R_\odot$ and $0.985R_\odot$; red crosses are averages of inverted adiabatic index between $0.99R_\odot$ and $0.998R_\odot$.

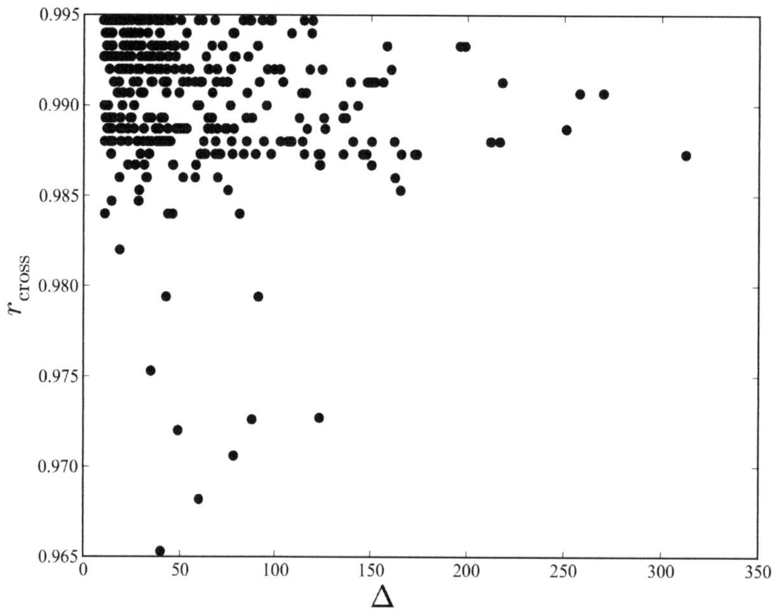

Figure 5. Crossing points for inverted sound speed profiles, plotted against Δ MAI.

to keep the averaging kernels as symmetric as possible and keep the cross-term contamination as small as possible. Figure 2 shows inversions for adiabatic index for the same regions.

Figure 3 shows averages of the inverted sound speed for all regions in our sample for different depth ranges. For sound speed averaged between $0.975R_{\odot}$ and $0.985R_{\odot}$, sound speeds are generally enhanced in the presence of magnetic fields, while in the region from $0.99R_{\odot}$ to $0.998R_{\odot}$, sound speeds decrease. In both regions, the magnitude of the change tends to increase with magnetic field strength, although the relationship seems to be more of an envelope than a linear relation. Further, there seems to be some saturation of the effect at very high magnetic field strengths. The inversion results for the adiabatic index are shown over the same range of depths. The behaviour of Γ_1 is largely the same as sound speed.

In Figure 5, we plot the boundary between the positive and negative perturbations as a function of ΔMAI. There is no obvious dependence of the boundary point on magnetic activity.

4. Conclusions and further work

Earlier works [6, 7] have studied the structure of sunspots using ring diagrams, but have had much smaller sample sizes. We confirm the results of these earlier works. We find that both sound speed and adiabatic index are enhanced in the layers between approximately $0.975R_{\odot}$ and $0.985R_{\odot}$, and are depressed in the shallower layers between $0.99R_{\odot}$ to $0.998R_{\odot}$. We do not extend our inversions beyond these regions, as the helioseismic data contains less information outside these layers, and getting reasonable inversions requires great care.

[7] found a linear correlation between magnitude of the sound speed change and strength of the active region. We find a similar relation, but with substantial scatter. Further, we find that the correlation appears to saturate at high field strengths. We do not find that the depths of the enhancements or suppression of sound speed and adiabatic index depend on the strength of the active region in any way.

The scatter we see is to a certain extent due to instabilities in the inversions themselves. Although we have expended significant effort on selecting appropriate inversion parameters, it is not possible to explore the parameter space as thoroughly as earlier authors could on smaller data sets. We are continuing work to improve the quality of the inversions. Nevertheless, we believe that a substantial amount of the scatter is in the data itself, rather than due simply to the inversions. We are working on quantifying the intrinsic scatter.

It should be noted that what we have interpreted as a sound speed change is more precisely a wave speed change. Since the magnetic fields in the active regions change the speed of wave propagation as well as the structure, not all of the change in inverted wave speed is necessarily due to a thermal perturbation. The difference between the inverted sound speed and the actual sound speed, given a number of assumptions about the magnetic fields, was shown by [10].

Work is continuing in an effort to improve the quality of the inversions, to verify them against complimentary inversion techniques, and to more precisely determine the relationship between the structure and the strength of the active region.

Acknowledgments
CB is supported by a NASA Earth & Space Sciences Fellowship NNX08AY41H. SB is supported by NASA grant NNG06GD139 and NSF Career grant ATM-0348837. The Solar Oscillations Investigation / Michelson Doppler Imager project on SOHO is supported by NASA grant NNX09AI90G to Stanford University. SOHO is a project of international cooperation between ESA and NASA.

References

[1] Hill, F 1988, *ApJ* **333**, 996
[2] Hindman, B, Haber, D, Toomre, J and Bogart, R 2000, *Sol. Phys.* **192**, 363
[3] Rajaguru, S. P., Basu, S and Antia, H M 2001, *ApJ* **563**, 410
[4] Rabello-Soares, M C, Bogart, R S and Basu, S 2008, *J. Phys.: Conf. Series* **118**, 012084
[5] Baldner, C S, Bogart, R S, and Basu, S, 2010, HELAS IV in press
[6] Basu, S, Antia, H M and Bogart, R S 2004, *ApJ* **610**, 1157
[7] Bogart, R S, Basu, S, Rabello-Soares, M C and Antia, H M 2008, *Sol. Phys.* **251**, 439
[8] Basu, S and Antia, H M 1999, *ApJ* **525**, 517
[9] Pijpers, F P and Thompson, M J 1994, *A&A* **281**, 231
[10] Lin, C-H, Basu, S and Li, L 2009, *Sol. Phys.* **257**, 37

Properties of Near-surface Flows around Active Regions from Helioseismic Holography

D. C. Braun & K. Wan

NWRA, CoRA Div., 3380 Mitchell Ln, Boulder CO 80301, USA

E-mail: dbraun@cora.nwra.com

Abstract. A variety of local-helioseismic analyses have shown ~50 m/s flows converging on active regions (ARs). We have examined the average properties of both the 75 strongest converging and 75 strongest diverging flows present in Carrington rotation CR1988 within the uppermost 3 Mm of the Sun. The flows, averaged over 5 days, were deduced from calibrated helioseismic holography measurements applied to MDI observations of CR1988. Inflows associated with ARs typically have maximum speed of between 20 and 60 m/s at about 3 heliocentric degrees from their centers and fall to zero by a radius of 7 degrees. Similar converging flows, however, are prevalent in the quiet Sun. Outflows of similar spatial extent, but signficantly larger speeds, are present diverging from sunspots (i.e. the moat flows). Many of the converging flows in ARs appear to simply mark the boundaries of the moats while others converge on plage regions. In general, large ARs containing sunspots contain a complex mixture of both inflows and outflows which, aside from sunspot moats, also appear similar in property to convective components of the quiet Sun.

1. Introduction

Shallow converging flows, on the order of ~50 m/s and centered on solar active regions (ARs), are a significant finding of local helioseismology [1] [2]. They are believed to extend down to 10 Mm below the photosphere and appear to lie above deeper diverging flows [3]. However, few systematic studies have been done to date of their general properties (a notable exception is [4]). Here we quantify the magnitude and horizontal extent of both near-surface inflows and outflows present over one solar rotation and obtained by applying helioseismic holography to Doppler observations obtained with MDI/SOHO [5]. We are particularly interested in addressing the following questions: 1) how do flows around ARs differ from quiet-Sun flows and 2) how do the properties of inflows compare with those of outflows (for example, moat flows around sunspots). Answering these questions is critical to deriving and interpreting models of these flows [6]. For this study we selected the time period 2000 Mar 29-Apr 26 (corresponding to Carrington rotation CR1988), a period close to the peak of solar cycle 23 and with almost 50 NOAA numbered ARs visible on the solar disk.

2. Methods and Results

Lateral-vantage helioseismic holography [7] [8] [9] is applied to one month of MDI observations spanning the Carrington rotation (CR1988) under study. Briefly, considering waves propagating down to a focus depth of 3 Mm, we measure acoustic (p-mode) travel times between opposite quadrants of an annular pupil The difference in the travel times, between waves propagating from one quadrant to its opposite and the travel times of waves propagating in the reverse direction, is sensitive to horizontal flows near the focus [8] [9].

The travel-time differences between the east and west quadrants, τ_{we}, and the differences between the north and south quadrants, τ_{ns}, are calibrated into eastward and northward vector components of a horizontal flow by applying different tracking rates to the same region of the Sun (the Carrington rotation rate and the Carrington rate plus a constant offset). The shift in τ_{we} between the two sets of measurements, divided by the known tracking offset rate, provides a calibration factor relating the travel time differences and the flow averaged over some (sensitivity) function of depth. The sensitivity function, computed under the Born approximation, for lateral-vantage holography at a 3 Mm focus-depth shows a strong peak within 1 Mm below the photosphere as well as somewhat weaker contributions at depths between approximately 2 and 5 Mm below the photosphere [9].

For each 15° span of longitude, daily maps of the horizontal flow ($\mathbf{v_h}$) are averaged over 5 days around the central-meridian passage to produce a synoptic flow map (e.g. see Fig. 4 of [8]). Supergranulation dominates the flow maps (even over 5 days), making the identification of AR-scale flows challenging. To do this, we find the local maxima and minima (hereafter "extrema") of maps of the horizontal component of the flow divergence smeared with a 2D Gaussian with FWHM of 7.5°. In heliographic coordinates (L, B), the horizontal component of the flow divergence is given by

$$\nabla_h \cdot \mathbf{v_h} = (1/\cos B)\,[\partial(v_B \cos B)/\,\partial B + \partial v_L/\,\partial L]\,, \tag{1}$$

where $\mathbf{v_h} = (v_L, v_B)$. For this study, we only consider centers of inflows or outflows within latitudes B between $-40°$ and $+40°$.

To look for correlations of the flows with solar activity, we use MDI magnetograms to assess the total (unsigned line-of-sight) magnetic flux within an angular distance of 7.5° (hereafter defined as "nearby") of each identified flow location. Fig. 1 shows a scatter plot of the flow divergence at the extrema against the nearby flux. It is apparent that there is a weak correlation of inflow (negative-divergence) strength with magnetic flux. A stronger correlation is observed for outflows, with the sunspot moats dominating the correlation at high flux and flow divergence values. We select for further study the strongest 75 inflows and outflows, using cutoffs in $\nabla_h \cdot \mathbf{v_h}$ indicated by the vertical lines in Fig. 1.

Fig. 2 shows probability distribution functions of the nearby magnetic flux for the strongest 75 inflows and outflows as well as a set of random locations within the same latitude range (±40°). Inflows have a significantly higher probability over chance of having nearby flux levels of between 3 and 10 $\times 10^{21}$ Mx. On the other hand, outflows have a higher probability than chance to be near the highest flux values. Somewhat surprising, however, is the large number of strong flows, of both types, in the quiet Sun. Thus, it is worth examining how flows associated with strong nearby magnetic flux differ from those associated with weaker nearby flux.

To explore further the dependence of the properties of the flows on solar activity, we define three subgroups of each of the sets of 75 strongest inflows and outflows based on the nearby magnetic flux (the horizontal lines in Fig. 1 separate these subgroups). First, we split the inflows into three subgroups of 25 members each (denoted $A-$, $B-$, and $C-$). The first 17 outflows, ranked by flux, are clearly identified with sunspot moats and define group $A+$. Assigning the 25 outflows with the weakest nearby flux to group $C+$ (as for the inflow subgroup $C-$) leaves 33 outflows associated with intermediate flux values ($B+$). The subgroups $C-$ and $C+$ are essentially associated with quiet-Sun.

For each of the 75 inflows and 75 outflows, the original (unsmeared) vector flow fields are projected onto Postel coordinates centered on the extrema of the divergence signal. A "radial velocity" is defined to be the component of the horizontal flow along a great circle passing through the center of the projection. Fig. 3 shows the averages of the radial velocity over

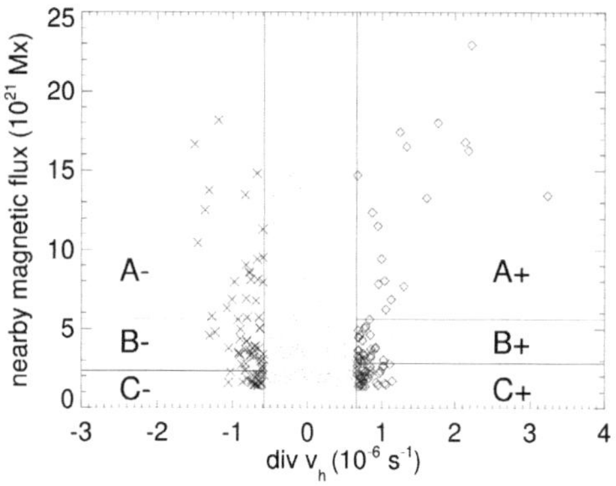

Figure 1. Scatter plot of the horizontal component of the flow divergence for each local maximum or minimum of a (Gaussian-smeared) synoptic map against the total unsigned magnetic flux within 7.5° of the given divergence extremum. The vertical solid lines isolate the 75 strongest divergence extrema, with inflows denoted by Xs and outflows by diamonds, from the weaker divergence extrema denoted by points. The horizontal lines separate six inflow or outflow subgroups (see text).

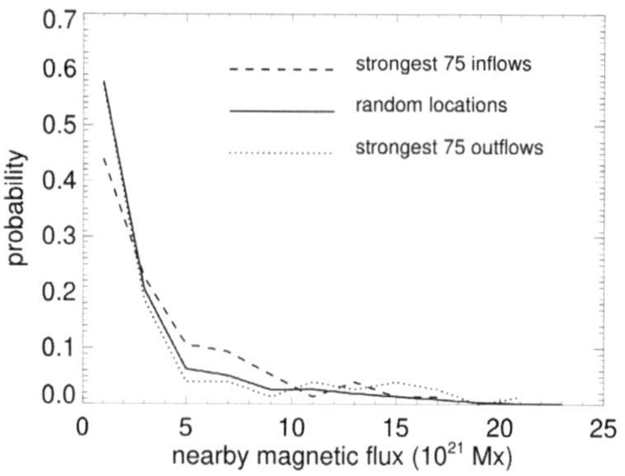

Figure 2. Probability distribution of the magnetic flux (in 2×10^{21}Mx bins) near the 75 strongest inflows (dashed line) and 75 strongest outflows (dotted line) compared with the probability distribution for random locations (solid line). Taken together, this and the preceding plot demonstrate that the association of inflows with active regions, while significant, is not an especially strong one considering the large number of similar flows in the quiet Sun.

the perpendicular (azimuthal) coordinate in the Postel projection for the inflows and outflows associated with the strongest nearby flux (left panels) as well as the quiet Sun (right panels). Plots of the same quantities for the intermediate flux groups $B-$ and $B+$ (not shown) are similar to those for $C-$ and $C+$. It is clear that the inflows all have similar properties, regardless of their proximity to active regions. Thus, inflows clearly associated with ARs appear to be quantitatively and qualitatively similar to those found in the quiet Sun. These inflows are significantly more compact (i.e. extending no more than about 7° from their centers) than previously noted [1]. The azimuthally-averaged flows peak at around 40 m/s. An examination of individual flows in group $A-$ shows that the inflows near ARs generally fall into two types: those that are centered near plage (~40%), and those that coincide with the edge of a sunspot moat flow (~60%). Outflows around sunspots (i.e. the moat flows) are substantially stronger than those outflows not associated with spots. Nevertheless, all inflows and outflows are similarly compact.

3. Conclusions

While occuring more often near ARs then expected by chance (Fig. 2) the inflows examined here are similar to those found in quiet Sun and thus may not require specialized physics [6], particularly those which may simply mark the boundaries of the sunspot moats. For the inflows centered on plage, there remains a "chicken and egg" question: do converging flows form because of some physical influence of the magnetic fields, or does field collect in the center of existing inflows? But by far the strongest AR-related near-surface flows seen (at least in CR1988) are associated with the diverging moat flows which, while having no significant counterpart in the quiet Sun, are an important component of the dynamics of ARs. More systematic studies of helioseismically-inferred flows [4] are needed.

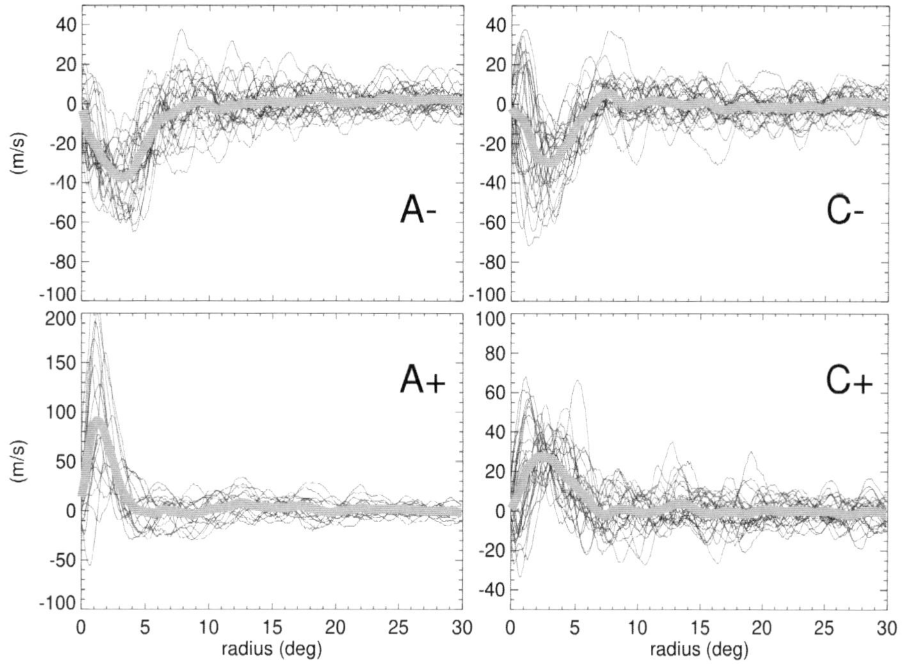

Figure 3. Azimuthal averages of the radial velocity component of the flows for four subgroups. The thin black lines indicate the radial velocities for each of the subgroup members (to give some indication of the variation within each group), while the thick grey line indicates the mean of the radial velocities over the entire subgroup. The left panels indicate the inflows (top) and outflows (bottom) associated with the strongest nearby flux, while the right panels indicate the inflows (top) and outflows (bottom) present in quiet Sun. Note the change in the vertical scale for group *A*+ (the sunspot moats).

Acknowledgments

This work is supported by the Astronomical Division of the NSF (AST-0406225)

References

[1] Gizon L, Duvall Jr T L and Larsen R M 2001 *Recent insights into the Physics of the Sun and Heliosphere: Highlights from SOHO and Other Space Missions* eds P Brekke, B Fleck and J Gurman (San Francisco, Astronomical Society of the Pacific) vol 203 of IAU Symposia pp 189-191

[2] Haber, D A, Hindman, B W, Toomre, J, Bogart, R S and Hill, F 2001 *Helio- and Asteroseismology at the Dawn of the Millennium* ed A Wilson (Noordwijk, ESA) vol SP-464 pp 209-218

[3] Gizon L and Birch A C 2005 *Living Rev. Solar Phys.* **2**, 6 URL (cited on 2010 Sep 10): `http://www.livingreviews.org/lrsp-2005-6`

[4] Hindman, B W, Haber, D A and Toomre, J 2009 *Astrophys. J.* **698** 1749

[5] Scherrer, P H *et al* 1995 *Solar Phys.* **162** 129

[6] Spruit, H C 2003 *Solar Phys.* **213** 1

[7] Lindsey C and Braun D C 2004 *Astrophys. J. Suppl.* **155** 209

[8] Braun D C, Birch A C and Lindsey C 2004 *Helio- and Asteroseismology: Towards a Golden Future* ed D Dancy (Noordwijk, ESA) vol SP-559 pp 337340

[9] Braun, D C, Birch, A C, Benson, D, Stein, R F and Nordlund, A 2007 *Astrophys. J.* **669** 1395

HMI ring diagram analysis I. The processing pipeline

R. S. Bogart[1], **C. Baldner**[2], **S. Basu**[2], **D. A. Haber**[3] **and M. C. Rabello-Soares**[1]

[1] Stanford University, Stanford, Calif. USA
[2] Yale University, New Haven, Conn. USA
[3] JILA, University of Colorado, Boulder, Colo. USA

E-mail: rbogart@stanford.edu

Abstract. The combination of high resolution, spatial coverage, and continuity of photospheric Doppler and other data from HMI has allowed us to embark on a program of systematic exploration of solar subsurface flows and thermal structure variations using the technique of ring-diagram analysis on an unprecedented scale. Two ring-diagrams pipelines exist: a synoptic program aimed at mapping the evolution of the circulation and local subsurface flows on a global scale from the surface to depths of down to $0.9R_sun$, and a targeted program designed to provide a comprehensive view of the thermal structure anomalies associated with loci of magnetic activity over the course of their life cycles. In this paper we describe the analysis techniques implemented in the processing pipelines.

1. Introduction

Two ring-diagram pipelines are being developed: one producing synoptic views of the flows and structure on regular spatial grids, and the other producing detailed flow and structure anomalies associated with specific target regions. We will refer to these as the "synoptic" and "target" pipelines, respectively. Both have common requirements for most of the analysis modules and data structures, so it is appropriate to consider them together.

The HMI ring-diagram analysis pipelines consist of several distinct elements. Each of these pipeline elements is described in a section below. Fuller documentation on the pipeline analysis modules and associated data products can be found on the web pages of the HMI Ring Diagrams Team, at *http://hmi.stanford.edu/teams/rings/*.

The primary analysis dataset for the standard pipelines is the set of full-resolution full-disc Dopplergrams produced by the HMI Observables Pipeline at a cadence of one per 45 sec. However, ring-diagram pipelines have been run on other observables as well, including continuum intensity and line depth.

2. Quality checks and coverage

HMI data coverage has been almost continuous since 1 May 2010 and is expected to remain so. Occasional images are unsuitable for inclusion in the processing pipeline for various reasons, usually because the Image Stabilization System was turned off for instrument calibrations, occasionally for other reasons such as transmission or processing errors. These do not generally amount to more than a fraction of a percent of the total, so low duty cycle is not an issue. The pipelines have, however, been used to process earlier test data sequences as part of the instrument calibration, so it was necessary to include data coverage checks. The module **rdcover** reports the available duty cycle for the input data using the same target interval and rejection criteria as the tracking. With MDI data it was established that data coverages of less than 0.85 led to degradation in the quality and number of fits, and those less than 0.70 led to serious degradation. These numbers may be conservative for HMI, but will be irrelevant except in the case of spacecraft emergencies or for shorter analysis intervals around the times of eclipses.

Sufficient data are available to the HMI observables processing for reliable and useful quality flags to be set on the individual output records. The tracking module can reject images from a hand-prepared list as well as on the basis of their quality flags. We have been visually inspecting the observable image statistics and updating a rejection list to trap cases in which automatic quality assessment failed. As those cases are understood and the observables code matures, this step should become unnecessary. The tracked data cubes include in their associated metadata log files identifying the input images which were rejected and interpolated over.

3. Data averaging

Spatial variations in the Doppler signal are removed before analysis by subtracting a long-term mean. These variations are primarily due to the solar rotation of course, but there are also effects due to the non-uniform response of the instrument; an example of the latter are the "fringes" which are actually in the lookup tables for the Doppler calibration based on integrated-light observing sequences in which the front window rather than the Sun is imaged. As regions are tracked across the field of view such spatial variations acquire a temporal component.

The Doppler data have a very large diurnal variation due to the orbital motion of the spacecraft. The orbital velocity must be removed from both the temporal averages and the individual images as they are tracked. The module **datavg** is used to produce such corrected averages over one-third of a Carrington rotation (about 9 days), sampled six times per rotation. Data series produced by this module such as **hmi.V_avg120** are available to the tracking module for background subtraction. The synoptic pipeline requires them, creating the appropriate records if they do not already exist.

4. Target generation

The synoptic data products for transverse velocities are sampled at a spatial scale corresponding to 2°.5 heliographic (~ 30 Mm), with analysis regions of diameter 5°, extending out to $\mu = 0.986$, about 80° from disc center. At this distance the area of foreshortened HMI pixels is comparable to that of the limit at which we were able to extract useful ring-diagram fits from full-disc MDI data at one-third the spatial resolution.

To invert for velocities and structure below the immediate sub-surface layers, we must sample larger areas for longer intervals as well, as the small regions can only resolve the lowest order modes, the spatial and temporal Nyquist frequency being comparatively high. We also analyse on 7°.5 and 15° sampling grids, with the regions again having diameters twice the grid spacing. Because we sample so close to the limb, the annual variation of heliographic latitude on the ecliptic provides an opportunity to measure at least the smaller tiles at higher latitudes during part of the year. We divide the year into four grid "seasons" at the times when $B_0 = \pm 3°.625$. (The two extreme cases are mirror images of one another in latitude.) Our ability to reach near the limb and a desire to keep the tiles approximately equally spaced regardless of latitude leads to a rather complicated selection of the target locations in heliographic latitude and Stonyhurst longitude (*i.e.* referred to geocentric central meridian) for the smaller tiles. The program **gentargs** generates the appropriate list of target latitudes and longitudes for a given time to be used as parameters for the tracking module.

The temporal sampling intervals correspond to the time it takes a region to rotate through its diameter, 5° of Carrington rotation (1/72 of a synodic rotation, 545 min) for the 5° tiles, and correspondingly longer for the larger tiles — 1/24 and 1/12 of a synodic rotation (1635 min and 54.5 hr), respectively. The actual tracking times are somewhat longer: 576, 1728, and 3456 min, respectively, to allow for modest apodization of the power spectra.

For the target pipeline, we compare results for the region of interest compared with those for one or more other regions observed under as close to the same conditions as possible. In practice, that means selecting comparison regions at the same latitude and within about 120° longitude. The comparison regions are selected on the basis of their Magnetic Activity Index (MAI, described in the next section) being as low as possible within the location constraints. Region sizes and tracking times for the target pipeline are still to be determined. We plan to begin with 15° regions tracked for 1/12 or 1/6 of a Carrington rotation, as has been done with MDI.

5. MAI determination

In the past the Magnetic Activity Index has been determined by mapping and tracking regions of the same geometry from MDI magnetograms and integrating the unsigned flux execeeding a certain threshold, based on the noise characteristics of the instrument. While MDI and HMI are co-observing that is still being done, but we will obviously have to convert to a new MAI based on HMI magnetograms. We expect to use the same method, but the selection of parameters for noise thresholds and sampling rates must be adjusted for the different observations. The module **maicalc** is still under development, but this only affects the implementation of the target pipeline.

6. Tracking

Region definition in time and space is at the heart of local-area helioseismology. The ring-diagram pipelines, as well as the time-distance pipeline, use the module **mtrack** to convert a time-sequence of solar images to a collection of three-dimensional data cubes, each representing the data for a particular target region. If the region is centered at fixed heliographic coordinates, this implies tracking at the Carrington rotation rate, and this is what is done for the HMI ring diagram pipelines. (Ring-diagram analysis of MDI and GONG data has been done traditionally with regions at different latitudes tracked at different rates, approximating the differential rotation rates in photospheric Doppler signal at the latitudes involved.) Image data are mapped from their surface spherical coordinates to the plane using Postel's azimuthal equidistant projection, interpolated from the image plane with a cubic convolution kernel. In the synoptic pipeline the map scale at center of each region is 0°.04 per map pixel for the 5° and 15° tiles, corresponding roughly to the image scale at disc center, 0°.08 for the 30° tiles. With spatial extents of 128, 384, and 384 pixels, the tracked cubes have a full extent of 5°.12, 15°.36, and 30°.72 respectively, allowing for circular spatial apodization of the power spectra to the target radii.

Region parameters in the target pipelines have not been defined, and may not be rigid. Initial tests have been with 15° tiles tracked with similar parameters as for the synoptic pipeline, but for twice the duration.

We plan eventually to track each of the target regions for their full disc passages, and then to extract from those long data cubes the intervals needed for each analysis period. This will reduce the number of times that the input data files need to be read. For now, each region is tracked from original data for each analysis period. The tracking module does create multiple cubes for different regions from the same input data set, but system limits constrain the number of regions that can be tracked concurrently to about 250. For the 5° synoptic series, in which over 2700 regions must be tracked every 9 hours, this is a particular issue.

7. Power spectra

Power spectra are calculated via a straightforward 3-d Fourier transform of the real data in the tracked cubes, using the **pspec3** module. The data are apodized with a 4-th order polynomial taper in both the spatial and temporal dimensions. The radial apodization varies from unity at the target radius to zero at the edge of the map. For the map sizes and scales used in the pipeline, that is a fractional distance of 0.0234375 of the map width. The temporal apodization is from 0 at the endpoints to 1 over a distance of 0.015625 of the total interval.

The power spectra are normalized so that the integrated power over all voxels in the power spectrum is equal to the variance of the original data cube. To convert the power spectral density to physical units, one must divide by the physical scales (CDELTi) of the power spectrum.

8. Ring fitting

We have two different algorithms for parametric fitting of the "rings", ridges in the 3-d power spectrum. Module **rdfitc** implements the approach of [1], in which fits to a 13-parameter model including multiple background and asymmetry terms are performed for each point in selected temporal frequency planes of the power spectrum, with generation of a full covariance matrix. This method provides more reliable parameter estimates, but is comparatively time-consuming. Module **rdfitf**, based on the approach of [4], fits only 6 parameters in the frequency dependence of the lowest order azimuthal Fourier terms of the azimuthally analysed spatial power spectrum. It is very fast and well-suited to the determination of the U_x and U_y terms needed for flow inversions; but the frequency determinations are not as reliable.

For the synoptic pipeline, the fast fitting is applied to all regions. For the 15° and 30° tiles comprehensive fits are done only for the tiles on central meridian. We are currently doing comprehensive fits for all of the 5° tiles, since their small size permits this (the fitting time varies roughly as the 4th power of the cube sizes). Once we have gained sufficient confidence in the results we will likely abandon one of these two otherwise redundant sets of fits.

Only the comprehensive fits are performed for the target pipeline, as the inversions depend sensitively on the mode frequency determinations.

9. Inversions

The two parameters U_x and U_y, representing horizontal displacement of the centers of the rings in the frequency plane corresponding to different modes, are inverted for the depth dependence of the transverse velocity of the observed field (relative to the tracking rate!) using the OLA procedure described in [1]. The inversion is implemented in module **rdvinv**.

For the structure pipeline frequency differences between comparison regions are inverted for sound-speed and adiabatic gradient variations with depth, as described in [3]. Inversions for thermal structure perturbations are not yet implemented in the synoptic pipeline, because of the lack of an appropriate set of quiet-Sun reference frequencies over the disc.

Full three-dimensional flow inversions from combinations of the synoptic analysis regions at all scales are planned, but these techniques are still under development.

Acknowledgments

This work was supported in part through NASA grants NAS5-02139, NNX09AB04G, and NNX07AH82G.

References

[1] Basu S and Antia H.M 1999 *Astrophys. J.* **525** 517

[2] Basu S, Antia H.M and Bogart R S 2004 *Astrophys. J.* **610** 1157

[3] Bogart R S, Basu S, Rabello-Soares M C and Antia H.M 2008 *Solar Phys.* **251** 439

[4] Haber D A, Hindman B W, Toomre J, Bogart R S, Thompson M J and Hill F 2000 *Solar Phys.* **192** 335

HMI ring diagram analysis II. Data products

R. S. Bogart[1], C. Baldner[2], S. Basu[2], D. A. Haber[3] and M. C. Rabello-Soares[1]

[1] Stanford University, Stanford, Calif. USA
[2] Yale University, New Haven, Conn. USA
[3] JILA, University of Colorado, Boulder, Colo. USA

E-mail: `rbogart@stanford.edu`

Abstract. The combination of high resolution, spatial coverage, and continuity of photospheric Doppler and other data from HMI has allowed us to embark on a program of systematic exploration of solar subsurface flows and thermal structure variations using the technique of ring-diagram analysis on an unprecedented scale. There are two ring-diagrams pipelines, as described in [1]. In this paper we discuss the synoptic pipeline execution and describe the data being processed and produced.

1. Data products

Input data for the ring-diagram pipelines are at present taken from the published and mirrored data series **hmi_test.V_45s** in the Data Record Management System (DRMS) as described in [2]. This is an unarchived prototype series for the forthcoming definitive HMI Level 1 observable data product **hmi.V_45s**. The structure of the definitive series is not expected to change significantly from that of the prototype series. The various output data series are correspondingly being put in the DRMS namespace **hmi_test**; when the input series is migrated to its official namespace the output series will be as well. The output data series are available for mirroring as requested by NetDRMS sites [2].

All ring-diagram pipeline data products are named according to the rule {*NS*}.**rd**{*OBS*}{*PROD*}_{*SER*}, where *NS* is the DRMS namespace (**hmi** for definitive products based primarily on HMI data), *OBS* is a code for the input observable (**V** for Doppler data), *PROD* is a product name describing the module producing the data series, and *SER* is a name specific to a particular pipeline. The product names for data produced by the various modules are: **track** for output of module **mtrack**; **pspec** for output of **pspec3**; **fitsc** for output of **rdfitc**; **fitsf** for output of **rdfitf**; and **flows** for output of **rdvinv**. The implemented pipeline series codes are: **fd05** for the synoptic 5° tiles; **fd15** for the synoptic 15° tiles; **fd30** for the synoptic 30° tiles; and **targ** for the target pipeline. The various products created in the synoptic 15° tile set, for example, are:

- hmi_test.rdVtrack_fd15
- hmi_test.rdVpspec_fd15
- hmi_test.rdVfitsf_fd15
- hmi_test.rdVfitsc_fd15
- hmi_test.rdVflows_fd15_frame

(The last has a **_frame** designator added because the inversion results from many DRMS records of ring fits are gathered together in a single inversions record. Inversions for all regions tracked at a certain time, a *frame*, are gathered together in this series. There will be other series for particular sets of locations collected over a rotation.)

A DRMS data series is organized by its metadata *prime keys:* the union of values of the prime keys is taken to uniquely define a record. Multiple records with the same values of their prime keys are assumed to refer to the same data, and only the one with the highest internal record number (the one most recently added to the database) is ordinarily provided by data requests. For an image series such as the input Dopplergrams it is normally sufficient to use a single prime key, the observation time; but for the pipeline data sets described here it is obviously necessary to provide at least three prime keys, designating both the time and the location of the tracked cubes or those to which the analysis products refer. In principle all of the products could be included in a single data series if we used more prime keys to designate for example the product type and the relevant pipeline or the parameters uniquely defining it, but that would be cumbersome, which is why we have chosen to create the 20 or more different series described above. Each of those series is also uniquely described by a number of non-prime keys whose values are constant in the database. This prevents for example a power spectrum from accidentally being ingested into a tracked cube series or a 5° tile from being ingested into a series reserved for 30° tiles.

Because the phasing of ring-diagram pipeline analyses is keyed to Carrington times, the product data series use two prime keys to distinguish the time: Carrington Rotation and Central Meridian Longitude at the midpoint of the analysis interval. Three prime keys are used to describe the spatial locations of the analysis regions: Heliographic Latitude, Carrington Longitude, and Central Meridian (Stonyhurst) Longitude. Although the three longitude keys are redundant, they are chosen for the convenience of being able to select certain groupings easily, for example all analyses for a given Carrington longitude, or all analyses on the central meridian for a certain time range.

For the products of the target pipeline, the prime keys are an Identifier (the active region number with an optional longitude offset for comparison regions), plus the Carrington Rotation and Carrington coordinates of the region. This assumes that all regions will be analyzed for intervals centred on the times of their central meridian crossings. If that constraint is relaxed at least one additional prime key will be required.

The synoptic pipeline has been run for all target intervals from Carrington time 2096:030 (17 May 2010) through 2100:035 (3 Sep 2010) — four full rotations. Since the data product series were for testing, no data were archived to tape, and the data on disc from times prior to 2098:330 (18 June 2010) have aged off. Continuous coverage in the analysis data series of the synoptic pipeline is currently (26 Sep 2010) available for the following Carrington frames:

- hmi_test.rdV*_fd30 : 2098:330 – 2100:060; 33 frames
- hmi_test.rdV*_fd15 : 2098:330 – 2100:045; 67 frames
- hmi_test.rdV*_fd05 : 2098:330 – 2100:035; 204 frames

Each frame includes a large number of analysis data cubes: 69 30° tiles, 284 or 281 15° tiles, and 2748 or 2727 5° tiles. The number and distribution of tiles change with B_0 as described in [1] and illustrated schematically in Figure 1.

Only one record set of the target pipeline exist so far — AR 11072 (2097:315.5,15.0S, 2010.05.23) — due to the lack of strong active regions at central meridian in the data to date.

2. Pipeline execution

In order to keep cadence, it is clear that the three components of the synoptic pipeline must be started once every 9, 27, and 54 hours. The pipeline has been running long enough to gain some

Figure 1. Schematic locations of the centres of the target grid points for the 15° and 5° synoptic data series. The figure on the left illustrates the 15° locations when $|B0| < 3°.625$, and the one in the center for times when $B0 \geq 3°.625$. The 5° grid on the tight is for $B0 \leq -3°.625$. The shaded dots correspond in aspect ratio to their projected appearance, but are smaller than the actual extent of the regions, which overlap their neighbors by one half their width.

experience of its running times under various load conditions. The typical latency (the time to complete the pipeline for a single analysis time) of the 30° pipeline is about 20 hours, meaning that under normal conditions it is running only about 1/3 of the time. The 15° pipeline has a typical latency of about 18 hours, so it must be running about 2/3 of the time. The 5° pipeline has a minimum latency (without the inversions) of about 12 hours, which is longer than its required cadence; an average of 1.3 instances of it must be running at all times. During times of extreme system load the timings for the 5° pipeline have increased by as much as a factor of two. At these times the system load was so severe that all activities were affected and problems identified and corrected. The ring-diagrams pipeline itself has occasionally run for over a week at triple cadence with no noticeable impact on the system load nor on its own latencies.

The real latency of availability of data products of course involves more than the processing times for the ring-diagrams pipeline itself. There is a delay of at least one day for production of the definitive observables. Because the pipeline requires background subtraction of a 9-day mean, it cannot commence until at least 5 days after the target time. If full disc crossings are used for the initial tracking sets, this will require a latency of about 12 days (although the latency for the background mean will be subsumed in that).

The various elements of the pipeline contribute very differently to the overall processing time for the different synoptic pipelines. For the 30° pipeline, the **rdfitc** module (for just the nine regions on central meridian or seventeen regions on meridian or equator, running on parallel processors) takes more than half the time, while the flow inversions take a negligible fraction. For the 5° pipeline on the other hand, the inversions (if they were being carried out) would consume the largest fraction, while the **rdfitc** fitting can be run for all 2700+ regions in only half the tracking time. Relative timings for the individual module components of the different synoptic pipelines are given in Table 1. The relative timings for the alternate 5° pipeline are for ones in which **rdfitc** fits are done only on the central meridian, but for which the flow inversions are performed.

3. Other data sets

The pipeline elements have been run on Doppler data concurrently observed with both HMI and MDI for Carrington times 2096:150–2097:225 (8–30 May 2010). A sample comparison, of the averaged U_x and U_y parameters from the ring fits, is shown in Figure 2. The agreement of

Table 1. Relative timings for individual modules in the synoptic pipelines.

Module	30° tiles	15°	5°	5°*
mtrack	0.29	0.55	0.44	0.28
pspec3	0.07	0.15	0.14	0.08
rdfitf	0.08	0.15	0.20	0.13
rdfitc	0.54	0.10	0.22	< 0.01
rdvinv	0.02	0.05	—	0.50

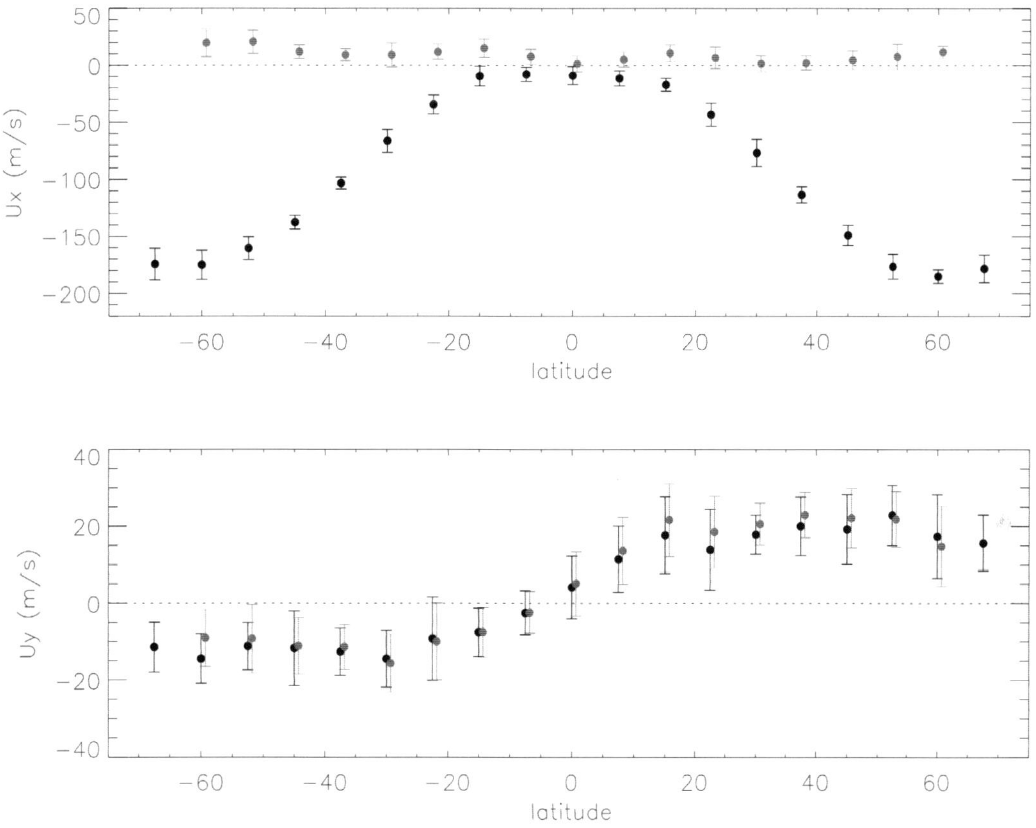

Figure 2. Average values of the U_x and U_y parameters (from rdfitc) for 15° tiles along the central meridian, averaged over all modes, as a function of latitude. The red symbols are for MDI data and the black symbols for HMI data for the same 10 time intervals. The difference in zonal velocity values reflects the difference in tracking rates for the MDI and HMI pipelines [1].

values and dispersion is excellent, as we would expect from the similarity of size and scale of the regions analyzed.

References

[1] Bogart R S, Baldner C, Basu S, Haber D A and Rabello-Soares M C 2011 *J. Phys. Conf. Proc.* this issue
[2] Bogart R S, 2007 *Astron. Nachrichten* **328** 352

The Need for Physics-based Inversions of Sunspot Structure and Flows

D. C. Braun[1], **A. C. Birch**[1], **A. D. Crouch**[1], **M. Rempel**[2]

[1]NorthWest Research Assoc, CoRA Div, 3380 Mitchell Ln, Boulder, CO, USA
[2]NCAR, HAO Div, 3080 Center Green Dr, Boulder, CO, USA

E-mail: `dbraun@cora.nwra.com`

Abstract. Current controversy exists in the interpretation and modeling of helioseismic signals in and around magnetic regions like sunspots. Unresolved issues include the dependence of the sign of both the inferred flows and wave speed on the type of filtering used, and the discrepancy between the relatively deep two-layer wave-speed models derived from standard time-distance methods and shallow, positive wave-speed models derived using forward models which include effects of mode conversion To make full use of the year-round, almost limb-to-limb, coverage provided by the *Solar Dynamics Observatory*, an efficient and reliable inversion method incorporating possible magnetic effects and the currently unexplained sensitivity to methodology is critical.

1. Introduction

A cool layer inferred below sunspots in many helioseismic (time-distance) inversions is largely derived from positive (slower) travel-time perturbations, relative to the quiet Sun, observed at small values of the mode phase speed [1] [2]. Controversy exists whether these positive travel-time perturbations may arise due the effects of filters applied to the observations [3] [4] [5]. Travel-time perturbations within sunspots show strong variations, including changes of sign, with frequency at fixed phase speed [3] [6]. This behavior has also been qualitatively (and in some cases remarkably quantitatively) reproduced with artificial data – derived from hydrostatic models [7], magnetostatic models [8], and MHD models [9] – in which no slower layer is present. In contrast to measurements made using phase-speed filters, travel-time perturbations derived using ridge filters are exclusively negative (implying faster wave speeds) in sunspots [4]. A number of forward models, including some with magnetic fields, provide evidence for shallow, positive wave speed perturbations below sunspots [10]. Thus, considerable uncertainty remains in the inference of the wave-speed structure below sunspots. However, an efficient, robust inverse method which includes the effects of magnetic fields has not yet been developed.

2. Travel-time measurements in spots

The measurement of travel-times in sunspots is subject to a fair amount of uncertainty and can be systematically sensitive to details of the measurement procedure. The effects of a reduced wave amplitude (or reduced excitation) in magnetic regions on the measurement of travel times have been recently explored [11] [12] [13] [14]. Additional complications arise due to differences, between sunspots and quiet-Sun, of the temporal-frequency content of the cross-covariance functions. Figure 1 shows some cross-covariance functions and their power spectra for both a real sunspot and the realistic sunspot simulation of Rempel and collaborators [16] [9]. The power spectra of the cross-covariance functions show that, in both the real and artificial data, there is a shift in power towards lower temporal frequencies in the sunspot umbrae as compared to the quiet Sun. In the temporal domain, this can be seen as an increase in the spacing between the peaks (or valleys) of the umbral-averaged cross-covariance functions, relative to those of the quiet-Sun cross-covariance functions. The result is that the difference between the phase travel-time shifts (spot minus quiet-Sun travel times), determined through wavelet fitting, will vary according to which peak (or valley) is used to fit the phase travel time.

As one might expect, tests have indicated that the sensitivity of the travel-time measurements to methodology decreases considerably with cross-covariance functions computed after narrow temporal-frequency bandpass filters are applied to the data. This type of filtering is useful for studying the frequency variation of travel-time perturbations in both real and artificial data [3] [9].

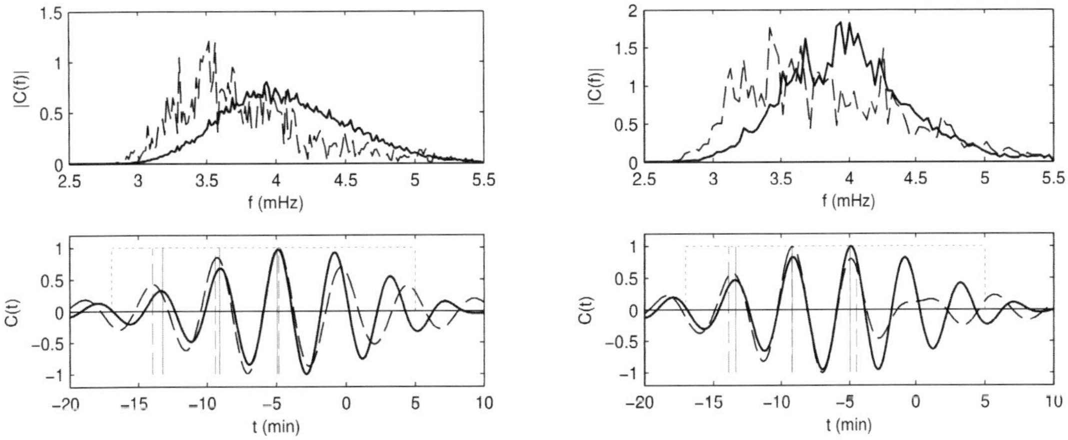

Figure 1. Cross-covariance functions and their power spectra determined for a region containing a sunspot observed with MDI [15] (left panels) and for the artificial sunspot modeled in a realistic magnetoconvection simulation [16] [9] (right panels). The bottom panels show the cross-covariance functions, averaged over the sunspot umbra (dashed curves) and a region of quiet Sun (solid curve) determined by helioseismic holography and corresponding to outgoing waves with the application of a standard phase-speed filter (filter "2" of [2]). The zero point of the (horizontal) time axis is arbitrary, and the amplitude of the umbral cross-covariance functions has been multiplied by a constant to facilitate comparison with the quiet Sun. The vertical lines indicate phase travel-times determined from wavelet fitting tuned to individual peaks in the cross covariances. The dotted lines illustrate the window used in the fitting. The top panels show the power spectra of the cross-covariance functions, with the line-types having the same meaning as the lower panel. It is evident, in both the real and simulated sunspot, that the power in sunspot umbrae is shifted to lower frequencies, which results in a larger spacing between peaks observed in the umbral-averaged cross-covariance functions.

3. Ridge-filtered Inversions for Flows

As for the sign of the near-surface wave-speed perturbation, the inferred direction of subsurface flows below sunspots also appears to be highly dependent on the analysis methodology. In particular, shallow inflows are inferred with phase-speed filters [17] [5] while outflows at the same depths are seen in inversions with ridge filters [4].

We carried out an inversion for subsurface flows from ridge-filtered travel-time measurements made using 7.8 hours of Doppler velocity measurements from the Helioseismic and Magnetic Imager (HMI) onboard the *Solar Dynamics Observatory.* A region containing several sunspots (AR 11057), observed by HMI starting 2010 March 28 was selected for analysis. This inversion is based on travel times measured for the modes n = 0,1,2, & 3. In addition to the ridge filters we also used filters that isolated 0.5 mHz wide bands in frequency. The kernels for the inversion were computed in the Born approximation. Figure 2 shows an example inversion result for flows in a layer extending from the surface to 3 Mm below the surface. Notice that there are outflows from the sunspots. In addition, the outflows associated with the supergranulation pattern can be clearly seen. The shallow sunspot outflows obtained with this ridge-filtered inversion differ substantially from flow inversions (which show shallow inflows) performed on the same set of HMI data after applying phase-speed filters [18]. On the other hand, this result is consistent with outflows surrounding sunspots seen in inversions of other sunspots using MDI observations as well as photospheric motions of magnetic features [4].

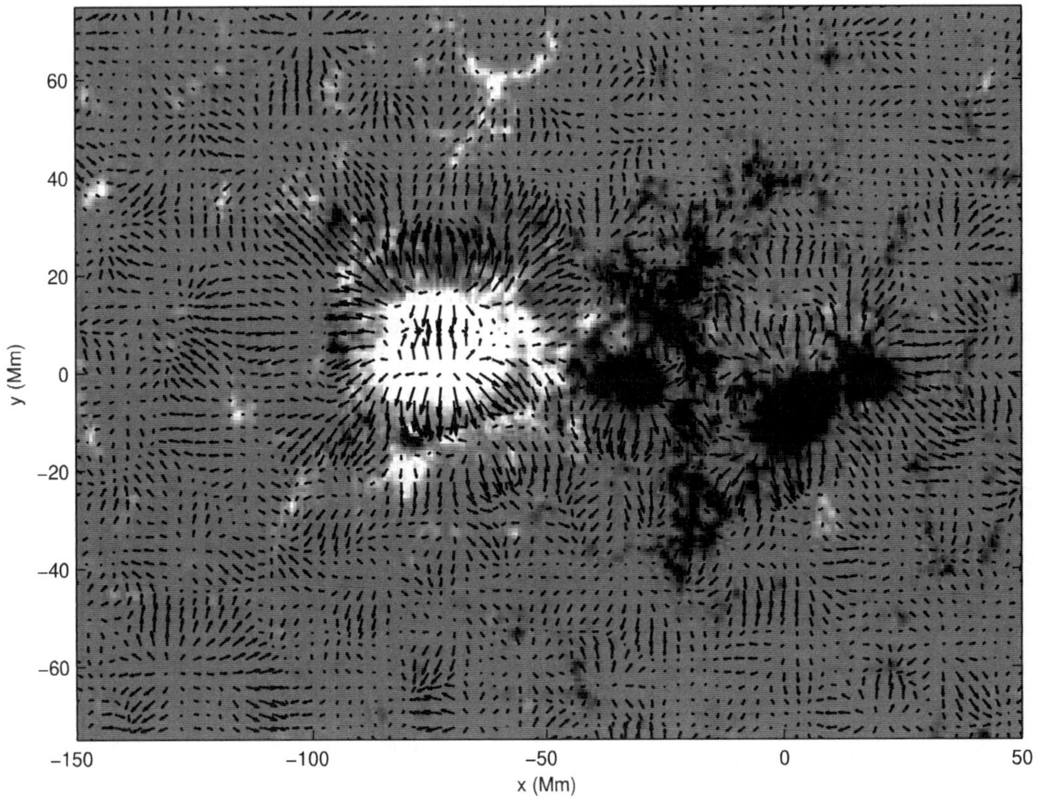

Figure 2. Results of an inversion for horizontal flows from helioseismic holography applied to SDO/HMI Dopplergrams for the active region AR 11057. The largest vectors represent velocities of a few hundred m/s. The background is the line-of-sight magnetic field saturated at $\pm$ 300 Gauss.

4. Magnetic Effects

Although anomalous sensitivities of measurements and inversions to the details of the methodology may be behind some of the discrepancies in the structural models of sunspots and active regions [10], the role of the magnetic field must also be considered. For example, it is possible that, under conditions likely for sunspots, travel-time perturbations may not be related to simple wave-speeds (either the sound speed or an isotropic fast-mode speed). This is illustrated by forward models [19] [20] which include the effects of mode conversion, and which are based on measurements of both travel-time perturbations (or equivalently, phase-shifts) and absorption [21]. Although no inversion method has yet been developed which includes magnetic effects, a promising start is the computation of translationally invariant inversion kernels which include the effects of magnetic fields [22].

5. Conclusions

The effort to understand and resolve these and other issues in sunspot seismology relies heavily on numerical computations of wave propagation through model sunspots, including realistic MHD models [16] as well as magneto-hydro-static (MHS) models (such as translationally invariant models [22]). MHS models are relatively efficient to construct, making them particularly useful to understand the physics of the interaction between waves and magnetic structures. They may also play a critical role as a background (reference) model for inversion methods. Realistic MHD models help to predict what conditions may actually be relevant for real sunspots, as well as provide the necessary validation of helioseismic methods in magnetic regions.

Acknowledgments

This work is supported by the NASA SDO Science Center Program (contract NNH09CE41C) and the Heliophysics GI program (contract NNG07EI51C).

References

[1] Kosovichev, A G, Duvall T L Jr, and Scherrer, P H 2000 *Solar Phys.*, **192**, 159
[2] Couvidat, S, Birch, A C, and Kosovichev, A G 2006 *Astrophys. J*, **640**, 516
[3] Braun, D C, and Birch, A C 2008, *Solar Phys.*, **251**, 267
[4] Gizon, L and 14 coauthors 2009 *Space Sci. Revs*, **144**, 249
[5] Zhao, J, Kosovichev, A G, and Sekii, T 2010, *Astrophys. J*, **708**, 304
[6] Couvidat, S, and Rajaguru, S P 2007, *Astrophys. J*, **661**, 558
[7] Birch, A C, Braun, D C, Hanasoge, S M, and Cameron, R 2009, *Solar Phys.*, **254**, 17
[8] Moradi, H, Hanasoge, S M, and Cally, P S 2009, *Astrophys. J*, **690**, L72
[9] Braun, D C, Birch, A C, Crouch, A D, and Rempel, M 2010, *Proceedings IAU Symposium 273 Physics of Sun and Star Spots*, submitted
[10] Gizon, L, Birch A C, and Spruit, H 2010, *Ann. Rev. Astron. Astrophys.*, **48**, 289
[11] Rajaguru, S P, Birch, A C, Duvall, T L Jr, Thompson, M J, and Zhao, J 2006, *Strops. J*, **646**, 543
[12] Hanasoge, S M, Couvidat, S, Rajaguru, S P, and Birch, A C 2008, *Mon. Not. Roy. Astron. Soc.*, **391**, 1931
[13] Parchevsky, K V, Zhao, J, and Kosovichev, A G 2008, *Isotropy's. J*, **678**, 1498
[14] Nigam, R, and Kosovichev, A G 2010, *Astrophys. J*, **708**, 1475
[15] Scherrer, P H *et al* 1995 *Solar Phys.* **162** 129
[16] Rempel, M, Schüssler, M, & Knölker, M 2009, *Astrophys. J*, **691**, 640
[17] Zhao, J, Kosovichev, A G, and Duvall, T L Jr 2001, *Astrophys. J*, **557**, 384
[18] *Nat. Geog.* published online at `http://news.nationalgeographic.com/news/2010/04/photogalleries/100421-nasa-sun-solar-dynamics-observatory-first-pictures`
[19] Crouch, A D, Cally, P S, Charbonneau, P, Braun, D C, and Desjardins, M 2005, *Mon. Not. Roy. Aston. Soc.*, **363**, 1188
[20] Cameron, R, Gizon, L, Schunker, H, and Pietarila, A. 2010 (arXiv:1003.0528)
[21] Braun, D C, 1995, *Astrophys. J*, **451**, 859
[22] Crouch, A D, Birch, A C, Braun, D C, and Clack, C T M 2010, *Proceedings IAU Symposium 273 Physics of Sun and Star Spots*, submitted

Preliminary velocity flows inside NOA AR 10720 derived by temporally evolving ring diagram analysis of SOHO/MDI dopplergrams

Keiran Burrows and Balázs Pintér

Solar Research Group, Institute of Mathematics and Physics, University of Aberystwyth, United Kingdom, SY23 1BZ

E-mail: kbb09@aber.ac.uk, b.pinter@aber.ac.uk

Abstract. Between 13th and 16th January 2005, NOA active region 10720 was the site of several large flares, one of which induced a solar quake. The expanding wave front of the quake was visible across the surface, causing both horizontal and vertical plasma displacements. Using a new temporal scanning technique for ring diagram analysis of SOHO/MDI (Michelson Doppler imager on board the Solar and Heliospheric Observatory) dopplergrams, we have calculated the horizontal and vertical velocity flows within the active region and the surrounding areas to a depth of 15Mm. We have been able to prove that it is possible to determine changes to a steadily varying subsurface flow, over time scales of hours.

1. Introduction

Solar flares are associated with surface sunspot activity and reorganization of magnetic flux within the lower atmosphere. Most flare eruptions are directed outwards, away from the sun and into the heliosphere. However, flares can also direct all the stored kinetic and thermal energy inwardly, causing a response within the photospheric layers. This kind of eruption can cause white light emission and visible ripples to appear on the solar surface, and is known as a solar quake. [1][2] have suggested rapid heating of the low photosphere as a possible mechanism for this type of event.

NOA active region 10720 rotated onto the visible solar disc on January 11th 2005 and evolved into a highly complex delta region. It produced many significant X and M class flares, and a large earth directed CME. On January 15th 2005 an eruption caused one of the largest flare induced quake events of cycle twenty three. The wavefront of the quake was visible on the surface, expanding away from the foot point of the flare [1][2]. Figure 1 is an SOHO/MDI (Michelson Doppler imager on board the Solar and Heliospheric Observatory) continuum image taken after the quake event, and shows the complex shape of the active region.

It is of interest to know how the subsurface plasma within the active region reacted to the sudden impulsive shock of the quake event. Temporally averaged plasma flow patterns can be achieved using the local helioseismic technique of ring diagram analysis. [3] realized that it was possible to extend global helioseismology to localized patches of the solar surface. By using the millions of oscillations generated within the convection zone, it was possible to disentangle information about the material which each wave passed through. From this information, interior plasma flows and structure can be determined in the horizontal and vertical directions.

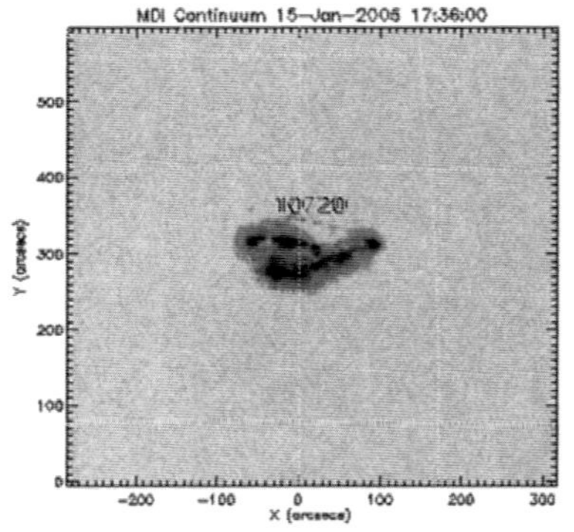

Figure 1. SOHO/MDI continuum image of NOA active region 10720 approximately seventeen hours after the quake event

Ring diagrams have also been very successful at determining subsurface global flows such as the meridional and zonal flows within the convection zone. These are likely to be controlled by the dynamo and could also be the anchor points of sunspot activity. This technique has been very successful in determining large scale subsurface weather patterns, which have been used as early forecast tools in space weather prediction. One major part of ring diagrams which has, until now, not been possible is analysis of temporal flows on time scales of hours or minutes. This has ment many short lived events such as flare, CME eruptions, and solar quakes have not been studied in terms of the subsurface structure and effects.

2. Method

The quake event occurred between 00:21 and 00:43 UT and was induced by a X1.2 flare which erupted above the active region. In order to study the effects which the quake had on the subsurface plasma flows, we have used the local helioseismic technique of ring diagram analysis. We have developed a method of analyzing the temporal variations of these flows, before, during and after the time of the quake.

We created an IDL code which is able to select all the relevant dopplergram images, based on the header information for each file, and a list of start and end dates and times to analyze. This method results in blocks of time which are independent of any other and so subsurface velocity flows can be easily calculated. The code can also be used to study events which occur over time scales of minutes, down to the cadence limit of the MDI instrument. We have designed the code to take an initial start date and time, the total length of time, the number of time blocks, and the length of each time block. It will then determine the relevant data files for each time range, again based on the file headers. The program will then perform ring diagram analysis on each block of time and for a pre-determined number of data cubes surrounding a specific region of interest. When complete, the code produces plots which display the horizontal and vertical components of the flow.

We have used several standard techniques to create ring diagrams from dopplergram data produced by the SOHO/MDI instrument. Firstly the dopplergrams were remapped onto great circles using Postel's projection, and subsections cut out surrounding regions of interest [4][5]. These patches of the surface were spaced evenly around the centre of mass of the active region. This was calculated from an initial continuum image, using contour lines of constant intensity.

For this analysis, the size of each patch was 16° x 16° or 128 x 128 pixels and the patch centers were separated by 7.5°. Individual patch centers were tracked at a rate equal to the doppler differential rotation rate given by [6], where the coefficients have been adjusted to match the rotation rate of the active region. The form of the rotation equation used here is within the quoted errors and is given by equation 1.

$$\omega_r(\vartheta_0) = 473.92 - 89.83(sin^2(\vartheta_0)) - 59.22(sin^4(\vartheta_0)) \text{ n}Hz \tag{1}$$

Data cubes were assembled by stacking together many images from the same patch site, and missing images were replaced by zeros to keep the cadence the same throughout the analysis. Each data cube was detended in time to remove the large scale effects of solar rotation and apodized with a cosine bell function, to remove Gibbs ringing, that can occur later in the analysis. A three dimensional Fourier transform was applied to the data, two in space and one in time. The resulting power spectrum which, when cut at constant frequency, shows rings nested inside each other, one ring for each radial mode n.

The ring diagrams were unwrapped and filtered to remove high order variations in the theta direction [7][8]. Each ring diagram was fit with a model of the form

$$P = [A/(\omega - \omega_0 + U_x k_x + U_y k_y)^2 + \Gamma^2] + b_0/k^3 \tag{2}$$

where the flow parameters U_x and U_y and central frequency ω of each oscillation mode were inverted using a 1D RLS (Regularised Least Sqaures) inversion procedure, similar to that used in the GONG pipeline. We have produced velocity flows to a depth of 15Mm below the photosphere for several patches of the surface.

3. Discussion

We have previously presented preliminary velocity flows and disturbances caused by the quake event. We did this for 2 patches of the surface, one centered on the active reigion and one centered 7.5° south west of the first. The ring diagram procedure used for the preliminary results was similar to that previously described, but the tracking rate was not uniform. This caused very high velocities to be imprinted into the subsurface flows, masking out the true velocities. Figure 2 shows the original results for a patch co-rotating with the active region, where the velocities are wildly larger than should be expected.

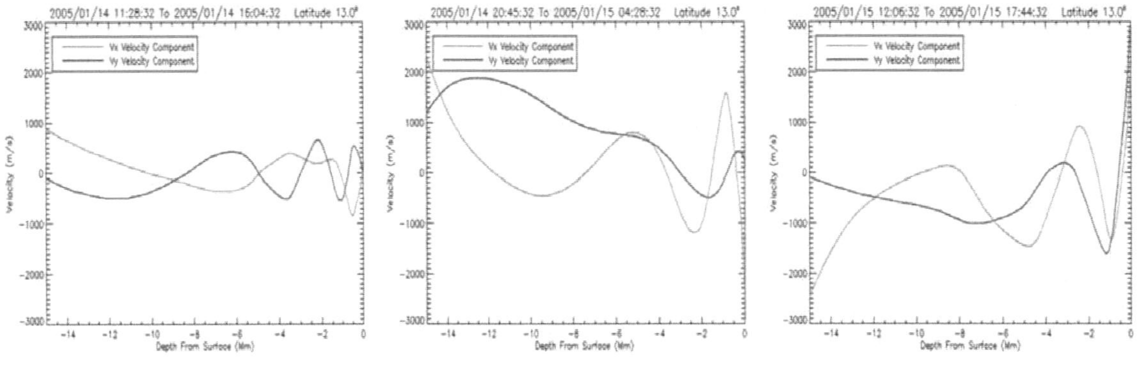

Figure 2. These plots show the horizontal and vertical velocity components of subsurface flows for a patch rotating with the active region. The large velocities are due to the non uniform tracking originally used for this research. The plots represent blocks of time, before, during and after the quake respectively.

We now only track at a uniform rate for each latitude. We recalculated the flows for several patches of the surface. Here we display only the velocity profile of the main patch rotating with the active region.

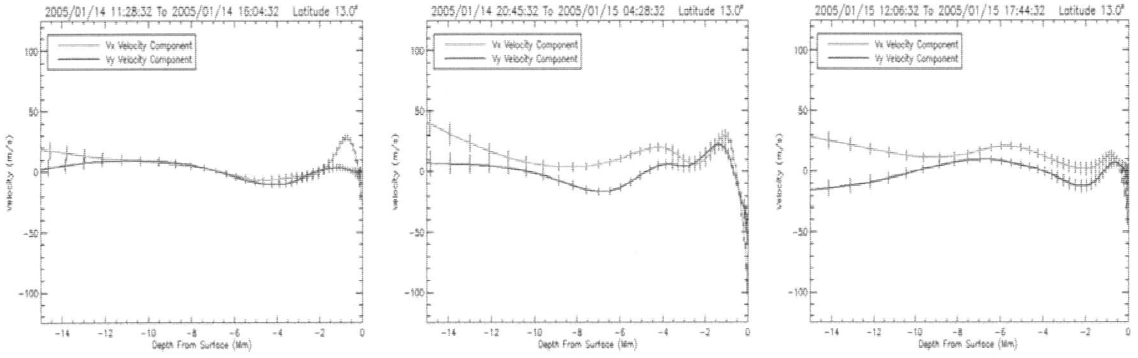

Figure 3. The plots here show the same patch of the surface, tracked at a uniform rate. Again blocks of time before, during and after the quake have been investigated. The near surface layers under the active region show a reaction to the quake, and a slow relaxation to a previous steady flow.

As can be seen in Figure 3 there is a steady plasma flow field before the quake event, which changes direction with depth and is typical of large active regions. During the quake event, large disturbances in the near surface layers can be seen, which indicate an inward radial path for some the energy of the quake event. The subsurface plasma appears to relax back towards the bulk flow of the active region and surrounding area. Further investigation is needed to determine how the quake is coupled to the subsurface flows seen with ring diagram anaylsis.

4. Conclusions

NOA active region 10720 was one of the largest and most complex magnetic structures to appear on the solar disk during solar cycle twenty three. It produced many large flares and a disruptive CME during its transit across the surface. The acoustically active flare of January 15th 2005 caused a solar quake event between 00:21 UT and 00:43 UT which caused visible ripples on the solar surface. We have shown that it is possible to detect temporally varying flows using a new method of ring diagram analysis. We need to extend this work to examine the sub-photospheric reaction to the quake and its links with the flare eruption.

Acknowledgments

The authors would like to thank Brad Hindman and Deborah Haber for their help with the ring fitting procedure. Also we would like to thank the UK Science and Technology Funding Council for funding this research.

References

[1] Moradi H, Donea AC, Lindsey C, Besliu-Ionescu D, Cally PS, 2007, *Mon. Not. R. Astron. Soc.*, **000**
[2] Martínez-Oliveros JC, Donea AC, Cally PS and Moradi H, 2008, *Mon. Not. R. Astron. Soc.*, **389**
[3] Hill F, 1988, *ApJ*, Part 1, **vol. 333**
[4] Corbard T, 2003, `http://www.mps.mpg.de/projects/seismo/NA4/SW/ring_analysis.html`
[5] Zaatri A, Corbard T, Roth M, González Hernández I, Von Der Luhe O, 2008, proc. second helas international conference, *Journal of Physics*, **118**
[6] Snodgrass HB & Ulrich RK, 1990, *ApJ*, **351**, 309.
[7] Schou J & Bogart RS, 1998, *ApJ*, **504**, L131
[8] Haber DA, Hindamn BW, Toomre J, Bogart RS, Larsen RM, Hill F, 2000, *Solar Phys*, **192**

Temporal Variations of High-Degree Solar p-Modes from GONG and MDI

Olga Burtseva[1], Sushant Tripathy[1], Rachel Howe[1], Kiran Jain[1], Frank Hill[1], Richard Bogart[2] and Maria Cristina Rabello-Soares[2]

[1] National Solar Observatory, Tucson, AZ, USA
[2] HEPL, Stanford University, CA, USA

E-mail: burtseva@noao.edu

Abstract. We study temporal variations in the amplitudes and widths of high-degree acoustic modes in the quiet and active Sun by applying ring-diagram technique to the GONG+ and MDI Dopplergrams during the declining phase of cycle 23. The increase in amplitudes and decrease in line-widths in the declining phase of the solar activity is in agreement with previous studies. A similar solar cycle trend in the mode parameters is also seen in the quiet-Sun regions but with a reduced magnitude. Moreover, the amplitudes obtained from GONG+ data show long-term variations on top of the solar cycle trend.

1. Introduction

The global and local analysis of solar acoustic modes has shown that the mode amplitudes and lifetimes are anti-correlated with the solar activity level, and depend strongly on the local magnetic flux [1, 2, 3]. Most of the analyses have been done for average quantities without discrimination between quiet and active areas, while the solar cycle behavior of the quiet-Sun regions is not well understood. High-degree *p*-mode lifetimes measured in the quiet-Sun regions using cross-correlation analysis show variations with the activity cycle [4]. In the quiet Sun, no large-scale magnetic field concentrations, one of the potential damping sources, are present on the surface. Another explanation could be the activity-related variation of the convective properties near the solar surface. The studies of the solar cycle variations of the size of the solar granules so far arrive at contradictory conclusions [5, 6]. An attempt to mask strong surface activity and analyze high-degree *p*-mode amplitudes in the quiet-Sun regions at solar minimum and maximum indicated that the amplitude at solar minimum is higher than that at solar maximum [7], however, the effect introduced by the mask needs to be better understood. In this work we apply ring-diagram analysis to eight years of data to characterize the high-degree acoustic mode amplitudes and widths in the quiet and active Sun during declining phase of the activity cycle.

2. Data analysis

The mode parameters of the solar acoustic oscillations analyzed in this work are obtained from Global Oscillation Network Group (GONG) and Michelson Doppler Imager (MDI) high-resolution Dopplergrams for the period from 2001 to 2009 using the standard ring-diagram technique [8] of the GONG and MDI pipelines. The size of a standard patch in the ring analysis is $15^o \times 15^o$. The most distant patches are centered at $52^o.5$ from the disk center. More recently the pipeline developed for the Helioseismic and Magnetic Imager (HMI) [9], applying both asymmetric [10] and symmetric [11] profiles fitting to derive the mode parameters, has also been used to analyze the MDI data, in the traditional MDI patch set. The results from both procedures are consistent, thus, here we show the symmetric profile fitting results only.

The Magnetic Activity Index (MAI) values are computed from MDI 96-minute magnetograms for the same time frame of Dopplergrams. The MAI is computed by averaging unsigned flux above 50 G and below 500 G to avoid contamination from noise in the measurements of the quiet-Sun flux values and saturation issues of the MDI magnetograms [12]. We define the quiet-Sun regions as those with MAI values below 5 G.

As discussed in [3], the amplitudes and widths fitted from the ring analysis need to be corrected for center-to-limb, residual B_0-angle variations, and duty-cycle dependences. In our analysis, GONG data were restricted to duty cycle of 70% and higher. In contrast, MDI data have high duty cycle and therefore no selection criteria were used to reject the fitted parameteres. In this work, only disk center patches were analyzed to avoid the foreshortening effects.

3. Results and discussion

The results for amplitudes and widths shown in this paper are for a multiplet $\ell = 440$, $n = 2$ ($\nu = 3.2$ mHz). We find that the correlation of this multiplet with other multiplets in the 2.5–3.5 mHz frequency range over solar activity cycle is higher than 70%.

3.1. Activity-related trend

The variations of amplitude and width as a function of time and MAI, computed from all patches and the quiet-Sun patches at the disk center using GONG data, are shown in Figure 1. According to the Mt.Wilson sunspot index data (see Figure 2), during most of the year 2008 the Sun was quiet with no or very few small sunspots appearing on the solar surface. Therefore, we took average of year 2008 consisting of 13 Carrington Rotations as a reference, and plotted the mode parameters relative to this mean value. The amplitudes increase by $\sim$ 10% from the high solar activity period in 2001 till 2004. After this period, we find a long-term variation, but no clear association with activity cycle. It is interesting to note that the mode amplitude, derived from Variability of Solar Irradiance and Gravity Oscillations (VIRGO) and GONG data using global helioseismology technique, shows similar wiggles though they are small in amplitude relative to the extent of the solar cycle trend [13]. If we ignore the unexplained long-term variations, we can conclude that the mode amplitudes obtained in our study increased from 2001 to 2008 by $\sim$ 22% in all patches and by $\sim$ 16% in the quiet-Sun regions. The widths decreased from 2001 to about 2008 by $\sim$ 9%, then started showing a rising trend.

The variations in amplitude and width from MDI data (see Figure 1, top panels) show a solar cycle trend similar to that from GONG data. The long-term variations in the amplitudes obtained from GONG data are not seen in the MDI amplitudes. However, this could be due to the limited amount of data (only 135 days) that were analyzed with the MDI Dopplergrams. We plan to improve statistics on MDI data to confirm these results.

The decrease in amplitudes and increase in widths with increase in MAI (bottom panels of Figure 1) are consistent with known results from global and local analysis. The linear correlation analysis shows that the correlation coefficient between amplitude and MAI is -0.49 for all patches and -0.29 for the quiet-Sun patches. The correlation coefficient between width and MAI is 0.73 for all patches and 0.48 for the quiet-Sun patches.

Figure 2 shows the variation of the magnetic activity index over time. The MAI values of all patches at the disk center are well correlated with Mt. Wilson sunspot index. A weaker solar-cycle trend is also visible in the magnetic indices of the quiet-Sun patches. The solar cycle variation of the MAIs of the quiet-Sun regions computed from MDI magnetograms was also

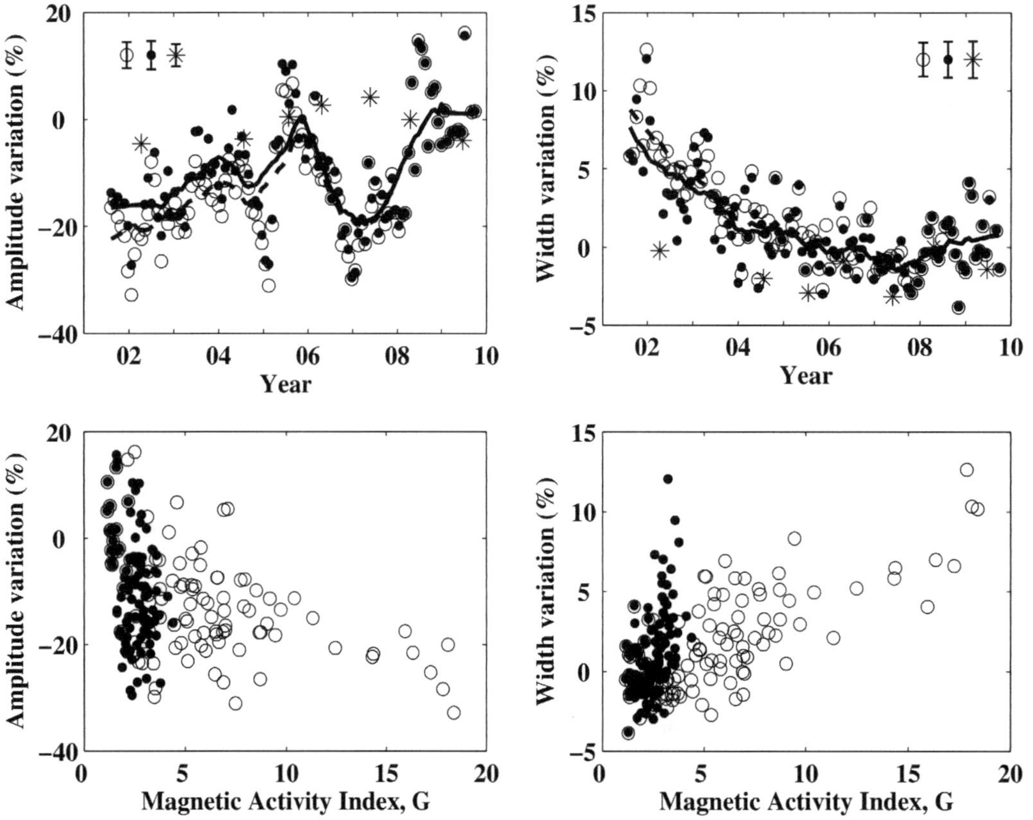

Figure 1. Variation of amplitude and width (relative to the mean value of 2008) as a function of time and MAI, computed in the quiet-Sun (closed circles) and all patches without MAI threshold (open cicrcles). The mode parameters are derived from disk center patches using GONG data. Each point is an average over one Carrington rotation. Solid and dashed curves represent the temporally smoothed values in the quiet-Sun and all patches, respectively. Stars show the amplitude and width variation as a function of time, computed from MDI data in all patches at the disk center. In this case, each point is an average over one year. The typical error bars for the measurements are shown in the top panels.

noted in [4] and could be due solar cycle variation of the strong-field component of the quiet-Sun network [14, 15]. This suggests that the magnetic field plays a role in the activity related variations of the acoustic mode parameters in the quiet Sun.

3.2. Long-term variations

In an attempt to understand the long-term variations, on top of the solar cycle trend, seen in the amplitudes computed from GONG data, we restrict our analysis to 2 G $\leq$ MAI $\leq$ 3 G patches. We notice the long-term variation in the amplitude but no dependence on magnetic activity. This leads to the conclusion that the variations are probably not related to solar activity cycle. Moreover, according to our preliminary estimates from the patches at higher latitudes, the variations are present up to the regions centered at $52^o.5$ in latitude. We plan to investigate this aspect in more detail after applying the necessary geometrical corrections.

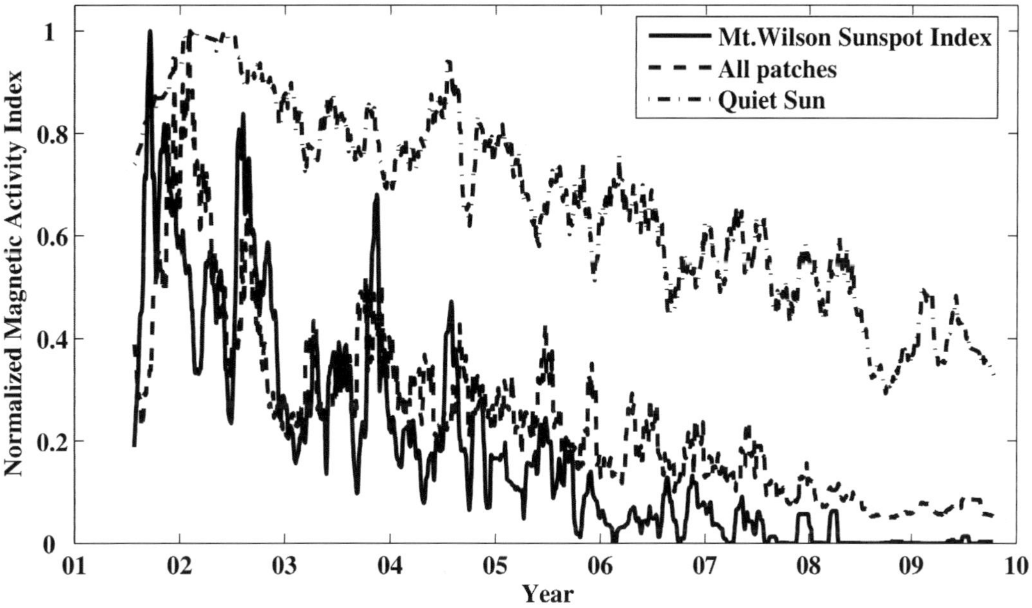

Figure 2. Normalized magnetic activity index, computed from MDI magnetograms in the quiet-Sun (dash-dot curve) and all regions (dashed curve) at the disk center, plotted as a function of time. The solid line represents the Mt.Wilson Sunspot Index. Note, that each curve is normalized to its maximum value to show the relative variation.

Acknowledgments

This work utilizes data obtained by the Global Oscillation Network Group (GONG) Program, managed by the National Solar Observatory, which is operated by AURA, Inc. under a cooperative agreement with the National Science Foundation. The data were acquired by instruments operated by the Big Bear Solar Observatory, High Altitude Observatory, Learmonth Solar Observatory, Udaipur Solar Observatory, Instituto de Astrofisica de Canarias, and Cerro Tololo Interamerican Observatory. SOHO is a mission of international cooperation between ESA and NASA.

References

[1] Chaplin W J, Elsworth Y, Isaak G R, Miller B A and New R 2000 *Mon. Not. Royal Astron. Soc.* **313** 32
[2] Rajaguru S P, Basu S and Antia H M 2001 *Astrophys. J.* **563** 410
[3] Howe R, Komm R W, Hill F, Haber D A and Hindman B W 2004 *Astrophys. J.* **608** 562
[4] Burtseva O, Hill F, Kholikov S and Chou D-Y 2009 *Solar Phys.* **258** 1
[5] Berrilli F, Ermolli I, Florio A and Pietropaolo E 1999 *Astron. Astrophys.* **344** 965
[6] Saldana-Munoz M, Muller R and Hanslmeier A 2004 *Hvar Obs. Bull.* **28** 9
[7] Burtseva O, Tripathy S C, Hill F, Kholikov S, Raouafi N-E and Lindsey C 2009 in GONG 2008/SOHO 21 ASP Conference Series, Solar-Stellar Dynamos as Revealed by Helio- and Asteroseismology **416** 293
[8] Hill F 1988 *Astrophys. J.* **333** 996
[9] Bogart R S, Baldner C, Basu S, Haber D A and Rabello-Soares M C 2011 *J. Phys. Conf. Proc.* this issue
[10] Basu S and Antia H M 1999 *Astrophys. J.* **525** 517
[11] Haber D A, Hindman B W, Toomre J, Bogart R S, Thompson M J and Hill F. 2000 *Solar Phys.* **192** 335
[12] Basu S, Antia H M and Bogart R S 2004 *Astrophys. J.* **610** 1157
[13] Salabert D, Garcia R A, Palle P L and Jimenez A 2011 *J. Phys. Conf. Proc.* this issue
[14] Hagenaar H J, Schrijver C J and Title A M 2003 *Astrophys. J.* **584** 1107
[15] Pauluhn A and Solanki S K 2003 *Astron. Astrophys.* **407** 359

Helioseismic Tests With the FLASH Simulation Code

Pedro A. González-Morales[1], Rekha Jain[2] and Michael J. Thompson[3]
[1,2]Dept. Applied Maths at University of Sheffield (UK),
[3]High Altitude Observatory, Boulder, CO (USA).
E-mail: [1]`pedro.gonzalez@sheffield.ac.uk`, [2]`r.jain@sheffield.ac.uk`, [3]`mjt@ucar.edu`

Abstract. We show our first results from local helioseismic simulations using the numerical code FLASH by testing its suitability for simulating subphotospheric wave motions in helioseismology. In order to check the capability of this code for different sources of waves, we have implemented a non-magnetic plane-parallel atmosphere adding a source term to the energy equation. We confirm the capabilities of FLASH code for investigating the propagation of sound waves into a realistically stratified solar interior.

1. Introduction

Helioseismology is the study of pressure and gravity waves which are excited stochastically below the solar surface, these oscillations are a key to understanding the structure and dynamics of the solar interior. Forward modelling using theoretical and simulation studies combined with observational data-analysis are being used for this purpose. In recent years, it has become increasingly obvious that this combined effort is now essential if we are to understand fully the magnetic regions such as sunspots and plages and their influence on these oscillations.

The FLASH code solves the fully compressible, reactive hydrodynamic equations. It was initially developed to model nuclear flashes on the surfaces of neutron stars and white dwarfs, and the interior of white dwarfs; but it has since been applied to modelling a wide variety of astrophysical flows, see also Fryxell et al. (2000) for more details. In this paper we report the first use of FLASH numerical code to understand the behaviour of wave-field propagation in the context of helioseismology (see also Moradi et al. (2010) for a summary of other numerical codes used in local Heliosesimology).

In section 2 we describe the construction of our background model and the form of acoustic source used. In section 3, we describe the numerical model setup. In section 4 we show the time-distance diagrams obtained from our simulations with FLASH. The results are briefly discussed in section 5. We then conclude followed by a short outline of our future work in section 6.

2. The Model

Following the standard procedure, we choose for our non-magnetic plane-parallel stratification the standard Solar Model (the S model) of Christensen-Dalsgaard (1996) and try to make it stable against convection. We consider the adiabatic approximation for an ideal gas, so the adiabatic index, Γ_1 $(= 5/3)$, and the mean molecular weight, μ $(=1.25)$, take constant values.

To eliminate the convective instability in the upper layers the temperature profile is slightly modified according to the convective stability criterion shown below by equation (3) to be less or equal zero but, keeping the condition of hydrostatic equilibrium as described by eq. (2). We also use the equation of state for an ideal gas given by eq. (1), to close the system of equations.

$$T = \frac{\mu p}{\mathbb{R}\rho} , \tag{1}$$

$$\frac{dp}{dz} = -g(z)\rho(z) , \tag{2}$$

$$\left|\frac{dT}{dz}\right| - \left(1 - \frac{1}{\Gamma_1}\right)\frac{T}{p}\left|\frac{dp}{dz}\right| = s(z) , \tag{3}$$

where p, ρ, and T are the pressure, density and temperature respectively and $\mathbb{R}$ is the gas constant. The initial pressure and density are recalculated to maintain the hydrostatic equilibrium and the convective stability (see also, Shelyag et al. (2006) for more details).

We solve Euler's equations for compressible gas dynamics in two spatial dimensions, adding a single acoustic source $Q(\mathbf{r}, t)$ to the energy equation as described by Rast (1999).

$$Q(\mathbf{r}, t) = Q_0 \left\{1 + \tanh\left[\frac{\ln(3)}{\sigma_1}(t - t_0)\right]\right\} \exp\left[-\frac{(\mathbf{r} - \mathbf{r}_0)^2}{\sigma_2^2}\right] , \tag{4}$$

where the amplitude is $Q_0 = -150$ erg s^{-1} cm^{-3}, the characteristic time length is $\sigma_1 = 120$ s, the full-width at half maximum is $\sigma_2 = 0.5$ Mm, the source shifting time is $t_0 = 1000$ s. This source imposes a cooling event that produces a localised drop in pressure and temperature which then results in an expanding outflow around the source. As a side effect, this source introduces convective instabilities around itself.

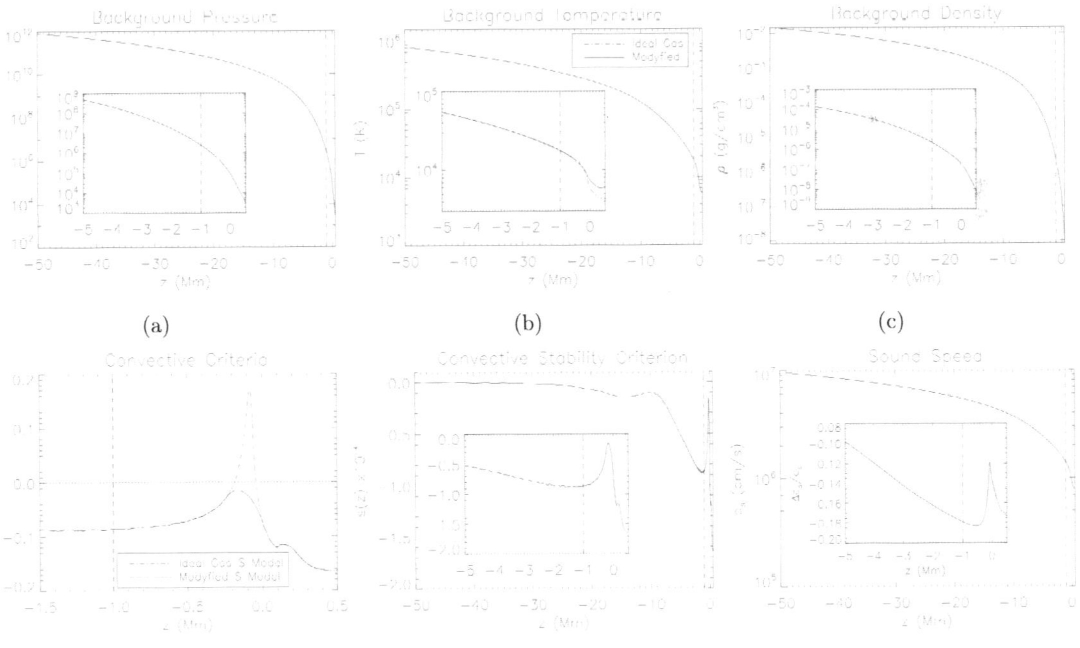

(a) (b) (c)

(d) (e) (f)

Figure 1: *Modified S Model profiles for our convectively stable background. The subplot in figure (f) shows, in detail, the relative differences between the sound speeds obtained from our model and the S Model. The rest of the subplots show the details in the upper layers of the corresponding parameter.*

3. Numerical Setup

Our numerical domain is represented schematically in figure 2. The simulation box has D Mm of thickness by L Mm of width. The source is located at $L/2$ at a depth z_s below the solar surface; the surface being located at $R_\odot$ or $z = 0$. We consider our atmosphere under a vertical gravitational field, constant in time i.e. $\mathbf{g}(\mathbf{r}, t) = \mathbf{g}(z)$.

We carried out three numerical experiments with different depth and grid points as well as different source positions to test the code, see table (1).

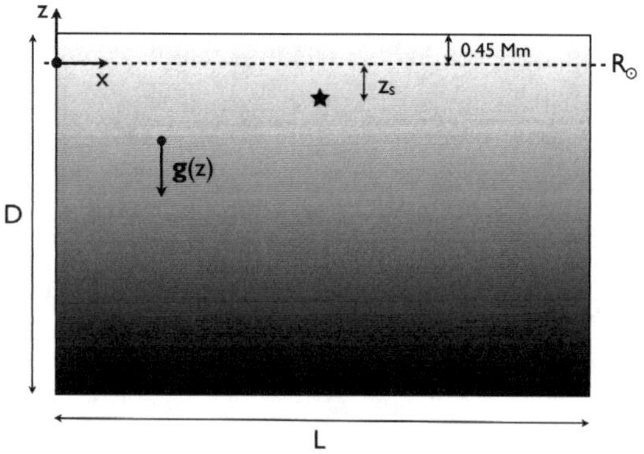

Figure 2: *Schematic representation of our numerical domain. The solar surface is represented by the dashed line at $R_\odot$. D is the thickness whereas L is the width. The gravitational field is represented by* $\mathbf{g}(z)$ *and the location of the source by an asterisk at* $(L/2,\ z_s)$.

Table 1: *Summary with the model parameters used in our simulations.*

Model Name	S10.1	S10.2	S50.1
D (Mm)	10	10	50
L (Mm)	150	150	150
z_s (Mm)	-1	-1.55	-1
Grid (HxV)	392x1200	392x1200	960x6000

The boundary conditions used in our simulations are outflow for left and right sides in the horizontal direction (ASC FLASH Center 2009). For the vertical direction we keep the hydrostatic equilibrium with outflow condition for the top and bottom boundaries.

4. Results

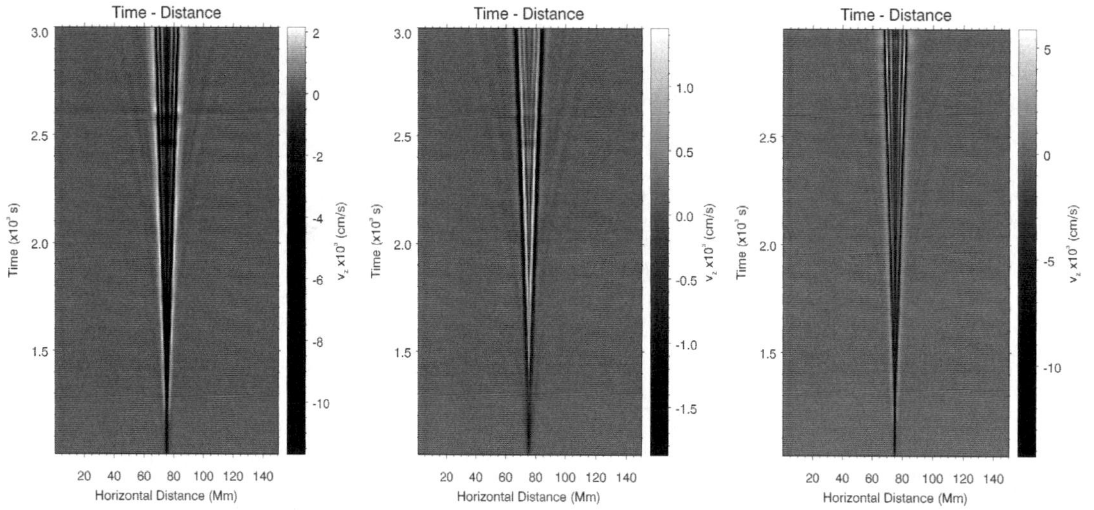

(a) S10.1 (b) S10.2 (c) S50.1

Figure 3: *Time-distance diagrams for the vertical velocity, v_z. The measured level is located at $R_\odot$. Figures (a) and (b) correspond to the source depth of* 10 *Mm and (c)* 50 *Mm respectively. We can see how the trace due to convective cells around the source is less prominent in model S10.2 than in model S10.1 due to the source position with respect to the measured level.*

5. Discussion and Conclusion

In figure 1 we show the profiles for our modified S Model and the position of the source with a vertical dashed line (shown only for experiments S10.1 and S50.1). The ideal gas consideration and the elimination of convection close to the surface have produced a slight increment in the temperature in this region when compared to the unmodified S Model, see subplot of figure 1b.

In figure 1f we show the relative differences of the sound speeds obtained from our model and the S Model, i.e. $\Delta = (c_s - c_0)/c_s$. The relative difference is greatest close to the upper layers due to the modification of temperature profile, therefore we are modifying the eigenfrequencies and eigenfunctions of the S Model close to the surface in an important way. The difficulties as well as the possibilities of finding a non-convective solar-like model has also been addressed in Schunker et al. (2010)

Figure 3 shows the time-distance diagrams for v_z. In figures 3a and 3b, we can see how the amplitude of v_z decreases by increasing z_s. Furthermore, we see sharper ridges for S10.1 model. This sharp ridges are due to the convective cells generated by the source. Figure 3c shows the time-distance diagram for model S50.1. We can see similar structure as in models S10.1 and S10.2, but with less numerical noise, due to higher number of points used in horizontal direction.

Comparison with figure 8 of Shelyag et al. (2006), suggests that although there are some differences in the boundary conditions treatment, numerical technique, source position and the atmospheric model, the results obtained with FLASH are similar to their simulation. These preliminary results have shown the capabilities of FLASH code for investigating the helioseismic studies, in particular the propagation of sound waves into a realistically stratified solar interior.

6. Future Work

The most important step will be to implement more accurate boundary conditions to reduce noise and reflections. We also hope to use the adaptive mesh capabilities of FLASH to improve the resolution and reduce the computational time. Clearly we need a more accurate modelling in $s(z)$ profile to get a more realistic convectively stable S Model. A new source to avoid the insertion of convective instability in our numerical domain and a study with different random sources would be also very interesting. Another important and necessary step is to introduce magnetic fields. Finally it would be interesting to analyse the simulated and the real observational data of wavefields with the same local helioseismic techniques.

Acknowledgements

This work is supported by a UK-STFC studentship in collaboration with IAC (Spain). The software used in this work was in part developed by the DOE-supported ASC/Alliance Center for Astrophysical Thermonuclear Flashes at the University of Chicago.

References

ASC FLASH Center. 2009, FLASH User's Guide, University of Chicago

Christensen-Dalsgaard, J. 1996, Nuclear Physics B Proceedings Supplements, 48, 325

Colella, P. & Woodward, P. R. 1984, Journal of Computational Physics, 54, 174

Fryxell, B., Olson, K., Ricker, P., et al. 2000, *ApJS*, 131, 273

Moradi, H., Baldner, C., Birch, A. C., et al. 2010, *Sol. Phys.*, 171

Rast, M. P. 1999, *ApJ*, 524, 462

Schunker, H., Cameron, R., & Gizon, L. 2010, ArXiv e-prints

Shelyag, S., Erdélyi, R., & Thompson, M. J. 2006, *ApJ*, 651, 576

The Influence of Tracking Rate on Helioseismic Flow Inferences

Swati Routh[1], Deborah A. Haber[1], Bradley W. Hindman[1,2], Richard S. Bogart[3], Juri Toomre[1,2]

[1] JILA, University of Colorado, Boulder, CO 80309, USA

[2] Department of Astrophysical and Planetary Sciences, University of Colorado, Boulder, CO 80309, USA

[3] Stanford University, Stanford, CA 94305,USA

E-mail: swro2964@lcd.colorado.edu

Abstract. Traditionally, most local helioseismic studies of subsurface flows have removed the large signal due to the Sun's rotation by tracking the analysis region across the solar surface. In order to work in a uniformly rotating reference frame, the ring-analysis pipeline of the recently launched Helioseismic and Magnetic Imager (HMI) will track all analysis regions at the solid-body Carrington rate. To test this tracking scheme, we compare flow determinations resulting from two different tracking schemes. In one scheme we use the HMI pipeline implementation which tracks at the Carrington rotation rate. In the other, the tiles are tracked at the local differential surface rotation rate as measured by Snodgrass (1984). We observe systematic differences between the flows obtained by the two schemes even after transforming them to a common frame (Snodgrass frame), with the zonal flows measured in the Carrington frame being faster by 5-20 m/s.

1. Introduction

When measuring convective flows through local helioseismology, one must deal with the fact that the sun's rotational velocity is often an order of magnitude larger than the flows being sampled (typical flow speeds for supergranulation are 300 m/s, compared to the equatorial rotational velocity of 2000 m/s). Therefore, in order to measure the more sedate convective flows, ring-analysis procedures have usually tracked data cubes at the local surface differential rotation rate of the plasma (Haber et al. 2002, Corbard et al. 2003). In practice, the Snodgrass rate (Snodgrass 1984) appropriate for the center of the analysis region has been used in order to remove as much of the rotational signature as possible and to keep the measurement procedure in a slow, linear regime.

For consistency in region labelling, to allow for tracking over long time intervals, and to avoid a shearing reference frame, the designers of the HMI ring-analysis pipeline have opted to track at the uniform Carrington rotation rate. If the ring-analysis flow-measurement procedure were purely linear, the choice of reference frame would be immaterial. However, large velocities can Doppler shift power out of the predefined frequency windows used to perform the analysis, thereby causing a potential nonlinearity in the measurement procedure. In order to test the proposed tracking scheme for HMI, in this paper we analyze the same data using both tracking

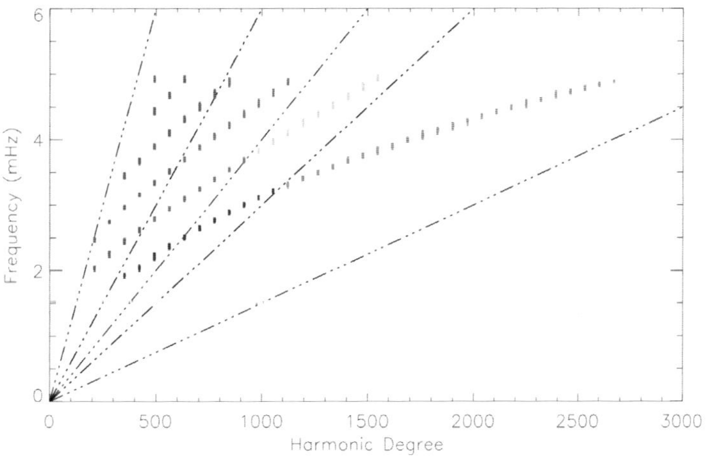

Figure 1. Dispersion diagram for the set of modes common to both data sets, one obtained by tracking at the uniform Carrington rate and the other at the shearing Snodgrass rate. The dash-dotted lines indicate the boundaries of bins in horizontal phase speed over which averaging of flows was performed. The phase speeds correspond to 1.0, 2.1, 2.8, 4.2 and 8.4 km/s. Each fitted mode is shown as a tiny asterisk and the spread in symbols shows the range of rest frequencies for tiles at different locations on the disk. The color of each asterisk indicates which phase speed bin the mode falls within. Note, the f mode has been binned separately from the p modes.

schemes: the Carrington frame and the shearing Snodgrass frame. The two sets of results are intercompared and systematic differences identified.

2. The Data

We have analyzed Dopplergram data from HMI consisting of a multitude of tracked data cubes each corresponding to a square region, 5.12° on a side, that has been tracked for 9.6 hr. Each cube lies at a different location on the solar disk, and the regions fill the solar disk with roughly equal density. The regions overlap by roughly 50% and, depending on the solar B angle, there are either 2727 or 2748 separate regions. Data from 18 separate 9.6 hr periods, spanning 6 days, have been analyzed and the results averaged together. Two different tracking schemes have been employed on the same HMI images. One set of data cubes has been tracked at the Carrington rate $\Omega_C/2\pi = 456.03$ nHz, and the other at the Snodgrass rate $\Omega_S/2\pi = (451.426 - 54.77\ \sin^2\theta - 80.17\ \sin^4\theta)$ nHz, where θ is the latitude of the center of the region. Each data cube is apodized in space and time and Fourier transformed to form a 3-D power spectrum that is a function of cyclic frequency ν and two horizontal wavenumbers, k_x and k_y, corresponding to the prograde, zonal direction (the x-direction) and northward (the y-direction).

3. Mode Fitting

Parameters for the f and p modes have been extracted from each power spectra by a maximum-likelihood fitting procedure with a fitting function given by,

$$P(\nu, k_x, k_y) = \frac{A\Gamma/2}{[(\nu - \nu_0 - U_x k_x - U_y k_y)^2 + (\Gamma/2)^2]} + b/k^3, \tag{1}$$

where the fitting is performed separately for each total horizontal wavenumber $k = (k_x^2 + k_y^2)^{1/2}$ and for each radial order n. The fitted parameters are the mode amplitude A, the line width Γ, the rest frequency ν_0, two flow speeds U_x and U_y, and a background power level b. For the entire set of data cubes, 60812 modes were common to both data sets. Figure 1 shows the dispersion diagram for these fits. Modes ranging in order from the f mode to p_5 were obtained.

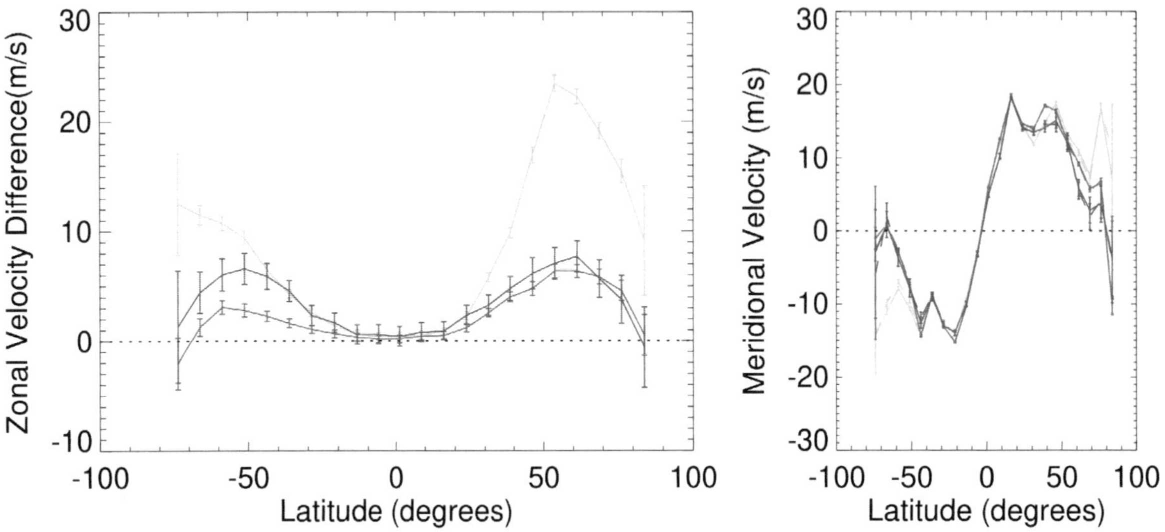

Figure 2. (a) Difference between the mean zonal flow measured in the two frames ($v_C - v_S$) and (*b*) mean meridional flow plotted separately for each frame. The means were computed by averaging over longitude and time, with further averaging over 7.5° bins in latitude and 3 bins in horizontal phase speed performed to increase the ratio of the signal to noise. All zonal flows have been transformed to a common reference frame (Snodgrass) for comparison. The different colors of the curves correspond to averages over p modes with different horizontal phase speeds, with green corresponding to modes with the shallowest acoustic cavity and violet to those with the deepest. The colors correspond to those used in Figure 1. The zonal flows measured in the Carrington frame are systematically faster than those measured in the Snodgrass frame. The meridional flows measured in the two frames are identical to within their measurement errors. Measurement made in the Carrington frame are shown with solid curves and those made in the Snodgrass frame with dashed curves.

4. Mean Flow Analysis

In order to compare the data obtained by the two different tracking rates, we have transformed the zonal flows U_x measured in the Carrington frame to the Snodgrass frame, by adding the difference between the surface velocity for the two frame rates. Error weighted means were then computed by averaging all flow speeds in longitude and time. The ratio of signal to noise was further increased by averaging over latitudinal bins that were 7.5° wide and by averaging all modes falling within specified ranges of horizontal phase speed $2\pi\nu/k$ (see Figure 1). Modes with a common phase speed have acoustic cavities with the same lower turning point; thus, the phase speed binning averages over modes with similar sampling in depth. Furthermore, since the f modes are surface gravity waves without an acoustic cavity, we have performed a separate analysis for the f modes and the p modes. Finally, only modes common to both data sets were included in the means.

5. Results

The mean flow speeds are shown in Figures 2 and 3. The mean zonal flows measured in the Carrington reference frame are systematically faster than those measured in the Snodgrass frame (shown as positive values in Figures 2*a* and 3*a*). The difference is small near the equator (less than 5 m/s), but grows toward the poles reaching values of 20 m/s or more. Furthermore, for modes of large harmonic degree ($l > 1000$) the discrepancy is greatly enhanced preferentially in the northerm hemisphere.

The meridional flows measured in the two reference frames are identical to within their measurement errors. The flows are poleward with a speed of 15-20 m/s at mid-latitudes, vanishing flow at the equator, and evidence for reduced flow near the poles. For the p modes there is a marked lack of dependence on phase speed indicating little variation with depth; however, the flows obtained with f modes are systematically higher by a factor of roughly 50%, perhaps suggesting a near-surface enhancement of the meridional flow

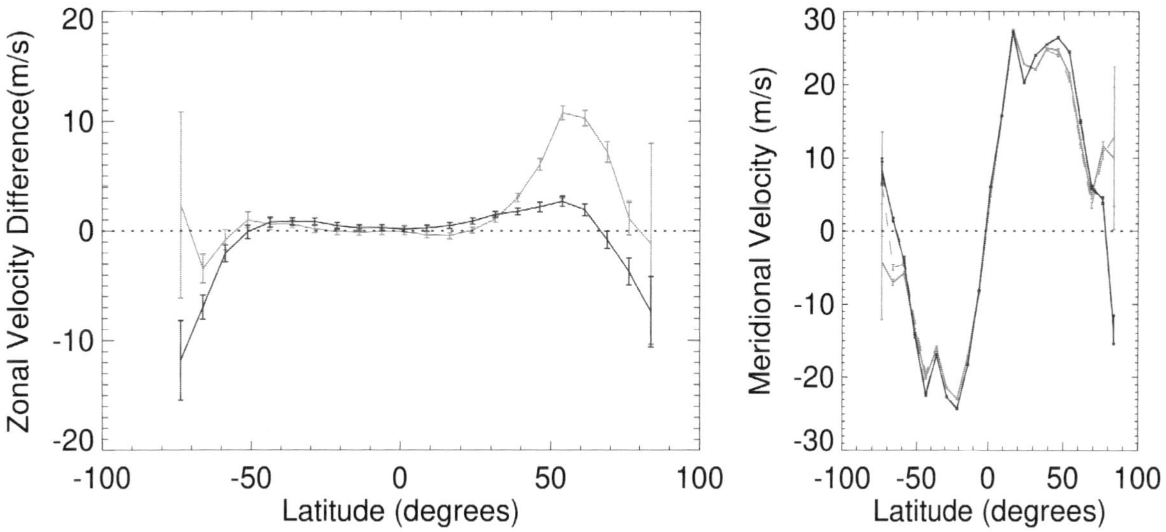

Figure 3. Same as Figure 2, except the analysis uses only f modes. Orange corresponds to high degree modes $l > 1000$

6. Discussion

The fact that the mean flows are systematically different depending on the tracking rate is an indication that the mode fitting procedure isn't linear with respect to tracking. One possible source for this systematic error arises because the fitting procedure is performed over fixed frequency windows and large flow velocities shift the power within this window. For small analysis regions, the p-mode power profiles are sufficiently broad that this redistribution of power may move the wings of the target mode ridge out of the fitting domain, and perhaps move the wings of nearby ridges into the domain as well. This causes a systematic mismeasurement of the flow speed. Such problems would be exacerbated by the use of the Carrington rate because the rotation rate of the frame and the local plasma rate have substantial mismatch. Further, this problem is more prominent when analyzing small tiles. The mode profiles for small analysis regions are broadened due to the spatial window function. Thus, a further test would be to repeat the analysis with tiles that are 15° on a side. The disagreement in the zonal velocity should be reduced for such a tile size, if the source is as suggested.

Acknowledgments

We thank NASA for its support through grants NNG06GD97G, NNX07AH82G, NNX08AJ08G, NNX08AQ28G, NNX09AB04G and NAS5-02139. HMI data courtesy of SDO (NASA) and the HMI consortium.

7. References

[1] Corbard, T, Toner, C, Hill, F, Hanna, K D, Haber, D A, Hindman, B W and Bogart, R S 2003 *ESA SP-517: SOHO12/GONG+ 2002: 'Local and Global Helioseismology: The Present and Future' (Big Bear Lake, CA)* ed H. Sawaya-Lacoste pp 255-8

[2] Haber, D A, Hindman, B W, Toomre, J, Bogart, R S, Larsen, R M and Hill, F 2002 *Astrophys. J.* **570** 855-64

[3] Snodgrass, H B 1984, *Solar Phys.* **94** 13-31

Ring-diagram parameter comparisons for GONG, MDI and HMI

R Howe,[1] S Tripathy,[1] I González Hernández,[1] R Komm[1], F Hill,[1] R Bogart[2] and D Haber[3]

[1] National Solar Observatory, 950 N Cherry Ave, Tucson, AZ 85719, USA
[2] HEPL, Stanford University, Stanford, CA 94305, USA
[3] JILA and Department of Astrophysical and Planetary Sciences, University of Colorado, Boulder, CO 80309, USA

E-mail: `rhowe@noao.edu`

Abstract. We examine the differences between ring-diagram mode frequency estimates from samples of Global Oscillation Network Group [GONG], Michelson Doppler Imager [MDI] and Helioseismic and Magnetic Imager [HMI] data, and find that different instruments and analysis pipelines do result in small systematic differences which may not be uniform across the solar disk.

1. Introduction

In the ring-diagram analysis technique of local helioseismology [1] peak parameters are obtained for small regions of the solar disk by fitting the three-dimensional power spectrum of a data cube, usually obtained by tracking a chosen region across the disk as the Sun rotates. This spectrum, when cut at constant temporal frequency, appears as a set of concentric rings, giving the technique its name. Flows cause the rings to shift in k-space, and changes in the acoustic cavity, such as those caused by magnetic activity, increase or decrease their size, changing the measured frequency of the peaks. The frequencies are also affected by geometric effects such as foreshortening as the patch moves away from disk center, and by instrumental distortions [2]. Here we consider the frequency parameters obtained from samples of GONG [3], MDI [4] and HMI data during the development of the HMI ring-diagram pipeline. Development is still ongoing, and the HMI results presented here are preliminary.

The data analyzed here are taken from square patches about 15° on a side, tracked for approximately the time it takes for the Sun to rotate 15°. For GONG and MDI, where the detector pixels are about 2 arcsec, patches of 128×128 pixels, covering $16° \times 16°$ on the disk, are tracked for 1664 minutes at one-minute cadence. For HMI, the detector pixels are about 0.5 arcsec; the patches are remapped to 384×384 pixels covering 15.36° on a side, and tracked for 1728 minutes at a cadence of 45s. The area covered by the HMI tiles extends farther from disk center, allowing up to 284 tiles to be tracked for each "day" instead of the 189 of the standard "dense pack" analysis [5] long used for GONG and MDI data. The HMI data are tracked at the Carrington rate for all latitudes in order to facilitate tracking over long periods, while for the MDI and GONG data the tracking rate varies with latitude following the Snodgrass rotation law.

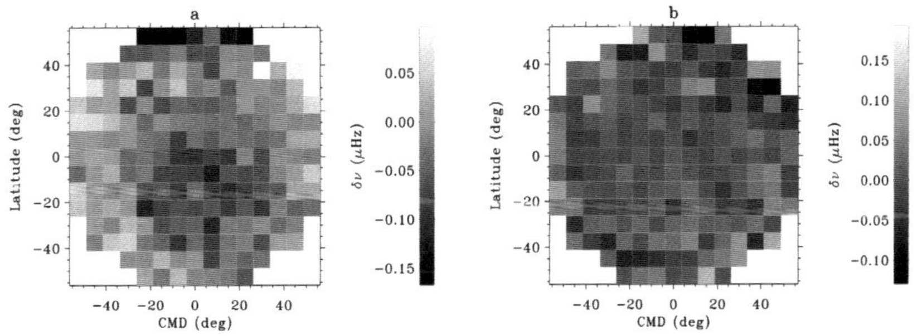

Figure 1. Mean differences in μ Hz between frequency estimates from Carrington and Snodgrass tracking, for (a) MDI CR 2068 and (b) GONG CR 2069.

2. Analysis techniques

The HMI ring-diagram pipeline includes two different fitting procedures: rdfitf [CO] and rdfitc [CT]. The rdfitf code that is used in this analysis is based on the standard ring-analysis fitting code originally developed [5] to quickly fit power spectra from many regions of the solar disk. The rdfitf procedure unwraps the given power spectrum onto a polar coordinate system (θ, k) and then filters out an average of the power in frequency at each θ and k in order to eliminate any stationary instrumental effects. The data are then subsampled before being fit with the function

$$P(\theta, k, \nu) = A\Gamma/2/[(\nu - \nu_0 + U_x * k\cos\theta + U_y * k\sin\theta)^2 + (\Gamma/2)^2] + bg/k^3 \qquad (1)$$

This is also the algorithm used in the routine analysis of GONG ring data. The rdfitc algorithm [6] uses a 13-parameter model with asymmetric peak profiles to fit at constant frequency:

$$P(k_x, k_y, \nu) = \exp[A_0 + (k - k_0)A_1 + A_2(k_x/k)^2 + A_3 k_x k_y/k^2] \qquad (2)$$
$$\times[S^2 + (1 + S_x)^2](x^2 + 1) - 1 + e^{z1}/k^3 + e^{z2}/k^4$$

The rdfitc method is more computationally expensive than rdfitf. Therefore, for the larger HMI tiles we have used this method only for patches on the central meridian.

3. Results

3.1. Preliminary checks

3.1.1. Carrington versus Snodgrass Using MDI and GONG data, we checked for the effects of different tracking rates on the frequency parameter. To look for systematic effects across the disk, frequency differences in each dense-pack position were averaged over a range of peaks and over a full rotation of observations. Figure 1 shows the results for frequency differences between Carrington and Snodgrass-rate tracking for GONG CR 2069 and MDI CR 2068. In each case the differences are small and do not show a systematic pattern.

3.1.2. GONG versus MDI We compared MDI data (tracked at the Snodgrass rate using the HMI pipeline and fitted with the CO algorithm) and GONG data analyzed with the standard GONG pipeline, for two rotations, 2068 and 2084 (Figure 2). In the second case the observations were made when SOHO was "upside down"; the distinctive pattern of the differences across the disk is rotated 180°, demonstrating that this is an effect of incompletely corrected instrumental distortion in MDI. (Earlier versions of the data with even less complete distortion correction showed more dramatic differences.)

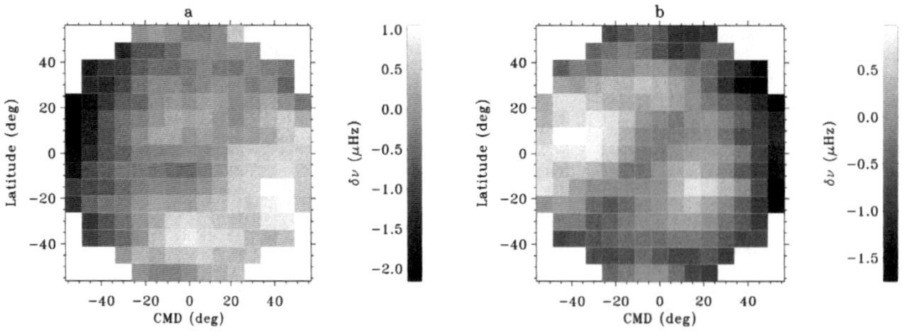

Figure 2. Mean differences in μ Hz between frequency estimates from CO fits of MDI and GONG data, for (a) CR 2068 and (b) CR 2084.

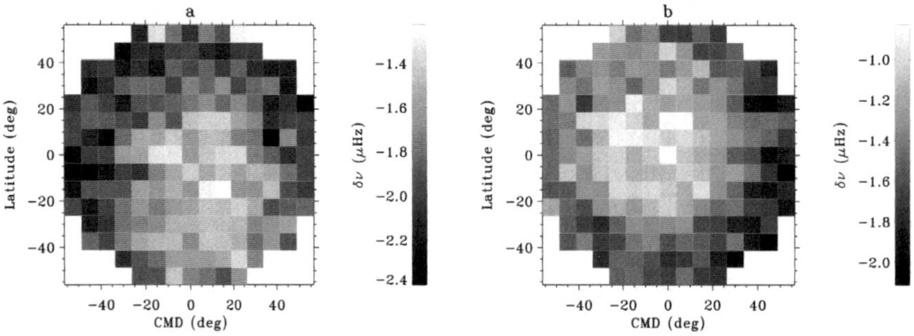

Figure 3. Mean differences in μ Hz between frequency estimates from CO and CT fits of MDI data, for (a) CR 2068 and (b) CR 2084.

3.1.3. CO versus CT As might be expected because of the different peak models, the CO and CT fits give slightly different frequency estimates when applied to the same data. This is illustrated in Figure 3 for two rotations of MDI data. There appears to be a trace of the MDI-distortion signature in the differences, perhaps because the different peak models respond differently to the distortion effects.

3.2. HMI and MDI observations of CR 2096 – 2097

We analyzed results from several days of observations from HMI and from the MDI Dynamics Run during May 2010, with CO fits covering the whole disk for 15 days of HMI and 18 of MDI, and CT fits on the central meridian for 21 days of HMI and 20 of MDI. The variation of the frequency of a particular peak across the disk can be roughly described by the equation $\nu = \nu_0 + a_1 x + a_2 x^2 + a_3 y + a_4 y^2 + a_5 B + a_6 B^2$, where x and y are the distance from disk center in the x and y directions, as a fraction of the apparent radius, and B is magnetic activity index (in Gauss) for each patch, calculated by integrating the unsigned longitudinal magnetic field strength from MDI magnetograms over the area and duration of the patch. When only the central meridian patches are available, we omit the terms in x. Figures 4 and 5 show the results of such an analysis with all four data sets plotted on the same axes.

4. Conclusions

Instrumental and geometric effects cause variations in the ring-diagram frequency estimates across the solar disk. The frequency estimates also, not unexpectedly, depend on the fitting

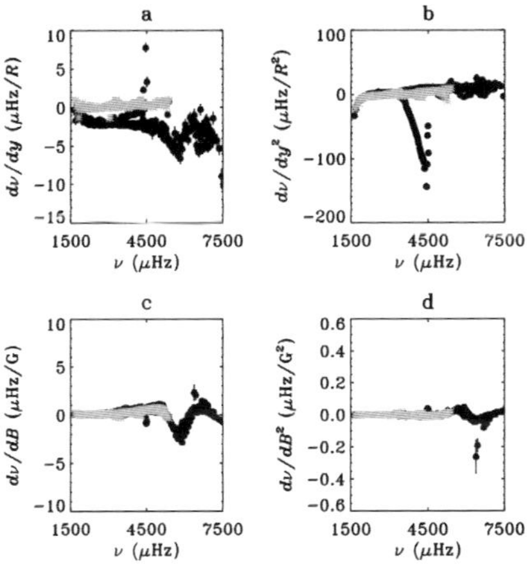

Figure 4. $d\nu/dy$ (a), $d\nu/dy^2$ (b), $d\nu/dB$ (c), and $d\nu/dB^2$ (d) terms for fits to HMI CO (black) and MDI CO (gray) frequency estimates.

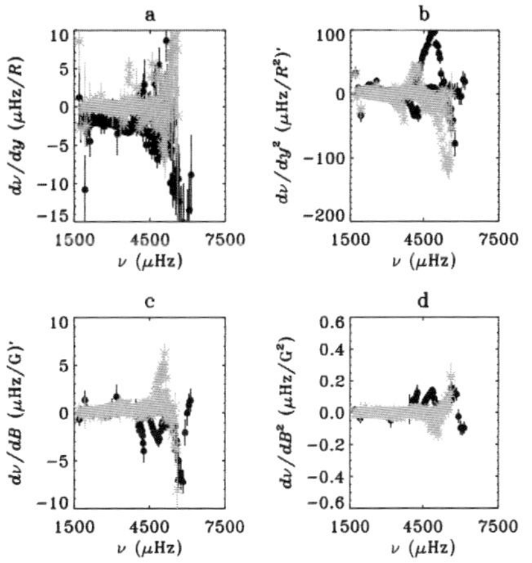

Figure 5. $d\nu/dy$ (a), $d\nu/dy^2$ (b), $d\nu/dB$ (c), and $d\nu/dB^2$ (d) terms for fits to HMI CT (black) and MDI CT (gray) frequency estimates.

algorithm and model; differences tend to be more pronounced at high degree and nearer the limb. Tests with artificial data may eventually throw more light on these issues. Tracking at the Carrington rate rather than the Snodgrass rate has very little effect on the frequency estimates.

Acknowledgments
This work utilizes data obtained by the Global Oscillation Network Group (GONG) Program, managed by the National Solar Observatory, which is operated by AURA, Inc. under a cooperative agreement with the National Science Foundation. The data were acquired by instruments operated by the Big Bear Solar Observatory, High Altitude Observatory, Learmonth Solar Observatory, Udaipur Solar Observatory, Instituto de Astrofísica de Canarias, and Cerro Tololo Interamerican Observatory. SOHO is a project of international collaboration between ESA and NASA. HMI data courtesy of SDO (NASA) and the HMI consortium. Haber's work was partially supported by NASA grants NAS5-02139, NNX07AH82G, and NNX09AB04G.

References

[1] Hill F 1988 *Astrophys. J.* **333** 996–1013

[2] Howe R, Komm R W, González Hernández I, Hill F, Haber D A and Hindman B W 2004 *ESA SP-559: SOHO 14 Helio- and Asteroseismology: Towards a Golden Future* ed Danesy D pp 484–7

[3] Harvey J W, Hill F, Hubbard R, Kennedy J R, Leibacher J W, Pintar J A, Gilman P A, Noyes R W, Title A M, Toomre J, Ulrich R K, Bhatnagar A, Kennewell J A, Marquette W, Patrón J, Saá O and Yasukawa E 1996 *Science* **272** 1284–6

[4] Scherrer P H, Bogart R S, Bush R I, Hoeksema J T, Kosovichev A G, Schou J, Rosenberg W, Springer L, Tarbell T D, Title A, Wolfson C J, Zayer I and MDI Engineering Team 1995 *Solar Phys.* **162** 129–88

[5] Haber D A, Hindman B W, Toomre J, Bogart R S, Larsen R M and Hill F 2002 *Astrophys. J.* **570** 855–64

[6] Basu S and Antia H M 1999 *Astrophys. J.* **525** 517–23

Axisymmetric absorption of p modes by an ensemble of thin, magnetic-flux tubes

R. Jain[1], A. Gascoyne[1] and B.W. Hindman[2]

1 Department of Applied Mathematics, University of Sheffield, UK
2 JILA, University of Colorado at Boulder, USA

E-mail: R.Jain@sheffield.ac.uk

Abstract. The buffeting action of the solar acoustic waves (p modes) on magnetic fibrils excites magnetohydrodynamic (MHD) tube waves. We model these fibrils as axisymmetric, untwisted, vertically oriented, thin, magnetic-flux tubes. The MHD tube waves propagate along the length of the tube and carry energy away from the p-mode cavity creating a source of p-mode absorption. We calculate the absorption arising from the excitation of *sausage* MHD waves within a model plage composed of many flux tubes with differing plasma properties. We find that for a collection of tubes with normally distributed plasma parameters β, the macroscopic absorption coefficient of the collection effectively depends on only the mean value of β.

1. Introduction

It is well established that the solar f and p modes are influenced by the properties of magnetic structures such as sunspots and plages. It is also known that the effect of magnetic field is not just confined within the boundaries of the magnetised region but extends beyond it into an "acoustic shadow". The work of Braun (1987, 1988), Bogdan and Braun (1995), and more recently Braun and Birch (2008) suggests that both sunspots and plages are ravenous absorbers of p-mode power. Thus, the importance of sub-surface field structure in modifying the properties of f and p modes has been realised and many theoretical investigations have concentrated on understanding the physical mechanism responsible for this absorption (e.g., Spruit 1991; Bogdan and Cally 1995; Bogdan et al. 1996; Crouch and Cally 2005; Jain et al. 2009).

Jain et al. (2009) (hereafter referred to as JHBB) studied f- and p-mode absorption in a plage region by modelling the plage as an *idealised* forest of thin, untwisted, vertical, magnetic-flux tubes. By idealised, we mean each flux tube in the plage has the same properties, i.e., the same magnetic flux and plasma β (the ratio of gas to magnetic pressure). However, the distribution of flux within a plage is generally quite heterogeneous and there is no reason to believe that a plage region is comprised of identical flux tubes. In this paper, we model the plage as consisting of many flux tubes each with the same magnetic flux but with a Gaussian distribution of β values. We then investigate the effect of such an ensemble on the absorption of p modes.

2. The Equilibrium

The fibril fields are embedded in a non-magnetic medium. We model the nonmagnetic medium as a plane-parallel gravitationally stratified polytropic atmosphere with gravity $\mathbf{g} = -g\hat{z}$. The pressure, density and sound speed vary with depth z as power laws with a polytropic index

a. Following Bogdan et al. (1996) and Hindman and Jain (2008) we truncate the polytrope at $z = -z_0$ which represents the model photosphere. Above the truncation depth $z > -z_0$ we assume the existence of a hot vacuum ($\rho_{\rm ext} \to 0$ with $T_{\rm ext} \to \infty$). For thin flux tubes, the lateral variation of the magnetic field across the tube is ignored and the plasma β is constant with depth (see Bogdan et al. (1996) for details).

3. The Governing Equation

The incident acoustic wavefield in the non-magnetic medium can be expressed as a single partial differential equation for the displacement potential Φ:

$$\frac{\partial^2\Phi}{\partial t^2} = c^2\nabla^2\Phi - g\frac{\partial\Phi}{\partial z}, \tag{1}$$

where c^2 is the square of the adiabatic sound speed. This equation can be transformed into an ordinary differential equation that supports plane wave solutions of the form:

$$\Phi(\mathbf{x},t) = \mathcal{A}\mathrm{e}^{-i\omega t}\mathrm{e}^{ik_n x}Q_n(z)\ ; \quad Q_n(z) = (-2k_n z)^{-(\mu+1/2)}W_{\kappa_n,\mu}(-2k_n z)$$

where $\mathcal{A}$ is the complex wave amplitude, ω is the temporal frequency and $k_n(\omega)$ the wavenumber eigenvalue; $Q_n(z)$ is the vertical eigenfunction proportional to Whittaker's W function (Abramowitz & Stegun 1964), with $\mu = (a+1)/2$, $\nu^2 = a\omega^2 z_0/g$ and $\kappa_n = \nu^2/(2k_n z_0)$. We calculate the eigenvalues and eigenfunctions for the truncated polytrope by requiring that the Lagrangian pressure perturbation vanishes at the truncation depth, i.e., $\nabla\cdot\xi = 0$. Mathematically this takes the form $W_{\kappa_n,\mu+1}(2k_n z_0) = 0$.

Thin flux tubes support both longitudinal (sausage) waves and transverse (kink) waves; here we only consider the sausage waves. These waves are driven by the external f and p modes. Using the formulation of Hindman and Jain (2008), the vertical fluid displacement due to the excitation of sausage waves within the tube can be described by

$$\left[\frac{\partial^2}{\partial t^2} + \frac{2gz}{2a+\beta(1+a)}\frac{\partial^2}{\partial z^2} + \frac{g(1+a)}{2a+\beta(1+a)}\frac{\partial}{\partial z}\right]\xi_\parallel = \frac{(1+a)(\beta+1)}{2a+\beta(1+a)}\frac{\partial^3\Phi}{\partial z\partial t^2}$$

4. Absorption coefficient for an ensemble of flux tubes

Following the procedure in JHBB, we calculate the absorption coefficient, α_n, for the nth-order p mode for a single tube. We wish to apply our theoretical results to model the absorption coefficient measured for a solar plage region by examining the effect of a large number of thin flux tubes. We, thus assume that the tubes are sufficiently separated that we can ignore their acoustic jacket modes (Bogdan and Cally 1995) and the effects of multiple scattering (Hanasoge and Cally 2009).

In JHBB, the absorption coefficient for an individual tube was linearly proportional to the magnetic flux ϕ contained by that tube,

$$\alpha_n = C_n(\beta,\omega)\ \phi\ . \tag{2}$$

The function $C_n(\beta,\omega)$ is a smooth but rather complicated function of the frequency ω and the β value of the tube. The absorption coefficient at position $\boldsymbol{x}$ measured by local helioseismic techniques such as ridge-filtered holography (Braun and Birch 2008), can be related to the distribution of tubes through a spatial weighting function or kernel, $K_n(\boldsymbol{x},\omega)$,

$$\alpha_n(\boldsymbol{x},\omega) = \sum_i C_n(\beta_i,\omega)\ \phi_i\ K_n(\boldsymbol{x}_i - \boldsymbol{x},\omega)\ ,$$

where each flux tube in the plage is labelled by an index i and $\boldsymbol{x}_i$, ϕ_i and β_i are the position, magnetic flux and plasma parameter for tube i.

In JHBB all the flux tubes were identical; thus, the quantity $C_n(\beta_i, \omega)$ could be taken outside the summation and the remaining sum simply became the kernel weighted magnetic flux, $\Theta_n(\boldsymbol{x}, \omega)$, which is a measurable quantity. Here, however, we will be considering a random distribution of flux tubes for a range of β values between 0 and 1. Thus, we compute an ensemble average of the absorption coefficient,

$$\langle \alpha_n(\boldsymbol{x}, \omega) \rangle = \left\langle \sum_i C_n(\beta_i, \omega) \sum_i \phi K_n(\boldsymbol{x}_i - \boldsymbol{x}, \omega) \right\rangle .$$

We will assume that the locations $\boldsymbol{x}_i$ and β values of the tubes are not correlated. Thus, if all the tubes have the same flux,

$$\langle \alpha_n(\boldsymbol{x}, \omega) \rangle = \langle C_n(\beta_i, \omega) \rangle \, \Theta_n(\boldsymbol{x}, \omega) \; ; \; \langle C_n(\beta_i, \omega) \rangle = \mathcal{C} \int_0^1 C_n(\beta, \omega) \mathrm{e}^{-\frac{(\beta - \beta_0)^2}{2\sigma^2}} \, d\beta, \qquad (3)$$

where σ is the variance, $\mathcal{C}$ is a distribution normalization, and

$$\Theta_n(\boldsymbol{x}, \omega) = \left\langle \sum_i \phi K_n(\boldsymbol{x}_i - \boldsymbol{x}, \omega) \right\rangle \approx \int d\boldsymbol{x}' \; |B(\boldsymbol{x}')| \; K_n(\boldsymbol{x}' - \boldsymbol{x}, \omega) .$$

We show the results for maximal flux boundary condition. (Hindman & Jain 2008).

5. Results and Discussion

We compute equation (3) for different distributions of β values. The magnetic flux Θ_n and the kernel functions K_n are identical to that used by JHBB. In Figure 1, we plot the absorption coefficient for a modelled plage as a function of frequency. The curves in this figure are for a modelled plage that is composed of a host of magnetic flux tubes whose β values have been drawn from a gaussian distribution limited to values between 0 and 1. The three horizontal panels correspond to different peak values of the distribution β_0, while the different line styles for the curves denote different distribution widths σ. While the curves for various β_0 and σ all differ, the primary dependence of the absorption coefficient is on the mean of the distribution $\bar{\beta}$. Note, $\bar{\beta}$ is not equal to the most probable value β_0 because of the fact that β is restricted to values within 0 and 1.

The Gaussian distribution used in Figure 1 is shown as curves, whereas the symbols correspond to a plage comprised of a multitude of identical flux tubes, all with the same value of β equal to the mean of the gaussian distribution $\beta = \bar{\beta}$. One can clearly see that two very different tube distributions produce essentially the same absorption coefficient as long as their mean β values are the same.

Also note from Figure 1 that the absorption reduces significantly above $\nu =$ 2.5-3 mHz for low radial orders. This is due to the fact that at low mode orders and high frequencies, waves have larger horizontal wavenumbers and thus the kernel function K_n (see JHBB) whose width depends on the horizontal wavelength samples relatively smaller spatial region, resulting in reduced total magnetic flux. Thus, from equation (3) the absorption coefficient α reduces for low radial order and higher frequencies.

6. Conclusions

We compute absorption coefficients for collections of vertical, axisymmetric, thin, magnetic-flux tubes representing a solar plage region. We have shown that the macroscopic absorption coefficient of the collection effectively depends only on the mean value of β for the distribution.

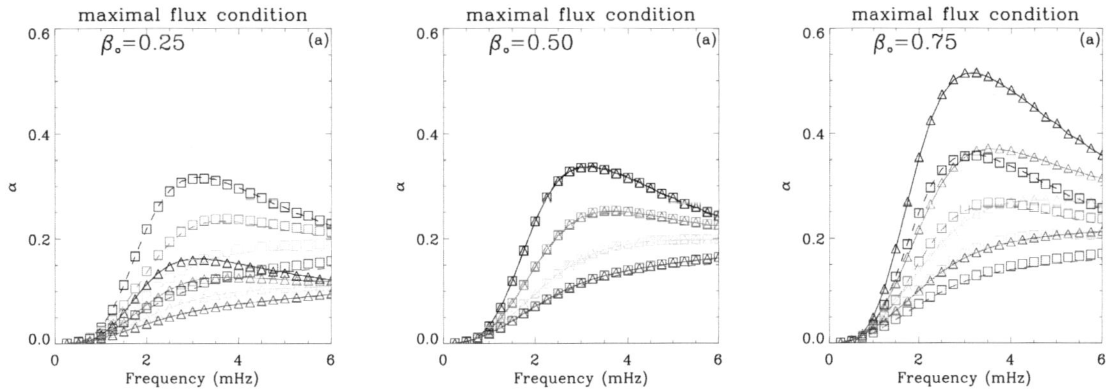

Figure 1. Absorption coefficient as a function of frequency for a modelled plage constructed from many thin, magnetic-flux tube with plasma β varying between 0 and 1. The distribution function is gaussian in form with a peak value of β_0 and a width of σ. The solid and dashed lines are for $\sigma = 0.1$ and 0.8, respectively. Various symbols represent the corresponding absorption coefficients for identical tubes, the β value for each of these tubes was selected such that it matched the mean value of the guassian distributions. Each mode order is denoted by a different colour: black (p1), red (p2), turqouise (p3) etc. One can easily see that the only relevant parameter appears to be the mean of the distribution.

6.1. Acknowledgments

RJ and AG acknowledge STFC (UK) for partial funding. BWH acknowledges NASA grants NNG05GM83G, NNG06GD97G, NNX07AH82G, NNX08AJ08G, NNX08AQ28G, and NNX09AB04G.

References

[1] Abramowitz, M. & Stegun, I. A. 1964, Handbook of Mathematical Functions (New York: Dover)
[2] Bogdan, T. J. and Braun, D. C. 1995 in Proc. 4th SoHO Workshop: Helioseismology (ed. J. T. Hoeksema), ESA SP-376, 31
[3] Bogdan T. J. and Cally, P. S. 1995, ApJ, 453, 919
[4] Bogdan, T. J., Hindman, B. W., Cally, P. S. & Charbonneau, P. 1996, ApJ, 465, 406
[5] Braun, D. C., Duvall, T. J., Jr and LaBonte, B. J. 1987, ApJ, 319, L27
[6] Braun, D. C., Duvall, T. J., Jr and LaBonte, B. J. 1988, ApJ, 335, 1015
[7] Braun, D. C. & Birch, A. C. 2008, Sol. Phys., 251, 267
[8] Crouch, A. D. and Cally, P. S. 2005, Solar Phys., 227, 1
[9] Hanasoge, S. and Cally, P. S. 2009, ApJ, 697, 651
[10] Hindman, B. W. & Jain, R. 2008, ApJ, 677, 769
[11] Jain, R., Hindman, B. W., Braun, D. C. & Birch, A. C. 2009, ApJ, 695, 325 (JHBB2009)
[12] Spruit, H. C. 1991 in Lecture Notes in Physics (eds: J. Toomre and D. O. Gough), Springer-Verlag 388)

Testing the GONG ring-diagram pipeline with HMI Dopplergrams

Kiran Jain[1], S. C. Tripathy[1], I. González Hernández[1], S. Kholikov[1], F. Hill[1], R. Komm[1], R. Bogart[2] and D. Haber[3]

[1] GONG, National Solar Observatory, Tucson, AZ 85719-4933, USA
[2]HEPL, Stanford University, Stanford, CA 94305-4085, USA
[3]JILA, University of Colorado, Boulder, CO 80309-0440, USA

E-mail: kjain@noao.edu

Abstract. The GONG ring-diagram pipeline was developed to analyze GONG+ Dopplergrams in order to extract information about solar subsurface flows and has been extensively tested for this purpose. Here we present preliminary results obtained by analyzing the HMI Dopplergrams with the GONG pipeline and compare them with those obtained from the HMI ring-diagram pipeline.

1. Introduction

The techniques of local-helioseismology, e.g ring-diagram, time-distance etc., have been widely used to study the structure and dynamics of localized regions on the solar disk [1]. These techniques utilize high-resolution and high-cadence observations. Dopplergrams from the Michelson Doppler Imager (MDI) and the Global Oscillation Network Group (GONG) during the past solar cycle have been extensively used in the analysis, but the technique of ring diagram has been limited by the size of the region ($\approx 15^\circ$) which is generally larger than the size of typical active regions. Now, with the availability of images from the Helioseismic and Magnetic Imager (HMI) with better resolution, the ring-diagram technique can be reliably applied to regions as small as 5°. A comparison of Dopplergrams obtained from all three instruments is discussed in [2]. Here, we present our preliminary efforts to use the GONG ring-diagram pipeline [3] for the HMI data. Although it was primarily developed for the use of GONG data, it has been reliably used for various other data sets [4]. We also compare our results with those obtained with the ring-diagram pipeline designed specifically for the HMI data [5]. The comparison not only validates independently developed pipelines but also tests the reliability of the inferred features.

2. HMI Dopplergrams and Ring-diagram Pipelines

We use HMI Dopplergrams from 24–25 April 2010 corresponding to Carrington Rotation (CR) 2096 and Central Meridian Longitude (CML) 330°. The spatial resolution and temporal cadence of these images are $0^\circ.04$ and 45 seconds, respectively. Before processing these Dopplergrams through ring-analysis pipelines, we corrected them for the orbital motion. Since the GONG pipeline is mostly used for 15° and 30° patches with spatial resolutions of $0^\circ.125$ and $0^\circ.25$ respectively, which is lower than the HMI resolution, we carried out an analysis for three different patch sizes. Details of various patches are given in Table 1.

Table 1. Parameters used for various patch sizes.

Patch size	Grid size	Grid spacing	Spatial frequency resolution(Mm^{-1})	# of images	Temporal frequency resolution (μHz)
5°	128×128	0°.04	0.1010	768	28.93
15°	384×384	0°.04	0.0337	2304	9.65
30°	384×384	0°.08	0.0168	2304	9.65

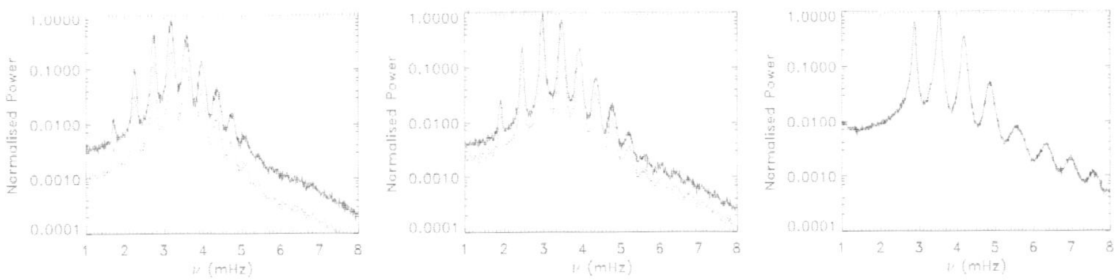

Figure 1. Azimuthally-averaged power spectra at (left) ℓ=280, (middle) ℓ=351, and (right) ℓ=819, for 15° patch at disk center. Solid and dotted lines correspond to the GONG and HMI pipelines respectively.

Each set of Dopplergrams was processed through the GONG as well as HMI ring pipelines to produce the 3-D power spectra. Finally the power spectra were fitted to obtain the mode parameters and sub-surface flows. The GONG pipeline uses a symmetric-profile model to fit the power spectrum, while the HMI pipeline uses both symmetric- and asymmetric-profile models. In order to make a direct comparison between mode parameters obtained from both pipelines, we confine this analysis to the symmetric-profile model. Although the functions used in both pipelines are the same, there are a few differences in the fitting procedure: (i) the use of different guess tables for fitting the mode parameters, (ii) a multiplication factor for enhancing the power to an optimal value for fitting, and (iii) the number of radial orders (n) that are fitted. In addition, although the GONG pipeline can fit ridges up to n=16, the standard pipeline fits up to n=6 only.

3. Analysis and results

Figure 1 illustrates the comparison between azimuthally-averaged power (P) from both pipelines for 15° patches at disk center. A Postel projection is used to remap the data and the tracking is done at the Carrington rate rather than at the Snodgrass rate. Since different normalizations have been used to calculate the 3D power spectra, we rescaled each spectrum to the interval [0,1] using a range normalization. Each value P in the spectrum is replaced by the value $(P - P_{min})/(P_{max} - P_{min})$ where P_{max} and P_{min} are the maximum and minimum values of power in the spectrum. As evident from the figure, the power obtained from the HMI pipeline is lower than that for the GONG pipeline at ℓ=280 and 351. This difference decreases with increasing ℓ and we do not find any noticeable difference at ℓ=819. The modes fitted for all three patch sizes at disk center are shown in Figure 2. As mentioned in Section 2, the maximum values of n fitted with GONG pipeline is 6 while the HMI pipeline fits modes for higher values of n. For a 5° patch, there are few modes fitted in both pipelines as compared to larger patches. While the GONG pipeline fits 41, 242 and 536 modes for 5°, 15° and 30° patches, these numbers

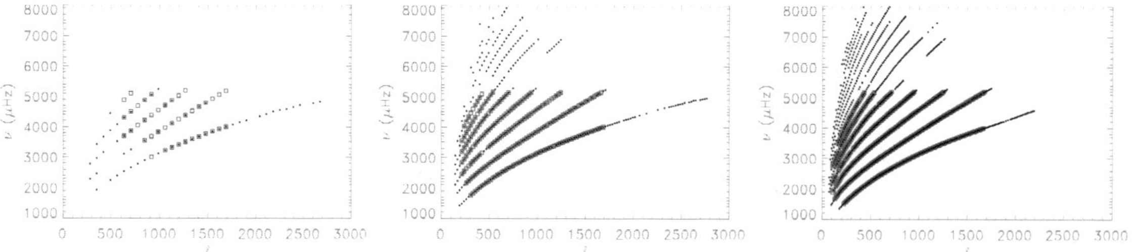

Figure 2. The ℓ - ν diagram at the disk center for (*left*) 5°, (*middle*) 15°, (*right*) 30° patches. The diamonds and filled circles are for GONG and HMI pipelines, respectively.

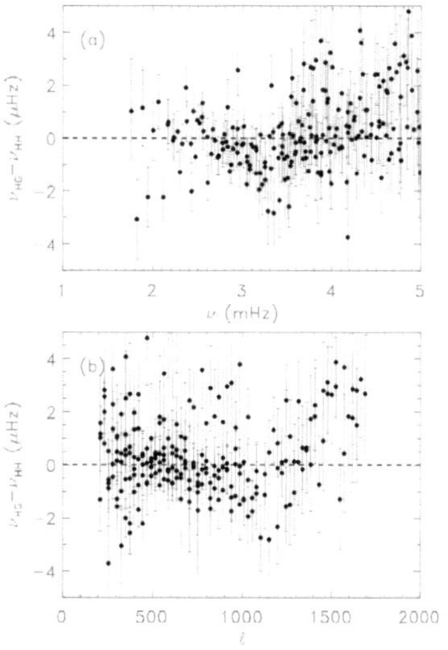

Figure 3. Difference between the frequencies obtained with the GONG (HG) and HMI (HH) pipelines for a 15° patch at disk center as a function of (a) frequency and (b) degree.

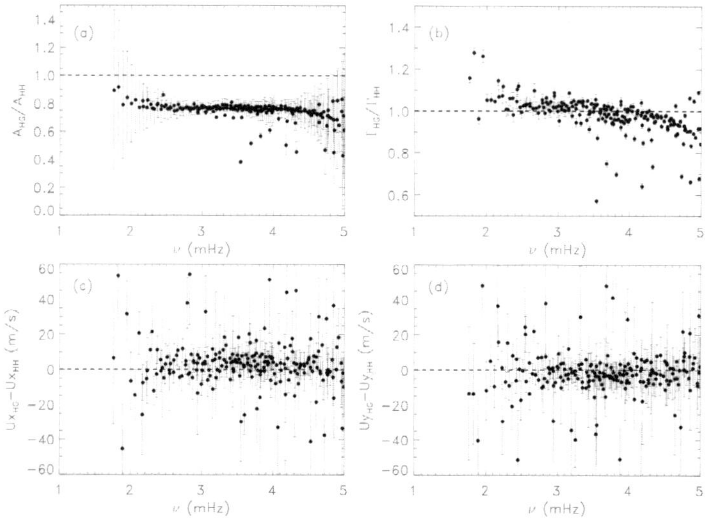

Figure 4. Differences between the mode parameters, (a) amplitude, (b) line width, (c) x-component of horizontal velocity, and (d) the y-component of velocity, at the disk center for a 15° patch obtained from analyzing the HMI data with GONG and HMI pipelines.

are higher in the HMI pipeline, i.e. 59, 387 and 536 respectively. The difference in fitted modes arises due to the use of different guess tables. In addition, the GONG pipeline fits modes up to $\ell \approx 1700$ and $\nu \approx 5$ mHz only. Although the n=0 ridge goes beyond $\ell = 2500$ in the HMI pipeline, the maximum ℓ values in other ridges corresponding to p modes are comparable to the GONG pipeline.

Figure 3 shows the difference between frequencies obtained from both pipelines for a 15° patch as a function of ν and ℓ. There is a large scatter in frequency differences and it increases with frequency. It appears that the frequencies from the GONG pipeline are lower around 3 mHz band while these are higher when we move away from this range. However, we do not find any clear trend with ℓ. The differences between other mode parameters are shown in Figure 4. Here the difference is more systematic, while the mode amplitude is lower in the GONG pipeline, the line width is larger at low frequencies and smaller at high frequencies. In both parameters, the difference increases at higher frequencies. The obtained velocities, U_x and U_y, are also found to be different and the difference is within ± 15 m/s in most cases. These preliminary results

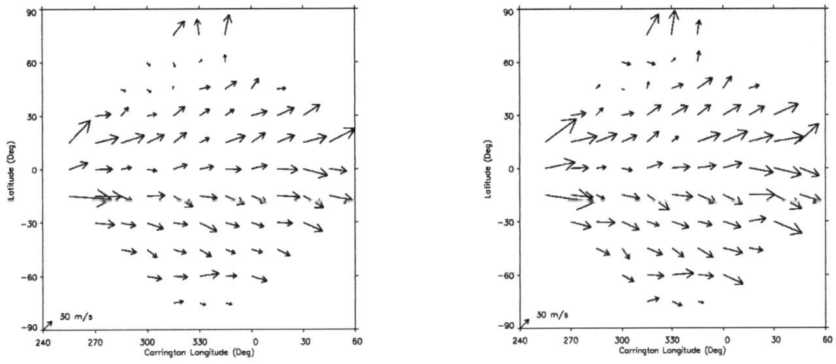

Figure 5. Sub-surface flow maps obtained with the GONG pipeline at the depth of (left) 7.1± 1.77 Mm and (right) 10.2± 2.72 Mm.

clearly indicate that the normalization of the power spectra can influence the magnitude of mode parameters significantly and there are other systematics that need to be investigated in detail. The results obtained from both pipelines can not be combined yet to study relative variations.

Figure 5 shows flow maps at two depths for 15° non-overlapping regions. These flows are obtained by tracking regions at Snodgrass rate with the GONG pipeline. The existing version restricts patches to ± 52.5° in both latitudes and longitudes from central meridian, we have expanded the limit to ±75° in order to check the advantage with the HMI Dopplergrams. We obtain a smooth variation in horizontal flow and, as expected, there is a clear poleward trend in each hemisphere. We also obtain the variation at higher latitudes within normal range. Although errors increase when we move towards higher latitudes, these are much smaller than those obtained from the processing of MDI and GONG Dopplergrams.

4. Summary

The current version of the GONG ring-diagram pipeline is able to process the high-resolution HMI data, however there is a need to make it more computationally efficient in order to process the huge amount of data from the HMI. Although both pipelines use the same function to fit the power spectra, there are differences in the fitted values with the GONG pipeline from those obtained from the HMI pipeline. The differences could be due to the use of different guess tables and different normalization procedures for calculating the power. There is a significant difference in the number of modes fitted with both pipelines. We find that higher-n modes are difficult to fit in the current GONG pipeline. A revision of the code is necessary to optimize the analysis of HMI data.

Acknowledgments

HMI data courtesy of SDO (NASA) and the HMI consortium. This work is supported in part by NASA grants NNG-08EI54I, 14583340-26967, NNX08AQ28G, NNX09AB04G and NAS5-02139.

References

[1] Gizon L and Birch A C 2005, Living Rev. Sol. Phys., 2, 6

[2] Howe R, Jain K, Hill F, Komm R, González Hernández I and Bogart R 2010, these proceedings

[3] Corbard T, Toner C G, Hill F, Hanna K D, Haber D A, Hindman B W and Bogart R S 2003 *Proc. Local and Global Helioseismology: The Present and Future* vol 517 ed H Sawaya-Lacoste (Noordwijk: ESA) p 255

[4] Jain K, Hill F, Tripathy S C, Antia H M, Armstrong J D, Jefferies S M, Rhodes Jr E J and Rose P J 2006 *Proc. Beyond the spherical Sun* vol 624 ed K Fletcher and M Thompson (Noordwijk: ESA) p 103

[5] Bogart R S, Baldner C, Basu S, Haber D A and Rabello-Soares M C 2010, these proceedings

Towards near real time high-resolution Dopplergrams from GONG

Kiran Jain, S. McManus, I. González Hernández, S. C. Tripathy, J. Bolding, F. Hill, T. M. Wentzel

National Solar Observatory, 950 N Cherry Av., Tucson, AZ 85719, USA

E-mail: kjain@noao.edu

Abstract. The GONG network, consisting of six sites around the globe, provides continuous observations of the Sun. The processing and merging of Dopplergrams from various sites usually takes several months before these are made available to the community for analysis. In this paper, we discuss our recent attempts to reduce the delay between observations and the availability of merged Dopplergrams. Our analysis indicates that the modified approach does not influence mode parameters and inferred helioseismic flows. However, the duty cycle plays a significant role in inferring the sub-surface flows and a low duty cycle, if less stations contribute, may lead to qualitatively different results.

1. Introduction

The Global Oscillation Network Group (GONG) upgraded its cameras in 2001 to provide high-resolution Dopplergrams for studying solar interior using various local-helioseismology techniques. The reliability of these techniques depends upon the availability of non-interrupted observations that require combining simultaneous velocity images recorded by the Network into a single, once per minute, set of registered merged images. The process of merging is performed in various steps [1,2] and takes up to several months before merged Dopplergrams are available for analysis. This time delay is primarily caused by the arrival of data tapes at the Data Management and Analysis Center (DMAC) in Tucson (USA), and various calibration procedures applied to the site images. Recently, DMAC has started receiving high-resolution Dopplergrams via Internet on a daily basis from five of its sites. We use these images to reduce the gap between observations and availability of merged Dopplergrams. In this paper, we describe how the quality of the "Quick Reduce" data affects the local helioseismology products, mainly on the results obtained from ring-diagram analysis [3]. We study the effect of angle determination and the duty cycle on various mode parameters and subsurface flows with depth.

2. Data and Procedure

We use sample Dopplergrams obtained via Internet for 8–10 May 2008 from five sites, namely Mauna Loa (ML), Big Bear (BB), Cerro Tololo (CT), El Teide (TD) and Udaipur (UD). Figure 1 shows the availability of Dopplergrams from individual sites for these sample days. Note that the Dopplergrams from the sixth site, i.e. Learmonth (LE), are not available due to poor bandwidth. The large gaps in observations, as shown in Figure 1, are due to the few images from UD and non-

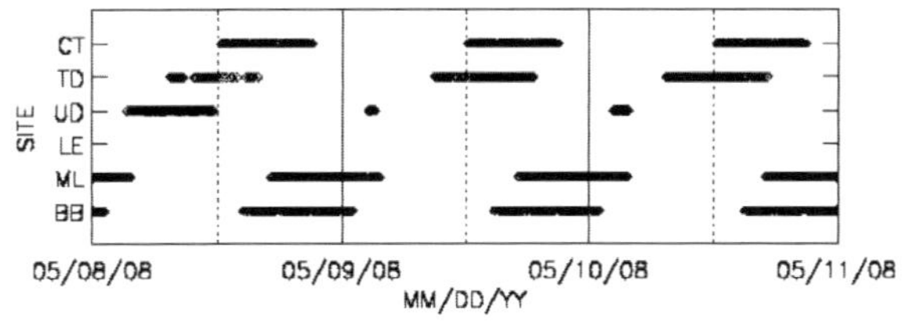

Figure 1. Coverage of GONG sites used in the QR data.

availability of images from LE. Following the standard procedure, before merging Dopplergrams from different sites, the site images were calibrated and the image rotation was calculated.

One of the important steps in combining images is the precise determination of Solar North in each site image. For our analysis, we calculate the image orientation using two different set of images:

- Standard: Calibrated Dopplergrams with existing method to determine optimized angle equations that include noon drifts and ±7 days of fully calibrated images,
- Quick Reduce (QR): Calibrated Dopplergrams with optimized angle equations determined from noon drifts and ±7 days of near-real-time QR images.

Figure 2 shows the difference between the image P-angle obtained from the standard and the QR individual site images for three sample days. Both methods provide comparable orientation more than 50% of the time. The maximum difference is about 0.03 degrees. The relative offset between the merged images for these two sets is given in Figure 3. As seen in individual site images, the difference is again less than 0.05 degrees. It clearly indicates that the QR images can be reliably used to determine the image alignment. However, we notice an increase in offset angle or a drop in correlation coefficient when images from particular sites are used. Since seeing conditions at individual sites are different, and the QR and the standard images also pass through different calibration processes, this appears to affect the calculated angles. We need to investigate it in detail before using QR images for any scientific purpose.

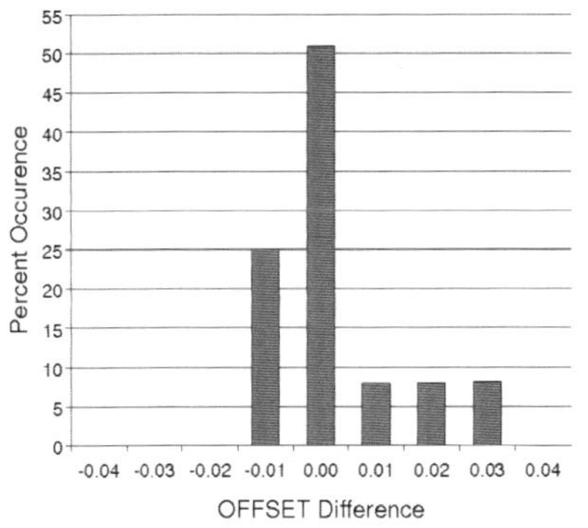

Figure 2. Histogram showing differences between the image rotation angles of individual site images obtained from the standard and the QR procedures for three sample days.

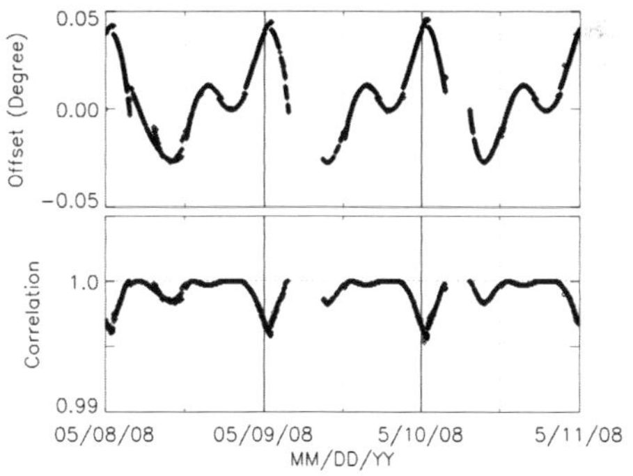

Figure 3. (Top) Relative offset between the merged images obtained with standard and QR angle determination procedures. The errors in these angles are of the order of 10^{-4} degrees. (Bottom) Cross-correlation coefficient between standard and QR images.

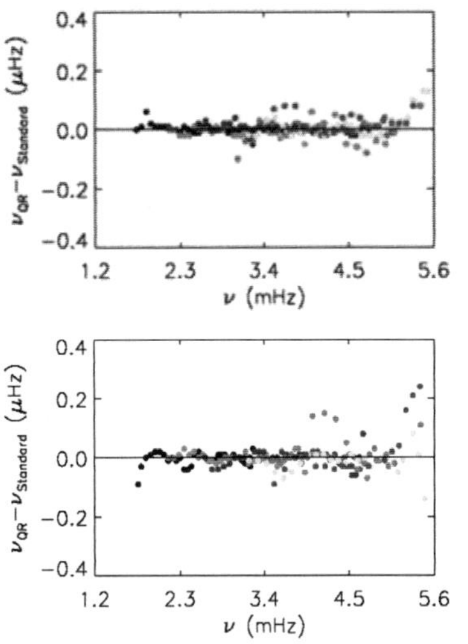

Figure 4. Difference in frequencies obtained from the QR and the standard Dopplergrams as a function of frequency for 8 May 2008 (top) and 9 May 2008 (bottom). Different colors are for different n values.

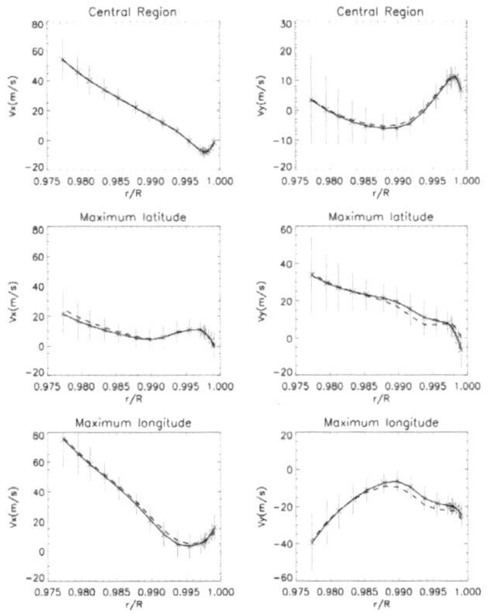

Figure 5. Variation of (left) x- and (right) y-components of the sub-surface flows with depth for 8 May 2008. The flows obtained with the standard and the QR Dopplergrams are shown by solid and dashed lines, respectively.

3. Influence on local helioseismic inferences

In order to understand the influence of near real time Dopplergrams on the helioseismic inferences, we processed three different sets of images through the GONG ring diagram pipeline [4] using 1440 minutes of merged data. These sets are (i) 5-site Standard, (ii) 5-site QR, and (iii) 6-site Standard Dopplergrams. The obtained power spectra are fitted to a Lorentzian profile model to estimate various mode parameters. The obtained velocity components are inverted to infer depth dependence of the horizontal flow.

3.1. 5-site standard vs. 5-site QR

In Figure 4, we plot the difference in mode frequencies at disk center obtained from the 5-site QR and the 5-site standard Dopplergrams. In all cases, the agreement is good at low frequencies while it starts to deviate at $\nu > 4.5$ mHz. Similar results are obtained for other mode parameters, e.g. amplitude and width. The variation of x- and y-components of the flow with depth at three locations on the disk is shown in Figure 5. The center of the maximum latitude/longitude patch is +52.5 degree from the disk center. The inferred flow profiles obtained for both data sets are identical and the values are in good agreement within errors in all cases. While work is in progress, our analysis demonstrates that the QR images provide comparable results to those obtained with a set of standard images.

3.2. 5-site standard vs. 6-site standard

To test the utility of using the merged images from a 5-site network, we have compared horizontal flows of a "Dense Pack" for standard images in Figure 6. The reduction in number of sites contributing to the merged images may also reduce the duty cycle. Since fewer images are available from UD for 9–10 May, and those from LE are completely absent, the duty cycles for

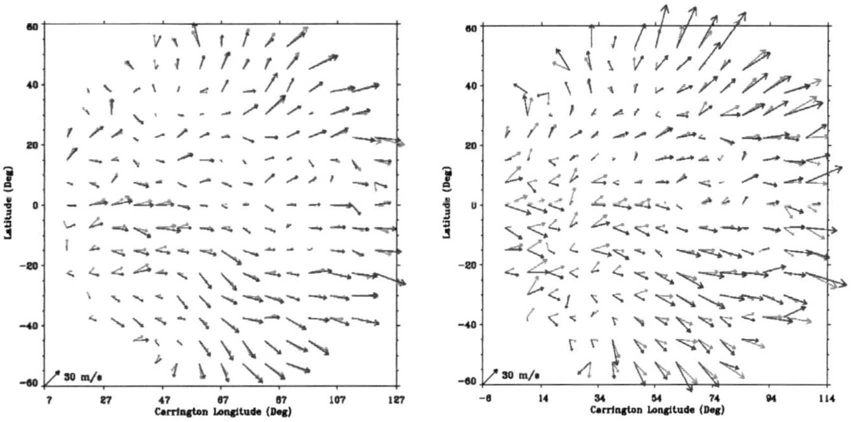

Figure 6. Sub-surface flow maps at the depth of 7.1± 1.5 Mm for (left) 8 May 2008 , and (right) 9 May 2008. The flows for 5-site standard and 6-site standard are shown by the blue (dark) and green (light) arrows, respectively.

these two days have substantially decreased. The maximum decrease is found on 9 May (22%) while it is marginal for 8 May (3%). The effect of duty cycle is clearly visible on flow vectors as both their magnitude and direction have changed significantly. These differences increase with larger changes in the duty cycle.

4. Summary

The modified method of determining the position of Solar North does influence mode parameters and inferred helioseismic flows, although all deviations are within the estimated errors. However, the duty cycle plays a significant role in inferring the sub-surface flows and a low duty cycle may lead to much different results. In order to increase the duty cycle for near real time Dopplergrams, it is crucial to get images from Learmonth via the Internet and we hope that the upcoming improvements in bandwidth at Learmonth will make this faster transfer possible in the near future. Furthermore, the rotation angle determination algorithm needs to be reviewed in order to shorten the number of trailing days needed.

Acknowledgments

This work utilizes data obtained by the GONG program, managed by the National Solar Observatory, which is operated by AURA, Inc. under a cooperative agreement with the National Science Foundation. The data were acquired by instruments operated by the Big Bear Solar Observatory, High Altitude Observatory, Learmonth Solar Observatory, Udaipur Solar Observatory, Instituto de Astrofsica de Canarias, and Cerro Tololo Interamerican Observatory. This work has been partially supported by NASA NNG-08EI54I to National Solar Observatory and the NASA Living with a Star-Targeted Research and Technology Program.

References

[1] Toner C G 2001 *Helio- and asteroseismology at the dawn of the millennium (Tenerife)* vol 464, ed A. Wilson and P. L. Pallé (Noordwijk: ESA) p 355

[2] Toner C G, Haber D, Corbard T, Bogart R and Hindman B 2003 *Proc Local and Global Helioseismology: The present and future (Big Bear Lake)* vol 517, ed H. Sawaya-Lacoste (Noordwijk: ESA) p 405

[3] Hill F 1988 *Astrophys. J.* **333** 996-1013

[4] Corbard T, Toner C G, Hill F, Hanna K D, Haber D A Hindman B W and Bogart R S 2003 *Proc. Local and Global Helioseismology: The Present and Future* vol 517 ed H. Sawaya-Lacoste (Noordwijk: ESA) p 255

Solar flares and temporal changes in subsurface vorticity measurements

R Komm[1], K Jain[1], A Reinard[2], R Howe[1] and F Hill[1]

[1] National Solar Observatory, Tucson, AZ 85719, USA
[2] University of Colorado, NOAA Space Weather Prediction Center, Boulder, CO, USA

E-mail: rkomm@nso.edu

Abstract. We derive the kinetic helicity density of subsurface flows applying ring-diagram analysis to Global Oscillation Network Group (GONG) data. Here, we focus on flows derived from times series of 8 hours and compare them to daily values for a high- and a low-activity sample. Compared with daily values, the horizontal flows derived from 8-hour time series are reasonable near disk center and less reliable near the limb. Also, the errors are larger for shorter time series. A dipolar helicity pattern is present in the flows derived from 8-hour and 24-hour time series of flare-productive active region 10808. For the quiet-Sun sample, the subsurface kinetic helicity is considerably smaller without any pattern.

1. Introduction

Previous studies have shown that the flare activity of solar active regions is intrinsically linked with the vorticity of solar subsurface flows on spatial scales comparable to the size of active regions and on temporal scales from days to the lifetime of active regions [1, 2]. We derive the vorticity vector and the kinetic helicity density of subsurface flows by applying ring-diagram analysis to Global Oscillation Network Group (GONG) Dopplergram data. We are studying the temporal evolution of these subsurface quantities and their relation to magnetic flux and flare activity of active regions. Here, we focus on flows derived from shorter times series of 8 hours duration and compare them to daily values to see if we can improve the temporal resolution. For this preliminary analysis, we have chosen observations during a high-activity period that includes a flare-productive active region and a low-activity period that includes only a remnant of an active region. The results of this analysis are encouraging.

2. Data and Analysis

We determine the horizontal components of solar subsurface flows with a ring-diagram analysis using the dense-pack technique [3]. Acoustic modes excited near the solar surface and traveling through the interior are advected by flows. To measure flows below the solar surface, we track a patch on the solar disk in GONG Dopplergrams for a given time (8 hours or one day) and calculate a three-dimensional power spectrum. Two-dimensional slices at a given temporal frequency show rings. The shift of these rings in spatial frequency is a measure of the subsurface flows. Using the measured horizontal velocities, we calculate the kinetic helicity density defined as the scalar product of the velocity and vorticity vector [4, 5].

We perform this analysis for a high- and a low-activity period. For the high-activity sample, we have selected GONG Dopplergram data obtained during 13 Sep. 2005, which includes active region 10808. This region was very flare-prolific producing several C-, M-, and X-class flares during the disk passage. The subsurface flows associated with such an active region are highly twisted. For the low-activity sample, we have selected GONG Dopplergram data obtained during 1 Aug. 2009, which includes only a remnant of decaying active region 11024 near 30° latitude in the southern hemisphere. Corresponding full-disk magnetograms are shown in Figure 1.

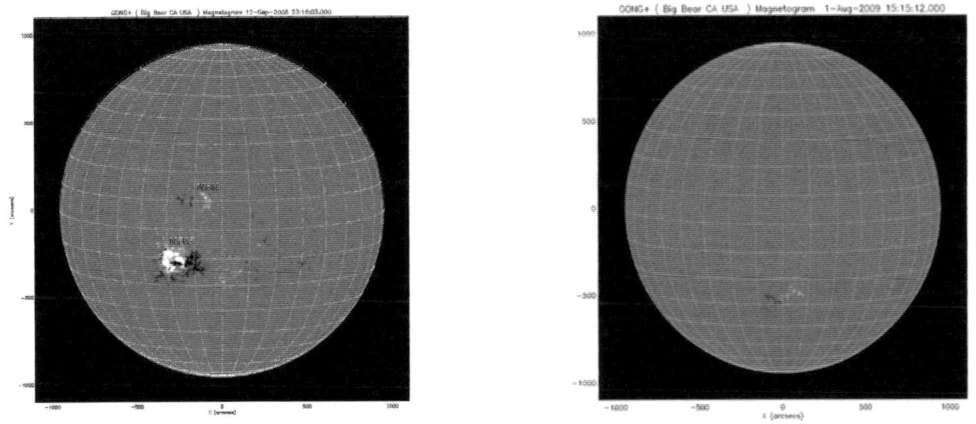

Figure 1. Daily GONG magnetograms showing the flare-producing active region 10808 (left) and the remnant of active region 11024 (right).

3. Results

Figure 2 shows the horizontal flow maps at a depth of 7 Mm or 1% of the solar radius for both sample periods. The daily flow maps (top) show large-scale horizontal flows such as the differential rotation and the poleward meridional flow. The average surface rotation rate has been subtracted. The errors are small near disk center and increase toward the solar limb due to geometric foreshortening, which increases projection errors making rings harder to fit. In the left panel, active region 10808 is noticeable as a location of enhanced error. The corresponding flows derived from an 8-hour subset show larger errors (bottom). Flows near disk center are similar to the ones of the daily flow maps, while flows near the limb are less reliably measured due to larger errors. Since active regions are restricted to mid- to low latitudes, this matters only near the eastern and western limb. Each of the three 8-hour time series leads to similar large-scale flows. The error values also depend on the fill factor; a decrease in fill factor leads to a smaller signal-to-noise ratio making rings harder to fit. One of the three 8-hour subsets with a fill factor of 86% has about 50% larger errors than the other subsets with a 100% fill factor.

Figure 3 shows north-south slices of kinetic helicity density as a function of depth and latitude. For the high-activity sample (left), we have chosen Carrington longitude 232.5°, which includes active region 10808 at 11° South in the unsigned magnetic flux. The kinetic helicity density at this location shows a "dipolar" pattern, a characteristic of flare-productive active regions. The pattern is most noticeable in the slice from the synoptic map (top) and in the one from the daily map (middle). However, despite the overall noisier result, the pattern is present in the 8-hour time series (bottom). The other two 8-hour subsets also show this helicity pattern. Helicity values at high latitudes are not significant due to large velocity errors. In addition, we have analyzed 8-hour time series of two other active regions (10759 and 10486) leading to similar helicity patterns. For the low-activity sample (right), we have chosen a longitude of 240°,

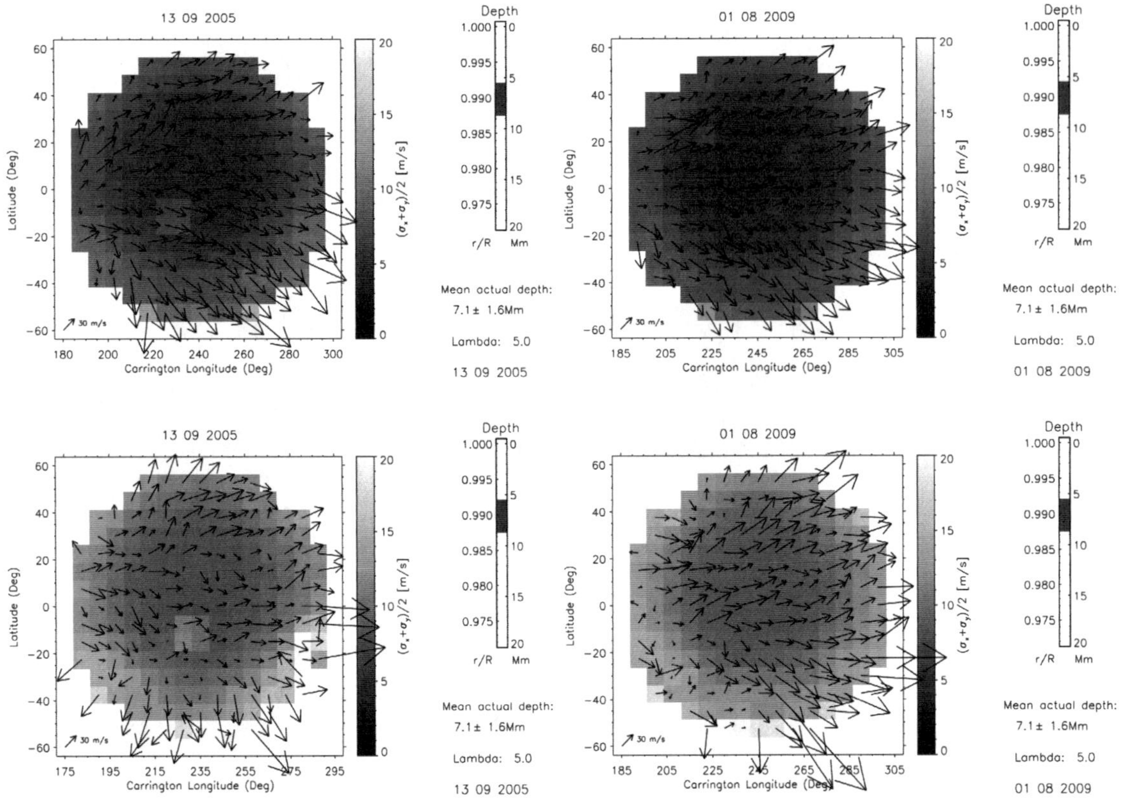

Figure 2. Full-disk maps of horizontal flows (black arrows) at a depth of 7 Mm derived from 24 hours (top) and 8 hours of data (bottom) with active region 10808 (left; 13 Sep 2005) and the remnant of active region 11024 (right; 1 Aug 2009). The background indicates the error values averaged over both velocity components.

which includes the remnant of active region 11024. The magnetic activity is about one order of magnitude smaller than that during the high-activity sample. The kinetic helicity density is much smaller and shows no distinct pattern compared to the active-region maps.

4. Discussion

The flows derived from 8-hour time series are reasonable near disk center and less reliable near the limb when compared to "standard" daily flow maps. The errors are larger for shorter time series, as to be expected. While the helicity values derived from the 8-hour data sets are noisier than the values derived from daily flows, the overall patterns remain the same. The high-activity sample shows a helicity pattern, that is expected from flare-productive regions, near active region 10808 in the 8-hour or 24-hour flow maps. The low-activity sample shows considerably smaller values of subsurface helicity without any pattern in the flow maps derived from 8-hour or 24-hour time series. With these encouraging results, we intend to analyze the disk passage of several active and quiet regions for a quantitative comparison.

Acknowledgments

This work utilizes data obtained by the Global Oscillation Network Group (GONG) program, managed by the National Solar Observatory, which is operated by the Association of Universities

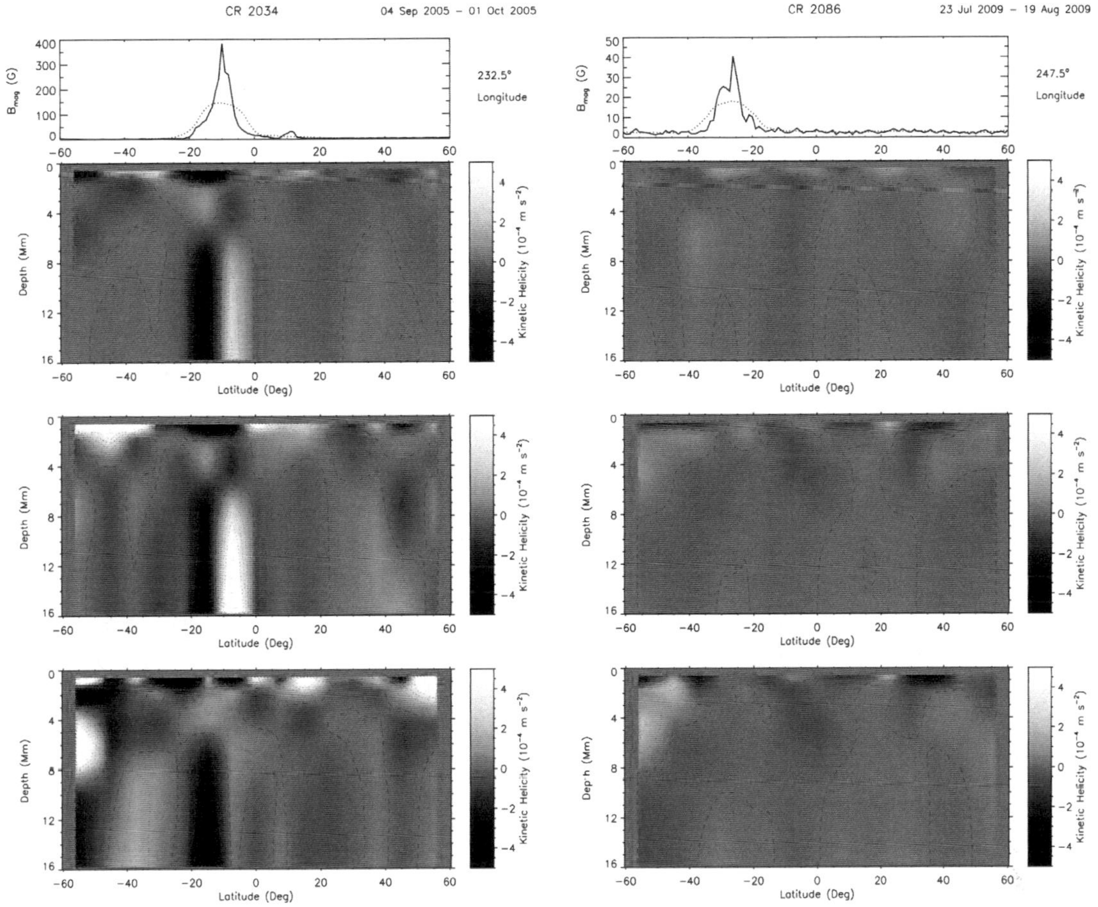

Figure 3. A north-south slice of kinetic helicity density as a function of latitude and depth with active region 10808 (left; 13 Sep 2005) and the remnant of active region 11024 (right; 1 Aug 2009). The line drawing indicates the location of activity. The helicity maps are derived from synoptic maps (top), 24-hour maps (middle), and 8 hours of data (bottom).

for Research in Astronomy (AURA), Inc. under a cooperative agreement with the National Science Foundation. The data were acquired by instruments operated by the Big Bear Solar Observatory, High Altitude Observatory, Learmonth Solar Observatory, Udaipur Solar Observatory, Instituto de Astrofísica de Canarias, and Cerro Tololo Interamerican Observatory. SOHO is a mission of international cooperation between ESA and NASA. This work was supported by NASA grant NNG08EI54I to the National Solar Observatory.

References

[1] Reinard A A, Henthorn J, Komm R and Hill F 2010 *ApJL* **710** L121–5
[2] Komm R, Ferguson R, Hill F, Barnes G and Leka K D 2010 *Solar Phys.* DOI 10.1007/s11207-010-9552-1
[3] Haber D A, Hindman B W, Toomre J, Bogart R S, Larsen, R M and Hill F 2002 *ApJ* **570** 855–64
[4] Komm R, Corbard T, Durney B R, González Hernández I, Hill F, Howe R and Toner C 2004 *ApJ* **605** 554–67
[5] Komm R 2007 *Astron. Nachr.* **328** 269–72

Analysis of peculiar penumbral flows observed in the active region NOAA 10930 during a major solar flare

Brajesh Kumar[1], **P. Venkatakrishnan**[1], **Savita Mathur**[2], **Sanjiv Kumar Tiwari**[1] **and R. A. García**[3]

1 Udaipur Solar Observatory, Physical Research Laboratory, Dewali, Badi Road, Udaipur 313004, INDIA

2 High Altitude Observatory, 3080 Center Green Drive, Boulder, CO 80302, USA

3 Laboratoire AIM, CEA/DSM-CNRS, Université Paris 7 Diderot, IRFU/SAp, Center de Saclay, 91191, Gif-sur-Yvette, FRANCE

E-mail: `brajesh@prl.res.in`

Abstract. It is believed that the high energetic particles and tremendous amount of energy released during the flares can induce velocity oscillations in the Sun. Using the Dopplergrams obtained by Global Oscillation Network Group (GONG) telescope, we analyze the velocity flows in the active region NOAA 10930 during a major flare (of class X3.4) that occurred on 13 December 2006. We observe peculiar evolution of velocity flows in some localized portions of the penumbra of this active region during the flare. Application of Wavelet transform to these velocity flows reveals that there is major enhancement of velocity oscillations in the high-frequency regime (5–8 mHz), while there is feeble enhancement in the p mode oscillations (2–5 mHz) in the aforementioned location. It has been recently shown that flares can induce high-frequency global oscillations in the Sun. Therefore, it appears that during the flare process there might be a common origin for the excitation of local and global high-frequency oscillations in the Sun.

1. Introduction

During the solar minimum phase between the activity cycles 23 and 24, a complex active region NOAA 10930 gave rise to a major flare (of class X3.4) on 13 December 2006 during its passage on the solar disk. The photospheric and chromospheric observations of this flare event taken in G-band and Ca II H, respectively, by Solar Optical Telescope (SOT) onboard *Hinode* spacecraft[1] have shown elongated flare ribbons which separate apart with the progress of the flare[2, 3]. The hard X-ray (HXR) images obtained by *RHESSI* spacecraft[4] show double-footpoint HXR sources located on the flare ribbons[5, 6]. There are also reports on the lateral motion of penumbral filaments in this active region during the flare[7].

Motivated by the aforesaid examples, we have analyzed the motions in this active region during the flare using the full-disk Dopplergrams obtained by the Global Oscillation Network Group (GONG)[8] telescope. Wavelet transform[9,10] is applied to velocity time series to examine whether there are any short-lived oscillations induced by the flare. Wavelet technique has the capability to show immediate consequence of any change in the physical parameter of a time series. This study will allow us to identify the seismic counterparts of transients events, such as flares, in the Sun. It will be of extreme importance to recognize the existence of such events in other stars through asteroseismology. Using this technique, today it is possible to measure magnetic activity cycles in other stars[11] from the high quality data obtained from CoRot[12] and Kepler[13] missions. Thus, it is a matter of time to identify high-energetic flares and other magnetic events using asteroseismology.

In the following Sections, we present description of the data and analysis, results, and finally the discussions.

2. Data and Analysis

The GONG instrument obtains full-disk photospheric Dopplergrams of the Sun using Ni I 6768 Å line with a spatial sampling rate of about 2.5 arcsec per pixel and at a cadence of one minute. We have used these velocity images between 01:00 UT and 03:02 UT on 13 December 2006 spanning the flare event. The sequence of images are derotated with respect to the first image of the series and then we choose a raster of 60×60 pixels centered around the active region for analysis. The sequence of the raster images are co-aligned with a sub-pixel accuracy. For the flare event of 13 December 2006, the flare intensity was highest at 02:25 UT as seen in the microwave emission observations from Nobeyama Radio Polarimeters (NoRP) (Reference). In the Figure 1, we show the difference Doppler image obtained by taking the difference between the Doppler image at 02:20 UT and 02:26 UT. This image clearly shows a bright patch (indicated by 'P' in this image) in the localized penumbral region of the active region, which is the site of flare induced large velocity flows. We have analyzed velocity flows in the rasters of 3×3 pixels in the centroid of bright patch (P) and one of the quiet regions (indicated by 'Q' in this image) far away from the active region. In the Figures 2(a) and 3(a), we show the Wavelet Power Spectra obtained for the flaring and quiet regions, respectively. The Figures 2(b) and 3(b) show the scale-average variance obtained from the corresponding Wavelet Power Spectrum in the frequency regimes: (0–2 mHz), (2–5 mHz), and (5–8 mHz). These results indicate major enhancement in high-frequency (5–8 mHz) velocity oscillations in penumbral flaring region just after the flare maximum as observed by NoRP. Besides, we can see that in the low frequency regime (0–2 mHz), the scale-average variance for both 'P' and 'Q' are below 50% significance level (top of panels of Figures 2(b) and 3(b)). In the frequency regime (2–5 mHz) where the p modes are expected, the power is suppressed in the penumbral region 'P' (middle panel of Figure 2(b)), whereas in a quiet region 'Q' of the Sun, the power of the p modes is above 50% and 90% significance level (middle panel of Figure 3(b)). Finally, at high frequency (5–8 mHz), the increase of power after the flare is observed in the average variance (bottom panel of Figure 2(b)) for the penumbral region 'P'. Our results are comparable with the earlier reports[14,15] on the detection of pronounced seismic emission from different flares into the solar interior in the 5–7 mHz frequency band.

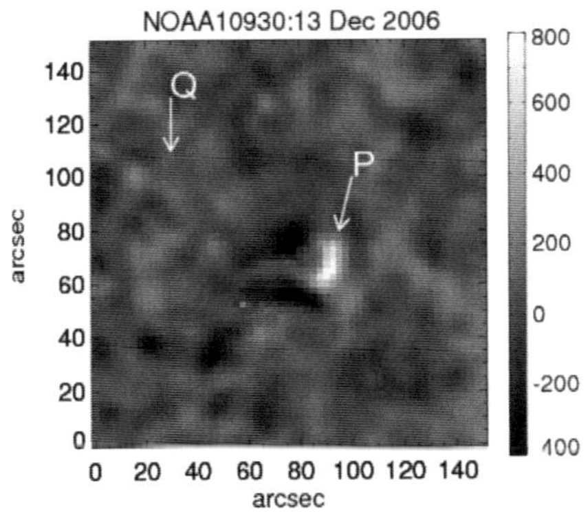

Figure 1. The difference image between the GONG Dopplergrams taken at 02:20 UT and 02:26 UT. The bright patchy region indicated by 'P' shows the location of the suddenly enhanced velocity oscillations in the localized penumbral region of the active region. The flare maximum was around 02:25 UT. One of the quiet regions of the Sun is indicated by 'Q' in this image.

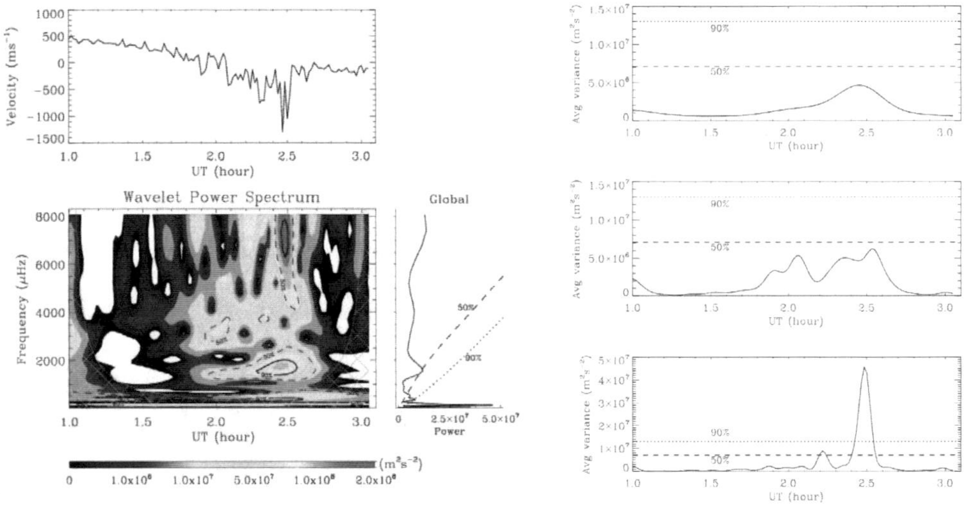

(a) (b)

Figure 2. (a) The upper panel shows the temporal evolution of the line-of-sight velocity averaged over a raster of 3×3 pixels in the centroid of the bright patchy penumbral region (P) as shown in the Figure 1. The lower panel shows the Wavelet Power Spectrum (WPS) and the Global-wavelet Power Spectrum (GWPS) computed from this time series. (b) The plots illustrate scale-average time series for the WPS in the frequency regime 0–2 mHz (top panel), 2–5 mHz (middle panel), and 5–8 mHz (bottom panel). In these plots, the dotted line corresponds to 90% significance level and the dashed line corresponds to 50% significance level.

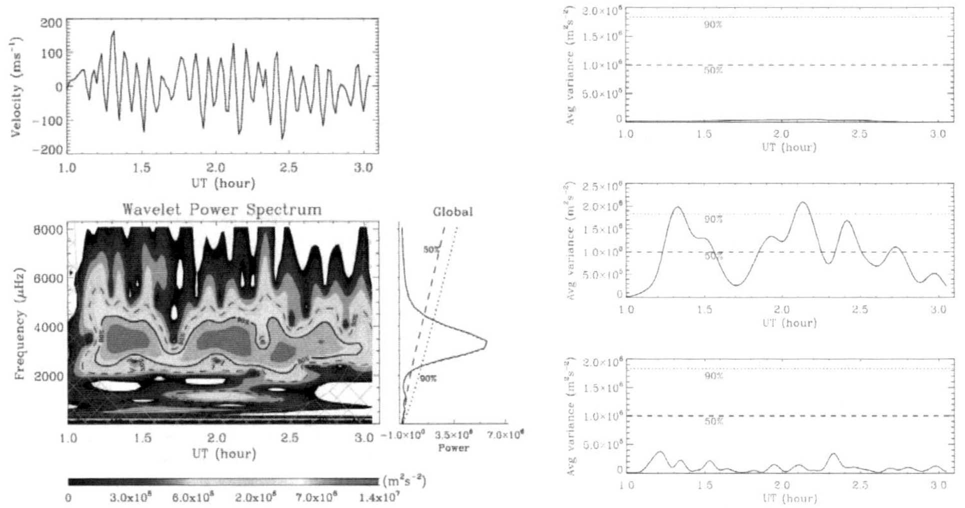

(a) (b)

Figure 3. Same as Figure 2, but for a quiet region (Q) as indicated in the Figure 1. Here, the dominant 5-min oscillations are clearly seen in the Wavelet Power Spectrum, Global-wavelet Power Spectrum and the scale-average time series.

3. Discussions

The impact of the X3.4 class flare that happened in the active region NOAA 10930 on 13 December 2006 was very much pronounced on the solar photosphere as it could produce flare ribbons seen in photospheric G-band 4305 Å images obtained by SOT on board *Hinode* spacecraft. The GONG Dopplergrams do show large velocity enhancements in the flaring region just after the impulsive phase of the flare. Recently, it has been shown that flares can induce high-frequency global oscillations in the Sun[16,17]. The enhancement of high frequency power in these localized velocity oscillations during the flare is comparable with the flare related enhancements seen in disk-integrated velocity and intensity observations of the Sun obtained from the Global Oscillation at Low Frequency (GOLF)[18] and Variability of solar IRadiance and Gravity Oscillations (VIRGO)[19] instruments, respectively, on board Solar and Heliospheric Observatory (*SOHO*) spacecraft[20]. These results might be indicative of a common origin for the excitation of local and global high-frequency oscillations in the Sun, that is somehow related to the flare process.

Acknowledgments

This work utilizes data obtained by the Global Oscillation Network Group (GONG) program, managed by the National Solar Observatory, which is operated by AURA, Inc. under a cooperative agreement with the National Science Foundation. The data were acquired by instruments operated by the Big Bear Solar Observatory, High Altitude Observatory, Learmonth Solar Observatory, Udaipur Solar Observatory, Instituto de Astrofísica de Canarias, and Cerro Tololo Interamerican Observatory. This work has been partially supported by the CNES/GOLF grant at the Service d'Astrophysique (CEA/Saclay). NCAR is partially funded by the National Science Foundation. We are thankful to John Leibacher, Frank Hill, Thierry Appourchaux, Paul Cally, Rudolf Komm, Irene González-Hernández and Sushanta Tripathy for useful discussions related to this work.

References

[1] Kosugi T *et al* 2007 *Solar Phys.* **243** 3
[2] Isobe H *et al* 2007 *Pub. Astron. Soc. Japan* **59** 807
[3] Jing J, Chae J and Wang H 2008 *The Astrophys. J. Lett.* **672** L73
[4] Lin R P *et al* 2002 *Solar Phys.* **210** 3
[5] Su Y *et al* 2007 *Pub. Astron. Soc. Japan* **59** 785
[6] Minoshima T *et al* 2009 *The Astrophys. J.* **697** 843
[7] Gosain S, Venkatakrishnan P and Tiwari S K 2009 *The Astrophys. J. Lett.* **706** L240
[8] Harvey J *et al* 1995, in GONG'94: Helio- and Astero-Seismology from the Earth and Space, eds. R. K. Ulrich, E. J. Rhodes, Jr., & W. Dappen, *Astron. Soc. Pac. Conf. Ser.* **76** 432
[9] Torrence C and Compo G P 1998 *Bull. Amer. Met. Soc.* **79** 61
[10] Liu Y, Liang X S and Weisberg R H 2007 *J. Atmos. & Ocean. Tech.* **24(12)** 2093
[11] García R A, Mathur S, Salabert D *et al* 2010 *Science* **In press**
[12] García R A, Régulo C, Samadi R *et al* 2009 *Astron. & Astrophys.* **506** 41
[13] Basri G, Borucki W J and Koch D 2005 *New Astron. Revs.* **49** 478
[14] Donea A -C and Lindsey C 2005 *The Astrophys. J.* **630** 1168
[15] Donea A -C *et al* 2006 *Solar Phys.* **239** 113
[16] Karoff C and Kjeldsen H 2008 *The Astrophys. J. Lett.* **678** L73
[17] Kumar B, Mathur S, García, R A and Venkatakrishnan P 2010 *The Astrophys. J. Lett.* **711** L12
[18] Gabriel A H *et al* 1995 *Solar Phys.* **162** 61
[19] Fröhlich C *et al* 1995 *Solar Phys.* **162** 101
[20] Domingo V, Fleck B and Poland A I 1995 *Solar Phys.* **162** 1

A study of torsional sunspot motion

Balázs Pintér, Keiran Burrows, Erin Connor and Karl Stokes

Institute of Mathematics and Physics, Aberystwyth University, Penglais Campus, Physics Building, SY23 3BZ, Wales, UK

E-mail: b.pinter@aber.ac.uk

Abstract. A fundamental understanding of the behaviour of sunspots is a powerful tool by which plasma dynamics and magnetism of the solar interior can also be better understood. Here a straightforward method is presented to investigate the temporal evolution of various sunspot features. Our particular aim is to detect rotations and search for oscillations in the torsional motions. We demonstrate the method by using the SOHO/MDI (Michelson Doppler Imager on board the Solar and Heliospheric Observatory) white-light intensity continuum images of a sunspot, which was visible on the solar disc between 17th and 28th July 2003.

1. Introduction

Sunspot dynamics have been observed and investigated for a long time, since Galilei turned his telescope to the sun in the early seventeenth century. Variations in sunspot structure, such as radial movements, have been studied (among others) by [1], sunspot rotations have been observed by [2] and [3] and others. The rotational motions have been interpreted theoretically, for example, by [4].

There are at least two basic techniques in the literature to measure sunspot rotations. A three-dimensional velocity vector of plasma motion from its observed radial component is derived in [5] and other publications by the authors, while [6] have used a method that provides high-resolution maps of rotational speeds as a function of time, sunspot radius and angular coordinate. The latter ones use SOHO/MDI (Michelson Doppler Imager on board the Solar and Heliospheric Observatory) white-light images, from which they create time slices along rings around the sunspot centre by uncurling the original images. Torsional motions appear in such uncurled plots as movements of small features (within and near the sunspot umbra and penumbra) along the axis for the angle. The rotational speeds can be derived from the gradient of the diagonal streaks. This technique can provide more detailed description of sunspot dynamics, as the spatial and temporal resolutions of the new generations of space-born solar observatories increase.

In the present paper we introduce another way of deriving the properties of torsional motions of sunspots. The technique works particularly well in case of isolated single sunspots, which have no dynamic interactions with nearby features of the photosphere. The obtained rotational speeds can be used to study the radial profile of sunspot motions and to search for short-time torsional oscillations by wavelet analysing the data.

2. Data processing

The data used in this study are taken from SOHO/MDI continuum (white-light) images. A sunspot, of simple shape and well isolated from its photospheric environment, was visible on the solar disc for the period of 17 to 28th July 2003. MDI produced 64 images of the sun (approximately five frames per day) as the sunspot swept across the disc at 7.5° latitude on the southern hemisphere of the sun. The spatial sampling of the MDI continuum images is 2 arcsecs, which correlates with 1.45 Mm.

The SolarSoft package was used to read and process the data. First, the effects of the photospheric differential rotation and limb darkening were removed from each image, then the selected sunspot was tracked. The sunspot centre was identified by finding the darkest position in the umbra by interpolation.

The boundaries of the sunspot umbra and penumbra can be defined as the set of positions where the image intensity has a large local increase as moving away from the centre. Those locations can be found by taking radial cross sections across the sunspot centre and finding the four peaks (umbra and penumbra boundaries at both sides of the centre) of the intensity gradient profile. Cuts can be taken across the centre at different angles, where the intensity at each gradient peak can be read. Average intensities at the peaks nearer and farther from the sunspot centre can be used to define the boundaries of the umbra and penumbra, respectively. The details are shown

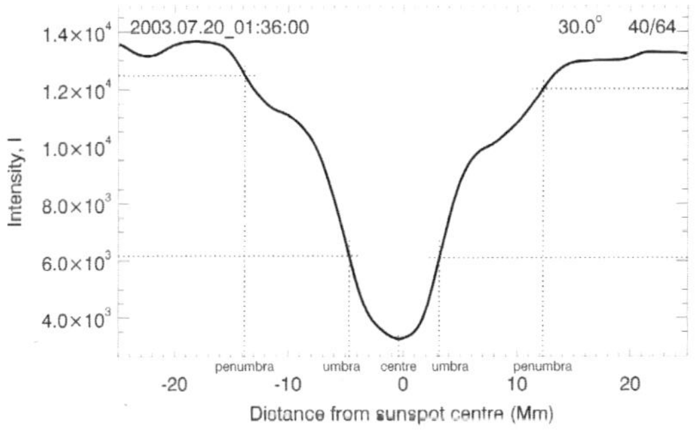

Figure 1. Example for the intensity profile along a slit, with angle 30° to the x axis, across the sunspot centre to determine the umbra and penumbra boundaries

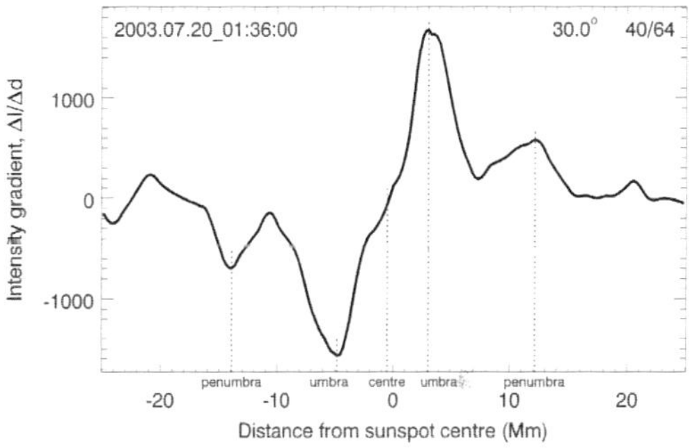

Figure 2. Same as in figure 1 but showing the derivative of the intensity along the slit

in figures 1 and 2, which display the variations of the intensity and its gradient along the line across the sunspot centre that has an angle of 30° with the positive x (equatorial) direction.

The sunspot with its derived umbra and penumbra boundaries (white and black dotted lines, respectively) can be seen in figure 3.

The basic idea of following the sunspot rotation around its centre is that an *orientation* can be attributed to the sunspot at every moment. When we take the penumbra contour line (determined by a constant averaged intensity), the distance of its points from a straight line that contains the sunspot centre and makes a certain angle to the positive x axis can be calculated. The integral of that distance along the boundary measures how 'squashed' the closed boundary line is in the direction perpendicular to that straight line. By iteration (and using discrete summation instead of an integral), the direction of the straight line across the sunspot centre for which the integral is the least (that is, the sunspot is the most 'squashed') can be obtained. This direction, called the *orientation* of the sunspot, is dependent of the shape of the sunspot. Orientation can be assigned to any two-dimensional shape but one with cylindrical (axial) symmetry. Assuming no large changes in the sunspot shape with time, any variation of

its orientation indicates a rotational movement of the sunspot.
An orientation angle can be found not only by using the penumbra boundary but also the umbra boundary or any other constant-intensity contour lines. The variation in orientation derived from loops of constant values from the intensity of the umbra centre to the intensity of the penumbra boundary describes the *radial* variation of the sunspot shape in a compact and measurable, quantitative way. The sunspot orientations derived from the umbra and penumbra boundaries

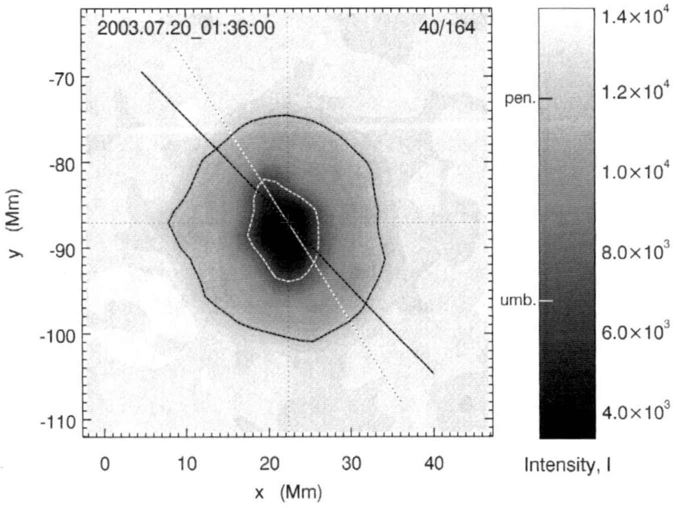

Figure 3. The sunspot in the 40th frame of the white-light images. The dotted straight lines indicate the orientation of the umbra and penumbra boundaries.

in frame No. 40 are shown in figure 3 by the white (at 123.32°) and black (at 135.06°) dotted straight lines, respectively.

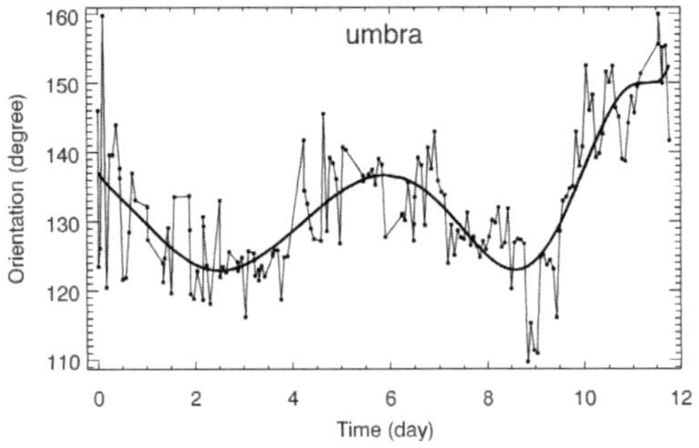

Figure 4. Temporal variation of the orientation of the umbra boundary

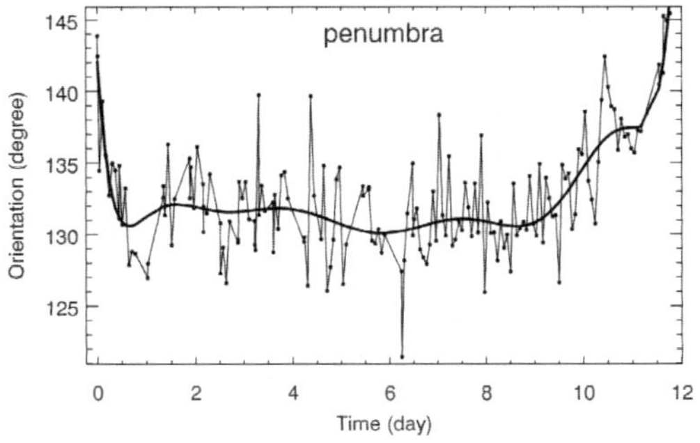

Figure 5. Temporal variation of the orientation of the penumbra boundary

3. Results

The time series of the orientation angles derived from the umbra and penumbra lines are given in figures 4 and 5, respectively. The variation on a long time scale is shown by plotting the fifth-order polynomial fitted to the raw data points. The obvious difference between the umbra and penumbra cases show that layers around the sunspot centre are not coupled strongly, there can be a strong velocity shear between neighbouring shells.

Once the smooth variation (fitted polynomial) is removed from the orientation data, torsional motions at short time scales (changes within a few hours) can be studied. To search for

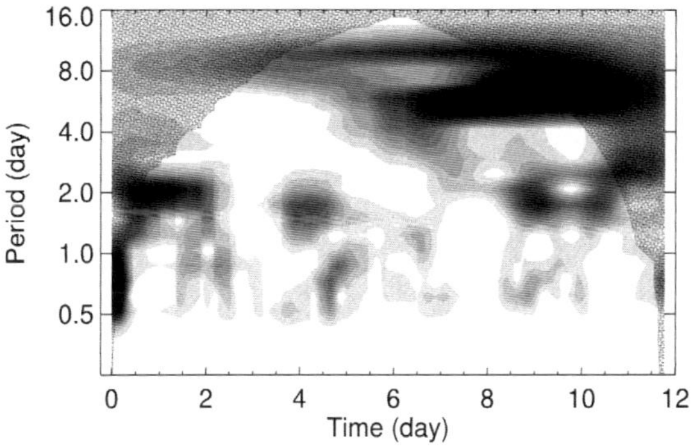

Figure 6. Wavelet power spectrum of the umbra orientation. Dark patches indicate enhanced powers at certain regions of time and oscillation period.

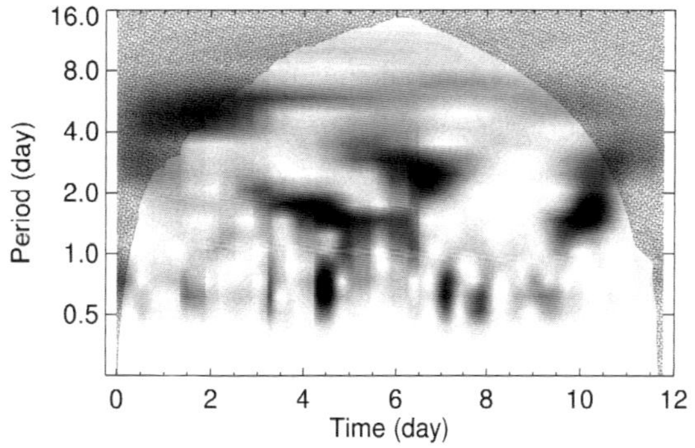

Figure 7. Same as in figure 4 but for the penumbra. The wavelet spectrum technique does not provide reliable data in the shaded marginal areas.

oscillatory motions, we can wavelet analyse the time series. The wavelet power spectra of the residual orientation time series derived from the umbra and penumbra contour lines are shown in figures 6 and 7. The oscillation power with certain periods (vertical axis) at a given time (horizontal axis) are indicated by the darkness of the gray colour scale. Both plots show somewhat strong oscillation power (dark patches) with about two-day and seven-day periods. The latter one seems more constant, while the oscillations with a two-hour period have smaller and larger amplitudes varying with time.

4. Conclusions

By defining an orientation to sunspots, we have introduced a technique to derive their torsional motion, which can be wavelet analysed to find torsional oscillations even at a short time interval. The technique works well for sunspots which have no complicated structures. As a follow-up, some of the sunspots analysed by other methods (see Introduction) will be considered and the derived torsional motions shall be compared. Torsional oscillations can be used as diagnostic tools, revealing the dynamics and magnetic properties of the sub-photospheric plasma at the sunspot. The technique will also be used to search for sudden changes in the behaviour of the torsional motion of a sunspot shortly before or after a nearby energetic event, such as a flare. This would indicate a coupling between the two phenomena.

Acknowledgments

The wavelet analysis program was written by using the wavelet software provided by C. Torrence and G. Compo (available at http://atoc.colorado.edu/research/wavelets/). K. Burrows and K. Stokes acknowledge the financial support from STFC and RAS, respectively.

References

[1] Evershed J 1910 *MNRAS* **70** 217-263
[2] Maltby P 1964 *Astrophisica Norvegica* **8** 205-269
[3] Gopasyuk S I and Moreton G E 1967 *Proc. of the Astronomical Society of Australia* **1** 8-9
[4] Stenflo J O 1969 *Solar Physics* **8** 115-118
[5] Gopasyuk O S and Gopasyuk S I 2008 *Bulletin of the Crimean Astrophysical Observatory* **104.1** 52-58
[6] Brown D S, Nightingale R W, Alexander D, Schrijver C J, Metcalf T R, Shine R A, Title A M and Wolfson C J 2003 *Solar Physics* **216** 79-108

Wave absorption and moat flow in AR9787

M Roth[1], **M J Thompson**[2]

[1] Kiepenheuer-Institut für Sonnenphysik, Schöneckstr. 6, Freiburg, Germany
[2] High Altitude Observatory, NCAR, Boulder, CO, USA

E-mail: `mroth@kis.uni-freiburg.de`

Abstract. We present the results of a study of the wave absorption and the moat flow around the sunspot in AR9787 (the "HELAS sunspot", so-called because it has been the object of in-depth study at a HELAS workshop) using Fourier-Hankel decomposition of the oscillatory wave field. We carry out inversions both for the flow speed and for the sound speed perturbation as a function of depth. The results are compared with results from time-distance helioseismology.

1. Introduction

Recently, the comparison of measurements of wave speed and flows in the sunspot in NOAA Active Region 9787 was the object of an in-depth study at a HELAS (European Helio- and Asteroseismology Network) workshop in Freiburg. The results were that the oscillation amplitudes are reduced in the sunspot at all frequencies [1]. In the surrounding plage region there seems to be enhanced acoustic power above the acoustic cutoff frequency of the quiet-Sun atmosphere [2], which could be caused by the conversion of p modes into magneto-acoustic waves. Detailed studies of the absorption of solar oscillations in this sunspot show that a significant fraction of the p and f modes passing through the sunspot are absorbed. One particularly puzzling result was a possible disagreement of standard ring-diagram analysis [3] and time-distance helioseismology [4] based on phase-speed filters when comparing the resulting relative wave speed perturbation caused by the active region [5].

We continue studying this sunspot in AR 9787 by probing the moat flow with Fourier-Hankel decomposition, i.e. we measure average frequencies of acoustic waves traveling inward and outward in an annular region around the sunspot. We implement this technique with a SOLA (Subtractive Optimally Localized Averages) inversion method [6], in order to determine the flow properties from the Doppler shift affecting the wave frequencies.

2. Data

The data and the data processing is described in [5]. The data consist of MDI full disk Dopplergrams covering the time span of 2002 January 20 – 28. The remapping of the images was done with a Postel projection with a map scale of 0.12°. The motion of the sunspot (Carrington longitude $\varphi \sim 133^\circ$ and latitude $\lambda = -8.3^\circ$) was tracked and the re-mapping employed a cubic convolution interpolation. Missing data were linearly interpolated. For each day a temporal average was subtracted from each Dopplergram. The resulting data cube of Doppler velocity measurements consists of 512 x 512 x 1440 data points per day. These final data sets are available on the European Helio- and Asteroseismology Network (HELAS) web site at http://www.helas-eu.org/ → NA4.

3. Fourier-Hankel Decomposition

The Fourier-Hankel decomposition was originally developed by [7] for the seismic probing of sunspots. The decomposition of the wave signal into incoming and outgoing waves is based on Hankel functions

$$\Phi(R_\odot, \theta, \phi, t) = \sum_{L,m,\nu} \mathrm{e}^{\mathrm{i}(m\phi-\nu t)} A_{L,m,\nu} H_m(L\theta) + B_{L,m,\nu} H_m^*(L\theta) \, , \tag{1}$$

where $R_\odot$ is the solar radius, θ and ϕ are the spherical polar coordinates with the sunspot centered at $\theta = 0$, $L = \sqrt{l(l+1)}$ is the spatial wavenumber, m the azimuthal order, ν the temporal frequency, and t is time. The complex amplitudes of the incoming and outgoing waves are given by $A_{L,m,\nu}$ and $B_{L,m,\nu}$, respectively. Hankel functions can be approximated in the far field by

$$H_m(L\theta) \approx \sqrt{\frac{2}{\pi L\theta}} \mathrm{e}^{\mathrm{i}(L\theta - m\pi/2 - \pi/4)} \, . \tag{2}$$

The further data reduction is carried out according to [8] and [9] on an annular region around the sunspot. The radii of the inner and outer boundary of the annulus are 30 and 137 Mm. The annulus properties were selected according to [9] in such a way that it is possible to measure a Doppler shift within an area restricted to the spatial extent of the moat flow only but also allows isolating the individual f and p mode ridges in the resulting diagnostic diagrams.

For our further studies p and f modes in the range of harmonic degree $l = 70, \ldots, 1500$ and azimuthal orders $m = -10, \ldots, 10$ are selected.

For studying the mode absorption by the sunspot we define an absorption coefficient according to

$$\alpha = \frac{\int W(P_{\mathrm{in}} - P_{\mathrm{out}}) \, d\nu dl}{\int W P_{\mathrm{in}} \, d\nu dl} \, , \tag{3}$$

where the window function W selects either a ridge with radial order n or a certain wavenumber, and the power of the ingoing and outgoing waves with frequency ν and degree l is given by P_{in} and P_{out}.

4. Results

One of the results is a ridge and wavenumber-dependent absorption of acoustic waves due to the sunspot. Figure 1 gives an overview on the resulting absorption coefficients. From ridge to ridge the absorption varies with the f mode being absorbed most. As a function of wavenumber we find that higher wavenumbers are absorbed strongest. This slope may be due to a change in the vertical structure of the sunspot. Investigating several other sunspots, [1] found a similar functional dependence of the wave absorption on the wavenumber.

Another result is an obvious shift of the frequencies due to the moat flow around the sunspot. Figure 2 displays two examples of p-mode power spectra, which were also shown by [9] for another sunspot. The power spectra of the outward propagating modes displayed were corrected for the effect of absorption. We note that the power spectra displayed are the average over the single spectra for modes with different azimuthal order m.

Compared to the inward propagating modes, the modes traveling away from the spot are shifted by approximately 10 μHz to higher frequencies due to the advection of the moat flow. This frequency shift can be described by [9; 10]

$$\Delta\nu_{nl} = \frac{l}{\pi R_\odot} \int \bar{U}(r) K_{nl}(r) \, dr \, , \tag{4}$$

where the frequency shift $\Delta\nu_{nl}$ is the difference between the frequencies of the outward and inward propagating modes with radial order n and harmonic degree l. The mean moat flow

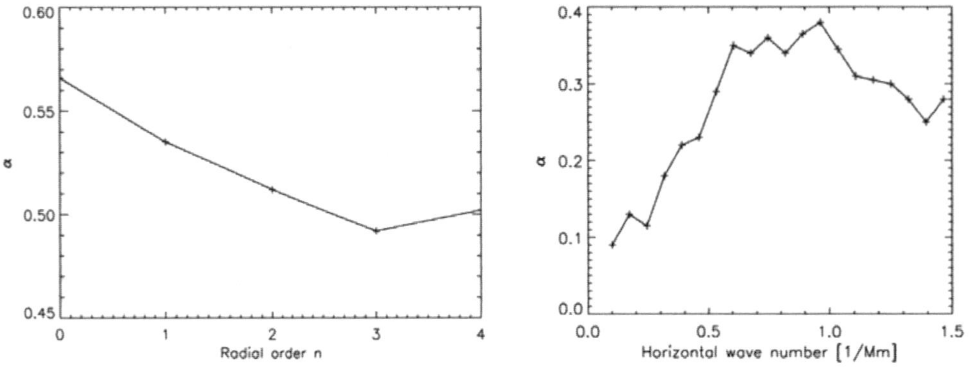

Figure 1. Absorption coefficient α as a function of radial order n (left) and as a function of wavenumber (right).

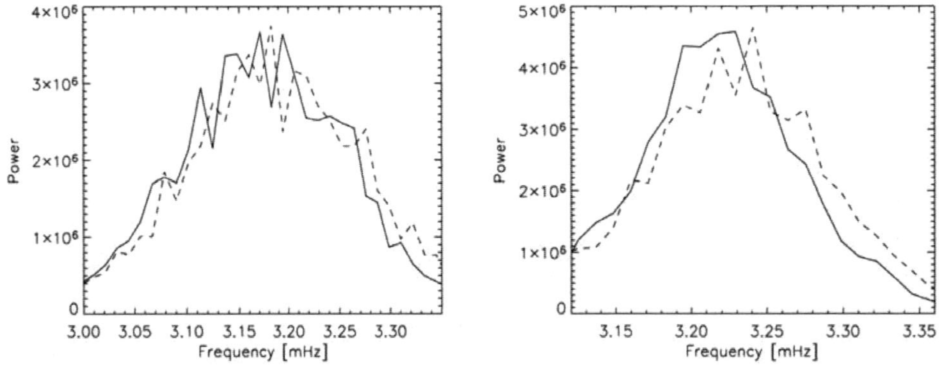

Figure 2. Power spectra in arbitrary units of inward (solid line) and outward (dashed line) propagating p modes. The two modes shown are $l = 288, n = 3$ (left) and $l = 452, n = 2$ (right) (compare with [9]).

over the annulus $\bar{U}(r)$ is directed outward and depends on the radius r inside the Sun. The sensitivity kernel $K_{nl}(r)$ is the energy density of a given mode as a function of radius.

We measure the frequency shift by fitting Lorentzian profiles to the individual power spectra and calculating the frequency differences between outward and inward propagating waves. From these frequency shifts we can estimate the moat flow around AR9787 by applying a standard 1D SOLA inversion [6]. The result is shown in Figure 3. The moat flow seems to have an almost constant amplitude of approximately 50 m/s in the first 4 Mm below the surface. Going deeper, the estimated flow amplitude shows indications for a higher amplitude at depths greater than 5 Mm. In these depths the amplitude is close to 60 m/s even though the error bars also increase. Time-distance measurements of the same sunspot as they are described in [5] result in a flow amplitude of approximately 30 m/s near the surface and approximately 50 m/s at a depth of 5 Mm.

In a final step, we apply an asymptotic inversion for the sound speed perturbations to the mean frequencies of the ingoing and outgoing waves. Based on our measurements, the sound speed perturbations in the moat around AR9787 are compatible with zero.

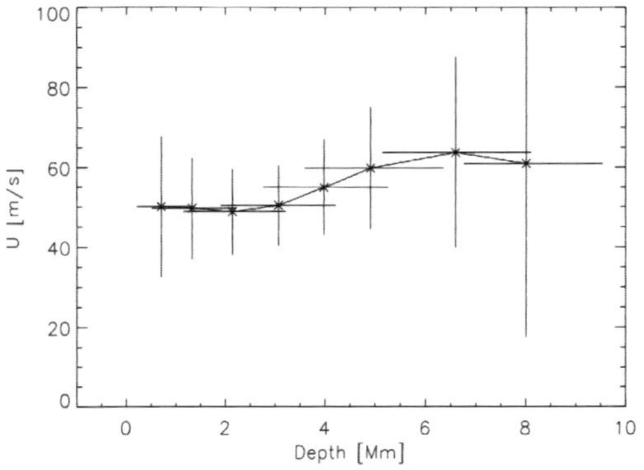

Figure 3. Flow profile of the moat flow around the sunspot in AR 9787 as determined by inversion. Vertical error bars are derived from the error magnification of the frequency shift via the SOLA inversion; horizontal bars correspond to the width of the averaging kernels.

5. Conclusions

We used the Fourier-Hankel decomposition technique to study the moat flow around AR9787 ("HELAS sunspot", [5]). The basis for our analysis are power spectra obtained for waves traveling toward and away from the sunspot. We conclude that in the moat around AR9787 waves are absorbed as less acoustic power is emitted in this area as received. Moreover the radially directed moat flow shifts the frequencies of the waves. Analyzing measured frequency shifts and by carrying out for the first time a SOLA inversion for the flow amplitude as a function of depth we find that the moat flow is in the order of 50 m/s down to 4 Mm depth. The flow shows indications of an increase between 4 – 8 Mm depth. These flow measurements are compatible with measurements from time-distance helioseismology. It seems that the sound speed is not significantly perturbed in the moat.

6. Acknowledgments

M.R. thanks Doug Braun and Laurent Gizon for useful discussions. The authors acknowledge support by the European Helio- and Asteroseismology Network (HELAS) which was funded as Coordination Action under the European Commission's Sixth Framework Programme (FP6). Data supplied courtesy of the SOHO/MDI consortia. SOHO is a project of international cooperation between ESA and NASA.

References

[1] Braun D C, Duvall Jr T L and Labonte B J 1988 *Astrophys. J.* **335** 1015–1025

[2] Braun D C, Lindsey C, Fan Y and Jefferies S M 1992 *Astrophys. J.* **392** 739–745

[3] Hill F 1988 *Astrophys. J.* **333** 996–1013

[4] Duvall T L, D'Silva S, Jefferies S M, Harvey J W and Schou J 1996 *Nature* **379** 235–237

[5] Gizon L, Schunker H, Baldner C S, Basu S, Birch A C, Bogart R S, Braun D C, Cameron R, Duvall T L, Hanasoge S M, Jackiewicz J, Roth M, Stahn T, Thompson M J and Zharkov S 2009 *Space Sci. Rev.* **144** 249–273 (*Preprint* `1002.2369`)

[6] Pijpers F P and Thompson M J 1994 *Astron. Astrophys.* **281** 231–240

[7] Braun D C, Duvall Jr T L and Labonte B J 1987 *Astrophys. J. Lett.* **319** L27–L31

[8] Braun D C 1995 *Astrophys. J.* **451** 859–+

[9] Braun D C, Fan Y, Lindsey C and Jefferies S M 1996 *ArXiv Astrophysics e-prints* (*Preprint* `arXiv:astro-ph/9603078`)

[10] Gough D O and Toomre J 1983 *Solar Phys.* **82** 401–410

A Ground Based Multiline Spectrometer for the Analysis of Solar Atmospheric Waves

J.Staiger

Kiepenheuer-Institut, 79104 Freiburg, Schöneckstrasse 6, Germany

joe@kis.uni-freiburg.de

Abstract. The presence of magnetic fields may influence the propagation characteristics of acoustic waves in the solar atmosphere. Investigating amplitude and phase properties may thus help to reveal the 3D geometry of magnetic fields above sunspots and activity regions. The height resolution of this type of measurements was limited in the past by the limitations of existing spectrometers.

We have developed a double-etalon based spectrometer allowing us to observe solar velocities with an unprecedented number of spectral lines. A test version of this instrument has been installed at the Vacuum Tower Telescope of the Kiepenheuer-Institut at Tenerife in July 2009 for the first time. 16 spectral lines were observed at a cadence of 60 secs over a period of 4 hours. The field of view was 100"x100".

The multiple diagnostic diagrams resulting from these observations clearly reveal the simultaneous p-mode ridge structure at every height level from the deep photosphere (538.0 nm C I) to the chromosphere (656.3 nm H-alpha). We are confident that individual acoustic modes may be vertically traced through the atmosphere at yet unsurpassed height resolution in the future.

1. Introduction

A consistent observational picture that would explain the energy transfer between the solar surface and the hot outer layers of the solar atmosphere currently does not yet exist. Tunneling of evanescent P-Modes through the photometric barrier and subsequent shock formation have long been suggested as a means to heat the solar chromosphere. Yet there is still a lack of experimental evidence for this type of conversion. Primary observational tools currently are etalon-based spectrometers as pioneered by Cavallini et al. [1] and Bonaccini et al. [2]. We suggest that overcoming the limitations of existing spectrometers with respect to their multiline capabilities may improve this situation.

We have designed a Fabry-Perot based device allowing us to observe velocities and intensities simultaneously across a large field-of-view with a high number of spectral lines. We hope that we will be able to contribute to a better understanding of chromospheric heating with this instrument.

2. Optical Layout

The core elements of of an etalon based spectrometer are a narrowband filter selection unit (see below, at image center) followed by two Fabry-Perot etalons arranged in a tandem configuration. During our experiments we used a collimated beam environment where the etalons were placed within the parallel beam section in front of the CCD camera (shaded box outlines spectrometer bounds).

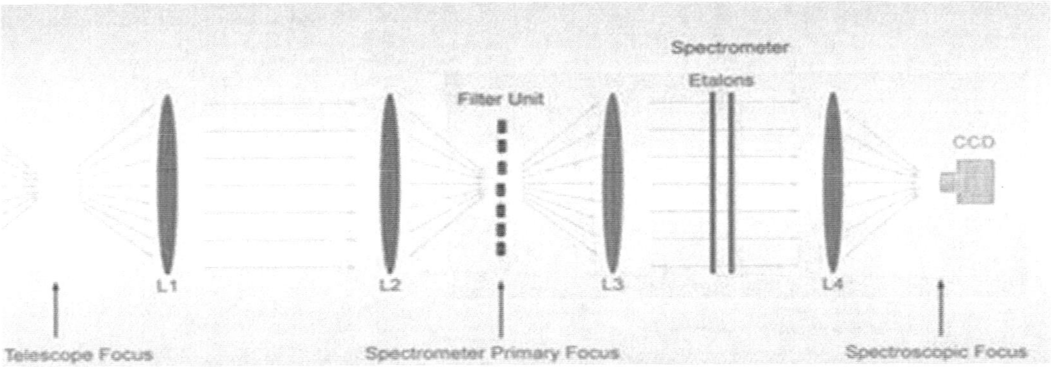

Figure 1. Spectrometer optical layout.

3. Matrix Filter Shifter

We have designed a new type of filter mount allowing us to change narrowband pre-filters with high speed and accuracy. This unit removes one of the primary temporal bottlenecks of existing spectrometers with respect to multiline capabilities. More spectral lines can be observed in a quasi-simultaneous mode than previously possible. 16 filters may be mounted to a grid-type pattern which will be shifted through the instrument's main beam by a Cartesian two-axis stepper drive.

Figure 2. The filter shifter unit is shown here with various narrowband interference filters mounted on the 4x4 grid matrix in preparation for a multiline observation.

4. VTT Experimental Setup

All experiments with the new spectrometer were carried out at the Vacuum Tower Telescope within the Teide Observatory at Izana, Tenerife. The etalons of the GFPI device (GFPI : Göttingen Fabry Perot Interferometer) which had been installed at the VTT in recent years were used as test etalons (see below, at image center).

Figure 3. VTT laboratory setup.

Setup Specifications:		
	CCD Camera:	DALSA 1M30
	Chip Size:	1024 x 1024
	Binning:	2 x 2
	Etalons:	IC Optical Ltd, 70 mm
	Spectral Range:	560 nm - 860 nm
	Field-of-View:	100" x 100"
	Exposure Times:	5 ms
	Resolution Limit:	0.2"

5. Test Observations

During an observational campaign in July 2009 the spectrometer was tested for the first time. 16 spectral lines were observed simultaneously. We achieved a cadence of 60 secs over a time span of 4 hours. Exposure times were of the order of 5 ms. The following table gives an overview over the spectral lines and the number of scansteps.

Wavelength	Element	Scansteps
517.2 nm		20
538.0 nm	C I	15
538.1 nm	Fe I	15
538.2 nm	Ti I	15
543.4 nm	Fe I	10
557.6 nm	Fe I	20
589.0 nm	Na D2	30
589.6 nm	Na D1	30
630.1 nm	Fe I	20
630.15 nm	Telluric	15
630.2 nm	Fe I	15
632.8 nm	He-Ne Laser	15
656.3 nm	H-Alpha	50
709.1 nm	Fe I	20
777.1 nm	Fe I	20
777.2 nm	Fe I	10

The operating software contains quick look tools which allow us to judge the quality of an observation shortly after the end of the measurements. The following screen dump displays such spectral data for 8 of the 16 observed absorption lines Shown here are simultaneous multiple p-mode diagrams taken from the low photosphere (upper left, 538.0 nm, C I) to the chromosphere (lower right, 656.3 nm, H-Alpha).

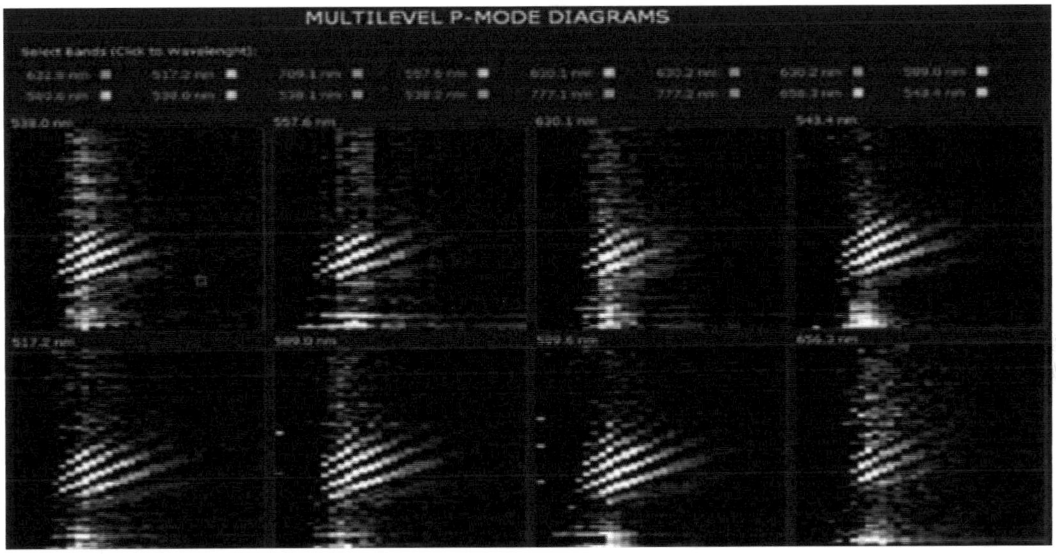

Figure 4. Multiple p-mode diagnostic diagrams.

7. Conclusions.

We have constructed and tested a multiline spectrometer which is able to handle 16 spectral lines in a simultaneous observation mode. We expect to get detailed information about amplitude and phase properties of solar atmospheric waves at a height resolution not achieved before.

8. Acknowledgements.

We want to thank Prof. Franz Kneer of IAG Göttingen for generously permitting us to make use of the GFPI etalons as test elements when they were still installed at the VTT. We want to thank Prof. Berdyugina of KIS for strongly supporting this project. We want to thank M. Roth of KIS for his scientific interest and participation in this project. Special thanks go to mechanical engineer T. Sonner of KIS responsible for the design and construction of the high quality mechanical components.

References

[1] Cavallini, F., Ceppatelli, G., Righini, A., Barletti, R.: 1979, A\&A 85, 255

[2] Bonaccini, D., Cavallini, F.,Ceppatelli, G., Righini, A.: 1989, A\&A 217, 386

Variation of high-degree mode frequencies during the declining phase of solar cycle 23

S. C. Tripathy, K. Jain and F. Hill

National Solar Observatory, Tucson, AZ 85719, USA

E-mail: `stripathy@nso.edu`

Abstract. We investigate the spatial and temporal variation of the high-degree mode frequencies during the declining phase of the solar cycle 23 and the extended minimum between the cycle 23 and 24. We find that the frequency shifts of high-degree modes obtained through the ring-diagram analysis in different phases of the solar cycle are not equally correlated with the local magnetic activity index.

1. Introduction

The acoustic frequencies of the Sun vary with the solar cycle and, in general, can be used as a tracer of magnetic activity [1]. However, recent studies using intermediate-degree global mode frequencies, both from GONG [2] and MDI [3], have shown complex variations during different phases of the solar cycle [4]. A similar study involving high-degree mode frequencies indicated that the correlation between frequency shifts and surface magnetic activity measured locally are significantly different during the two phases of the solar cycle [5]. Analyses of global frequencies during the extended minimum phase also indicate anti-correlation with the measurement of solar activity indices [6, 7]. In this context, we investigate the spatial and temporal variation of the high-degree mode frequencies during the declining phase of solar cycle 23 and the extended minimum phase between cycles 23 and 24.

2. Results

We use the ring-diagram technique [8, 9] to calculate the high-degree ($200 \leq \ell \leq 1100$; $0 \leq n \leq 6$) mode frequencies during nine-equally spaced Carrington Rotations (CRs) covering the period from 2002 to 2009 (CR1985 to CR2081). The high-degree mode frequencies are determined for each ring-day (1664 minutes) over 189 regions on the solar disk covering up to 60 degrees in latitude from the disk center. For each wavenumber and radial order, the frequency shifts are computed relative to the spatial average obtained from the 189 tiles; these shifts are then averaged over all multiplets to yield an average frequency shift, $\delta\nu$. We also estimate the magnetic activity associated with each tile by calculating a local magnetic activity index (MAI) that represents the average over all of the pixels in a given magnetogram tile. Here, we use the 96-min MDI magnetograms mapped and tracked in the same way as the Dopplergrams, so these MAIs represent contributions from about 18 images.

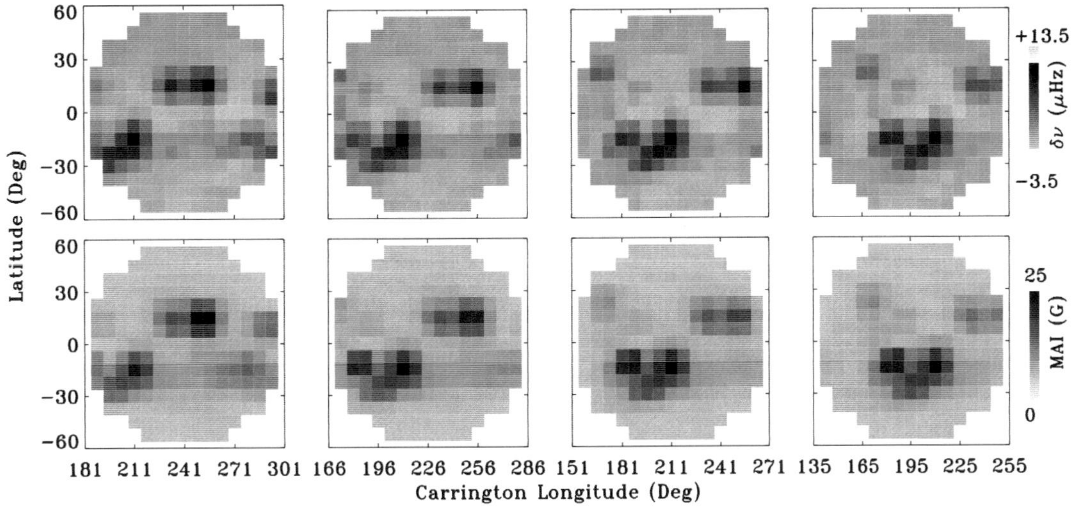

Figure 1. A sequence of images showing the frequency shifts (top panels) and Magnetic Activity Index (bottom panels) corresponding to 189 dense pack tiles for four ring-days (1–4 November, 2003) during the high activity period of CR 2009. The shifts represent average values over all modes present in a given tile.

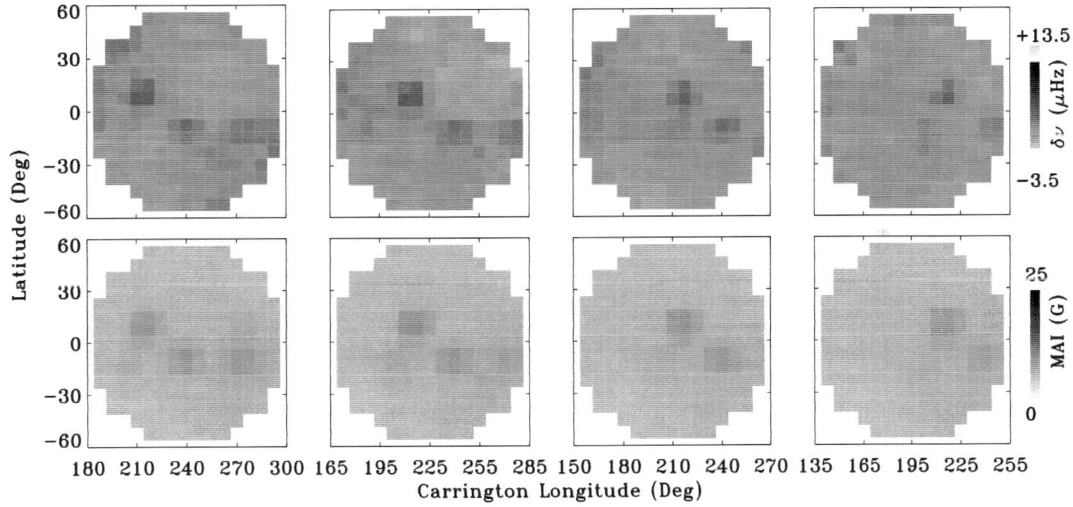

Figure 2. Same as Figure 1 but for low activity period of CR 2069 (24–28 April, 2008).

2.1. Spatial Variation

A sequence of images of the average frequency shift (top panel) and coeval magnetic activity index (bottom panel) for high- and low-activity periods corresponding to 1–4 November, 2003 and 24–28 April, 2008 are shown in Figure 1 and Figure 2, respectively. For each of the images, each pixel corresponds to a single tile in the 189 dense-pack mosaics. Once again the images confirm that during the high activity phase the local frequency shifts act as a tracer of magnetic activity while during the low-activity period, the agreement is weaker.

For a quantitative analysis, we perform a straight-line least-squares fit between the average

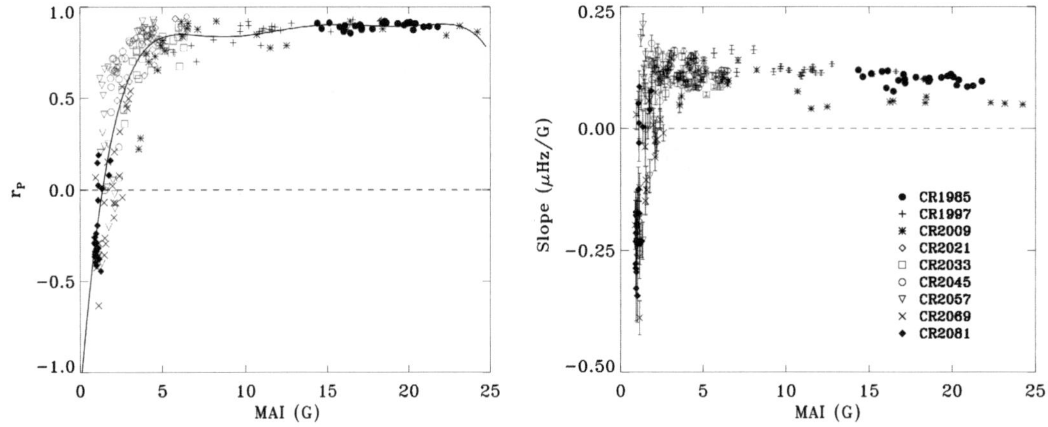

Figure 3. Linear correlation coefficient r_p (left) and the sensitivity in frequencies per unit change in activity i.e. slope of the linear regression correlation coefficient (right) between the frequency shift and MAI for each ring day. Different symbols represent different CRs and are marked in the figure.

frequency shifts and the MAI and also evaluated the linear Pearson's correlation coefficient, r_p for each ring day. These are shown in Figure 3 as a function of MAI for all the CRs. We find that for MAI above approximately 6–7 G, the correlation is larger than 80% and then shows a steep decrease as the MAI decreases. We also note negative coefficients for CR 2057–2081 which corresponds to the extended minimum phase between the cycle 23 and 24 implying that the frequencies do not follow the strong magnetic field component all the time. It is also interesting to note negative slopes for low MAI values.

2.2. Temporal Variation

We have also investigated the temporal variation of the frequency shifts over each CR, where the frequency shifts of each multiplet is calculated by subtracting an average frequency over the entire CR for each of the disk position and then averaged over all the multiplets. An average MAI is also obtained by averaging the MAI of each location over the entire CR. Figure 4 shows the linear correlation coefficients between frequency shifts and MAI. It is evident that as the solar cycle progresses towards the minimum phase, the correlation steadily decreases. We also note a significant drop during the extended minimum period and a negative correlation when the solar activity was near the minimum phase. As discussed earlier, such a negative correlation is also seen in the case of global modes [6, 7].

3. Summary

The frequency shifts of high degree modes obtained through the ring-diagram analysis in different phases of the solar cycle illustrate that the shifts are not equally correlate with the local magnetic activity index. During the descending phase, the frequency shifts are strongly correlated, while during period of extended minimum phase both the spatial and temporal shifts are weakly correlated with MAI implying that the shifts can not be accounted by the regions of observed strong field component of the magnetic field alone. Thus we conjecture that the changes in acoustic high-degree mode frequencies may be associated with two components: strongly localized active regions and the quiet-sun magnetic field. The former explains the larger shifts while the later may be responsible for the smaller shifts in all phases of solar activity.

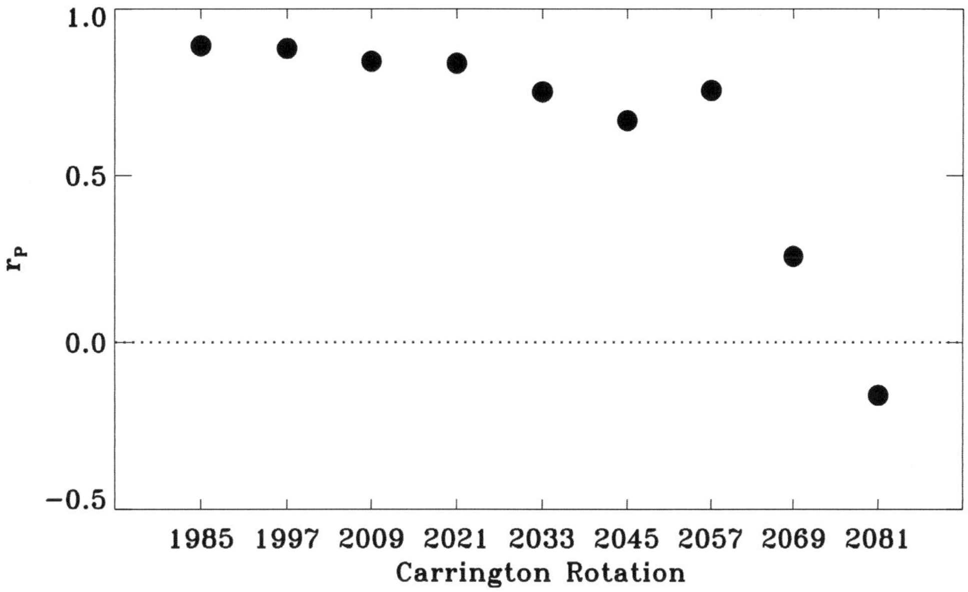

Figure 4. Linear correlation coefficients between frequency shifts and MAI for different CR's.

Acknowledgments

This work utilizes data obtained by the Global Oscillation Network Group (GONG) program, managed by the National Solar Observatory, which is operated by AURA, Inc. under a cooperative agreement with the National Science Foundation. The data were acquired by instruments operated by the Big Bear Solar Observatory, High Altitude Observatory, Learmonth Solar Observatory, Udaipur Solar Observatory, Instituto de Astrofísico de Canarias, and Cerro Tololo Interamerican Observatory. It also utilizes data from the Solar Oscillations Investigation/Michelson Doppler Imager on the Solar and Heliospheric Observatory. SOHO is a mission of international cooperation between ESA and NASA. This work was supported by NASA grant NNG 08EI54I.

References

[1] Hindman B, Haber D A, Toomere J and Bogart R S 2000, *Solar Phys.* **192** 363-372
[2] Harvey J W et al 1996 *Science* 272 1284-86
[3] Scherrer P H et al 1995 *Solar Phys.* 162 129-88
[4] Jain K, Tripathy S C and Hill F 2010 *Astrophys. J.* **695** 1567-76
[5] Tripathy S C, Jain K and Hill F 2010 *Proc. Magnetic Coupling between the Interior and Atmosphere of the Sun, Astrophys. and Space Science Proc.*, ed S. S. Hasan and R. J. Rutten (Berlin Heidelberg: Springer-Verlag) p 273
[6] Salabert D, Garcia R A, Palle P L and Jimenez-Reyes S J 2009 *A&A* **504** L1-4
[7] Tripathy S C, Jain K and Hill F 2010 *Astrophys. J.* **711** L84-8
[8] Hill F 1988 *Astrophys. J.* **333** 996-1013
[9] Corbard T, Toner C G, Hill F, Hanna K D, Haber D A Hindman B W and Bogart R S 2003 *Proc. Local and Global Helioseismology: The Present and Future* vol 517 (Noordwijk: ESA) p 255

Are short-term variations in solar oscillation frequencies the signature of a second solar dynamo?

Anne-Marie Broomhall[1], Stephen T. Fletcher[2], David Salabert[3,4], Sarbani Basu[5], William J. Chaplin[1], Yvonne Elsworth[1], Rafael A. García[6], Antonio Jiménez[3], and Roger New[2]

[1] School of Physics and Astronomy, University of Birmingham, Edgbaston, Birmingham B15 2TT, UK
[2] Faculty of Arts, Computing, Engineering and Sciences, Sheffield Hallam University, Sheffield S1 1WB, UK
[3] Instituto de Astrofísica de Canarias, E-38200 La Laguna, Tenerife, Spain
[4] Departamento de Astrofísica, Universidad de La Laguna, E-38205 La Laguna, Tenerife, Spain
[5] Yale University, P.O. Box 208101, New Haven, CT 06520-8101, USA
[6] Laboratoire AIM, CEA/DSM-CNRS-Université Paris Diderot, IRFU/SAp, Centre de Saclay, 91191 Gif-sur-Yvette, France

E-mail: amb@bison.ph.bham.ac.uk

Abstract. In addition to the well-known 11-year solar cycle, the Sun's magnetic activity also shows significant variation on shorter time scales, e.g. between one and two years. We observe a quasi-biennial (2-year) signal in the solar p-mode oscillation frequencies, which are sensitive probes of the solar interior. The signal is visible in Sun-as-a-star data observed by different instruments and here we describe the results obtained using BiSON, GOLF, and VIRGO data. Our results imply that the 2-year signal is susceptible to the influence of the main 11-year solar cycle. However, the source of the signal appears to be separate from that of the 11-year cycle. We speculate as to whether it might be the signature of a second dynamo, located in the region of near-surface rotational shear.

1. Introduction

The Sun is a variable star, whose magnetic activity shows systematic variations. The most conspicuous of these variations is the 11-year solar cycle [1]. However, over the past twenty years it has become apparent that significant (quasi-periodic) variability is also seen on shorter timescales, between 1 and 2 years [e.g. 2–4]. Fletcher et al. [5] investigated the origins of this so-called "mid-term" periodicity by looking at variations in the frequencies of solar oscillations. Fletcher et al. used the Sun-as-a-Star observations made by the Birmingham Solar Oscillations Network [BiSON; 6; 7] and the Global Oscillations at Low Frequencies [GOLF; 8–10] instrument onboard the *SOlar and Heliospheric Observatory* (*SOHO*) spacecraft. In this paper we extend the work of Fletcher et al. by examining data observed by the Variability of solar IRradiance and Gravity Oscillations [VIRGO; 11] instrument, which is also onboard SOHO. VIRGO consists of three sun photometers (SPMs), that observe at different wavelengths, namely the blue channel (402 nm), the green channel (500 nm), and the red channel (862 nm). We have examined each set of VIRGO data individually and find that the results are similar for each channel. Therefore, here we concentrate on the results found using the blue VIRGO data.

The frequencies of solar p modes vary throughout the solar cycle with the frequencies being at their largest when the solar activity is at its maximum [e.g. 9; 12–16]. By examining the changes in the observed p-mode frequencies throughout the solar cycle, we can learn about solar-cycle-related processes that occur beneath the Sun's surface.

We use oscillations data collected by making Sun-as-a-star observations, which are sensitive to the p modes with the largest horizontal scales (or the lowest angular degrees, l). Consequently, the observed frequencies are of the truly global modes of the Sun [e.g. 17, and references therein]. These modes travel to the Sun's core but, because the sound speed inside the Sun increases with depth, their dwell time at the surface is longer than at the solar core. Consequently, p modes are most sensitive to variations in regions of the interior that are close to the surface and so are able to give a global picture of the influence of near-surface activity.

Recently García et al. [18] observed signatures of a stellar activity cycle in asteroseismic data obtained by the Convection Rotation and Planetary Transits [CoRoT; 19] space mission. With the prospect of longer asteroseismic data sets (~ 5 years) becoming available through, for example, Kepler [20] there will be opportunities to observe activity cycles in other stars. These observations will provide constraints for models of stellar dynamos under conditions different from those in the Sun.

2. Uncovering the mid-term periodicity

The observations made by BiSON, GOLF and VIRGO were divided into 182.5-day-long independent subsets. BiSON has now been collecting data for over 30 years. The quality of the early data, however, is poor compared to more recent data because of poor time coverage. Here, we have analyzed the mode frequencies observed by BiSON during the last two solar cycles in their entirety i.e. from 1986 April 14 to 2009 October 7. GOLF and VIRGO have been collecting data since 1996 and so we have been able to analyze data covering almost the entirety of solar cycle 23, i.e., from 1996 April 11 to 2009 April 7. After 1996 April 11, when all three sets of data were available, we ensured that the start times of the subsets from each observational program were the same.

Estimates of the mode frequencies were extracted from each subset by fitting a modified Lorentzian model to the data using a standard likelihood maximization method. Two different fitting codes have been used to extract the mode frequencies, both giving the same results. For clarity, we only show the results of one method, which was applied in the manner described in [21]. A reference frequency set was determined by averaging the frequencies in subsets covering the minimum activity epoch at the boundary between cycle 22 and cycle 23. It should be noted that the main results of this paper are insensitive to the exact choice of subsets used to make the reference frequency set. Frequency shifts were then defined as the differences between frequencies given in the reference set and the frequencies of the corresponding modes observed at different epochs [22].

For each subset in time, three weighted-average frequency shifts were generated, where the weights were determined by the formal errors on the fitted frequencies: first, a "total" average shift was determined by averaging the individual shifts of the $l = 0$, 1, and 2 modes over fourteen overtones (covering a frequency range of 1.88 – 3.71 mHz); second, a "low-frequency" average shift was computed by averaging over seven overtones whose frequencies ranged from 1.88 to 2.77 mHz; and third, a "high-frequency" average shift was calculated using seven overtones whose frequencies ranged from 2.82 to 3.71 mHz. The lower limit of this frequency range (i.e., 1.88 mHz) was determined by how low in frequency it was possible to accurately fit the data before the modes were no longer prominent above the background noise. However, we note here that each of the fitted frequencies was checked for accuracy and this resulted in many of the low-n fitted frequencies from the VIRGO data being discarded. The lower signal-to-noise of the oscillations in the VIRGO data means that accurate fits to the data are only possible above approximately 2.3 mHz. The upper limit on the frequency range (i.e., 3.71 mHz) was determined by how high in frequency the data could be fitted before errors on the obtained frequencies became too large due to increasing line widths causing modes to overlap in frequency.

The left-hand panels of Figure 1 show mean frequency shifts of the p modes observed by BiSON, GOLF and blue VIRGO, respectively [also see 5; 22; 23]. The 11-year cycle is seen clearly and its signature is most prominent in the higher-frequency modes. This is a telltale indicator that the observed

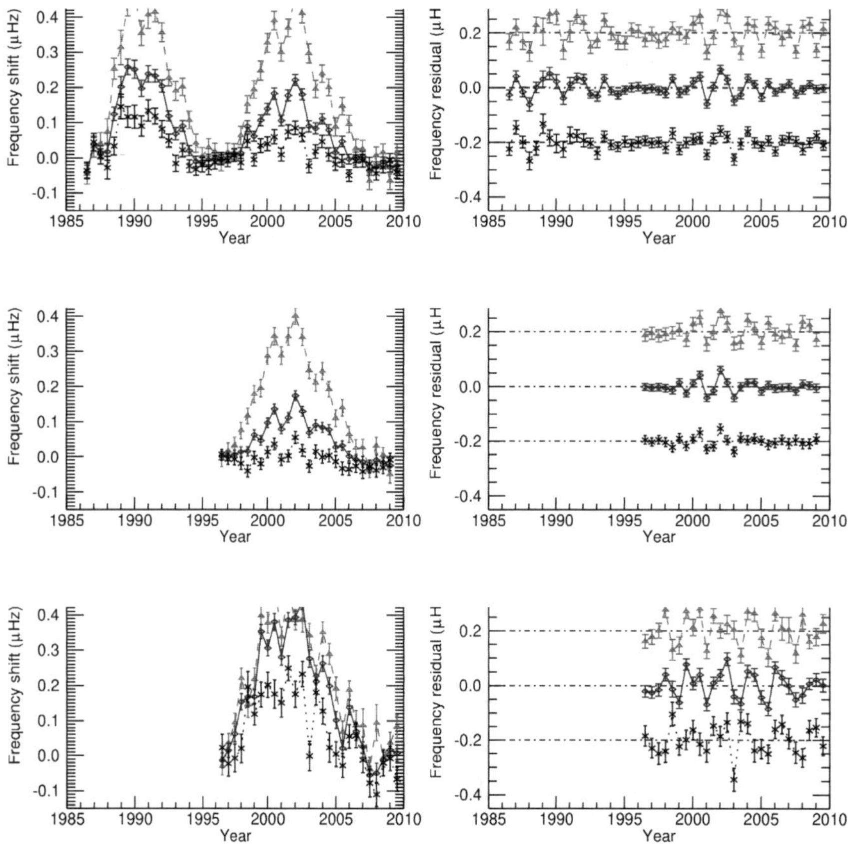

Figure 1. Left column: average frequency shifts of "Sun-as-a-star" modes with frequencies between 1.88 and 3.71 mHz (total-frequency band, solid line, and diamond symbols); 1.88 and 2.77 mHz (low-frequency band, dotted line, and cross symbols); and 2.82 and 3.71 mHz (high-frequency band, dashed line, and triangle symbols). Right column: residuals left after dominant 11-year signal has been removed (dotted and red dashed curves are displaced by −0.2 and +0.2, respectively, for clarity).

11-year signal must be the result of changes in acoustic properties in the few hundred kilometres just beneath the visible surface of the Sun, a region that the higher-frequency modes are much more sensitive to than their lower-frequency counterparts because of differences in the upper boundaries of the cavities in which the modes are trapped [24; 25]. Note that the difference between the low- and high-frequency range shifts is less for the blue VIRGO data, compared to the BiSON and GOLF data. This is because the low-frequency range for the blue VIRGO data does not extend as low in frequency as for the BiSON and GOLF data. Despite the low- and high-frequency bands showing different sensitivities to the 11-year cycle there is a significant correlation between the observed frequency-shifts. The correlations between the low- and high-frequency band shifts are 0.82 for the BiSON data, 0.67 for the GOLF data, and 0.78 for the VIRGO data. The errors on the correlations indicate that there is less than a 1% chance that each of these correlations would occur by chance. The signal is reassuringly similar in the different data sets. The correlation between the BiSON, GOLF, and blue VIRGO frequency shifts was found to be highly significant in all three frequency bands with less than a 1% chance that these correlations would occur randomly.

In order to extract mid-term periodicities, we subtracted a smooth trend from the average total shifts by applying a boxcar filter of width 2.5 years. This removed the dominant 11-year signal of the solar

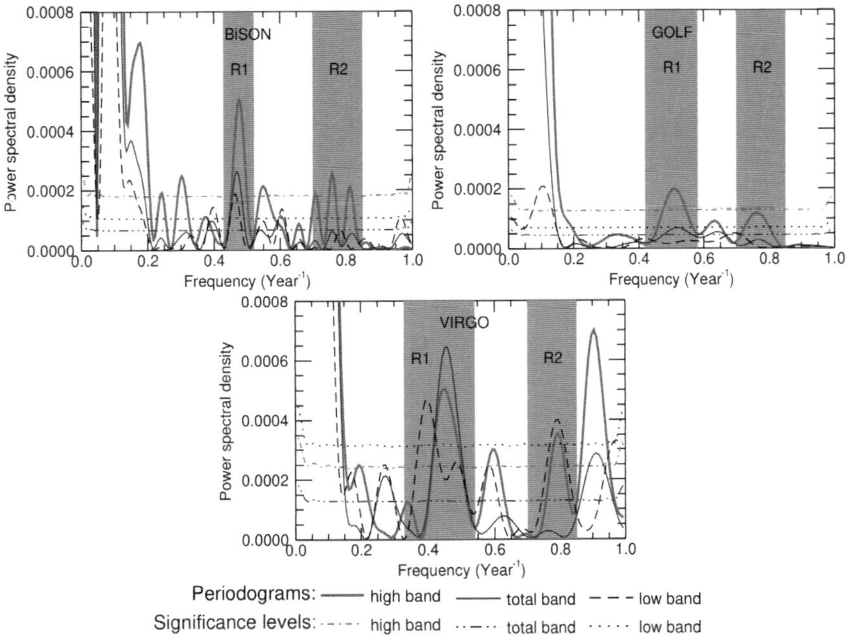

Figure 2. Periodograms of the frequency shifts observed in the different frequency bands (see legend). The 1% "false alarm" significance levels for the respective frequency ranges are also plotted. Shaded regions, denoted R1 and R2, are included in each panel to guide the eye towards the significant regions of interest.

cycle. Although the width of this boxcar is only slightly larger than the periodicity we are examining here, wider filters produce similar results. The resulting residuals, which can be seen in the right-hand panels of Figure 1, show a periodicity on a timescale of about 2 years.

There is a significant correlation between the low- and high-frequency band residuals for the BiSON (0.46), GOLF (0.55), and blue VIRGO (0.76) data and there is less than a 1% probability of these correlations occurring by chance. The correlations between the BiSON and GOLF residuals were found to be significant in all three frequency bands with less than a 1% probability of observing such correlations by chance. The BiSON and blue VIRGO residuals are also reasonably well correlated in all three frequency bands, with less than a 2% probability of observing the correlations by chance. However, the GOLF and blue VIRGO residuals are less well correlated.

The periodograms of the raw frequency shifts shown in the left-hand panels of Figure 1 were computed to assess the significance of the 2-year signal. Figure 2 shows the periodograms, oversampled by a factor of 10. Also plotted in Figure 2 are the 1% false alarm significance levels [26], which were determined using Monte Carlo simulations based on the size of the errors associated with the raw frequency shifts (see Figure 1). The large peak at 0.09 yr^{-1} is the signal from the 11-year cycle. There are also large peaks at approximately 0.5 yr^{-1} (indicated by the shaded regions denoted R1 in each panel of Figure 2). Statistical analysis of the BiSON periodograms established that the apparent 2-year periodicity was indeed significant, in the low-, total-, and high-frequency bands, with a false alarm probability of 1%. A peak at the same frequency is also significant in the high- and total-frequency bands in the GOLF and blue VIRGO data. Note that there is also a significant peak at a slightly lower frequency in the low-frequency band blue VIRGO data. The fact that the peaks in the GOLF and blue VIRGO data are not as prominent as the equivalent peaks observed in the BiSON data is expected because fewer GOLF and blue VIRGO data are available, particularly during periods of high activity when the 2-year signal is most prominent.

The examination of the VIRGO data supports the theory that there is a 2-year signal present in the frequency shifts. One possible explanation for this signal is a dynamo action seated near the bottom of the layer extending 5% below the solar surface [see 5, for details]. The amplitude envelope of the 2-year signal observed in the BiSON and GOLF data appears to be modulated by the 11-year cycle (see the right-hand panels of Figure 1). This is particularly true for the low-frequency band. Interestingly this does not appear to be the case for the signal observed in the blue VIRGO data, which could be because fewer very low-frequency modes were used to calculate the blue VIRGO frequency shifts, thus indicating that the signal shows some frequency dependence. Note that although asymmetries in the Sun's magnetic field have been used to explain the 2-year signal observed in other proxies of solar activity this would not explain why the amplitude of the signal observed in the p-mode frequency shifts is so similar in all frequency bands.

A prominent peak is observed at $\sim$ 0.9 yr^{-1} in the blue VIRGO high- and total-frequency bands. However, there is no signal present at the same frequency in either the BiSON or GOLF data. This peak could, therefore, be instrumental in origin.

3. Evidence for a 1.3-year periodicity

We also draw attention to a significant peak at a frequency of approximately 0.8 yr^{-1} or a period of $\sim$ 1.3 yr (indicated by the shaded regions denoted R2 in Figure 2). This peak is visible most strongly in the blue VIRGO data but an excess of power is also visible in the high-frequency range in the BiSON and GOLF data (although the GOLF peak is only significant at a 2% level). Notice that the 1.3-year signal observed in the VIRGO data is significant in both the low- and high-frequency bands but almost fully suppressed in the total-frequency band. This is because the signal is out of phase in the two different regions of the frequency-spectrum.

A 1.3 yr periodicity has been observed in other solar data. For example, Howe et al. [27] observed variations in the rotation profile of the Sun, most predominately at low latitudes and with a period of 1.3 yr. However, the signal was found to be intermittent and has not been observed since 2001 [28]. Jiménez-Reyes et al. [9] observed a 1.3 yr modulation in the energy supply rate. Wang and Sheeley [29] observe a 1.3 yr quasi-periodicity in the Sun's dipole magnetic moment and open magnetic flux. Wang and Sheeley [29] attribute this to the stochastic processes of active region emergence and a decay time of about 1 yr, which is determined by differential rotation, meridional flow and supergranule diffusion. The presence of excess power at this frequency in all three sets of data means that this feature warrants further investigation, as does the fact that the signal is out of phase in the low- and high-frequency band blue VIRGO data, and so the 1.3-year periodicity is the subject of ongoing work.

Acknowledgments

This paper utilizes data collected by the Birmingham Solar-Oscillations Network (BiSON), which is funded by the UK Science Technology and Facilities Council (STFC). We thank the members of the BiSON team, colleagues at our host institutes, and all others, past and present, who have been associated with BiSON. The GOLF instrument on board *SOHO* is a cooperative effort of many individuals, to whom we are indebted. *SOHO* is a project of international collaboration between ESA and NASA. S.T.F., A.M.B., W.J.C., Y.E., and R.N. acknowledge the financial support of STFC. S.B. acknowledges NSF grants ATM-0348837 and ATM-0737770. D.S. acknowledges the support of the grant PNAyA2007-62650 from the Spanish National Research Plan. R.A.G. thanks the support of the CNES/GOLF grant at the CEA/Saclay.

References

[1] Hathaway D H 2010 *Living Reviews in Solar Physics* **7** 1

[2] Benevolenskaya E E 1995 *Sol. Phys.* **161** 1–8

[3] Mursula K, Zieger B and Vilppola J H 2003 *Sol. Phys.* **212** 201–207

[4] Valdés-Galicia J F and Velasco V M 2008 *Advances in Space Research* **41** 297–305

[5] Fletcher S T, Broomhall A, Salabert D, Basu S, Chaplin W J, Elsworth Y, García R A and New R 2010 *ApJL* **718** L19–L22

[6] Elsworth Y, Howe R, Isaak G R, McLeod C P, Miller B A, New R and Wheeler S J 1995 *A&AS* **113** 379

[7] Chaplin W J, Elsworth Y, Howe R, Isaak G R, McLeod C P, Miller B A, van der Raay H B, Wheeler S J and New R 1996 *Sol. Phys.* **168** 1–18

[8] Gabriel A H, Grec G, Charra J, Robillot J, Roca Cortés T, Turck-Chièze S, Bocchia R, Boumier P, Cantin M, Cespédes E, Cougrand B, Crétolle J, Damé L, Decaudin M, Delache P, Denis N, Duc R, Dzitko H, Fossat E, Fourmond J, García R A, Gough D, Grivel C, Herreros J M, Lagardère H, Moalic J, Pallé P L, Pétrou N, Sanchez M, Ulrich R and van der Raay H B 1995 *Sol. Phys.* **162** 61–99

[9] Jiménez-Reyes S J, García R A, Jiménez A and Chaplin W J 2003 *ApJ* **595** 446–457

[10] García R A, Turck-Chièze S, Boumier P, Robillot J M, Bertello L, Charra J, Dzitko H, Gabriel A H, Jiménez-Reyes S J, Pallé P L, Renaud C, Roca Cortés T and Ulrich R K 2005 *A&A* **442** 385–395

[11] Fröhlich C, Romero J, Roth H, Wehrli C, Andersen B N, Appourchaux T, Domingo V, Telljohann U, Berthomieu G, Delache P, Provost J, Toutain T, Crommelynck D A, Chevalier A, Fichot A, Däppen W, Gough D, Hoeksema T, Jiménez A, Gómez M F, Herreros J M, Cortés T R, Jones A R, Pap J M and Willson R C 1995 *Sol. Phys.* **162** 101–128

[12] Woodard M F and Noyes R W 1985 *Nature* **318** 449–450

[13] Pallé P L, Régulo C and Roca Cortés T 1989 *A&A* **224** 253–258

[14] Elsworth Y, Howe R, Isaak G R, McLeod C P and New R 1990 *Nature* **345** 322–324

[15] Chaplin W J, Elsworth Y, Miller B A, Verner G A and New R 2007 *ApJ* **659** 1749–1760

[16] Jiménez-Reyes S J, Chaplin W J, Elsworth Y, García R A, Howe R, Socas Navarro H and Toutain T 2007 *ApJ* **654** 1135–1145

[17] Christensen-Dalsgaard J 2002 *Reviews of Modern Physics* **74** 1073–1129

[18] García R A, Mathur S, Salabert D, Ballot J, Régulo C, Metcalfe T S and Baglin A 2010 *Science* **329** 1032

[19] Baglin A, Auvergne M, Barge P, Deleuil M, Catala C, Michel E, Weiss W and The COROT Team 2006 *ESA Special Publication* (*ESA Special Publication* vol 1306) ed M Fridlund, A Baglin, J Lochard, & L Conroy p 33

[20] Borucki W, Koch D, Batalha N, Caldwell D, Christensen-Dalsgaard J, Cochran W D, Dunham E, Gautier T N, Geary J, Gilliland R, Jenkins J, Kjeldsen H, Lissauer J J and Rowe J 2009 *IAU Symposium* (*IAU Symposium* vol 253) pp 289–299

[21] Fletcher S T, Chaplin W J, Elsworth Y and New R 2009 *ApJ* **694** 144–150

[22] Broomhall A, Chaplin W J, Elsworth Y, Fletcher S T and New R 2009 *ApJL* **700** L162–L165

[23] Salabert D, García R A, Pallé P L and Jiménez-Reyes S J 2009 *A&A* **504** L1–L4

[24] Libbrecht K G and Woodard M F 1990 *Nature* **345** 779–782

[25] Christensen-Dalsgaard J and Berthomieu G 1991 *Theory of solar oscillations.* (Solar Interior and Atmosphere) pp 401–478

[26] Chaplin W J, Elsworth Y, Isaak G R, Marchenkov K I, Miller B A, New R, Pinter B and Appourchaux T 2002 *MNRAS* **336** 979–991

[27] Howe R, Christensen-Dalsgaard J, Hill F, Komm R W, Larsen R M, Schou J, Thompson M J and Toomre J 2000 *Science* **287** 2456–2460

[28] Howe R, Christensen-Dalsgaard J, Hill F, Komm R, Schou J, Thompson M J and Toomre J 2007 *Advances in Space Research* **40** 915–918

[29] Wang Y and Sheeley Jr N R 2003 *ApJ* **590** 1111–1120

Characterization of solar-cycle induced frequency shift of medium- and high-degree acoustic modes

M. Cristina Rabello-Soares
Stanford University, Stanford, CA 94305, USA
E-mail: csoares@sun.stanford.edu

Abstract.
Although it is well known that the solar acoustic mode frequency increases as the solar activity increases, the mechanism behind it is still unknown. Mode frequencies with $20 < l < 900$ obtained by applying spherical harmonic decomposition to MDI full-disk observations were used. First, the dependence of solar acoustic mode frequency with solar activity was examined and evidence of a quadratic relation was found indicating a saturation effect at high solar activity. Then, the frequency dependence of frequency differences between the activity minimum and maximum was analyzed. The frequency shift scaled by the normalized mode inertia follows a simple power law where the exponent for the p modes decreases by 37% for modes with frequency larger than 2.5 mHz.

1. Introduction and Data

It seems that the responsible mechanism for the mode frequencies change with solar activity is restricted to the outer layers of the Sun. However, at the moment, there is no general agreement in the precise physical cause. The frequency variations of medium and high-l modes with solar cycle were analyzed to help determine its physical origin. The solar-radio 10.7-cm daily flux (NGDC/NOAA) was used as the solar activity proxy.

The mode frequencies were obtained by applying spherical harmonic decomposition to MDI full-disk observations during the Dynamics and Structure observing modes. The first one has higher spatial resolution and is available every year for two or three months of continuous data. Data for 1999 to 2008 were used. The second one consists of 72-day time series from early 1996 to April 2010 (Fig. 1). They will be called the Dynamics and Structure sets from now on.

The mode frequencies were obtained using two distinct methods. The medium-l method ([5] and [7]) and the high-l method ([6]) which is applied only to the Dynamics set (Fig. 1 left panel). The second method is used when the spatial leaks of the modes overlap with the target mode making it more difficult to estimate the mode frequency.

2. Mode frequency variation with the solar activity cycle

There is a very high linear correlation of the mode frequency variation with several solar activity indices. However, deviations from a simple linear relation has been reported (see, for example, [2]). In this analysis, there is some indication of a quadratic relation:

$$\nu_q(n,l,F_r) = c_0(n,l) + c_1(n,l) \cdot F_r + c_2(n,l) \cdot F_r^2 \qquad (1)$$

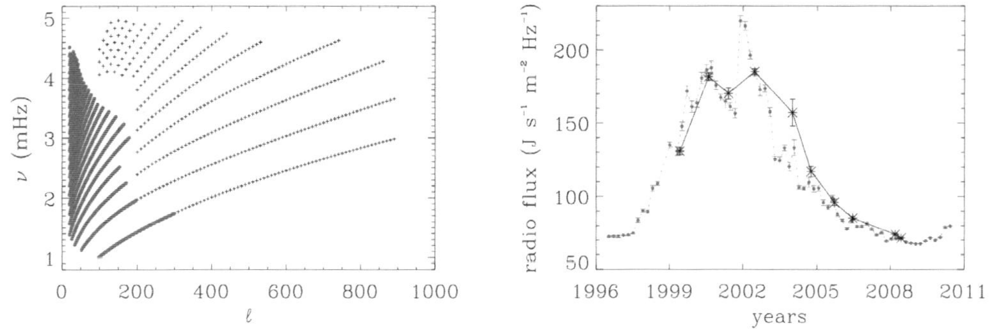

Figure 1. Right panel: Mode coverage. Modes obtained by the medium-l method are in red and by the high-l method in black. Left panel: Mean radio flux for the observational periods in the Dynamics set (in black) and in the Structure set (in red).

where F_r is the relative activity index defined as the mean solar-radio flux of a given observational period divided by the maximum mean flux of the set of observations (Fig. 2). The quadratic polynomial seems to indicate a saturation effect at high solar activity. A similar effect has been seen in frequencies at activity regions with a large surface magnetic field using ring analysis ([1]). The saturation occurs when $c_1 + 2 \cdot c_2 \cdot F_r^s = 0$, i.e.: $F_r^s = -c_1/(2 \cdot c_2)$.

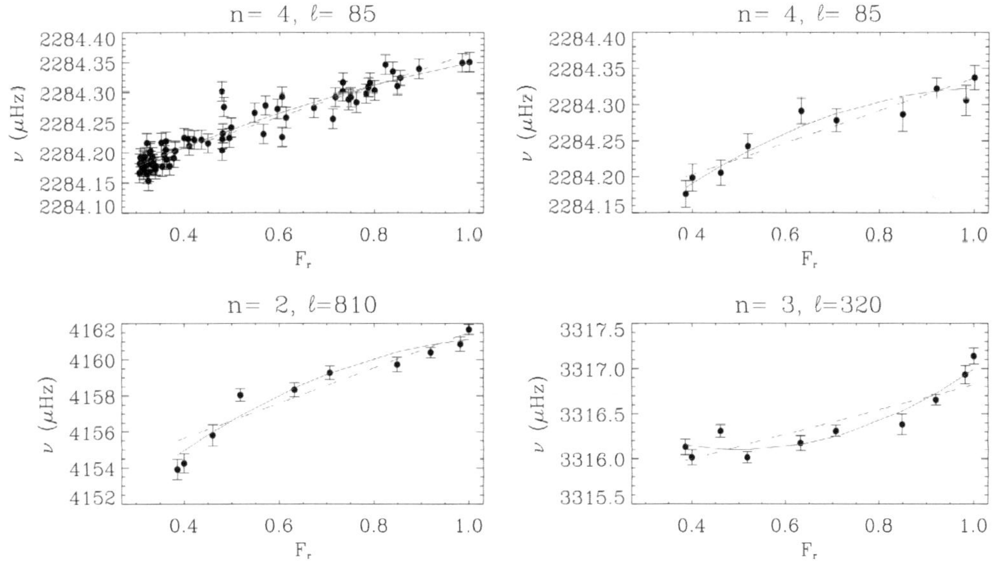

Figure 2. Examples of frequency variation with the relative activity index for different modes. The top left panel is using the Structure set and the others using the Dynamics set.

Fig. 3 shows the coefficient $c_2(n,l)$ as a function of $c_1(n,l)$. For the Structure set, the slope of the $c_2 - c_1$ linear relation is -0.3077 ± 0.0025 with a zero y-intercept (blue line in both panels of Fig. 3), indicating a saturation at $(357.2 \pm 2.9) \times 10^{-22}$ J s^{-1} m^{-2} Hz^{-1}. For the Dynamics set, the slope is -0.2813 ± 0.0083 with a very small y-intercept (-0.0169 ± 0.0052 μHz). For comparison, the linear regression was performed on the same c_1 range as the Structure set: $0 < c_1 < 4.34$ μHz. Modes obtained with both medium- and high-l methods were included in the fitting. Only 12% (133) of the modes in the Dynamics set were obtained with the high-l method. The saturation occurs at $(329.2 \pm 9.7) \times 10^{-22}$ J s^{-1} m^{-2} Hz^{-1}. This is more or less in agreement (differing by 2.8σ) with the value estimated by the Structure set. The slope increases as c_1 (and $|c_2|$) increases (Fig. 3). For $c_1 > 4.34$ μHz, the slope is -0.455 ± 0.028 and the y-intercept 0.50 ± 0.29 μHz, giving a saturation at $(204 \pm 12) \times 10^{-22}$ J s^{-1} m^{-2} Hz^{-1}.

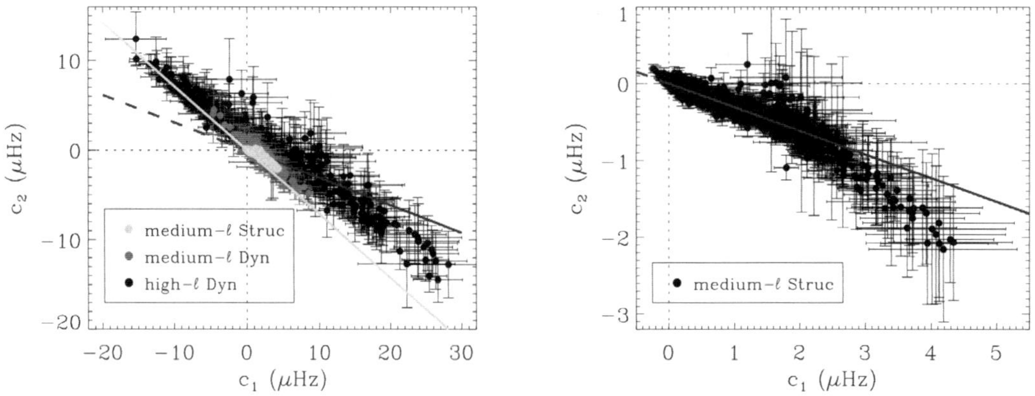

Figure 3. Left panel: Dynamics and Structure sets. Right panel: Structure set. The green circles in the left panel are the same as the black circles in the right panel. The red and black circles in the left panel are for the Dynamics set using the medium- and high-l methods respectively. The light-blue line is the mean slope of the simulations for the Dynamics set.

To test the significance of the quadratic relation in the data, the observed frequencies were fitted using a linear relation, $\nu_l(n, l, F_r) = d_0(n, l) + d_1(n, l) \cdot F_r$, and noise was added to the fitted frequencies. To each fitted frequency $\nu_l(n, l, F_r)$, the noise was generated as normally distributed random numbers with the same standard deviation as the error in the observed frequencies. For each (n, l) mode and each realization, a quadratic polynomial was fitted. Fig. 4 shows the coefficients obtained for one realization and the ones obtained for the Structure set for comparison. A straight line was fitted to $c_2(c_1)$ for each realization over the entire range of c_1 of each set. The slope averaged over 1000 realizations is -0.01437 ± 0.00011 and -0.712222 ± 0.000060 for the Structure and Dynamics set respectively. The first one is represented by the black line in Fig. 4 and the second one by the light-blue line in the left panel of Fig. 3. These slopes are very different from the ones obtained by the corresponding set indicating that the quadratic relation is not due to noise but to a real signal.

For the Dynamics set, there are several modes with a negative coefficient c_1 (Fig. 3). Nine percent (195) of the modes have $c_1 < -e_{c_1}$, where e_{c_1} is the fitting uncertainty. For the Structure set, there are two percent (27) of the modes with $c_1 < -e_{c_1}$. An example of the frequency variation with solar cycle for one of these modes is in the bottom-right panel of Fig. 2. The frequency of this mode is approximately constant for $F_r < 0.8$. To see if this is true for all the modes with a negative c_1, the mean relative frequency at a given time t is defined as:

$$\langle R_\nu(t) \rangle = \frac{1}{N} \sum_{n,l} \frac{\nu_{n,l}[t] - \nu_{n,l}[min(F_r)]}{\nu_{n,l}[max(F_r)] - \nu_{n,l}[min(F_r)]}, \tag{2}$$

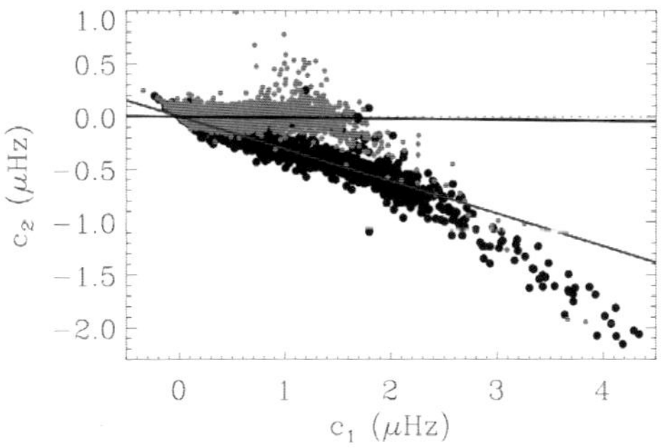

Figure 4. Coefficient c_2 versus c_1 for one realization (red circles) and from the observations using the Structure set (black). The black circles are the same as the black circles in the right panel of Fig. 3. The black and blue lines are the linear fit to the simulation and to observations respectively. The blue line is the same as the blue line in Fig. 3.

where N is the number of modes. $\langle R_\nu(t)\rangle$ was estimated over all modes with $c_1 > e_{c_1}$ and $c_1 < -e_{c_1}$ given by the black and red circles in Fig. 5 respectively. For the Structure set, $N = 1573$ for $c_1 > e_{c_1}$ and $N = 27$ for $c_1 < -e_{c_1}$. For the Dynamics set, $N = 1272$ for $c_1 > e_{c_1}$ and $N = 60$ for $c_1 < -e_{c_1}$. Only modes with a Spearman's correlation coefficient larger than 0.75 were used in the average of the Dynamics set to decrease the scatter. Modes with a negative c_1 seem to be less sensitive to the solar cycle, more so away from the solar maximum. Most of these modes have a small ν/L, between 3 and 25 μHz. In the Dynamics set, most of them that are p modes have $3 < \nu < 3.6$ mHz and $l > 200$ while the f modes have $l > 600$. In the Structure set, most that are p modes have $\nu < 1.6$ mHz and $l < 170$ while the f modes have $100 < l < 140$. It is not clear why these modes are less affected by the solar cycle. If this is proved to be true and not an artifact of the data, it will help in the understanding of the mechanism or mechanisms responsible for the mode frequency variation with solar activity.

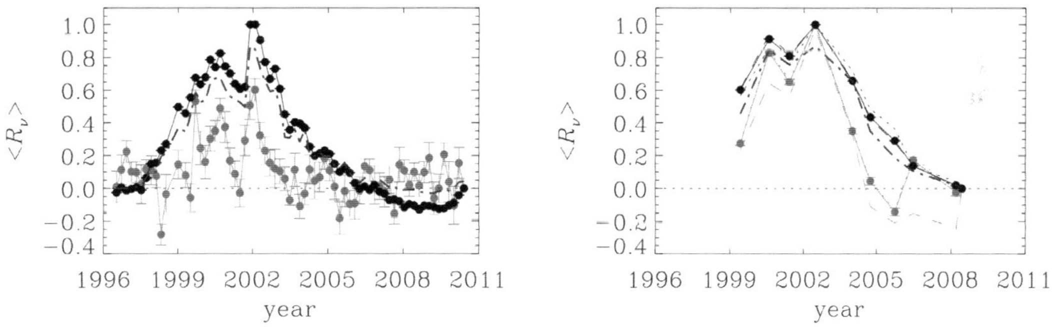

Figure 5. Left panel: Structure set. Right panel: Dynamics set. Modes with $c_1 > e_{c_1}$ are in black and modes with $c_1 < -e_{c_1}$ in red. The relative activity index is in blue. In the right panel, the medium- and high-l modes are given by the dashed and dotted lines respectively.

From Fig. 5 (left panel), the minimum between solar cycle 23 and 24 happened around August 2008. The frequencies were at a minimum from November 2007 to August 2009. As pointed out by other authors (see [4] and references within), the mode frequencies were smaller than the previous minimum. The frequency averaged over all medium-l modes is $\sim 10\%$ of the minimum-to-maximum variation smaller than in the previous minimum: $\langle R_\nu(\text{July 1996})\rangle - \langle R_\nu(\text{August 2008})\rangle \approx 0.1$.

3. The minimum-to-maximum frequency shift

Here, the minimum-to-maximum frequency shift $\delta\nu(n,l)$ is defined as the difference between the fitted frequency given by Eq. (1) at the maximum and minimum solar activity. The minimum and maximum activity was chosen to be those of the Dynamics set, which are around April 2008 and May 2002 respectively (Fig. 1 right panel). The differences in $\delta\nu(n,l)$ using a linear regression instead of Eq. (1) are small. The standard deviation of the differences is 0.0031 μHz and 0.25 μHz for the Structure and Dynamics set respectively. Most of the difference is due to the high-l modes in the Dynamics set. The effect of these differences in the estimated α and γ coefficients described below is in their 4-th significative digit or higher.

Most of, if not all, the perturbation to the structure of the Sun associated with the solar cycle is believed to be located in a thin layer near the solar surface which leads to a frequency perturbation roughly proportional to the inverse of the mode inertia $I_{n,l}$: $\delta\nu_{n,l} \propto \nu_{n,l}^{\alpha}/I_{n,l}$. The value of α depends upon the physical mechanism responsible for affecting the mode frequencies during the solar cycle. Some of the predicted values are -1, 1 and 3 (see [3] and references within). Fig. 6 (right panel) shows $\delta\nu(n,l)$ scaled by $I_{n,l}$ for both sets. Only modes with $c_1 > 0$ were plotted. Fitting only the p modes, α varies from approximately -4 for $\nu < 1.6$ mHz, to 0. for $1.66 < \nu < 2.5$ mHz, to 1.43772 ± 0.00018 for $2.4 < \nu < 3.55$ mHz, to 6.417 ± 0.023 for $\nu > 4$ mHz. The variation of α indicates a more complex mechanism than anticipated.

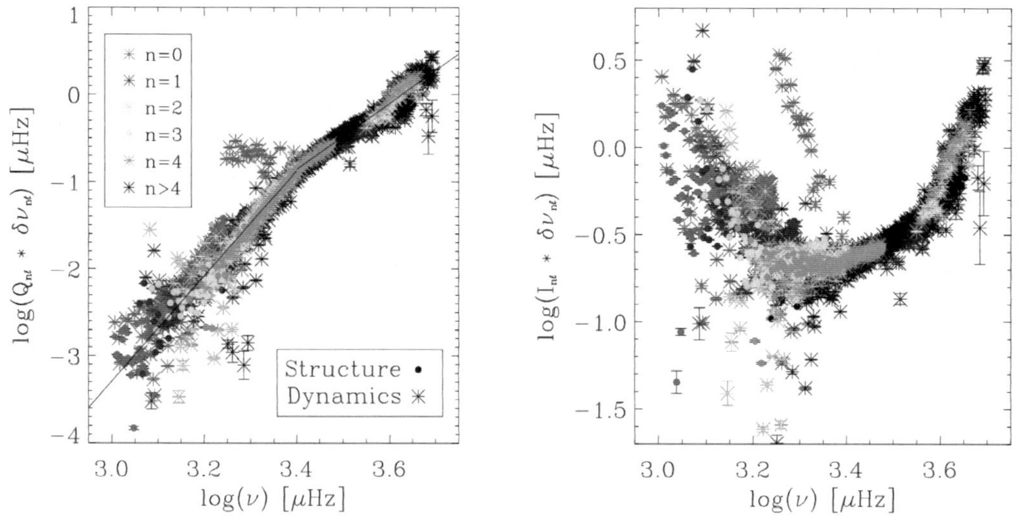

Figure 6. Only modes with $c_1 > 0$ are plotted.

Recently, [6] observed that scaling the frequency shift with the mode inertia normalized by the inertia of a radial mode of the same frequency ($Q_{n,l}$) also follows a simple power law: $\delta\nu_{n,l} \propto \nu_{n,l}^{\gamma}/Q_{n,l}$, but with a constant exponent. Extending this previous analysis to include data from the end of solar cycle 23 gives similar results (Fig. 6 left panel). However, there is a clear change in the coefficient γ around 2.5 mHz. For the p modes, $\gamma_p = 6.04466 \pm 0.00030$ for $\nu < 2.5$ mHz and $\gamma_p = 3.78657 \pm 0.00019$ for $\nu > 2.5$ mHz. In the previous analysis, it was estimated $\gamma_p = 3.60 \pm 0.01$ at all frequencies and it was argued that around 2.3 mHz there was a step in the p-mode frequency. This analysis show clearly that there in fact a change in γ_p.

The p modes obtained by the high-l method show a slightly different slope than those obtained by the medium-l method. In Fig. 6, the p modes obtained by the high-l method are concentrated at $\log[\nu(\mu\text{Hz})] > 3.55$ where they account for 70% of the modes (including the low n modes indicated by the colored stars). Although an artifact from the analysis can not be discarded due to the difficulty in estimating unbiased mode frequencies at high-l and/or high-frequency, it could be an indication of a degree dependence in the scaled frequency shift.

It is expected that the f modes are affected by the solar cycle in a different way than the p modes since they have very different properties. Fig. 7 shows the f-mode frequency shift. The high-l f modes in the Dynamics set has a distinctive behavior (red stars in Fig. 6 with $3.25 < \log[\nu(\mu\text{Hz})] < 3.4$). Fitting only the modes obtained by the medium-l method, $\gamma_f = 7.0880 \pm 0.0019$ which is 17% larger than γ_p at the corresponding frequency range.

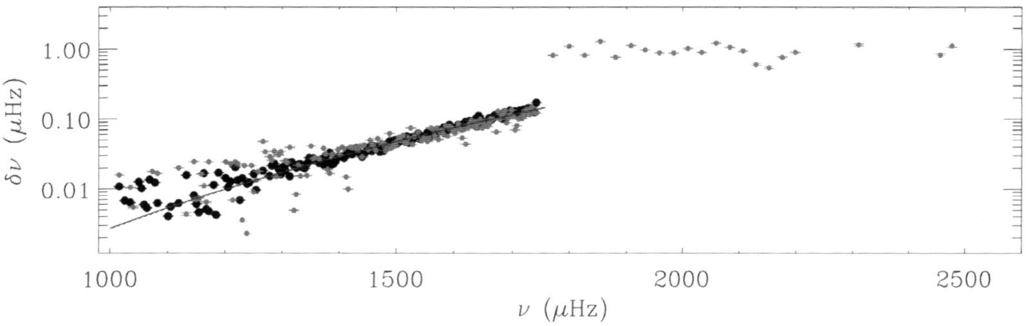

Figure 7. Only f modes. In red are the Dynamics set and in black the Structure set. The high-l modes are those with $\nu > 1760$ μHz. The blue line is for $\gamma_f = 7.0880 \pm 0.0019$.

4. Summary

There is some evidence of a quadratic relation of the frequency shifts with the solar-radio flux indicating a saturation effect at high solar activity. The saturation occurs at $\sim 350 \times 10^{-22}$ J s^{-1} m^{-2} Hz^{-1} and decreases as c_1 increases.

The minimum-to-maximum frequency shift scaled by the mode inertia is proportional to ν^{α} where α_p increases from -4 at low frequency to 6.4 at high frequency for the p modes. Scaling by the normalized mode inertia, there is a sharp change in the p-mode exponent at 2.5 mHz.

Acknowledgments

The frequencies obtained by the medium-l method were calculated by J. Schou and T. Larson.

References

[1] Basu, S., Antia, H. M. and Bogart, R. S. 2004, *ApJ*, **610**, 1157
[2] Chaplin, W.J., Elsworth, Y., Miller, B.A. and Verner, G.A. 2007, *ApJ*, **659**, 1749
[3] Gough, D.O. 1990, *Lecture Notes in Physics*, **367**, 283
[4] Jain, K. et al. 2010, *ASP Conf. Ser*, **428**, 57
[5] Larson, T. and Schou, J. 2009, *ASP Conf. Ser.*, **416**, 311
[6] Rabello-Soares, M. C., Korzennik, S. G. & Schou, J. 2008, *Solar Phys.*, **251**, 197
[7] Schou, J. 1999, *ApJ*, **523**, L181

Automated region detection and sunspot analysis using GONG imagery

Owen Giersch

Learmonth Solar Observatory, Exmouth, Western Australia
Centre for International Radio Astronomy, Bentley, Western Australia

owen.giersch@postgrad.curtin.edu.au

Abstract. I describe the methods for automatically reporting magnetic regions from GONG magnetograms and sunspot area and extent from GONG white light images. The magnetic region detection involves the sun being gridded into small boxes and finding adjacent boxes that have significant field. The sunspot analysis uses gradient changes in the cumulative histogram to find dark pixels in a region. This is a variation on the method described by Pettauer and Brandt[1]. Both of the methods are reasonably simple to implement and have produced good results for regions in the early part of cycle 24. Automated Sunspot areas were plotted against USAF/NOAA daily sunspot areas for the whole disk and produced a correlation co-efficient of 0.89. Future work with GONG H-alpha imagery will be described briefly.

1. Data

The data used in this project was from the GONG[2] site at Learmonth, Western Australia. White light and magnetogram images are acquired every hour from the GONG instrument between sunrise and sunset, giving between 8 and 12 images per day depending on time of year and weather. The data in this report spans from September 2009 to May 2010. Sunspot numbers were obtained from the SWPC website in the form of the Daily Activity Reports supplied by USAF/NOAA[3].

2. Magnetic Region Detection

Each magnetogram is processed in order to find active regions. There are five steps involved in this process:

1. Calculate the standard deviation for the field strength of pixels whose absolute field is less than 100 Gauss and within 0.95 solar radii (Rv).
2. Grid the magnetogram into 20 by 20 pixel boxes.
3. Find boxes that contain pixels with absolute field strength greater than 12 standard deviations. This is typically between 80 and 120 Gauss.
4. Combine all adjacent boxes and find the total width and height of the combined boxes.
5. Adjust the box so that the magnetic field is in the centre of the box.

After finding each region, the algorithm checks that there is at least 20 millionths of the solar hemisphere (20 μh) of coverage of significant field and that the region has both positive and negative field. This means that some small regions will be missed. The box parameters are stored in a text file:

Table 1. Region file for regions found on the 17 December 2010.

```
MJD:03675.539074074011296
DATE:2010/01/24 00:56:16
RGN ID      LAT    LON    WIDTH HEIGHT
L10012300 -24.79 -33.93 0304  0162
L10012301 +20.82 +42.15 0132  0112
```

The latitude and longitude are in degrees, and the box height and widths are in seconds of arc (Solar Diameter is ~1900 arcsec).

In addition, once per day magnetic parameters for each region are obtained. This is usually done near 0100 UT, weather permitting.

Table 2. Magnetic Parameters for the two regions found on the 24 January 2010.

```
MJD:03675.539074074011296

          MAX    MAX    MAX   MIN    MIN    MIN   TOTAL               GRAD
RGN       LAT    LNG    GAUSS LAT    LNG    GAUSS FLUX      GRADIENT  DIR
L10012300 -25.24 -31.33 +0747 -25.44 -39.28 -0566 163542937 -00122.38 89
L10012301 +19.78 +41.38 +0292 +19.86 +44.88 -0588 061861396 -00182.96 89.
```

Max Lat and Max Lng defines the location of the maximum field strength. Max Gauss is the maximum field strength. Min Lat, Min Lng and Min Gauss defines the position and field strength for the minimum field. Total flux is the total absolute flux in the region in units of Gauss-μh. The Gradient is the field gradient between the max and min Gauss values in units of Gauss/degree. The Grad Dir is the orientation from north of the line between the max and min Gauss Values.

The values for the positions and the sizes of the regions are then used to do an automated sunspot analysis once per day (usually around 0100 UT).

3. Limb Correction

Before a region can be analysed for sunspots, the white light image needs to be corrected for limb darkening. The pixels in the white light image are 8 bit (0 to 255). The average brightness of pixels at 20 different radii are obtained and a cubic spline fit is made to find brightness as a function of radius (Figure 1). This is normalised and inverted. Each pixel is multiplied by this function to obtain a limb corrected image.

4. Sunspot Detection

Once the image has been limb corrected, a cumulative histogram is generated for each region. In order for a spot analysis to be performed on a region, the histogram for that region must span more than 55 intensity bins. This avoids small gradient variations in the region being detected as sunspots. A sunspot threshold is found by finding where the slope of the histogram changes by an average of 10 counts/bin over four bins (Figure 2). Only pixels within 0.95 Solar Radii are included in the sunspot search to avoid problems near the limb. This does distort sunspot area values for regions near the limb. Figure 3a shows a limb corrected sunspot region and figure 3b shows the spot detected areas for the same region. The program outputs the sunspot area, extent and position for each sunspot region. Table 3 shows sample sunspot analysis output.

Table 3. Output of automated spots analysis for 24 January 2010.

```
MJD:03675.539074074011296
DATE:2010/01/24 00:56:16
REGION    LAT LNG EWEXT NSEXT AREA
L10012300 -26 -35  010   002   00230
L10012301 +21 +43  006   004   00214
```

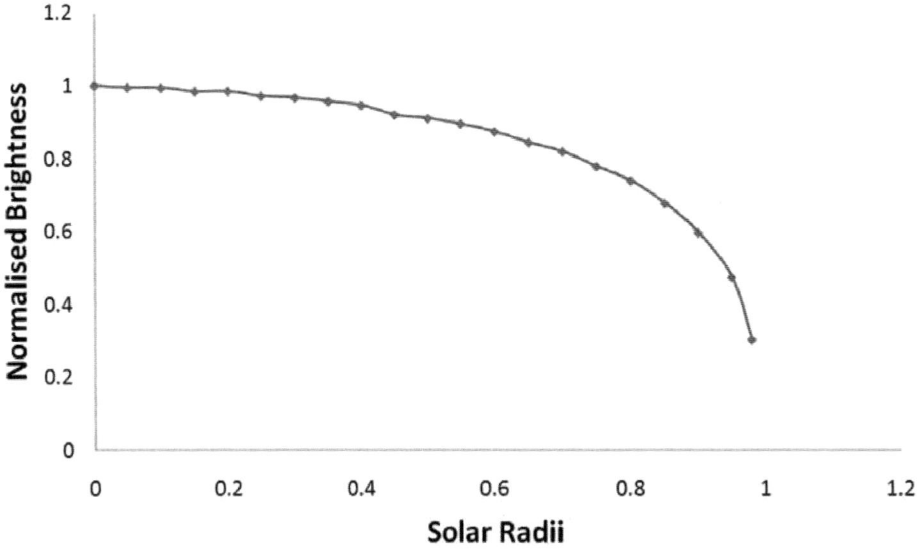

Figure 1. Normalised Brightness vs Solar Radii. The brightness curve for a white light image on 24 January 2010.

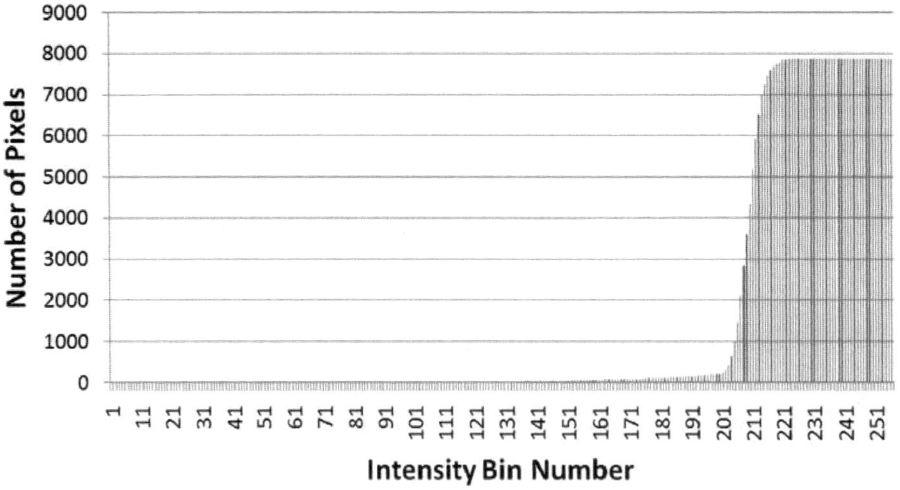

Figure 2. Cumulative Histogram for Region L10012300 on 24 January 2010. The threshold for sunspot detection is at bin 200.

(a) 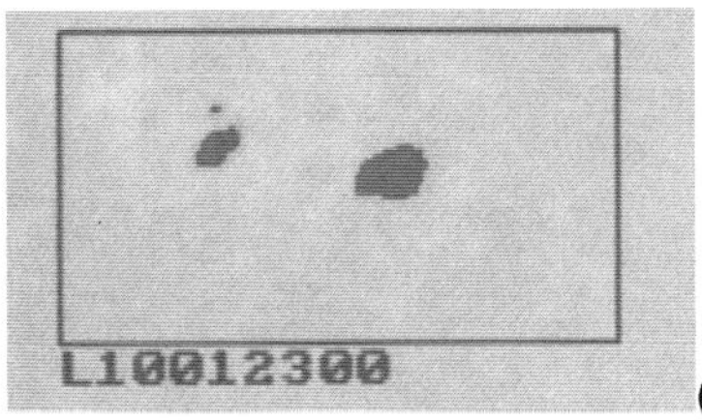

(b)

Figure 3. A sunspot region on 2010/01/24. (a) is the limb corrected region and (b) shows the automated detected area.

4. Analysis

The sunspot areas calculated by the automated analysis were compared to those reported by NOAA/AFWA. One hundred individual regions were compared from September 2009 to April 2010. Figure 4 summarises these results.

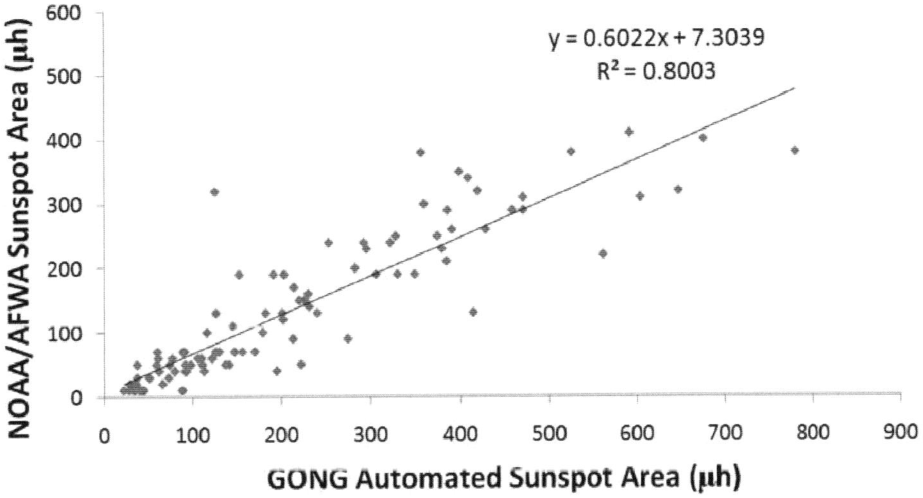

Figure 4. NOAA/AFWA Sunspot Area vs GONG Automated Sunspot Area

There is a strong co-relation between the NOAA/AFWA sunspot area and the automated sunspot area. However, the automated area is 60% higher than the NOAA/USAF area. Visual inspection of the areas seems to confirm that the automated procedure works well except for those regions very close to the limb. These results are similar to those reported by Pettauer and Brandt[2].

5. Future Work

The introduction of GONG H-alpha provides a good opportunity for automated flare and solar feature detection. Combined with sunspot and magnetic region information, this should lead to a consistent set of solar parameters to aid in space weather prediction.

References

1. Pettauer T and PN Brandt, *On Novel Methods to Determine Areas of Sunspots from Photoheliograms*, Solar Physics, 175, 197-203 (1997)
2. Global Oscillation Network Group (http://gong.nso.edu)
3. Space Weather Prediction Center Website (http://www.swpc.noaa.gov)

The far-side solar magnetic index

Irene González Hernández[1], Kiran Jain[1], W. Kent Tobiska[2] and Frank Hill[1]

[1]National Solar Observatory, Tucson, AZ, 857619, USA
[2]Space Environment Technologies (Space Weather Division), Pacific Palisades, CA, 90272, USA

E-mail: `irenegh@nso.edu`

Abstract. Several magnetic indices are used to model the solar irradiance and ultimately to forecast it. However, the observation of such indices are generally limited to the Earth-facing hemisphere of the Sun. Seismic maps of the far side of the Sun have proven their capability to locate and track medium-large active regions at the non-visible hemisphere. We present here the possibility of using the average signal from these seismic far-side maps as a proxy to the non-visible solar activity which can complement the current front-side solar activity indices.

1. Introduction

Photospheric features of solar activity account for a large portion of the Total Solar Irradiance (TSI) variation, with a superimposed modulation due to the solar cycle [1]. Magnetic indices related to photospheric activity, such as the Mount Wilson Plage Strength Index (MPSI) and the Mount Wilson Sunspot Index (MWSI), as well as magnetic activity proxies such as F_{10}, Mg II and Ly-α have been traditionally used as input to the modeling and prediction tools for TSI [2,3] and ultraviolet and extreme ultraviolet (UV/EUV) irradiance [4,5]. However, at each particular point in time, these indices contain information of the magnetic activity only for the solar hemisphere facing the Earth.

In this paper, we introduce the idea of using a Far-Side Magnetic Index (FSMI) to complement the front side ones. The FSMI is an integrated value over the full non-visible hemisphere calculated by averaging the seismic signal (phase-shift or travel time difference) produced by surface magnetic activity, namely plages and sunspots, on the waves. That is, a proxy to the photospheric magnetic activity in the far-side of the Sun.

2. Helioseismic far-side maps

Maps of the non-visible disk (far side) of the Sun calculated using the seismic holography technique [6,7] are able to detect medium-large active regions before they rotate onto the visible hemisphere. The Global Oscillations Network Group (GONG) project has been producing daily far-side maps of magnetic activity since 2005 that are available at http://gong.nso.edu/data/farside. In the last few years, new research has been incorporated resulting in an improved signal to noise in the far-side maps as well as the calibration of the signal in terms of magnetic field strength. Far-side maps showing a particular active region on the 5 and 9 of September 2010 are presented in Figure 1. These maps contain a single candidate, automatically highlighted by the process when the probability of appearance [8] is greater than

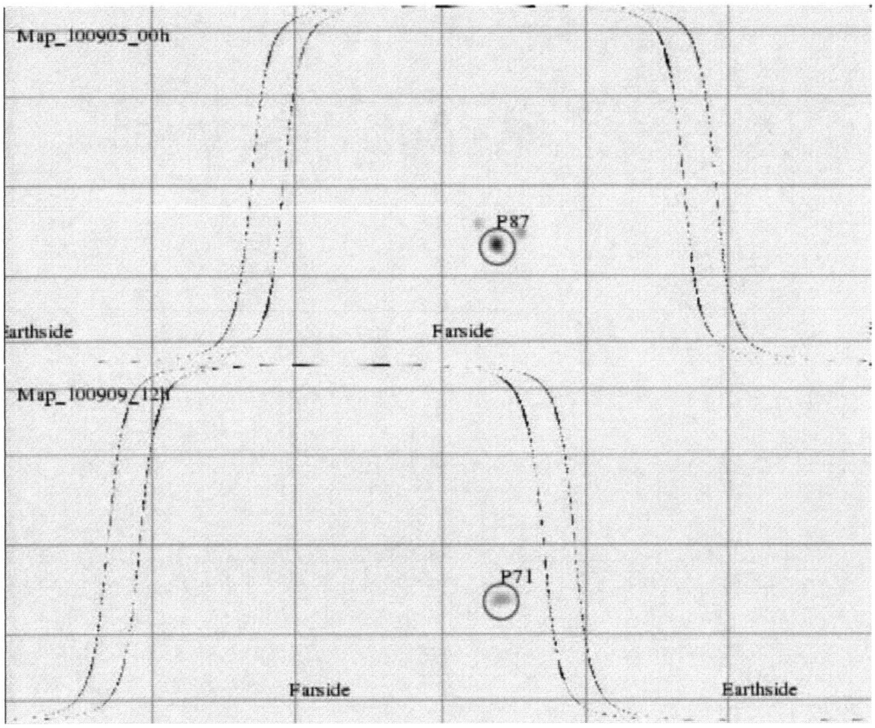

Figure 1. GONG far-side maps showing a single active region in the same heliographic coordinates on September 5 and 9 2010. Active region NOAA 11106 appeared on the East limb of the Sun on September 11 2010 at the same location.

70%. Active region NOAA 11106 appeared on the East limb of the Sun on September 11 2010 at the predicted location.

3. The far-side magnetic index (FSMI)

An integrated index of Far-Side Magnetic Activity has been calculated as an average of each individual helioseismic far-side map. We have arbitrarely ignored the positive phase shift in the maps and integrated only over the negative values, associated with magnetic features. Figure 2 shows the temporal variation for the period covering October 2002 to December 2006, the descending phase of solar cycle 23. The FSMI is shown as a running mean of 10 days to decrease the noise. A large increase in the FSMI can be seen associated to the large active regions that produced the Halloween flares in 2003.

The seismic far-side maps have been shown to be affected by a solar cycle variation, similar to the superimposed cycle modulation found in TSI measurements. In the case of the seismic maps, this cycle modulation has been associated to either global or localized structural changes in the solar convection zone [9]. To remove this variation each individual map is corrected by removing a 60-days average.

Figure 3 shows simultaneous observations of Ly-α (lasp.colorado.edu) and F_{10} (www.ngdc.noaa.gov), the calculated Mount Wilson Sunspot Index (www.astro.ucla.edu) and the Far Side Magnetic Index. The sequence spans from October 2002 to December 2006. The different indices have been scaled and shifted to aid the comparison.

The 27-day modulation due to the solar rotation is seen in both the front-side indices and the

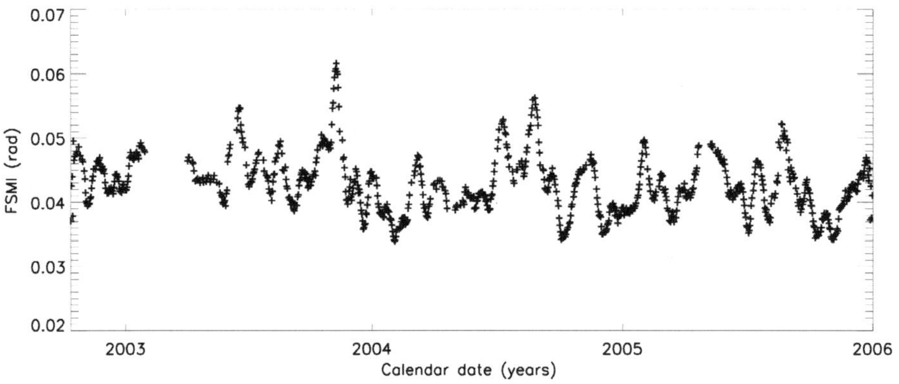

Figure 2. FSMI from October 2002 to December 2006. The far-side calculated index is shown as a running mean of 10 days to decrease the noise. There is a clear spike associated to the October-November 2003 large active regions that produced the Halloween flares.

far-side one. Two particular scenarios have been highlighted (A and B) for which the far-side index shows an increase after (A) and before (B) the front-side indices reflect it. This time lag between the front-side indices and the farside one is expected, as the active regions move from the front to the far side of the Sun and vice-versa.

4. Towards a global solar magnetic index

Since active regions emerge and decay in a semi-random manner on both the front and far side of the Sun, a correlation between the two data sets to infer a time lag and ultimately calibrate the index is not trivial, since it will be positive for certain periods, negative for others, and a complex mix for periods of high activity. In general, we expect to find a larger correlation between the FSMI and those indices that are proxies to large magnetic areas in the photosphere and lower chromoshere. To be able to compare these indices directly with the seismic signal due to photospheric activity, we need to calculate farside-like maps of the front side of the Sun, that is, use the same technique to map active regions in the front-side. That way, the integrated indices can be obtained simultaneously.

Although the seismic holography technique maps magnetic fields at the photosphere, a direct comparison of the far-side seismic maps with the chromospheric images (HeII 30.4 nm) of the EUVI instrument (STEREO-SECCHI), once the AHEAD and BEHIND spacecrafts cover the full far-side of the Sun would be interesting to complement the analysis. Future work also includes the calibration of the FSMI in terms of Ly-α, by comparing a front-side hemisphere seismic index with Ly-α simultaneous observations.

It has been shown that short term forecasting of UV irradiance can be improved when adding far-side seismic maps [10]. We expect that the integrated Far Side Magnetic Index will also help to complement the front-side indices for more accurate forecasting of solar irradiance.

4.1. Acknowledgments

The authors thank C. Lindsey and D. Braun for their extensive contribution to the GONG far-side pipeline. This work utilizes data obtained by the Global Oscillation Network Group (GONG) program, managed by the National Solar Observatory, which is operated by AURA, Inc. under a cooperative agreement with the National Science Foundation. The data were acquired by instruments operated by the Big Bear Solar Observatory, High Altitude Observatory,

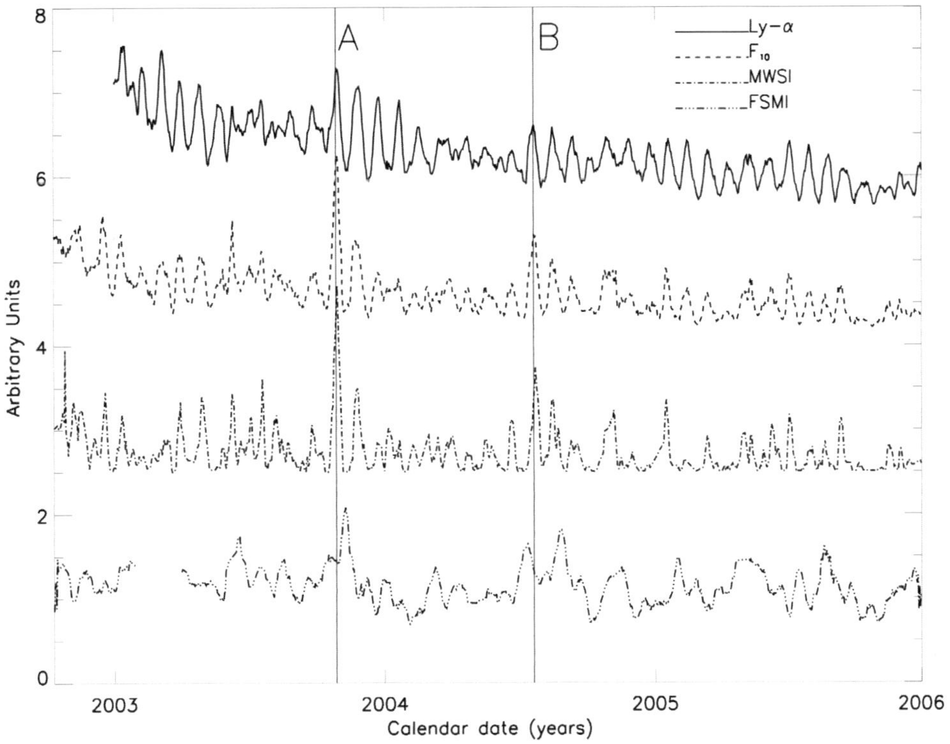

Figure 3. Simultaneous observations of Ly-α (lasp.colorado.edu), F_{10} (www.ngdc.noaa.gov), the calculated Mount Wilson Sunspot Index (www.astro.ucla.edu) and the Far Side Magnetic Index. The sequence spans from October 2002 to December 2006. The different indices have been scaled and shifted in the Y axis to aid the comparison.

Learmonth Solar Observatory, Udaipur Solar Observatory, Instituto de Astrofísica de Canarias, and Cerro Tololo Interamerican Observatory. SoHO is s project of international collaboration between ESA and NASA. This work has been supported by the NASA Living with a Star - Targeted Research and Technology program.

[1] Frohlich C. 1993 *AdSpR* **13** 429
[2] Jain K. and Hasan S. S. 2004 *A&A* **425** 301
[3] Tobiska W. K. 2002 *AdSpR* **29** 1969
[4] Dudok de Wit T., Kretzschmar M., Aboudarham J., Amblard P.-O., Auchre F. and Lilensten J. 2008 *AdSpR* **42** 903
[5] Viereck R., Puga L., McMullin D., Judge D., Weber, M. amd Tobiska W. K. 2001 *GeoRL* **28** 1343
[6] Lindsey C. and Braun D. C. 2000 *Solar Phys* **192** 261
[7] Braun D. C. and Lindsey C. 2001 *ApJ* **560** L189
[8] González Hernández I., Hill F. and Lindsey C. 2007 *ApJ* **669** 1382
[9] González Hernández I., Scherrer, P. and Hill F. 2009 *ApJ* **691** L87
[10] Fontenla J. M., Quémerais E., González Hernández I., Lindsey C. and Haberreiter M. 2009 *AdSpR* **44** 457

Temporal changes in the frequencies and widths of the solar p-mode oscillations

E J Rhodes, Jr [1,2], J Reiter [3], J Schou [4], T Larson [4], P Scherrer [4], J Brooks [1], P McFaddin [1], B Miller [1], J Rodriguez [1], J Yoo [1]

[1] Department of Physics & Astronomy, University of Southern California, Los Angeles, CA 90089-1342, U.S.A.

[2] Space Physics Research Element, Jet Propulsion Laboratory, Pasadena, CA 91101, U.S.A.

[3] Zentrum Mathematik, Technische Universität München, D-80333 Munich, Germany

[4] HEPL 4085, Department of Physics, Stanford University, Stanford, CA 94305-4085, U.S.A.

E-mail: erhodes@usc.edu

Abstract. We present a study of the temporal changes in the sensitivities of the frequencies and widths of the solar p-mode oscillations to corresponding changes in the levels of solar activity during Solar Cycle 23. From MDI and GONG++ full-disk Dopplergram three-day time series obtained between 1996 and 2008 we have computed a total of 221 sets of m-averaged power spectra for spherical harmonic degrees ranging up to 1000. We have then fit these 221 sets of m-averaged power spectra using our WMLTP fitting code and both symmetric Lorentzian profiles for the peaks as well as the asymmetric profile of Nigam and Kosovichev to obtain 442 tables of p-mode parameters. We then inter-compared these 442 tables which comprise in excess of 5.3 million p-mode parameters, and we performed linear regression analyses of the differences in p-mode frequencies and widths as functions of the differences in as many as ten different solar activity indices. From these linear regression analyses we have discovered new signatures of the frequency shifts of the p-modes and a similar, but slightly different, signature of the temporal shifts in the widths of the oscillations.

1. Introduction

At the present time, there is still disagreement over the nature of the physical mechanism which causes the frequencies of the solar p-mode oscillations to change during each solar cycle. Our main motivation for this study was our desire to provide more-detailed evidence of the frequency dependence of the responses of the oscillation frequencies to changes in solar activity than has been available in the past. Our hope has been that we could provide additional constraints on the possible mechanism that causes these frequency shifts.

2. Results

From MDI and GONG++ full-disk Dopplergram three-day time series obtained between 1996 and 2008 we have computed a total of 221 sets of m-averaged power spectra for spherical harmonic degrees ranging up to 1000. We have then fit these 221 sets of m-averaged power spectra (193 sets from MDI observations and 28 sets from GONG++ observations) using our WMLTP fitting code (Rhodes et al., 2001; Reiter et al., 2002) and both symmetric Lorentzian profiles for the peaks as well as the asymmetric profile of Nigam and Kosovichev (1998) to obtain

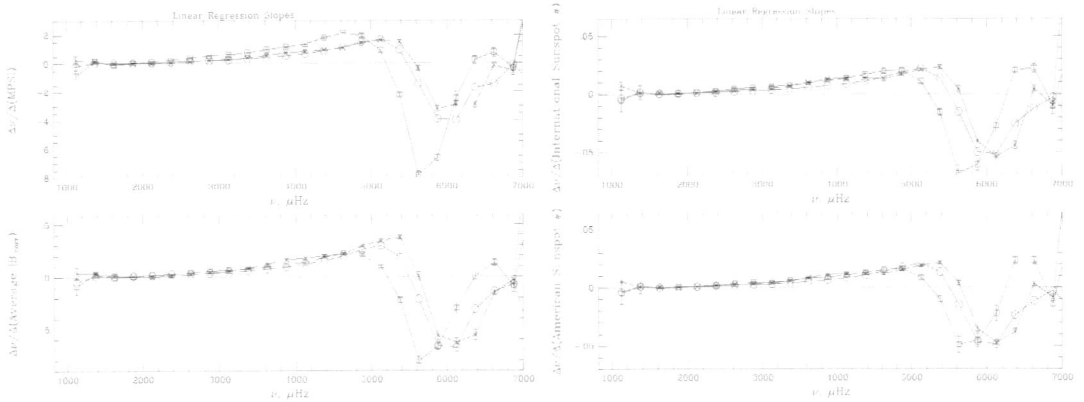

Figure 1. a.(top-left) The frequency dependence of the slopes of 75 separate linear regression analyses of symmetric-fit p-mode frequency shifts (which were binned in 250 μHz-wide bins) upon differences in Magnetic Plage Strength Index, MPSI. The late-2001 GONG+ (solar maximum) results are shown as the Xs, while the 1996 MDI (solar minimum) results are shown as the squares, and the early-2001 MDI results are shown as the circles. The average level of activity during the 2001 MDI time interval was slightly below that of the GONG++ time interval. The three $\pm$ zero-crossing frequencies can be seen to have increased systematically with increasing solar activity. Similarly, the three $\mp$ zero-crossing frequencies also increased systematically between 1996 and late 2001, as can be seen at the far right side of the panel. b.(lower-left) The frequency dependence of the 75 sets of linear regression slopes for which changes in the average of the unsigned magnitude of the corrected photospheric magnetic field, $|B|_{\rm corr}$, served as the independent variable. c.(top-right) Same as in panels a and b except that here differences in the three-day averages of the International Sunspot Number were used as the independent variable in the regression analyses. d.(bottom-right) In this case, differences in the three-day averages of the American Sunspot Number were used as the independent variable.

442 tables of p-mode parameters. We then inter-compared these 442 tables which comprise in excess of 5.3 million p-mode parameters, and we performed linear regression analyses of the differences in p-mode frequencies, widths, amplitudes, and asymmetries as functions of the differences in as many as ten different solar activity indices. Examples of the frequency dependence of some of the slopes of these linear regression comparisons are shown here in Figure 1. In Figure 1a, we show the frequency dependence of the 25 different slopes which we obtained from the linear regression analyses of the GONG++ frequency shifts upon the differences in the Magnetic Plage Strength Indicator (MPSI) as the Xs, while the 25 similar MDI regression slopes from early-2001 are shown as the circles, and 25 additional regression slopes computed from MDI observations obtained during 1996 are shown as the squares. In this panel we can see that the regression slopes did in fact change sign at three different frequencies in the case of the GONG++ and the two sets of MDI analyses. Based upon these results, we have defined the frequency at which the sign of the regression slopes change from being positive to being negative as the $\pm$ zero-crossing frequency. Figure 1a also shows that both sets of regression slopes later changed sign again from negative to positive at much higher frequencies. We refer to the locations of these changes in sign as the $\mp$ zero-crossing frequencies. In fact, both the $\pm$ and the $\mp$ zero-crossing frequencies were shifted toward higher values for the GONG++ observations from solar maximum in comparison with the zero-crossing frequencies of the MDI observations from solar minimum conditions. We believe that the most intriguing new feature that can be seen in Figure 1a is the obvious displacement of the two 2001 curves (i.e., the Xs and the circles) toward the right side of the panel. The location where the frequency

shifts change sign from positive to negative was moved to a higher frequency when the level of solar activity increased from the minimum levels of 1996 to the intermediate levels of early-2001 and was shifted even higher when the level of activity was the highest in late-2001. To our knowledge, this frequency shift in the zero-crossing frequency of the temporal frequency shifts has never been seen before by any other helioseismology studies. Not only are the zero-crossing frequencies shifted toward higher frequencies with increasing activity, but the entire high-frequency portions of the GONG++ and the 2001 MDI curves are shifted toward the right side of the plot. The same behavior can be seen in Figure 1b where the independent variable used in the regression analyses was the difference in average value of the unsigned magnitude of the corrected photospheric magnetic field, $|B|_{\rm corr}$. The same behavior can also be seen in Figures 1c and 1d, where we have plotted the regression slopes which we obtained when we employed the differences in the International and American Sunspot Numbers as the independent variables, respectively.

In Figure 2a we show the dependence of both the $\pm$ and $\mp$ zero-crossing frequencies upon the average International Sunspot Number of each of the 16 time intervals. These zero-crossing frequencies were all computed using symmetric Lorentzian profiles. In Figure 2b we show the similar solar cycle dependence of both sets of zero-crossing frequencies which we computed using the asymmetric profile of Nigam and Kosovichev (1998). Both of these panels show that the frequency where the temporal frequency shifts change from being correlated to being anti-correlated with solar ativity changes with the mean level of activity. Both panels also show that the frequency at which the frequency shifts go back to being positively correlated with the activity changes also changes with the mean level of activity.

In addition to studying the sensitivity of the temporal frequency shifts to changes in solar

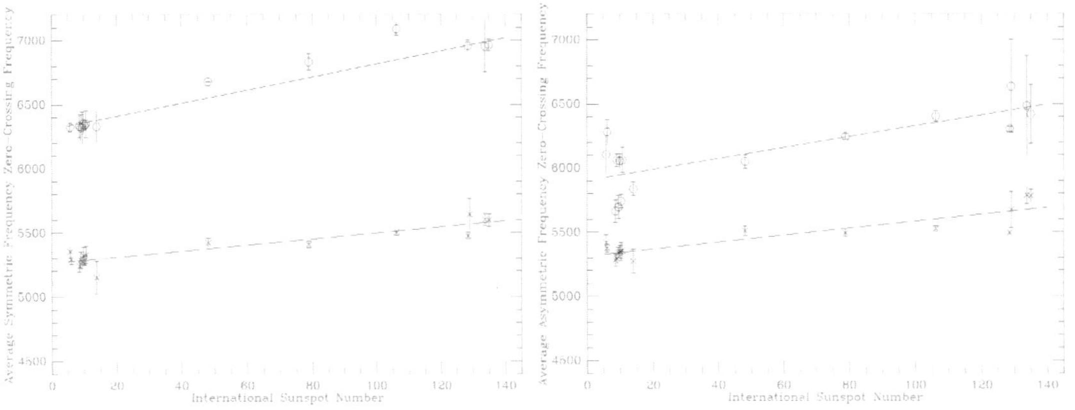

Figure 2. a.(left) Dependence of the mean $\pm$ zero-crossing frequencies (the xs) for all 16 different time intervals upon the mean value of the International Sunspot Number during each of those 16 time periods. The straight line is the linear regression fit to the 16 data points. It shows that the differences in the zero-crossing frequencies which were evident in all four panels of Figure 1 were three different examples of the solar cycle dependence of those zero-crossing frequencies. The 16 corresponding mean $\mp$ zero-crossing frequencies are shown as the 16 open circles in the upper portion of the panel. The linear regression fit to the 16 mean $\mp$ zero-crossing frequencies is shown as the upper solid line. b.(right) Similar results to those shown in the left-hand panel except that the asymmetric fitting profiles were used instead of the symmetric Lorentzian profiles. The $\pm$ zero-crossing frequencies and regression line were very similar to the results obtained using the symmetric profiles. However, the 15 $\mp$ zero-crossing frequencies computed using the asymmetric profiles were systematically lower than the corresponding points in panel a.

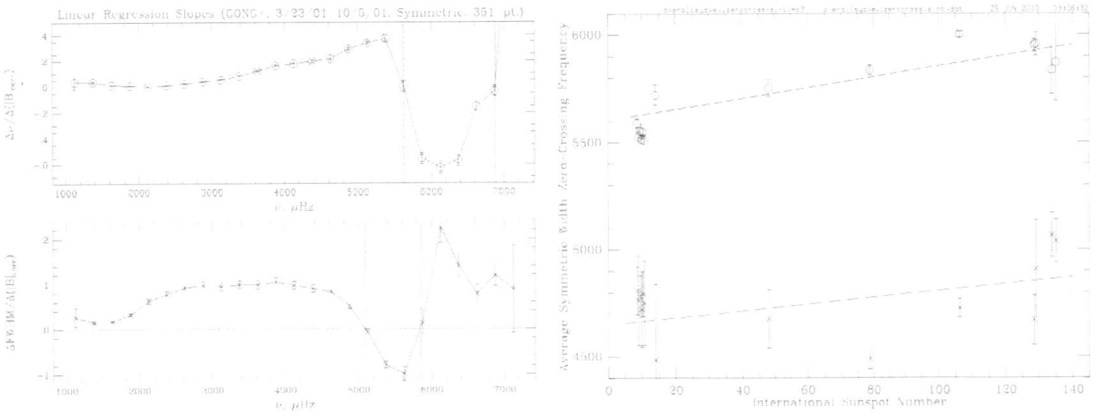

Figure 3. a.(upper-left) The frequency dependence of the 25 linear regression slopes from our study of the frequency shifts in which differences in the $|B|_{\mathrm{corr}}$ served as the independent variable. The $\pm$ zero-crossing frequency is marked by the vertical short-dashed line located near 5625 μHz, while the corresponding $\mp$ zero-crossing frequency is shown by the long-dashed line at the far right edge of the panel. b.(lower-left) Frequency dependence of the 25 linear regression slopes from our study of the width shifts using the same independent variable. The two zero-crossing frequencies are again marked with similar dashed lines near 5100 μHz and 5800 μHz, respectively. It is clear that both of these zero-crossings are located at lower frequencies than are the corresponding zero-crossings for the frequency shifts. c.(right) Dependence of both the $\pm$ and $\mp$ zero-crossing frequencies upon the mean levels of the International Sunspot Number as computed from our study of the temporal shifts of the p-mode widths. The lower solid line is the linear regression fit to the 16 $\pm$ zero-crossing frequencies. The upper solid line is the linear regression fit to the 16 $\mp$ zero-crossing frequencies. The slopes of these two regression fits are similar to, but somewhat smaller than, the slopes of similar regression fits that we obtained from our sensitivity analyses of the temporal frequency shifts.

activity, we repeated all of our regression analyses with the temporal shifts in the full-width-at-half-maxima (FWHM) taking the place of the temporal shifts in the frequencies. In Figure 3 we show the frequency dependences of the linear regression slopes that we obtained for both the temporal frequency shifts of the GONG++ observations in 2001 and for the FWHM shifts of those same observations. Specifically, we show in Figure 3a the frequency dependence of the 25 linear regression slopes from our study of the temporal frequency shifts in which differences in the $|B|_{\mathrm{corr}}$ served as the independent variable, while in Figure 3b we show the frequency dependence of the 25 linear regression slopes from our study of the width shifts using the same independent variable. In both panels we have marked the frequencies where the regression slopes change sign with dashed vertical lines. It is obvious that both the $\pm$ and the $\mp$ zero-crossing frequencies are smaller for the width shifts than they are for the frequency shifts. In Figure 3c we show the dependence of both the $\pm$ and the $\mp$ zero-crossing frequencies upon the mean levels of the International Sunspot Number as computed from our study of the temporal shifts of the p-mode widths. The two solid lines are the linear regression fits to the two sets of zero-crossing frequencies. Both lines have slopes which are similar to, although slightly smaller than, the slopes of the two similar regression lines that we showed in Figure 2 from our analysis of the p-mode frequency shifts.

3. Conclusions

We believe that the changes in the locations of the zero-crossing frequencies can be used as a new tool for probing changes in the sub-surface shear layer since some of the previous theoretical models which have been developed to explain the temporal frequency shifts have found that various combinations of temperature and magnetic field changes can result in a change in the sign of the frequency shifts at high frequencies (Jain & Roberts, 1993; Jain, 1995; Johnston et al., 1995). Hence, we believe that this new observational signature will allow estimates to be made of the changes in both temperature and field strength which occurred between 1996 and 2001. To summarize this important set of results, we have demonstrated

- that the p-mode oscillations demonstrate a complicated pattern in which their frequencies are correlated with changes in solar activity at low frequencies and then become anti-correlated with those changes in activity at intermediate frequencies before once again being positively correlated with the changes in activity at the highest frequencies;
- that the frequencies where the frequency shifts change from being correlated with changes in activity to being anti-correlated with those changes in solar activity (the so-called zero-crossing frequencies) actually change as the mean level of activity changes from month to month and from year to year;
- that the frequencies at which the frequency shifts go back to being positively correlated with the activity changes also changes with the mean level of activity;
- that the frequency dependence of the width shifts is similar to that of the frequencies, but the frequencies at which the sensitivity of the width shifts change from being positive to being negative and vice versa are lower than for the frequencies themselves.

Taken together, all of these new signatures of the temporal shifts in the p-mode frequencies and widths should give the theorists new information to include in their refined models of the causes of these temporal frequency and width shifts.

4. Acknowledgements

In this work we utilized data from the Michelson Doppler Imager (MDI) on the Solar and Heliospheric Observatory (SOHO). SOHO is a project of international cooperation between ESA and NASA. We also utilized data obtained by the Global Oscillation Network Group (GONG) project, managed by the National Solar Observatory, which is operated by AURA, Inc. under a cooperative agreement with the National Science Foundation. The data were acquired by instruments operated by the Big Bear Solar Observatory, High Altitude Observatory, Learmonth Solar Observatory, Udaipur Solar Observatory, Instituto de Astrofisico de Canarias, and Cerro Tololo Interamerican Observatory. The Stanford component of this work was supported by NASA Awards NNX07AK36G and predecessors and NAS5-02139. The portion of this research which was conducted at USC and at the Technical University of Munich was supported by the following grants to USC: NASA Grants NNX08AJ24G, NNX06AC24G, NAG5-13510, NAG5-11582, and NAG5-11001, NSF Grant AST-0307934, and by the following sub-awards from Stanford University to USC: Number 1503169-33789-A and Number 14405890-26967. Some of the computations for this work were carried out at USC using the HPCC and some of them were carried out with the use of the JPL Origin 2000 supercomputers.

5. References

[1] Jain R 1995 *ESA SP-376* 69
[2] Jain R and Roberts B 1993 *ApJ* **414** 898
[3] Johnston A, Roberts B and Wright A N 1995 *ASP Conf. Ser.* **76** 264
[4] Nigam R and Kosovichev A G 1998 *ApJ* **505** L51
[5] Rhodes E J Jr, Reiter J, Schou J, Kosovichev A G and Scherrer P H 2001 *ApJ* **561** 1127
[6] Reiter J, Rhodes E J Jr, Kosovichev A G, Schou J and Scherrer P H 2002 *ESA SP-508* 87

Global p-mode oscillations throughout the complete solar cycle 23 and the beginning of cycle 24

D Salabert[1,2], R A García[3], P L Pallé[1,2] and A Jiménez[1,2]

[1] Instituto de Astrofísica de Canarias, E-38200 La Laguna, Tenerife, Spain
[2] Departamento de Astrofísica, Universidad de La Laguna, E-38206 La Laguna, Tenerife, Spain
[3] Laboratoire AIM, CEA/DSM-CNRS, Université Paris 7 Diderot, IRFU/Sap, Centre de Saclay, F-91191 Gif-sur-Yvette, France

E-mail: `salabert@iac.es, rgarcia@cea.fr, plp@iac.es, ajm@iac.es`

Abstract. The parameters of the p-mode oscillations vary with solar activity. Such temporal variations provide insights for the study of the structural and dynamical changes occurring in the Sun's interior throughout the solar cycle. We present here a complete picture of the temporal variations of the global p-mode parameters (excitation, damping, frequency, peak asymmetry, and rotational splitting) over the entire solar cycle 23 and the beginning of cycle 24 as observed by the space-based, Sun-as-a-star helioseismic GOLF and VIRGO instruments onboard SoHO.

1. Introduction

The solar p-mode parameters have been demonstrated to vary with time and to be closely correlated with the solar activity proxies (Chaplin et al. [1]; Salabert et al. [2, 3]; Komm et al. [4]). These temporal variations provide insights to infer the interior of the Sun and its structural and dynamical changes throughout the solar cycle. However, clear differences between p-mode frequencies and solar activity during the unusually minimum of cycle 23 have been reported (Broomhall et al. [5]; Salabert et al. [6]). The origin of the p-mode variability is thus far from being properly understood and a better comprehension of its relationship with solar (and stellar, García et al. [7]) activity cycles will help us in our understanding of the dynamo processes.

2. Observations and analysis

We analyzed observations collected by the space-based, Sun-as-a-star instruments Global Oscillations at Low Frequency (GOLF, Gabriel et al. [8]) and Variability of Solar Irradiance and Gravity Oscillations (VIRGO, Fröhlich et al. [9]) onboard the *Solar and Heliospheric Observatory* (SoHO) spacecraft. GOLF measures the Doppler velocity in the D1 and D2 sodium lines (García et al. [10]). VIRGO is composed of three Sun photometers (SPM) at 402 nm (blue), 500 nm (green), and 862 nm (red). A total of 5202 and 5154 days of GOLF and VIRGO observations respectively covering more than 14 years between 1996 and 2010 were analyzed, with respective duty cycles of 95.4% and 95.2%. These datasets were split into contiguous 365-day subseries, with a four-time overlap. The power spectrum of each subseries was fitted to extract the mode parameters (Salabert et al. [11]) using a standard likelihood maximization function. Subseries with duty cycles less than 90% (basically around the SoHO vacation, 1998–1999) were removed. Each mode component was parameterized using an asymmetric Lorentzian profile,

Table 1. Changes (%) from maximum-to-minimum of the solar cycle in mode amplitudes $\langle \Delta h \rangle$, linewidths $\langle \Delta\gamma \rangle$, acoustic power $\langle \Delta p \rangle$, and energy supply rate $\langle \Delta\dot{e} \rangle$ measured from VIRGO data.

$\langle \Delta h \rangle$	$\langle \Delta\gamma \rangle$	$\langle \Delta p \rangle$	$\langle \Delta\dot{e} \rangle$
-37.9 ± 2.1	19.9 ± 1.6	-18.1 ± 1.0	1.6 ± 1.7

Table 2. Changes (%) from maximum-to-minimum of the solar cycle of the peak asymmetry $\langle \Delta b \rangle$ and their correlations with activity (using independent points only).

Instrument	$\langle \Delta b \rangle$	Correlation
GOLF blue period	-0.45 ± 0.17	-0.57
GOLF red period	-0.30 ± 0.33	-0.74
VIRGO blue	-0.69 ± 0.22	-0.55
VIRGO green	-0.52 ± 0.22	-0.50
VIRGO red	-0.44 ± 0.21	-0.38

including the $l = 4$ and 5 modes when visible. The temporal variations of the mode parameters were defined as the difference between reference values (taken as the average over 1996–1997) and the parameters of the corresponding modes observed at different dates. Their weighted averages over the central part of the 5-minute oscillation power ($\simeq$2200 – 3400 μHz) were then calculated. Mean values of daily measurements of the 10.7-cm radio flux were used as a proxy of the solar surface activity. Linear regressions were performed between the temporal variations of the mode parameters and the radio flux using independent points only. The color code in the following figures corresponds to the VIRGO blue, green, and red channels respectively.

3. Results

3.1. Mode excitation and damping

The temporal variations of the mode excitation and damping parameters from VIRGO were averaged over the $l = 0$, 1 and 2 modes. The changes in mode amplitudes $\langle \Delta h \rangle$ and linewidths $\langle \Delta\gamma \rangle$ are shown on Fig. 1. Note that due to absolute calibration problems and the changes of the observing wings, the GOLF amplitudes and linewidths are not exploitable for the moment. A proper calibration is currently underway. In order to compare with VIRGO, we performed, as described in Sec. 2, a preliminary analysis of the Global Oscillation Network Group (GONG, Harvey et al. [12]) $l = 0$ data (`gong.nso.edu/data/`). Similar fluctuations from the year 2006 in mode amplitudes (open circles on Fig. 1) and linewidths are observed as in VIRGO. The changes from maximum-to-minimum of the solar cycle are given in Table 1 and are consistent with previous work (e.g., Jiménez-Reyes et al. [13]; Salabert and Jiménez-Reyes [14]).

3.2. Frequencies of individual low-degree modes

Different temporal variations of the Sun-as-a-star p-mode frequencies are observed between individual angular degrees l = 0, 1, and 2 (Salabert et al. [6]) and are consistent between the observations from GOLF and the 3 VIRGO/SPMs (Fig. 2). They might be interpreted as different geometrical responses to the spatial distribution of the solar magnetic field beneath the surface of the Sun. Moreover, the peculiar behavior of the individual modes around the long and deep activity minimum of cycle 23 is of particular interest.

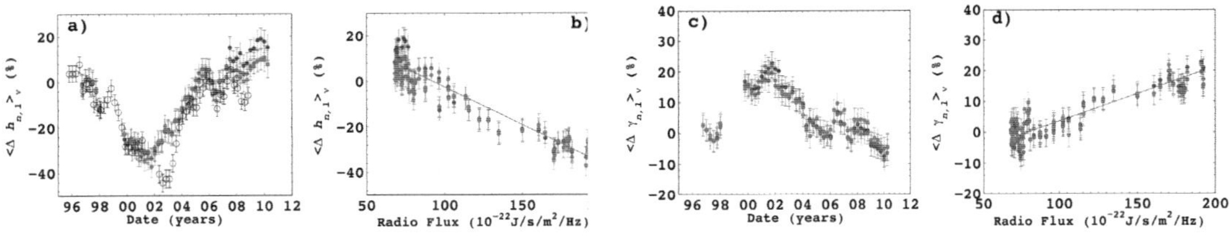

Figure 1. Temporal variations (%) in mode amplitudes $\langle \Delta h \rangle$ (a, b) and linewidths $\langle \Delta \gamma \rangle$ (c, d) measured from VIRGO data as a function of time and radio flux. The solid lines in panels b and d represent the best fits from weighted linear regressions.

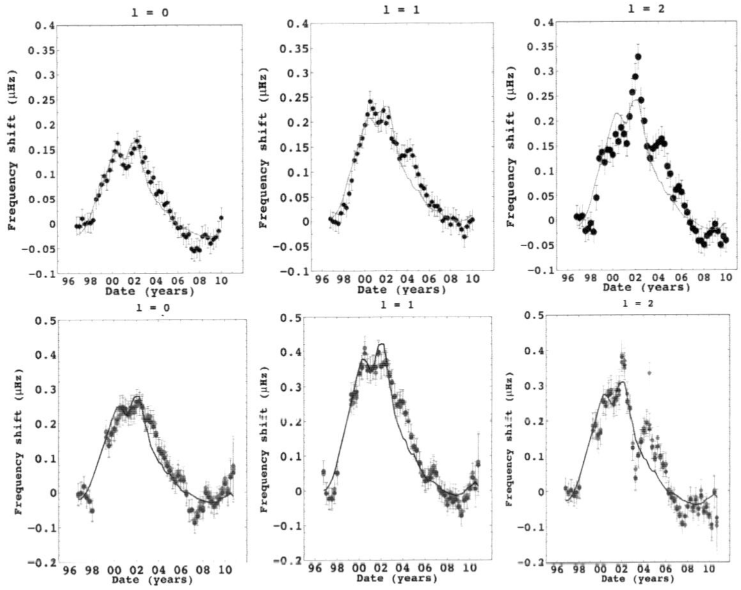

Figure 2. Temporal variations (μHz) of the $l = 0$, 1, and 2 p-mode frequencies measured from GOLF (top) and VIRGO (bottom) data. The solid lines correspond to the scaled radio flux.

3.3. Peak asymmetry

The peak asymmetry $\langle \Delta b \rangle$ of the pairs l = 0-2 and l = 1-3 modes shows significant temporal variations with solar activity (Fig. 3 and Table 2). Similar changes are obtained between velocity (GOLF) and intensity (VIRGO) measurements, while Jiménez-Reyes et al. [15] found opposite variations. No significant correlation between asymmetry and signal-to-noise ratio is observed.

3.4. Rotational splittings

No correlation is observed between the temporal variations of the mean l = 1 and 2 rotational splittings and the 11-year solar cycle (Fig. 3). However, some similar fluctuations between GOLF and VIRGO are present. Note also that the GOLF splittings show a jump between the red-wing period and the second blue-wing period starting end of 2002.

4. Conclusions

We analyzed more than 14 years of radial velocity and intensity helioseismic Sun-as-a-star data collected by the space-based GOLF and VIRGO instruments respectively to study the

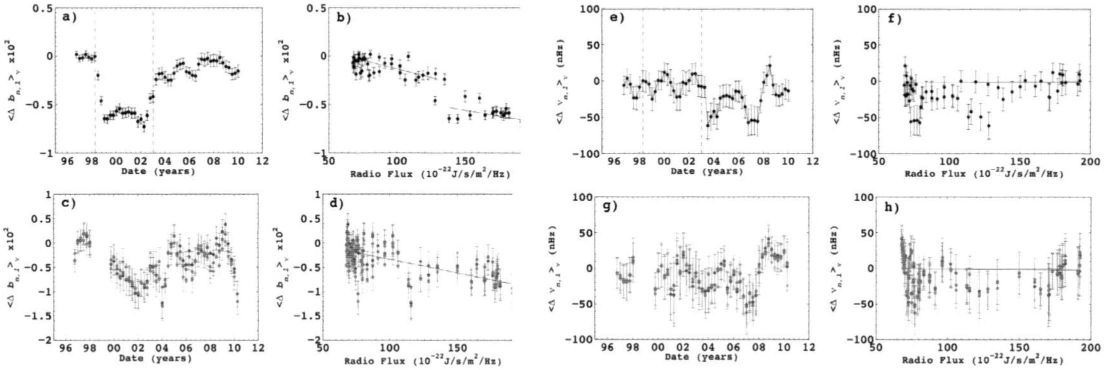

Figure 3. (Top row) Temporal variations ($\times 10^2$) in mode asymmetry $\langle \Delta b \rangle$ (a, b) and in rotational splitting (nHz) (e, f) measured from GOLF data as a function of time and radio flux. The vertical dashed lines separate the blue-, red-, and blue-wing GOLF periods. (Bottom row) Temporal variations ($\times 10^2$) in mode asymmetry $\langle \Delta b \rangle$ (c, d) and in rotational splitting (nHz) (g, h) measured from VIRGO data as a function of time and radio flux. The solid lines in panels b, d, f, and h represent the best fits from weighted linear regressions.

temporal variations of the low-degree p-mode parameters (excitation, damping, frequency, peak asymmetry, and rotational splitting) with solar activity. The observed changes in excitation and damping parameters confirm previous results. The frequency shifts present differences between individual l modes, for instance showing different minima for cycle 23. The peak asymmetry shows significant and similar variations between radial velocity and intensity measurements. The rotational splittings do not show correlations with the 11-year solar magnetic cycle.

Acknowledgments

The GOLF and VIRGO instruments onboard SoHO are a cooperative effort of many individuals, to whom we are indebted. SoHO is a project of international collaboration between ESA and NASA. The 10.7-cm radio flux was obtained from the National Geophysical Data Center. DS acknowledges the support of the grant PNAyA2007-62650 from the Spanish National Research Plan. This work has been supported by the CNES/GOLF grant at the SAp CEA-Saclay.

References

[1] Chaplin W J, Elsworth Y, Miller, B A, Verner G A and New R 2007 *ApJ* **659** 1749
[2] Salabert D, Fossat E, Gelly B, Kholikov S, Grec G, Lazrek M and Schmider F X 2004 *A&A* **413** 1135
[3] Salabert D, Jiménez-Reyes S J and Tomczyk S 2003 *A&A* **408** 729
[4] Komm R W, Howe R and Hill F 2000 *ApJ* **531** 1094
[5] Broomhall A-M, Chaplin W J, Elsworth Y, Fletcher S T and New R 2009 *ApJ* **700** L162
[6] Salabert D, García R A, Pallé P L and Jiménez-Reyes S J 2009 *A&A* **504** L1
[7] García R A, Mathur S, Salabert D, Ballot J, Régulo C, Metcalfe T and Baglin A 2010 *Science* **329** 1032
[8] Gabriel A H et al. 1995 *Solar Physics* **162** 61
[9] Fröhlich C et al. 1995 *Solar Physics* **162** 101
[10] García R A et al. 2005 *A&A* **442** 385
[11] Salabert D, Chaplin W J, Elsworth Y, New R and Verner G A 2007 *A&A* **463** 1181
[12] Harvey J W et al. 1996 *Science* **272** 1284
[13] Jiménez-Reyes S J, Chaplin W J, Elsworth Y and García R A 2004 *ApJ* **604** 969
[14] Salabert D and Jiménez-Reyes S J 2006 *ApJ* **650** 451
[15] Jiménez-Reyes S J, García R A, Jiménez A and Chaplin W J 2003 *ApJ* **595** 446

Open issues in probing interiors of solar-like oscillating main sequence stars
1. From the Sun to nearly suns

MJ Goupil[1], **Y. Lebreton**[1], **J.P. Marques**[1], **R. Samadi**[1], **F. Baudin**[2]

[1] Observatoire de Paris, UMR 8109, Paris, France
[2] Université Paris -Sud, Orsay, France

E-mail: mariejo.goupil]@obspm.fr

Abstract. We review some major, open issues in the current modelling of low and intermediate mass, main sequence stars based on seismological studies. In the present paper, the solar case is discussed together with current problems that are common to the Sun and stars with a structure similar to that of the Sun. Several additional issues specific to main sequence stars other than the Sun are reviewed and illustrated with a few stars observed with CoRoT in a companion paper.

1. Introduction

After more than two decades of helioseismology, almost four years of asteroseismology with CoRoT (1) and almost two years of intensive asteroseismology with Kepler (2), we review some major, current open issues about the internal structure of the Sun and solar-like oscillating stars. We discuss here the solar case, this also applies to oscillating stars that have a similar internal structure. For sake of brevity, we decided to include only unpublished figures and to cite published figures in the text. Several recent reviews exist on the topic, for instance (3), (4), (5).

2. The Sun

As it is well known, the Sun is a particular case. It is the closest star, and as a result we know with a high precision the luminosity, mass (through the product $GM_\odot$), radius, age and individual surface abundances of chemical elements[1]. Furthermore, a wealth of very accurate seismic constraints are available and have been successfully used. Inversion of a large set of mode frequencies has provided crucial information on the structure of the Sun, see for instance (5); (6). Accordingly, the following constraints must all be satisfied by any calibrated solar model: radius at the base of the convective envelope r_{bzc}, surface helium abundance Y_s, sound speed profile $c(r)$, internal rotation profile and location of ionization regions. To some extent, these constraints are found to be independent of the reference model (7). The current major challenges and open issues in the solar case then are:

- what are the values of the surface abundances, more specifically the oxygen abundance?

[1] although some of these latter are still debated, see below

Table 1. Value of the solar photospheric metallicity from 1993 to 2010 from Table 4 in Asplund *et al* (10) complemented by Caffau *et al* (11)

	GN93	GS98	AGS05	AGS09	Lod09	Caff10
Z/X	0.0244	0.0231	0.0165	0.0181	0.0191	0.0209

- what is the origin of the discrepancy between the seismic sound speed and that given by models below the convection zone?
- what are the dominant physical mechanisms responsible for a uniform rotation in the radiative region and a differential rotation in the convective zone? It is worth noting that this is the opposite in current calibrated 1D solar models: the convection zone is assumed to rotate uniformly and the rotation profile in the radiative zone is found to vary with radius if waves and/or magnetic field are not taken into account (see below) !
- how can we model properly near surface layers and the convection-pulsation interaction?
- how to succeed in probing the core?
- how to model oscillation mode line widths and amplitudes?

Some of these uncertainties about the Sun have consequences on the modelling of stars other than the Sun. On the other hand, some problems that are encountered with seismological studies of stars other than the Sun can be studied first with our well known Sun. These two points of view are discussed below.

2.1. Initial abundances: the solar mixture

In the late nineties, with the GN93 solar abundances (8), the seismic sun and calibrated solar models were in agreement by 1 to 5% for the sound speed profile and location of the base of the convective zone (9). However between 1993 and 2010 several revisions of the photospheric solar mixture were performed. Noteworthy, 3D model atmospheres including NLTE effects and improved atomic data have led to a substantial decrease of the C, N, O, Ne, Ar abundances and in turn to an important decrease of the solar metallicity (Z/X) (see Table 1).

Today, an internal consistency of the abundance determination from different ionization levels of a given element seems to have been obtained and a consensus between independent determinations seems to be reached (i.e. this shows the unvaluable benefit of independent approaches). However the revised modern solar chemical composition (10) leads to a strong disagreement between the sound speed of a calibrated solar model and that of the seismic Sun. This is also true for the depth of the convection zone and the envelope helium abundance (see (5), for a review). Fig.1 shows the difference of the sound speed profile between Basu's seismic model (12) and several models computed with Cesam2k (13) using some of the abundance mixtures listed in Tab.1. This figure shows that the discrepancy at the base of the convection zone is strongly increased when the revised abundance mixture AGS05 is used. This is mainly due to the decrease in oxygen and neon and in $(Z/X)_s$ which consequence is a decrease of the radiative opacities. With the newly derived abundance mixture AGS09 which has achieved some consensus, the discrepancy slightly decreases but remains significant.

2.1.1. Possible origins of the discrepancy First there could be some errors in the opacity derivation. As a check, the comparison between OPAL and OP opacities for a calibrated solar model shows that OP opacities give a (slightly) better fit than OPAL. However a change in opacity of about 30% at the base of the convection zone, at a temperature of 2 10^6 K would be necessary to compensate for the effects of the change in mixture (15; 3) and according to (16) there appears to be no way to change the OP opacity by such an amount (see also (17)

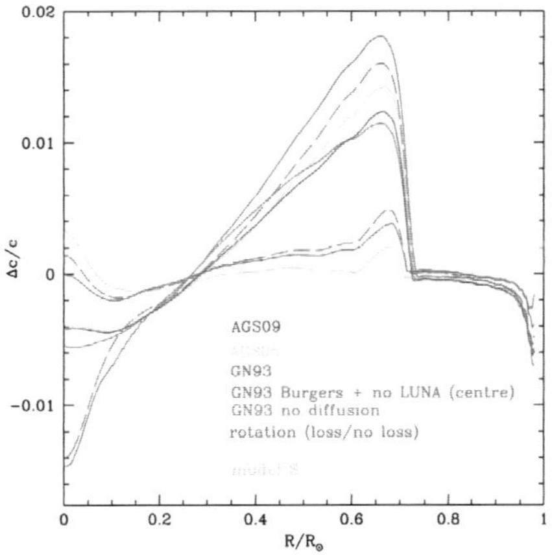

Figure 1. Differences in the sound speed profiles for the Sun between Basu's seismic model and calibrated solar models using various abundance mixtures and different assumptions for the physical description. All models have been calculated with the Cesam2k code except model S (14) which is based on the old solar mixture.

and (18)). Other possible causes have been discussed but none of the related improvements can reconcile simultaneously all the seismic constraints listed above (19),(20), (21).

In order to reconcile the seismic and theoretical sound speed, one needs higher opacities or higher helium below the upper convective zone (UZC) i.e. a higher He gradient. Any mixing below the UZC which smooths the gradient therefore would go in the wrong direction. In that case, one would rather need an advection process. It is not clear whether internal waves and/or which hydrodynamical instability could act in the sense of increasing helium below the UZC. Other further developements are underway such as taking magnetic effects into account in the derivation of solar abundance corrections. An investigation based on 3D magnetoconvection simulations shows that corrections to the solar abundance can be significant (22).

Abundances of other stars are determined by reference to the Sun, hence all stars are affected by errors or by inaccuracies in the solar mixture. Can other stars be discriminating in that issue? Note that one must wonder about the impact of the inconsistency which arises when modeling other stars using solar mixtures inferred from 3D model atmospheres if the stellar [Fe/H] itself has not been determined from a dedicated 3D model atmosphere.

2.2. Nuclear reaction rates

In stars, (most commonly charged induced) nuclear reactions occur at low energies (10- 300 keV) in the Gamow peak -which corresponds to the maximum probability of the reaction. The cross sections that are governed by Coulomb barriers and resonances show a strong and complex dependance on energy and globally decrease steeply towards low energy.

Cross sections are usually written in the form

$$\sigma(E) = \frac{S(E)}{E}\, e^{-2\pi\eta} \tag{1}$$

where $S(E)$ -the astrophysical factor- contains everything that concerns the nuclei and nuclear physics and varies slowly with E. The exponential term is related to the Coulomb barrier and relative velocity of the nuclei. Experiments which provide measurements of the cross sections in the laboratory generally occur at higher energy than the Gamow peak. Extrapolation of $S(E)$

to lower E is then necessary. This is difficult to achieve and cannot take into account possible unknown resonances occuring at low E. When the S-factor cannot be measured, the reaction rates are obtained from pure theory. Recent significant progress in laboratory and theory (hence better determination of S-factors down to the Gamow peak) has been achieved as discussed in the comprehensive reviews by (23), (24), see also (25).

2.2.1. Hydrogen burning reaction rates: Uncertainties still exist for the pp chain and CNO cycle reaction cross sections in the Sun. They are due to the difficulty to estimate the S-factor at low energy and to determine the exact role of the electron screening (both in the laboratory and in the star). For most of the concerned reactions, S-factors are determined from extrapolation of experimental data to low E but now, for some key reactions, energies corresponding to the Gamow peak are accessible through measurements provided by the LUNA experiment at Gran Sasso (26). This is the case of the $^3He(^3He, 2p)^4He$ reaction (see Fig. 1 in (26)) for which the $S(E)$-value has not been significantly modified but the error bars are largely decreased. Furthermore the cross section of the $N^{14}(p, \gamma)O^{15}$ reaction -the leading reaction in the CNO cycle- has now been measured down to energies relevant for $1M_\odot$ stars on the red giant branch (see Fig. 19 in (26)) and a resonance is observed. This represents a significant advance although an extrapolation is still needed towards solar conditions. Noteworthy, the revision is important with a decrease of $S(E)$ by 50%. This has crucial consequences for solar and stellar structure. In the solar core, the CNO cycle efficiency is reduced from 1.6 to 0.8% of the total energy. For main sequence stars slightly more massive than the Sun the occurrence of a convective core depends on the energy production. With the new $N^{14}(p, \gamma)O^{15}$ rate the convective core appears at higher stellar mass and is less massive for a given mass (see for instance the case of a 1.2 $M_\odot$ low metallicity star in Fig. 14 by (27)). Finally cross sections that are obtained from pure theory can be constrained by helioseismology (28),(29),(23).

2.2.2. Electron Screening Electron screening is based on the Salpeter's formula (30) with the underlying physical picture: the cloud of electrons decrease the repulsive Coulomb effect between interacting nuclei with the result of a decrease of the Coulomb potential and an enhancement of the reaction rate. This is a static description. It is currently not clear whether dynamic effects of the interacting ions can significantly change the impact of the screening in nuclear interactions for stars and wherther it must therefore be taken into account. The energy that initially fast moving interacting ions have when they get close enough to interact can be lower than the mean value of the medium; accordingly their reaction rate is reduced compared to Salpeter's prescription. This effect is difficult to quantify and relies on results from numerical simulations (31), (32), (33), (34) and the issue is not settled yet. (29) looked at the impact of changing the electron screening compared with the classical Salpeter's formula on the solar sound speed profile. They found that the solar seismic constraints do not allow variations larger than 1% when using GS98. (3) computed a solar model equivalent to the S model but assuming the extreme case of no screening at all. Their Fig.12 shows the difference in the sound speed profile between the seismic sun (12) and model S but switching off e- screening. A decrease of the reaction rate by swithcing off the e-screening clearly increases the discrepancy between observation and model. This is in agreement with previous results of (23) who showed that the large sound speed discrepancy with the use of AGS05 mixture is reduced to the previous level obtained with GS98 when the pp-reaction and the screening factor are increased up to 15% (Fig.3 in (23)) In that case, the surface helium and the depth of the convective envelope are in agreement with seismic determinations but the sound speed profile in the core significantly deviates from the seismic solar sound speed.

2.3. *Transport of chemicals and angular momentum*

Rotationally induced transport The physical origin of the uniform rotation profile in the radiative zone of the Sun unravelled by helioseismology is still debated. Rotationally induced transport of angular momentum resulting from a competition between shear-induced turbulence and meridional circulation driven by surface angular momentum losses is not able to make the rotation uniform in the radiative region of the Sun. For details, the reader is refered to (35). The impact of rotationally induced transport on the solar sound speed profile has been investigated by (36); (37). (38) have computed a solar model including rotationally induced transport with two different assumptions about the initial velocity (slow or 'fast' sun). Their conclusion is that for an initially slow rotation, the microscopic diffusion dominates whereas for an initially rapid enough rotation, meridional circulation dominates over turbulent shear. However in both cases, the discrepancy below the UCZ increases (see Fig.9 of (38)). This is also illustrated in Fig.1 which shows the differences in sound speed profile for calibrated solar models computed with AGS05 mixture computed assuming either no rotationally induced transport, or rotationally induced transport with no surface angular momentum loss or assuming rotationally induced transport with surface angular momentum loss. Adding rotationally induced transport makes the discrepancy with the observed sound speed worse. Indeed this process smoothes the helium gradient below the UZC whereas the observations seem to require an increase of helium in the radiative zone below the UCZ. Several prescriptions have been derived for the turbulent transport coefficients involved in rotationally induced transport but validitation of these prescriptions remain to be done (39).

Internal wave induced transport in radiative zones must exist in stars as these waves are generated at the interface between convective and radiative regions. They have been shown to transport angular momentum efficiently enough to make the rotation of the Sun in its radiative part rigid (40). One the main uncertainties is related to the abscnce of a viable quantitative description of the generation of waves.

Magnetic induced transport Instabilities driven by the interaction between rotation and a magnetic field could be responsible for the transport of angular momentum and the rigid solar rotation (41). This was shown by (37) in the solar case.

Whether all 3 processes work together to shape the solar rotation profile in the radiative zone or only one or two are dominant is not settled yet.

Investigations of the impact of rotationally and wave induced transport on the structure of stars other than the Sun have been performed for instance to explain the Li dip (42); (43). (44) studied the impact of including both rotationally induced transport and magnetic field on the structure of solar like stars while (45) estimate the seismic consequences and find that with the type of dynamo they assume in the radiative zone, the efficiency of rotational mixing in a radiative zone is significantly decreased and seismic parameters are then similar to those of a non-rotating, non-magnetic star.

2.4. *Near surface layers*

A direct comparison of the observed and numerical frequencies of the Sun shows systematic differences that remain small at low frequency but increase with increasing frequency (46). Several causes contribute to this offset with more or less importance (47). They are collectively referred to as near-surface effects (for reviews, see for instance (48), (4)).

2.4.1. Surface turbulent convection One important contribution to the differences between the observed and calculated frequencies comes from current modelling of the outer turbulent convective layers of the Sun. The description of the convective outer layers of the Sun in 1D stellar models remains quite approximate due to our inability to represent and implement in a 1D code a 3D multiscale nonlocal process such as turbulent dynamics (49). At the solar surface,

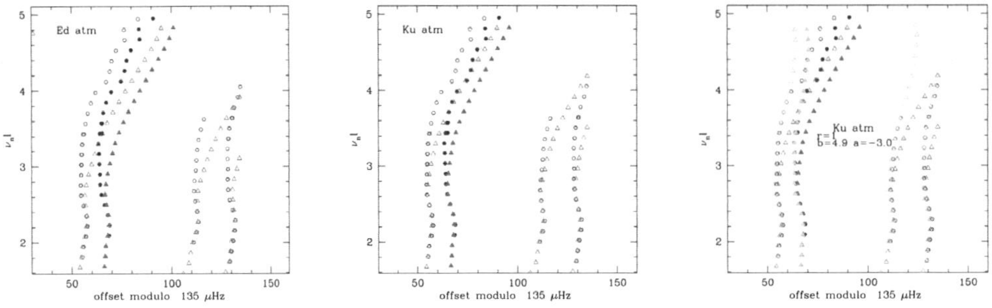

Figure 2. Echelle diagrams computed with GOLF solar data (blue dots) and using a calibrated solar model computed with either an Eddington (**left panel**) or a Kurucz (**middle panel**) $T(\tau)$ law (red dots) . Filled symbols correspond to radial modes. **Right panel:** An additional echelle diagram obtained from frequencies of the solar model (red dots) but corrected for near surface effects (green dots) agrees with the observations (blue dots) over the fitted interval.

turbulent convection is inefficient and therefore strongly dependent on the free parameters entering the local, 1D formulation for convection as well as many other assumptions in the formulation. A comparison between frequencies computed with two of the available formulations (MLT and CM (50)) for instance shows that the frequency differences increase with frequency and reach up to 0.15% at a frequency $\nu = 4.$ mHz i.e. a frequency offset of about $6 - 10\mu$Hz ((9)).

Patched models, that is 1D stellar models where the outer layers have been replaced in a proper way by those obtained with 3D numerical simulations, lead to frequencies in much better agreement with the observations (51), (52), (53), (54). For reviews, see (55), (56). This approach is valuable for studies of individual stars provided great care is taken in the patching procedure, but it cannot be used in a systematic investigation of a large number of stars. Indeed 3D numerical simulations with the required quality are quite numerical time consuming and therefore not available in the whole range of effective temperature, gravity and chemical composition.

Atmosphere as boundary condition: Another longstanding problem is the description of the atmosphere as boundary condition when computing a stellar model (57). In the solar case, one can use an empirical atmosphere derived from observations (HSRA) (58) which represents accurately enough the Sun atmospheric properties. But in other stars, one must rely on atmospheres built using a temperature-optical depth, $T(\tau)$, law. Commonly used $T(\tau)$ laws are the Eddington law or more realistic Kurucz model atmospheres (59). In order to illustrate the impact of the atmosphere boundary condition, Fig.2 (left panel) shows an echelle diagram built from observed solar oscillation frequencies based on SOHO data. An echelle diagram computed for a calibrated solar model where the temperature stratification of the atmosphere is assumed to follow a Eddington $T(\tau)$ law is also shown (middle panel). Largest differences are seen at high frequency. They are significantly larger than the observational errors. The discrepancy at high frequency decreases (roughly by half) when a more realistic Kurucz model is used (middle panel) . However one needs extensive grids of such models along an evolutionary track and for different masses. Furthermore, model atmospheres suffer from physical imperfections and are not applicable over the entire range of needed masses and ages (i.e. gravity, effective temperature) (60).

Nonadiabatic effects: Another source of uncertainty comes from what is usually referred to as nonadiabatic effects. They include the effects of interaction of the wave with the radiation and with the convection (see for instance (48)).

All these imperfections concur to generate significant errors on the computed oscillation frequencies which in the solar case amount up to 10 μHz at high frequency. As near surface effects cannot yet be reliably included in theoretical frequency computations, an alternative has been proposed which consists in removing these effects from observed frequencies. In the solar case, this has been quantitatively assessed with the comparison of theoretical and observed frequencies of modes with small to large degrees.

2.4.2. Correcting frequencies for near surface effects In the solar case, due to the large number of different modes, near surface effects can be removed from the frequencies. However, other procedures must be found for stars other than the Sun. Several approaches have been proposed. One consists in comparing the ratio of frequency combinations rather than frequencies themselves for low degree modes. In that case indeed near surface effects are at least partially cancelled (61). On the other hand, in order to be able to use absolute frequencies, (62) proposed a means to correct observed frequencies for near surface effects. They first showed that the systematic offset between the observed and theoretically computed frequencies of the Sun is well fitted with a power law

$$\nu_{mod}(n) = \nu_{obs}(n) - a(\nu_{obs}(n)/\nu_{max})^b \qquad (2)$$

with a, b fit to the data and $\nu_{max} = 3100$ Hz is a reference frequency that corresponds to the frequency of maximum power in a power spectrum. The frequencies $\nu_{obs}(n)$ and $\nu_{mod}(n)$ respectively represent the observed and model frequencies of radial modes with radial order n. For the Sun, such a correction leads for instance to small separations d_{02} which yield a solar age consistent with the meteoritic age (63).

Fig.2 (right panel) illustrates the effect of correcting $l = 0$ frequencies from near surface effects according to Eq.2 with $r = 1, a = -3.0, b = 4.9$ for the same solar model (with Kurucz $T(\tau)$ law) as in the middle panel. The corrected frequency echelle diagram coincides with the observations over the frequency interval that was used to derive the parameter values a, b. The next question is: how much the parameters a,b, ν_0 do depend on the adopted (Kurucz or another) model atmosphere? Of course one must also keep in mind that the values of the fitted parameters are valid only over the fitted observed frequency domain (see Fig.2 left).

Whether the above procedure can be applied to stars that are different from the Sun is an open question. Elements of the answer can be obtained with the use of available 3D numerical simulations and 1D patched models. Frequency differences for radial modes between a solar patched model and a non patched model are displayed in Fig.3 (left) as a function of the radial mode frequency of the patched model scaled to the frequency $\nu_{max,\odot}$ given according to (64) scaling law. The curve $a(\nu_{patch}/\nu_{max})^b$ is also plotted and coincides with the observations over the frequency interval that was used to derive the a, b parameter values over an interval up to $\nu_{patch}/\nu_{max} \approx 1.2$. Similar plots are displayed in Fig.3 (right) for 3 other models with different effective temperature, gravity or chemical abundance. For these models, we take again ν_{max} according to (64) and $T_{eff,\odot}$ and $\nu_{max,\odot}$ from our solar patched model. The parameters a, b are adapted to fit the frequency differences over the largest ν_{patch}/ν_{max} interval. Again a power law can fit the frequency differences over an interval up to $\nu_{patch}/\nu_{max} \approx 1.2 - 1.4$. We note that the frequency differences for the 3 models behave differently at high frequency than the solar ones. This is likely due to their lower gravity, higher effective temperature. Indeed, for these models, the ratio of turbulent to total pressure is higher than for the solar model (the higher the effective temperature, the lower the gravity, the higher the ratio P_{turb}/P_{tot}) and this results in larger differences between patched and non patched models. Note that for the three models

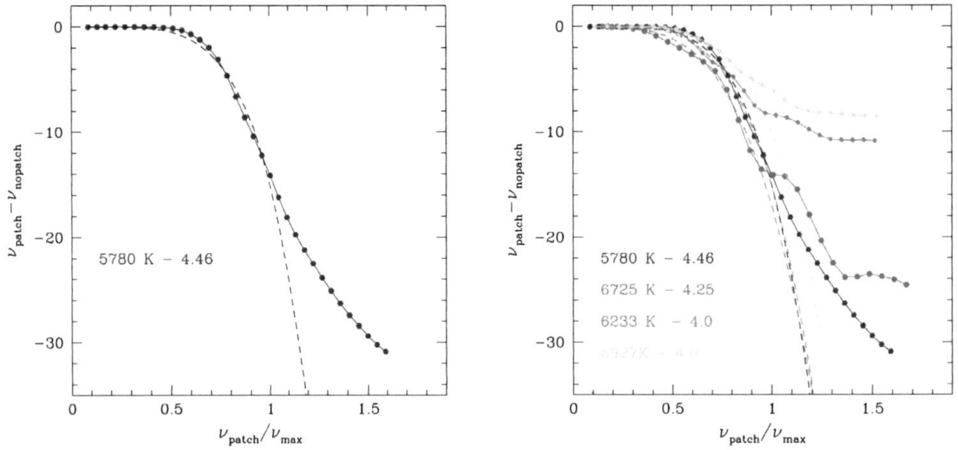

Figure 3. left: Frequency differences for radial modes $\nu_{patch} - \nu_{nopatch}$ between a patched and a non patched solar models (blue dots) as a function of ν_{patch}/ν_{max}. The black dashed line represents the power law according to Eq. 2 with $\nu_{max} = 3.207$mHz,$r = 1, a = -15, b = 4.9$ (For the solar models: $T_{eff} = 5780, g = 4.46$). **right:** Same as left for stellar models with different effective temperature, gravity as labelled in the left corner. Magenta and red models have a solar metallicity whereas the cyan models have a lower metallicity $[Fe/H] = -1$. Coloured dashed lines represent the power law according to Eq. 2 for each corresponding frequency difference. Parameter values are $\nu_{max} = 1.462$mHz, $r = 1, a = -17, b = 3.9$ (red); $\nu_{max} = 1.098$mHz, $r = 1, a = -10, b = 4.9$ (violet); and $\nu_{max} = 1.071$mHz, $r = 1, a = -15, b = 4.5$ (cyan)

(all three hotter than the solar model), the frequency differences show an oscillating behavior in function of the scaled frequency. Whether this behavior is real or artificially introduced by the patching process is not known yet. In any case, at lower frequency, the mean variation with frequency is well reproduced by a power law up to $\nu_{patch}/\nu_{max} \approx 1.2 - 1.4$

3. Conclusion:

Despite the huge amount of information provided by helioseismology about the internal structure of the Sun, several important open issues remain that we have reviewed. They can also impact our understanding and modelling of stars that have a similar structure to that of the Sun. An increasing set of such stars are being observed by CoRoT and Kepler, and seismic tools developped for studying the Sun are now being adapted to study other stars.

Acknowledgments

We gratefully thank our colleagues S. Talon, T. Corbard and A. Baglin for providing useful information and fruitful discussions when preparing this review. We also thank P. Morel and J. Christensen-Dalsgaard for providing public evolutionary and oscillation codes respectively that were used in the present review. We acknowledge financial support from CNES and the ANR SIROCO

References

[1] Baglin A, Auvergne M, Barge P, Deleuil M, Catala C, Michel E, Weiss W and The COROT Team 2006 *ESA Special Publication* (*ESA Special Publication* vol 1306) ed M Fridlund, A Baglin, J Lochard, & L Conroy p 33

[2] Borucki W J, Koch D G, Lissauer J and et al 2007 *Transiting Extrapolar Planets Workshop* (*Astronomical Society of the Pacific Conference Series* vol 366) ed C Afonso, D Weldrake, & T Henning p 309

[3] Christensen-Dalsgaard J 2009 *HELAS Workshop on New insights into the Sun* ed M S Cunha and M J Thompson (*Preprint* 0912.1405)

[4] Christensen-Dalsgaard J and Houdek G 2010 *AP&SS* **328** 51–66

[5] Basu S and Antia H M 2008 *Phys. Rep.* **457** 217–283

[6] Christensen-Dalsgaard J 2002 *RvMP* **74** 1073–1129

[7] Basu S, Pinsonneault M H and Bahcall J N 2000 *ApJ* **529** 1084–1100

[8] Grevesse N and Noels A 1993 *Origin and Evolution of the Elements* ed N Prantzos, E Vangioni-Flam, & M Casse pp 15–25

[9] Basu S 2008 *14th Cambridge Workshop on Cool Stars, Stellar Systems, and the Sun* (*Astronomical Society of the Pacific Conference Series* vol 384) ed G van Belle p 10

[10] Asplund M, Grevesse N, Sauval A J and Scott P 2009 *ARA&A* **47** 481–522

[11] Caffau E, Ludwig H, Steffen M, Ayres T R, Bonifacio P, Cayrel R, Freytag B and Plez B 2008 *A&A* **488** 1031–1046

[12] Basu S, Christensen-Dalsgaard J, Chaplin W J and et al 1997 *MNRAS* **292** 243

[13] Morel P and Lebreton Y 2008 *AP&SS* **316** 61–73

[14] Christensen-Dalsgaard J, Däppen W, Ajukov S and et al 1996 *Science* **272** 1286–1292

[15] Bahcall J N, Basu S, Pinsonneault M and Serenelli A M 2005 *ApJ* **618** 1049–1056

[16] Badnell N R, Bautista M A, Butler K, Delahaye F, Mendoza C, Palmeri P, Zeippen C J and Seaton M J 2005 *MNRAS* **360** 458–464

[17] Basu S 2010 *Proc. of GONG 2010 - SoHO 24 conference: A new era of seismology of the Sun and solar-like stars* (*ESA Special Publication* vol in press) ed Appourchaux T

[18] Turck-Chièze S 2010 *Proc. of GONG 2010 - SoHO 24 conference: A new era of seismology of the Sun and solar-like stars* (*ESA Special Publication* vol in press) ed Appourchaux T

[19] Guzik J A 2006 *Proceedings of SOHO 18/GONG 2006/HELAS I, Beyond the spherical Sun* (*ESA Special Publication* vol 624)

[20] Guzik J A 2008 *MmSAI* **79** 481

[21] Basu S, Chaplin W J, Elsworth Y, New R, Serenelli A M and Verner G A 2007 *ApJ* **655** 660–671

[22] Fabbian D, Khomenko E, Moreno-Insertis F and Nordlund Å 2010 *ArXiv e-prints* (*Preprint* 1006.0231)

[23] Weiss A 2008 *PhST* **133** 4025

[24] Adelberger E G, Balantekin A B, Bemmerer D and et al 2010 *ArXiv e-prints* (*Preprint* 1004.2318)

[25] Lebreton Y 2010 *ELSA conference 2010, Gaia: at the frontiers of astrometry, Sèvres* ed C Turon, F Arenou, F Meynadier, EAS Series, EDP Sciences, in press

[26] Costantini H, Formicola A, Imbriani G, Junker M, Rolfs C and Strieder F 2009 *RPPh* **72** 086301

[27] Lebreton Y and Montalbán J 2010 *AP&SS* **328** 29–38

[28] degl'Innocenti S, Fiorentini G and Ricci B 1998 *PhLB* **416** 365–368

[29] Weiss A, Flaskamp M and Tsytovich V N 2001 *A&A* **371** 1123–1127

[30] Salpeter E E 1954 *AuJPh* **7** 373

[31] Shaviv G and Shaviv N J 2000 *ApJ* **529** 1054–1069

[32] Shaviv G 2010 *MmSAI* **81** 77

[33] Mussack K and Däppen W 2010 *AP&SS* **328** 153–156

[34] Mao D, Mussack K and Däppen W 2009 *ApJ* **701** 1204–1208

[35] Maeder A 2009 vol ISBN 978-3-540-76948-4 (Springer Berlin Heidelberg)

[36] Palacios A, Talon S, Turck-Chièze S and Charbonnel C 2006 *Proceedings of SOHO 18/GONG 2006/HELAS I, Beyond the spherical Sun* (*ESA Special Publication* vol 624) p 38

[37] Yang W M and Bi S L 2006 *A&A* **449** 1161–1168

[38] Turck-Chièze S, Palacios A, Marques J P and Nghiem P A P 2010 *ApJ* **715** 1539–1555

[39] Talon S 2004 *Stellar Rotation* (*IAU Symposium* vol 215) ed A Maeder & P Eenens p 336

[40] Talon S and Charbonnel C 2005 *A&A* **440** 981–994

[41] Maeder A, Meynet G, Georgy C and Ekström S 2009 (*IAU Symposium* vol 259) pp 311–322

[42] Palacios A, Talon S, Charbonnel C and Forestini M 2003 *A&A* **399** 603–616

[43] Charbonnel C and Talon S 2005 (*EAS Publications Series* vol 17) ed G Alecian, O Richard, & S Vauclair pp 167–176

[44] Eggenberger P, Maeder A and Meynet G 2005 *A&A* **440** L9–L12

[45] Eggenberger P, Meynet G, Maeder A, Miglio A, Montalban J, Carrier F, Mathis S, Charbonnel C and Talon S 2010 *A&A* **519** A116

[46] Christensen-Dalsgaard J, Dappen W and Lebreton Y 1988 *Nature* **336** 634–638

[47] Christensen-Dalsgaard J and Thompson M J 1997 *MNRAS* **284** 527–540

[48] Goupil M J and Dupret M A 2007 (*EAS Publications Series* vol 26) ed C W Straka, Y Lebreton, & M J P F G Monteiro pp 93–110

[49] Kupka F 2009 *Interdisciplinary Aspects of Turbulence* (*Lecture Notes in Physics, Berlin Springer Verlag* vol 756) ed W Hillebrandt & F Kupka p 49

[50] Canuto V M and Mazzitelli I 1991 *ApJ* **370** 295–311

[51] Rosenthal C S, Christensen-Dalsgaard J, Nordlund Å, Stein R F and Trampedach R 1999 *A&A* **351** 689–700

[52] Li L H, Robinson F J, Demarque P, Sofia S and Guenther D B 2002 *ApJ* **567** 1192–1201

[53] Samadi R, Ludwig H, Belkacem K, Goupil M J and Dupret M 2010 *A&A* **509** A15

[54] Kupka F, Belkacem K, Goupil J and Samadi R 2009 *Co.Ast.* **159** 24–26

[55] Samadi R 2009 *ArXiv e-prints*

[56] Houdek G 2010 *AP&SS* **328** 237–244

[57] Morel P, van't Veer C, Provost J, Berthomieu G, Castelli F, Cayrel R, Goupil M J and Lebreton Y 1994 *A&A* **286** 91–102

[58] Gingerich O, Noyes R W, Kalkofen W and Cuny Y 1971 *Solar Phys.* **18** 347–365

[59] Kurucz R L 2005 *MmSAI* **8** 14

[60] Kurucz R L 2005 *MmSAI* **8** 73

[61] Roxburgh I W and Vorontsov S V 2003 *A&A* **411** 215–220

[62] Kjeldsen H, Bedding T R and Christensen-Dalsgaard J 2008 *ApJ Lett.* **683** L175–L178

[63] Doğan G, Bonanno A and Christensen-Dalsgaard J 2010 *ArXiv e-prints* (*Preprint* `1004.2215`)

[64] Bedding T R and Kjeldsen H 2003 *PASA* **20** 203–212

Open issues in probing interiors of solar-like oscillating main sequence stars: 2. Diversity in the HR diagram

MJ Goupil[1], **Y. Lebreton**[1], **J.P. Marques**[1], **S. Deheuvels**[2], **O. Benomar**[3], **J. Provost**[4]

[1] Observatoire de Paris, UMR 8109, Paris, France
[2] Astronomy Department, Yale University, New Haven CT, U.S.A.
[2] Sydney Institute for Astronomy (SIfA), School of Physics, University of Sydney, NSW 2006, Australia
[3] Université de Nice-Sophia Antipolis, CNRS UMR 6202, Observatoire de la Cte d'Azur, Nice, France

E-mail: `mariejo.goupil]@obspm.fr`

Abstract. We review some major open issues in the current modelling of low and intermediate mass, main sequence stars based on seismological studies. The solar case was discussed in a companion paper, here several issues specific to other stars than the Sun are illustrated with a few stars observed with CoRoT and expectations from Kepler data.

1. Introduction

After more than two decades of helioseismology, almost four years of asteroseismology with CoRoT (1) and almost two years of intensive asteroseismology with KEPLER (2), we review some current open issues about the internal structure of solar-like oscillating stars. We focus on low and intermediate mass, main sequence stars. We started with the Sun and stars that have a similar structure than the Sun in a companion paper. Here open issues not encountered with the Sun will be discussed and illustrated with a few individual stars observed by CoRoT and from ground. We end with a brief discussion about expectations from KEPLER data. For sake of shortness, we decided to include only unpublished figures and to cite published figures in the text. Several reviews exist on the topic, for instance (3), (4).

2. From the Sun to solar-like oscillating MS stars:

We focus on low and intermediate mass, main sequence stars that-is stars with masses up to 1.5 $M_\odot$ corresponding to F, G, K spectral types. These stars have an external convective region and can oscillate like the Sun with high frequency p modes. For these stars, one encounters the same problems as for the Sun, namely surface effects when comparing absolute values of the frequencies. But although we refer to these stars as solar-like stars in the present framework for shortness, these stars can differ from the Sun by several aspects: mass, age, surface metallicity and helium abundances, initial conditions for the chemical abundances, the rotation and magnetic properties. For masses larger than about 1.2 $M_\odot$, they have a convective

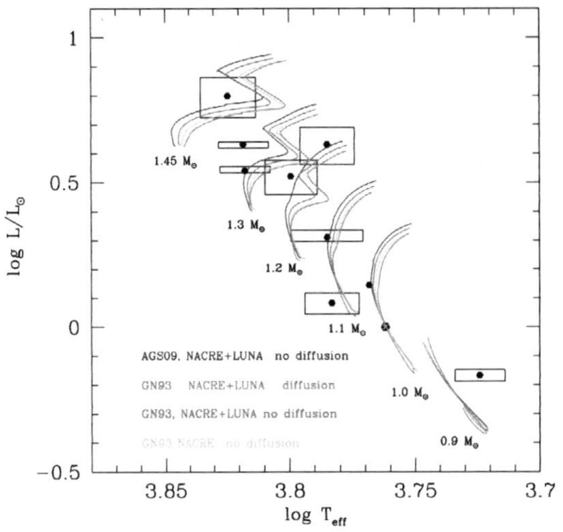

Figure 1. Evolutionary tracks of models built with several assumptions as indicated in the plot. All nuclear reaction rates are from NACRE except for the reaction $^{14}N(p,\gamma)^{15}$ of the CNO cycle which is from LUNA when specified. The other nuclear rates are all from NACRE. The helium and metallicity are solar.

core unlike the Sun. These differences therefore lead to additional open questions about their modelling. The major problem concerns dynamical processes occuring inside stars that have a significant impact on the mean structure of the stars and their ages. In particular, the 3D multiscaled turbulent convective transport and related instabilities is taken into account by means of a 1D crude formulation that involves free parameters which cannot be derived from first principles. As a consequence they are calibrated with observations (the Sun, binaries) but these values have no predictive quality.

The input parameters such as mass, age, initial chemical composition $Y_0, (Z/X)_0$, rotation profile are usually not well known. A first order of magnitude for mass and age is obtained via the location of the star in the HR diagram based on photometric and spectroscopic information. Uncertainties on this location and the chemical abundances give rise to a large number of possible models which makes difficult to probe in detail the internal structure of the star. This number is significantly reduced when seismic diagnostics such as the large separation and the small spacings are used. However free parameters used to describe the convective transport and related instabilities increase the family of acceptable models. The resulting solution therefore remains input physics dependent.

2.1. Observational constraints and seismic diagnostics

Efforts are therefore currently put at obtaining seismic constraints that are discriminating and as model independent as possible. A first information that is looked for is the large separation, its mean value as well as its variation with frequency (5), (6). Other combinations of low degree mode frequencies are built to locate the base of the upper convective zone and obtain properties about He ionization regions (7); (8); (9); (10) and surface helium content (11). Theoretical developements are carried also out to devise diagnostics and appropriate methods to determine the age ((12) and references therein); to probe the core (13), (14),(15); (16); (9), (17) with specific attention to tiny convective core properties (18) of low mass stars using only low degree modes;

to investigate mixing beyond the convective core (19) or semiconvection (20). For instance, the spacing d_{01} is sensitive to convective core properties (21), (22); ; (23); (24). (24) shows the variations of the seismic quantity d_{01} with frequency over a large range of frequency for a model representing the star HD 203608. The slope of the observed d_{01} over the observed frequency range is directly related to the slow period of oscillation seen on the theoretical d_{01} that corresponds to a time scale related to the acoustic radius of core convective radius.

2.2. CoRoT solar-like stars

Most CoRoT solar-like stars differ from the Sun either because they are more massive and faster rotators such as the F2 star HD181420 (25); (26) and the F5 star HD170987 (27) or because they are evolved and have an isothermal core such as the G0 star HD49385 (28),(29) or they have a mass similar to that of the Sun but differ in their metallicity such as HD49933 (30), HD 52265, or the young magnetic star HD 46375 (31). These stars can also differ significantly from the Sun by their level of magnetic activity and their magnetic cycles (32).

2.2.1. Which star for which diagnostic? Due to their differences, some of these stars are better suited to probe different physical processes.

Initial abundances and chemical mixture: Fig.1 shows some of these stars in a HR diagram together with evolutionary tracks of models built assuming no microsocopic diffusion; nuclear reaction rates are from NACRE completed with recent LUNA determinations. Two mixtures have been used : AGS09 and GN93. The effect of changing the mixture has a clear impact on a star such that the F star HD181906 (33) which has a convective core in the first case and no convective core in the second case. As the existence of a convective core manifests itself with a large slope of d_{01}, this can be in principle determined with this seismic diagnostic.

Nuclear reaction rate:CNO cycle and $^{14}N(p,\gamma)^{15}$ *burning:* The change from NACRE (34) to LUNA (35) reaction rate led to reduce the CNO cycle efficiency. As a result, the tracks for the most massive stars (i.e. with central temperature high enough for CNO to dominate) are slightly shifted upward. This of course coincides with stars having a convective core. For a $1.2M_\odot$ and $Z = 0.01$ model, the convective core is smaller at given mass and appears at higher mass (34).

Microscopic diffusion, surface helium abundance and initial metallicity: Due to gravitational settling and atomic diffusion in the radiative region below the convective envelope, the surface helium decreases with time. This decrease is larger for higher mass stars because the convective envelope is thiner. The decrease is also larger for lower metallic stars which are more compact and - due to smaller opacities and therefore smoother radiative temperature gradient- have also a thiner convective envelope. The disparition of helium in such thin convective zones is therefore very rapid. If microscopic diffusion acts alone, the envelope of HD49933 for instance would be fully depleted of helium. Mechanisms opposite to diffusion must therefore be at work such as turbulent diffusion and/or radiative acceleration, rotationally induced mixing and must therefore be included in the modelling.

2.2.2. Mode degree identification Seismic inferences assume that the modes are identified that-is that the observed frequencies can be attributed to modes with given degree l and azimuthal number m values in a spherical harmonics description. However in some cases, the $l = 1$ split multiplets can be mistaken with overlapping $l = 0$ and $l = 2$ modes in the fitting process. Hence some ambiguity can exist in the determination of $l = 0$ and $l = 1$ ridges in an echelle diagram. This is particularly true for the hottest (F type) stars because of their large mode linewidths and their relatively fast rotation. Exemples are HD49933 (36), HD181420 (25) and HD181906 (33) for which 2 scenarii are proposed for the identification of the $l = 0$ and $l = 1$ modes. In one case, fitting the data imposes a quite large core overshoot whereas in the other case a usual intermediate core overshoot amount is sufficient. This ambiguity that exists when the data sets

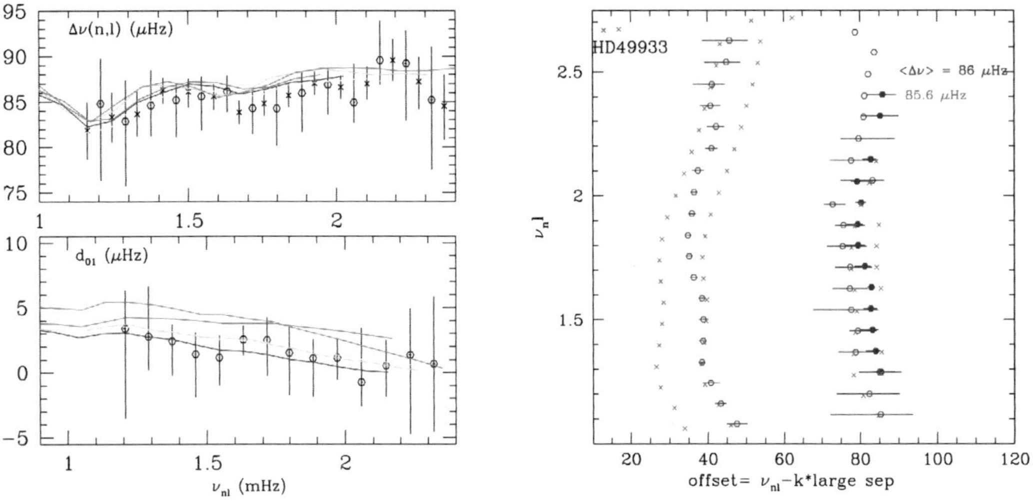

Figure 2. **(Left top:)** Large separation $\Delta\nu_{n,l}$ in function of the frequency $\nu_{n,l}$ and **(left bottom:)** the small spacings $d_{01} = \nu_{n,0} - (\nu_{n,1} + \nu_{n-1,1})/2$ and $d_{10} = -\nu_{n,1} + (\nu_{n+1,0} + \nu_{n,0})/2$ in function of the frequency $\nu_{n,0}$ for HD49933. Data (open circles and crosses) are from (38). Models (solid curves) are built assuming AGS05, no diffusion; $\alpha_{cgm} = 0.6$, $\alpha_{ov} = 0.27$ (magenta); AGS05, diffusion, $\alpha_{cgm} = 0.6$, $\alpha_{ov} = 0.21$, $Y = 0.27$, $Z/X = 0.0079$ (yellow) AGS09, diffusion (green); AGS05, diffusion,$\alpha_{ov} = 0.2$, rotationally induced transport (blue); AGS05, diffusion, $\alpha_{ov} = 0.$, rotationally induced transport, $Y = 0.27$, $\alpha_{cgm} = 0.6$ (chocolate).(**Right:**)Echelle diagram for HD49933. Blue dots represent the observations and red ones the model.

are too short can be lifted when using sophisticated data analysis treaments (37), (38), (39), (5). To lift the ambiguity in mode identification when it exists, (40) have proposed to use scaling relations (ν and $\Delta\nu$ scale as $< \Delta\nu >$) to build scaled echelle diagrams with reference to a star similar to the studied one. The authors tested this procedure with two sets of twin stars Sun and 18 Sco - τ Ceti and α Cen B and used it to determine the most probable scenario for two CoRoT stars HD181420 and HD181906 using HD49933 as the reference star. This scaling procedure has then been recently used on the ground based observed star, HD203608 (41). Indeed for this F8V star, two possible scenarii have been found with again important consequences on conclusions that can be drawn from the mode identification (28); (24). Echelle diagrams built assuming the scaling according to (40) coincide with the observed echelle diagram in the case of one of the two possible scenarii and definitely rejects the alternative scenario (41). This scenario favours a mild overshoot and the survical of convective core despite the small mass and old age of the star due its low metallicity (29). Mode identification based on scaling relations therefore appears as a potential interesting method that must nevertheless be studied further on theoretical ground before it can be used as a proper decision method.

2.2.3. HD49933: a low metallicity star With 180 days of observation with CoRoT, all seismic data analyses agree to provide the same scenario for the $l = 0, 1$ mode identification for this star (30). Seismic modelling of HD49933 illustrates the difficulty one encounters because of the degeneracy in input parameter space composed of mass, age, $(Z/X)_0, Y_0$, $\alpha_{cgm}, \alpha_{ov}$, initial rotation and transport coefficients. The free parameters $\alpha_{cgm}, \alpha_{ov}$ represent the convective mixing length for the adopted CGM formulation (42) and the convective core overshoot parameter respectively. All models discussed below are calibrated so that the mean large separation $< \Delta\nu >$ and the mean small spacings d_{01}, d_{10} agree with the observations as well as the observed location in a HR diagram. This lifts only partially the degeneracy. The phase of oscillation of the large separation is found to be quite sensitive to values of α_{cgm} and Y_0 and therefore such a constraint must be added in the minimisation process to reduce the number of acceptable models although this has not been done here. Individual frequencies have not been fitted (see below). The surface metallicity of the star is taken in the observed range $[Fe/H] = -0.4 \pm 0.1$, significantly lower than the Sun (43).

Fig.2 compares the mean large separation and the small spacings $d_{01}, d_{1,0}$ for models built assuming either AGS05 or AGS09 mixtures and various assumptions about microscopic diffusion, rotationally induced transport, convective core overshoot as listed in the caption. All models are computed with CESAM2k (44). Models including rotationally induced transport of angular momentum as implemented by J. Marques in the code CESTAM (a modified version of cesam2k) have been computed assuming no loss of angular momentum. The initial angular rotation on the PMS has been set in order to fit the observed rotation period of the star, $P = 3.4$ days at the age of HD49933. Details on the modelling of this star will be published elsewhere.

For all these different assumptions, one can find a mass and age that fit the mean large separation oscillation by ajusting α_{cgm} and Y_0 although this has not been done yet for our rotating models. The mass is found in the range 1.05-1.18 $M_\odot$ and the age in the range 2900-3900 Myr depending on the assumptions in the physical description and the chemical abundances. As a result of these various calculations, we find that a) when the AGS05 mixture is assumed, it is difficult to find a model satisfying all the observational constraints when $(Z/X)_0$ is on the smaller part of the authorized interval. This is less the case with the less extreme AGS09 mixture. It is important to stress that the star being metallic deficient compared with the Sun, the above mixtures taken from the Sun might not well be suited; b) because this star is low metallic, its thin convective envelope is rapidly devoid of helium when microscopic diffusion is included if one starts with solar initial helium abundance when one assumes AGS05 mixture. One then needs to start with a large initial helium abundance Y_0 or one must include some turbulence in the radiative zone. Starting then with a large $Y_0 = 0.35$, one still obtains a small $Y_{surf} = 0.10$ value for HD49933. One obtains a less extreme surface helium abundance $Y_s = 0.18$ when using the less extreme AGS09 mixture; c) without including microscopic diffusion nor rotationally induced transport, no model fits d_{01} for an overshoot smaller that $0.25 - 0.3$ Hp whatever the mixture; d) for AGS05 mixture, when diffusion is included with or without rotationally induced transport, some intermediate amount of overshoot (≈ 0.2 Hp) still remains necessary. As a conclusion, microscopic diffusion and rotationally induced transport as modelled here are not enough to render count of the slope of d_{01} variation with AGS05. One needs to include also some amount of convective core overshoot as a proxy for true overshoot and/or additional mixing process.

Absolute frequencies and echelle diagram The echelle diagram displayed in Fig.2 shows a systematic shift between the observed and theoretical $l = 1$ ridges roughly independent of the frequency when the mean separation is taken to be the same in building the observed and theoretical echelle diagrams. This discrepancy comes from the fact that absolute values of the frequencies have not been included in the optimisation procedure to find an optimal model. Increasing slightly the mean separation for building the echelle diagram for the model enables to perfectly match the low frequency part of the observed ridge as seen in Fig.2. The deviation of the theoretical ridge with the observed ones remains at high frequencies and might reflect surface effects that are not properly taken into account in the numerical frequencies.

2.2.4. HD181420: a fast rotator Analyses of the light curve of this star has been performed by (25) and (39) who found two possible scenarii for the mode identification. Results based on a Bayesian approach favors scenario 1 Benomar (2010, priv.com). This is also the conclusion of

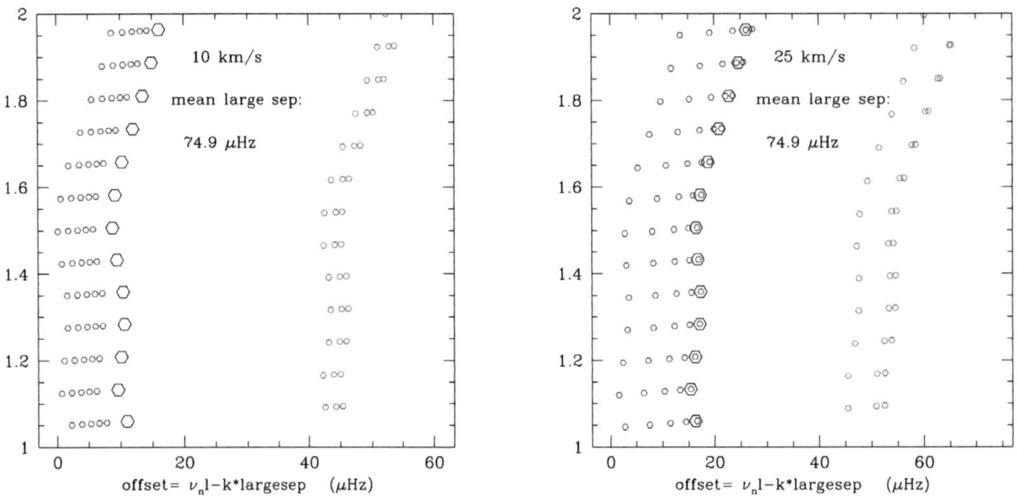

Figure 3. Echelle diagrams for a model of HD181420: rotational velocities $v = 10$ km/s **(left)** and 25 km/s **(right)** are assumed when computing the frequencies. From left to right in each pannel, ridges for $l = 2, 0, 1$ respectively appear.

(40) using scaling properties. Focusing then on scenario 1, one reproduces the large separation and the small spacing d_{01} with a $1.36M_{\odot}$ stellar model assuming a core overshoot of $0.2Hp$ as well as with a $1.37M_{\odot}$ stellar model assuming no overshoot, everything else being the same in particular microscopic diffusion and rotation are not included (45). A secondary oscillation component is seen in the observed large separation that is not reproduced by any models (45), Michel (2010 priv.com). The 'period' of this oscillation corresponds to the base of the convective zone of the above models, but the reality of this secondary component is not confirmed (26), Mosser (2010, priv. comm). This star is a 'rapid' rotator compared to the Sun. Indeed with a radius $R = 1.66R_{\odot}$ and the observed mean rotational splitting $\nu_{split} = (3. \pm 1)\mu Hz$ (25), one obtains a rotational velocity $v = 21.97.3$ km/s that corresponds to a ratio of the centrifugal to the gravitational accelerations of $\epsilon = \Omega^2/(GM/R^3) = 320\epsilon_{\odot}$! Perturbation methods to compute the effect of rotation on the frequencies nevertheless remain valid for this rotation rate (46). The non-spherically centrifugal distortion causes asymetries of split multiplets that are seen in echelle diagrams already for $v = 10$ km/s at high frequency (Fig.3, frequencies including rotating effects have been performed with the WarM oscillation code). Assuming a uniform rotation velocity of 25 km/s, the $m = 2$ components of the $l = 2$ modes coincide with the $l = 0$ mode for the lowest frequencies whereas at high frequencies both $m = 1$ and $m = 2$ components are mixed with the $l = 0$ frequency. Asymetries increase with frequency and are therefore larger at high frequencies where they contribute to surface effects! The difference between the rotation frequency measured at low frequency in a power spectrum (47) and the mean splitting (25) is found ot be compatible with a latitudinal dependence of the surface rotation of the star (Ouazzani, 2010 priv. com).

2.2.5. HD49385: an evolved star The light curve of this star has been analysed by (48) and (29). (29) clearly put in evidence the existence of an $l = 1$ avoided crossing and showed that it produces a characteristic distortion of the $l = 1$ ridge in the neighbour of this mode in an echelle diagram. A clear explanantion of the deformation of the ridge by the presence of an avoided crossing for a $l = 1$ mode has been provided by (48). The distorsion is caused by the

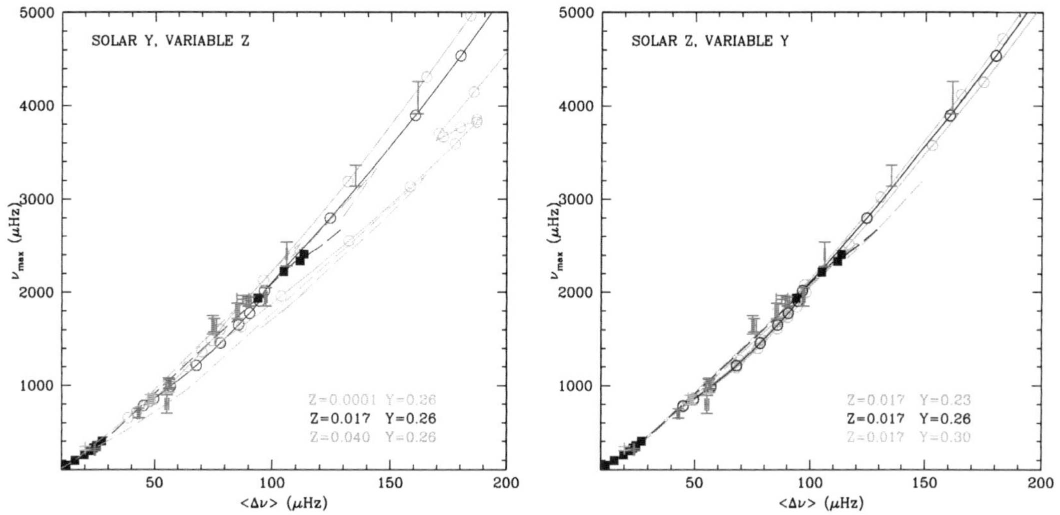

Figure 4. Frequency ν_{max} in function of the mean large separation $< \Delta\nu >$. Models (solid curves) have been computed with cesam2k code. Scaling laws are computed according to (49) and (50). Data for a few ground based observed stars and some CoRoT targets plotted with their errors bars represented by crosses.

fact that the modes propagate as p mode in a surface cavity and as g mode in a central cavity. For each mode, both cavities are separated by an evanescent region which acts as a coupling between the two regions. The magnitude of the ridge deformation is related to the properties of the evanescent region. The deformation of the ridge is then fitted to constrain the stellar model. A model satisfying all the constraints simultaneously then is difficult to obtain. Only a change in the chemical mixture from AGS05 to AGS09 is able to affect the evanescent region so as to provide a correct ridge deformation (41).

3. Kepler data and ensemble seismic investigations

The seismic part of the space mission Kepler produces seismic data for a huge number of stars for which stellar parameters are in general not well known. Ensemble investigations to derive stellar masses and radii then rely on scaling seismic properties (51). Fig.4 shows ν_{max} (frequency at maximum oscillation power in a power spectrum) in function of the mean large separation $< \Delta\nu >$ for series of stellar models (open circles) assuming different chemical compositions along evolutionary tracks from ZAMS (top right corner) to TAMS (down left corner). Data for a few ground based observed stars are overplotted as well as CoRoT targets. Due to NASA data policy, Kepler data are not shown. large The error bars indicate that one cannot distinguish between metallicities $Z = 0.017$ and $Z = 0.04$ (with $Y = 0.26$) for instance, nor between $Y = 0.23$ and $Y = 0.30$ ($Z = 0.017$). This leads to small uncertainties on masses and radii derived from these scalings. This is a crucial issue for the ESA project PLATO (52) that aims at detecting and studying earth-type exoplanets and therefore requires the determination of stellar masses of exoplanet host stars with an accuracy of about 10-15%. This means that it will be mandatory to determine the metallicity and helium abundances Z/X and Y by other means; for instance Z/X by spectroscopy and Y by the properties of the frequency dependence of the large separation. Another important issue is to establish the physical origin of the ν_{max} scaling which sofar has only been conjectured (53) but has been validated with observations (50). A better understanding of this origin could provide the explicit dependence of the scaling relations on metallicity and chemical composition, rotation etc ...

4. Conclusion:

With the wealth of data from CoRoT and Kepler, we are confronted with seismologial studies of a rich variety of low, and intermediate mass, main sequence stars that differ from the Sun in their stellar parameters and internal structure. Although this will enrich considerably our understanding of stellar physics and evolution, this also generates several difficulties not encountered with the Sun. Some were expected such as inaccurate determination of stellar parameters, degeneracy in parameter space and resulting non unicity of stellar models but some other problems were not really expected such as ambiguities in the mode identification due to broad linewidths and fast rotation (compared to the Sun) for F stars. So as far as probing internal structure by means of seismology of solar-like stars is concerned, we are therefore still at the beginning of the learning phase. Significant advances in stellar physics and solid conclusions about the open issues discussed in the present paper will require homogeneous detailed seismic studies for a larger number of individual stars that has been done so far. New roads also develope such as ensemble studies and scaling procedures and additional observational seismic constraints start to exist such as accurate mode amplitudes and linewitdhs and their variations with frequency that were not available from ground. Eventually studying a star and its planets as a global system is certainely the issue that must adressed in the future. In that framework, the perspective offered by the ESA project PLATO which has precisely this aim is a strong motivation for the seismic community to pursue its efforts and in turn PLATO will grandly benefit from all the forthcoming advances in the field.

Acknowledgments

We gratefully thank our colleagues K. Belkacem, J. Ballot, B. Mosser, C. Barban, T. Corbard, D. Reese, O. Creevey, A. Baglin, E. Michel, T. Appourchaux, C. Catala, A. Mazumdar for providing useful information and fruitful discussions when preparing this review. We also thank P. Morel and J. Christensen-Dalsgaard for providing public evolutionary and oscillation codes respectively that were used in the present review. We ackowledge financial support from CNES and the ANR SIROCO

References

[1] Baglin A, Auvergne M, Barge P, Deleuil M, Catala C, Michel E, Weiss W and The COROT Team 2006 *ESA Special Publication* (*ESA Special Publication* vol 1306) ed M Fridlund, A Baglin, J Lochard, & L Conroy p 33

[2] Borucki W J, Koch D G, Lissauer J, Basri G, Brown T, Caldwell D A, Jenkins J M, Caldwell J J, Christensen-Dalsgaard J, Cochran W D, Dunham E W, Gautier T N, Geary J C, Latham D, Sasselov D, Gilliland R L, Howell S, Monet D G and Batalha N 2007 *Transiting Extrapolar Planets Workshop* (*Astronomical Society of the Pacific Conference Series* vol 366) ed C Afonso, D Weldrake, & T Henning p 309

[3] Christensen-Dalsgaard J 2009 *ArXiv e-prints* (*Preprint* `0912.1405`)

[4] Christensen-Dalsgaard J and Houdek G 2010 *Ast. Sp. Sci* **328** 51–66

[5] Mosser B and Appourchaux T 2009 *Astr. Astroph* **508** 877–887

[6] Roxburgh I W 2009 *Astr. Astroph* **506** 435–444

[7] Monteiro M J P F G, Christensen-Dalsgaard J and Thompson M J 2000 *MNRAS* **316** 165–172

[8] Mazumdar A and Antia H M 2001 *Astr. Astroph* **368** L8–L12

[9] Mazumdar A and Antia H M 2001 *Astr. Astroph* **377** 192–205

[10] Mazumdar A and Michel E 2010 *ArXiv e-prints* (*Preprint* 1004.2739)

[11] Basu S, Mazumdar A, Antia H M and Demarque P 2004 *MNRAS* **350** 277–286

[12] Houdek G and Gough D O 2009 *Communications in Asteroseismology* **159** 27–29

[13] Provost J, Mosser B and Berthomieu G 1993 *Astr. Astroph* **274** 595–611

[14] Roxburgh I W and Vorontsov S V 1999 *Stellar Structure: Theory and Test of Connective Energy Transport* (*Astronomical Society of the Pacific Conference Series* vol 173) ed A Gimenez, E F Guinan, & B Montesinos p 257

[15] Roxburgh I W and Vorontsov S V 2002 *Stellar Structure and Habitable Planet Finding* (*ESA Special Publication* vol 485) ed B Battrick, F Favata, I W Roxburgh, & D Galadi pp 349–351

[16] Roxburgh I W and Vorontsov S V 2004 *Stellar Structure and Habitable Planet Finding* (*ESA Special Publication* vol 538) ed F Favata, S Aigrain, & A Wilson pp 403–406

[17] Mazumdar A, Basu S, Collier B L and Demarque P 2006 *Proceedings of SOHO 18/GONG 2006/HELAS I, Beyond the spherical Sun* (*ESA Special Publication* vol 624)

[18] Cunha M S and Brandão I M 2010 *ArXiv e-prints* (*Preprint* 1008.1299)

[19] Popielski B L and Dziembowski W A 2005 *Acta Ast.* **55** 177–193

[20] Silva Aguirre V, Ballot J, Serenelli A and Weiss A 2010 *Ast. Sp. Sci* **328** 129–133

[21] Provost J, Berthomieu G, Bigot L and Morel P 2005 *Astr. Astroph* **432** 225–233

[22] Miglio A and Montalbán J 2005 *Astr. Astroph* **441** 615–629

[23] Roxburgh I W and Vorontsov S V 2007 *MNRAS* **379** 801–806

[24] Deheuvels S, Michel E, Goupil M J, Marques J P, Mosser B, Dupret M A, Lebreton Y, Pichon B and Morel P 2010 *Astr. Astroph* **514** A31

[25] Barban C, Deheuvels S, Baudin F, Appourchaux T, Auvergne M, Ballot J, Boumier P, Chaplin W J, García R A, Gaulme P, Michel E, Mosser B, Régulo C, Roxburgh I W, Verner G, Baglin A, Catala C, Samadi R, Bruntt H, Elsworth Y and Mathur S 2009 *Astr. Astroph* **506** 51–56

[26] Benomar O 2010 *PhD thesis*

[27] Mathur S, García R A, Catala C, Bruntt H, Mosser B, Appourchaux T, Ballot J, Creevey O L, Gaulme P, Hekker S, Huber D, Karoff C, Piau L, Régulo C, Roxburgh I W, Salabert D, Verner G A, Auvergne M, Baglin A, Chaplin W J, Elsworth Y, Michel E, Samadi R, Sato K and Stello D 2010 *Astr. Astroph* **518** A53

[28] Mosser B, Deheuvels S, Michel E, Thévenin F, Dupret M A, Samadi R, Barban C and Goupil M J 2008 *Astr. Astroph* **488** 635–642

[29] Deheuvels S, Bruntt H, Michel E, Barban C, Verner G, Régulo C, Mosser B, Mathur S, Gaulme P, Garcia R A, Boumier P, Appourchaux T, Samadi R, Catala C, Baudin F, Baglin A, Auvergne M, Roxburgh I W and Pérez Hernández F 2010 *Astr. Astroph* **515** A87

[30] Benomar O 2010 *Astronomische Nachrichten, HELA IV proceedings* ed T Roca Cortés, P Pallé, S Jiménez Reyes,

[31] Gaulme P, Deheuvels S, Weiss W W, Mosser B, Moutou C, Bruntt H, Donati J, Vannier M, Guillot T, Appourchaux T, Michel E, Auvergne M, Samadi R, Baudin F, Catala C and Baglin A 2010 *ArXiv e-prints* (*Preprint* 1011.2671)

[32] García R A, Mathur S, Salabert D, Ballot J, Régulo C, Metcalfe T S and Baglin A 2010 *Science* **329** 1032–

[33] García R A, Régulo C, Samadi R, Ballot J, Barban C, Benomar O, Chaplin W J, Gaulme P, Appourchaux T, Mathur S, Mosser B, Toutain T, Verner G A, Auvergne M, Baglin A, Baudin F, Boumier P, Bruntt H, Catala C, Deheuvels S, Elsworth Y, Jiménez-Reyes S J, Michel E, Pérez Hernández F, Roxburgh I W and Salabert D 2009 *Astr. Astroph* **506** 41–50

[34] Angulo C, Arnould M, Rayet M and et al 1999 *Nuclear Physics A* **656** 3–183

[35] Formicola A, Imbriani G, Costantini H, Angulo C, Bemmerer D, Bonetti R, Broggini C, Corvisiero P, Cruz J, Descouvemont P, Fülöp Z, Gervino G, Guglielmetti A, Gustavino C, Gyürky G, Jesus A P, Junker M, Lemut A, Menegazzo R, Prati P, Roca V, Rolfs C, Romano M, Rossi Alvarez C, Schümann F, Somorjai E, Straniero O, Strieder F, Terrasi F, Trautvetter H P, Vomiero A and Zavatarelli S 2004 *Physics Letters B* **591** 61–68

[36] Appourchaux T, Michel E, Auvergne M, Baglin A, Toutain T, Baudin F, Benomar O, Chaplin W J, Deheuvels S, Samadi R, Verner G A, Boumier P, García R A, Mosser B, Hulot J, Ballot J, Barban C, Elsworth Y, Jiménez-Reyes S J, Kjeldsen H, Régulo C and Roxburgh I W 2008 *Astr. Astroph* **488** 705–714

[37] Benomar O, Appourchaux T and Baudin F 2009 *Astr. Astroph* **506** 15–32

[38] Benomar O, Baudin F, Campante T L, Chaplin W J, García R A, Gaulme P, Toutain T, Verner G A, Appourchaux T, Ballot J, Barban C, Elsworth Y, Mathur S, Mosser B, Régulo C, Roxburgh I W, Auvergne M, Baglin A, Catala C, Michel E and Samadi R 2009 *Astr. Astroph* **507** L13–L16

[39] Gaulme P, Appourchaux T and Boumier P 2009 *Astr. Astroph* **506** 7–14

[40] Bedding T R and Kjeldsen H 2010 *Communications in Asteroseismology* **161** 3–15

[41] Deheuvels S 2010 *PhD thesis*

[42] Canuto V M, Goldman I and Mazzitelli I 1996 *ApJ* **473** 550

[43] Bruntt H 2009 *Astr. Astroph* **506** 235–244

[44] Morel P and Lebreton Y 2008 *Ast. Sp. Sci* **316** 61–73

[45] Goupil M e a *HELAS Workshop on 'New insights into the Sun'* ed M Cunha, M Thompson

[46] Suárez J C, Goupil M J, Reese D R, Samadi R, Lignières F, Rieutord M and Lochard J 2010 *ApJ* **721** 537–546

[47] Mosser B 2010 *ArXiv e-prints* (*Preprint* `1006.5530`)

[48] Deheuvels S and Michel E 2010 *ArXiv e-prints* (*Preprint* `1006.3178`)

[49] Kjeldsen H and Bedding T R 1995 *Astr. Astroph* **293** 87–106

[50] Bedding T R and Kjeldsen H 2003 *PACA* **20** 203–212 (*Preprint* `arXiv:astro-ph/0305425`)

[51] Chaplin W J, Appourchaux T, Elsworth Y and et al 2010 *ApJ Lett.* **713** L169–L175

[52] Catala C, Arentoft T, Fridlund M, Lindberg R, Mas-Hesse J M, Micela G, Pollacco D, Poretti E, Rauer H, Roxburgh I, Stankov A and Udry S 2010 *Astronomical Society of the Pacific Conference Series* (*Astronomical Society of the Pacific Conference Series* vol 430) ed V Coudé Du Foresto, D M Gelino, & I Ribas p 260

[53] Brown T M, Gilliland R L, Noyes R W and Ramsey L W 1991 *ApJ* **368** 599–609

Low-Z solar model: sound speed profile under the convection zone

S V Ayukov and V A Baturin

Sternberg Astronomical Institute, Universitetsky pr. 13, Moscow, Russia

E-mail: asv@sai.msu.ru

Abstract. With the low solar heavy elements abundances published in 2004 discrepancy in sound speed profile between models and inversions became much larger than with older, higher abundances. We consider structure and source of the discrepancy. The most important source of discrepancy is shallow convection zone but making it deeper with localized opacity modifications does not lead to desired sound speed profile. Special opacity changes can produce models with sound speed profile which is close to inversion but these changes look fairly artificial. Enhanced heavy element diffusion improves agreement on He abundance but not on sound speed profile. We present solar models which include aforementioned effects or their approximations.

1. Introduction

Helioseismology provides means to verify models of the solar internal structure. The most important data used for this purpose are the sound speed profile, the convection zone depth R_{cz} and the helium abundance in the convection zone Y_{cz}. These are results of helioseismic inversion; in this work we're using compilation [1].

Since 1996 ([2]) until 2004 an agreement between models and inversions was quite remarkable. In 2004 solar composition was redetermined using new atmosphere models ([3], [4]); data were revised in 2009 ([5]). New abundances of many elements heavier than helium appeared to be significantly lower. Solar models computed with low abundances showed much worse agreement with inversion. We study low-Z problem and possible ways to obtain helioseismically consistent solar model with low-Z abundances.

Details of our model calculations are described in [6]. All low-Z models use AGSS09 ([5]) abundances and are calibrated to value of $Z/X = 0.0181$ which corresponds AGSS09.

2. Artificially enhanced diffusion

Enhanced diffusion was suggested by Asplund as a solution for low-Z problem ([3]) and later was investigated by many authors, including Basu and Antia ([7]), Montalban et al. ([8]), Bahcall et al. ([9]), Guzik et al. ([10]) and others. It was found that sound speed and convection zone (CZ) depth discrepancies can be significantly reduced in models with enhanced diffusion, but such models have rather low Y_{cz}. Even selective changes in diffusion rates ([10]) cannot produce model which gives good agreement on sound speed, Y_{cz} and R_{cz}.

In this work we enhance Z diffusion rate and don't change He diffusion rate. Let us remind that final model of the evolutionary track is adjusted to have predetermined Z/X (0.0181 for [5]). If we assume helioseismic calibration for $Y_{cz} \approx 0.25$, composition of the present Sun's envelope

Table 1. Solar models

Model	Mixture	Y_{cz}	Z_{cz}	Z/X	R_{cz}	
721-0001	GN93	0.242674	0.018045	0.024409	0.71509	high-Z, standard
721-0100	AGSS09	0.232267	0.013666	0.018123	0.72664	low-Z, standard
721-0113	AGSS09	0.249826	0.013353	0.018123	0.73095	enhanced Z diffusion
721-0102	AGSS09	0.224431	0.013805	0.018123	0.72506	no Z diffusion at all
721-0208	AGSS09	0.259377	0.013183	0.018123	0.72421	monotonic opacity change
721-0237	GN93	0.230775	0.018328	0.024409	0.72508	low-Z opacity in high-Z model
721-0238	AGSS09	0.243617	0.013464	0.018123	0.71770	high-Z opacity in low-Z model
721-0319	AGSS09	0.233292	0.013647	0.018123	0.71293	opacity correction near boundary CZ
721-0320	AGSS09	0.254679	0.013267	0.018123	0.71034	opacity corrections
721-0347	AGSS09	0.248932	0.013369	0.018123	0.71302	opacity corrections

becomes fixed; equation $X + 0.25 + X \cdot 0.0181 = 1$ gives $X = 0.7367$, $Z = 0.0133$. With so low Z in convection zone, enhanced Z diffusion increases Z below CZ boundary and partially restores model's structure. Our standard model with AGSS09 abundances (721-0100) has $X = 0.7541$, $Y = 0.232$, $Z = 0.0136$ in the CZ (Table 1). Increasing Z diffusion rate leads to decrease of Z in the envelope and, since Z/X is fixed, X is decreased and therefore Y_{cz} is increased. To obtain model with $Y_{cz} = 0.25$ we have to increase Z diffusion rate by 200% (721-0113 on Fig. 1).

Enhanced Z diffusion partially compensates overall decrease in Z caused by switching to low-Z abundances. To reach proper Y_{cz} we have to go about halfway to full compensation in Z below convection zone (Fig. 2).

Sound speed profile is not greatly affected by enhanced diffusion in the model calibrated to Z/X (Fig. 3; all sound speed comparisons are made against inversion [1]). The visible effects are probably caused by change of CZ depth. Therefore model with enhanced Z diffusion can have good Y_{cz} but not the proper sound speed profile and R_{cz}. In fact, such diffusion makes CZ even shallower than in standard low-Z model. This may look strange because we're increasing Z

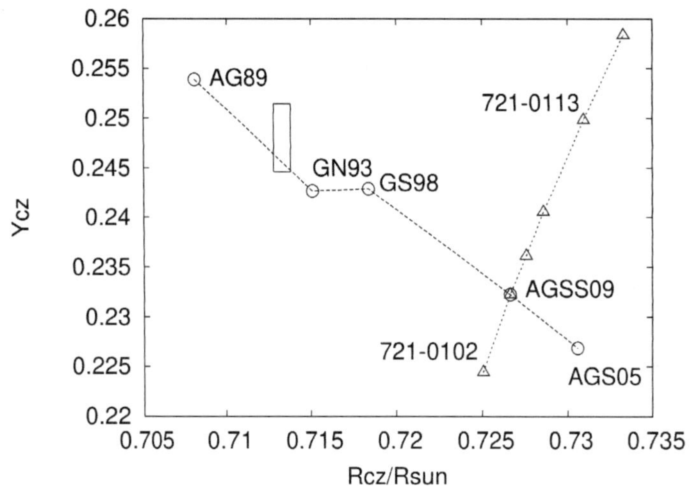

Figure 1. Helium abundance and convection zone boundary position. Circles—standard models with different composition, triangles—models with enhanced Z diffusion. Rectangle—results of helioseismic inversion ([1])

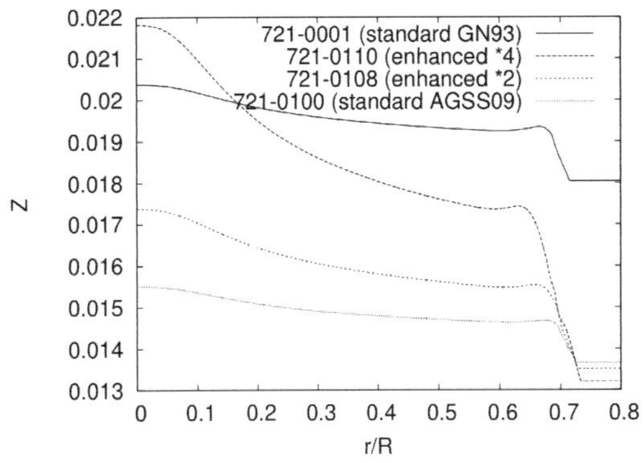

Figure 2. Models with enhanced Z diffusion; Z near convection zone boundary

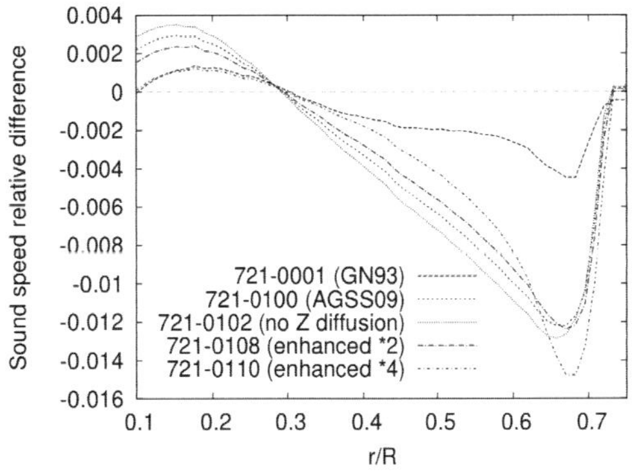

Figure 3. Sound speed comparison for models with enhanced heavy element diffusion

(and opacity) directly below CZ. However increasing opacity in the center makes CZ shallower (see 3.4 below), and since Z increase in the center is even larger than in the radiative zone this effect apparently suppresses Z increase in the radiative zone.

On the other hand, if we decrease Z diffusion rate or even disable Z diffusion, we get better (but still small) CZ depth and improper Y_{cz} and sound speed profile (Fig. 1, 3).

3. Opacity modifications

Opacity change was one of the first ideas to solve the low-Z problem. Basu and Antia ([7]) have found that opacity increase of about 19% near the CZ boundary improves density profile. Montalban ([8]) has suggested 14% opacity increase. Bahcall et al. ([9]) have made an extensive study and found that 11% increase in opacity in the region $T = 2 - 5 \cdot 10^6$ K helps to restore agreement. However not all of these works used models calibrated to exact value of Z/X, and we are not aware of any model with low-Z abundances which reproduces sound speed, Y_{cz} and R_{cz}.

Our models are calculated with OPAL opacities, 19-component interpolation scheme ([6]) and AGSS09 mixture. Introduced opacity modification factors are simple temperature-dependent functions.

3.1. Effect of low-Z abundances on opacity

To estimate the effect of low-Z composition on opacity, we have calculated opacity values for the solar conditions. Two calculations were made and compared. Both use the same temperature and density values taken from model with AGSS09 composition. X and total Z values, however, are taken from the two models with different mixtures. Thus we eliminate temperature and density dependence and consider only effect of chemical composition on the opacity at the conditions in the present Sun. Fig. 4 shows that low-Z effect on opacity is 3% decrease in the center and 17% near the bottom of the convection zone. This result is close to the plot published in [9].

We can introduce opacity correction resembling this profile. This correction is plotted on Fig. 4 as a function of $\lg T$; simple polynomial approximation is used. Correction in the atmosphere is excluded, but detailed computations confirm that it does not affect sound speed profile below the convection zone (only convection theory parameter α is affected by opacity in this region). Model with high-Z abundances and such opacity correction (721-0237) can help to understand the role of the opacity in model. It has sound speed profile very close to the proper

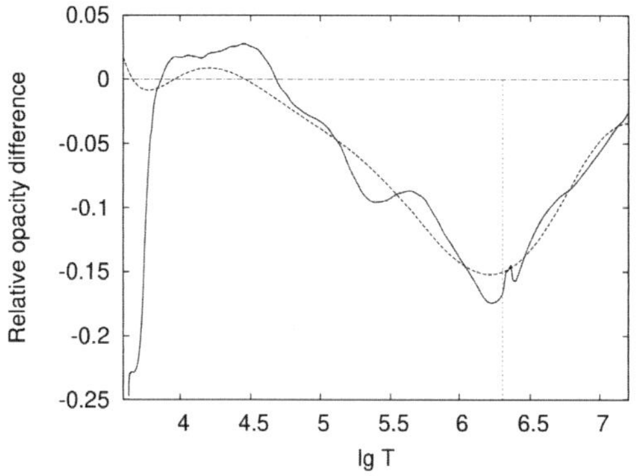

Figure 4. Opacity difference between low-Z and high-Z abundances (solid) and approximation of that difference (dashed)

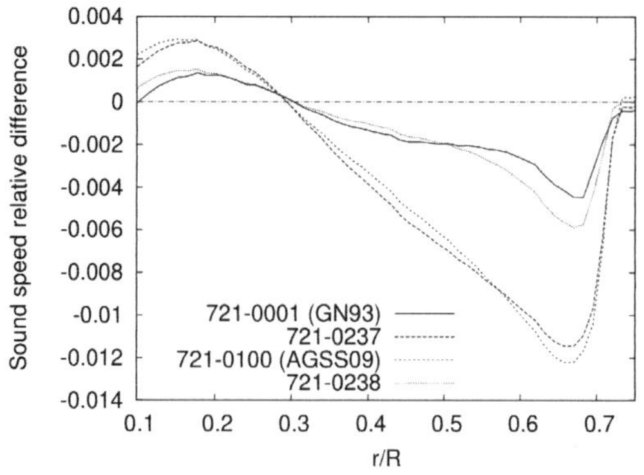

Figure 5. High-Z opacities in low-Z model (721-0238), low-Z opacities in high-Z model (721-0237)

low-Z model (721-0100). Density comparison plot is similar.

Comparison of opacity itself in these models (interpolated to the same radius points) shows differences about 4% (compare with correction amplitude of about 17%), so we did not achieve full compensation. Given that opacity correction can reproduce sound speed profile it is instructive to check how these models (721-0100 with AGSS09 and 721-0237 with GN93+opacity correction) differ in other parameters. The radiative interior did not change much. What we see is change in Z and change of convection parameter α. X and Y are almost fully compensated by opacity change, unlike Z. In other words, if we decrease opacity according low-Z abundances we get a model which is very close to original but has low Z everywhere.

This correction can be also applied in opposite direction, i.e. increase of opacity in the low-Z model (721-0238). This gives sound speed profile close to the high-Z sound speed profile (Fig. 5). Fit to heliosesmic sound speed profile can improved further by adding more local corrections (721-0320; Fig. 10).

3.2. Global opacity correction

Several models were computed with opacity uniformly increased everywhere by 5%, 10%, 15%, 20%. The effect on R_{cz} is noticeably small, and sound speed profile is almost unchanged even with 20% opacity increase (Fig. 6). This is somewhat surprising. It seems global opacity increase leads to overall decrease in Z (to keep opacity approximately at the same values) and, since Z/X is fixed in CZ, X also decreases. This in turn increases Y. Therefore global correction can help with low Y in low-Z models, but does not change R_{cz} and sound speed profile. This is illustrated by Fig. 7; models with such opacity correction have nearly the same R_{cz}.

3.3. Opacity correction localized at lower CZ boundary

Opacity correction localized near CZ boundary was also analyzed. Such correction affects only region near CZ boundary (Fig. 8). Model 721-0319 (AGSS09 composition; 24% increase in opacity; Fig. 9) has proper convection zone depth, but sound speed looks good only directly below CZ; in the deeper layers it is unchanged. Therefore such correction can produce model with proper CZ depth but sound speed in the radiative zone is not helioseismically consistent. Helium abundance in the envelope is almost unaffected by such correction.

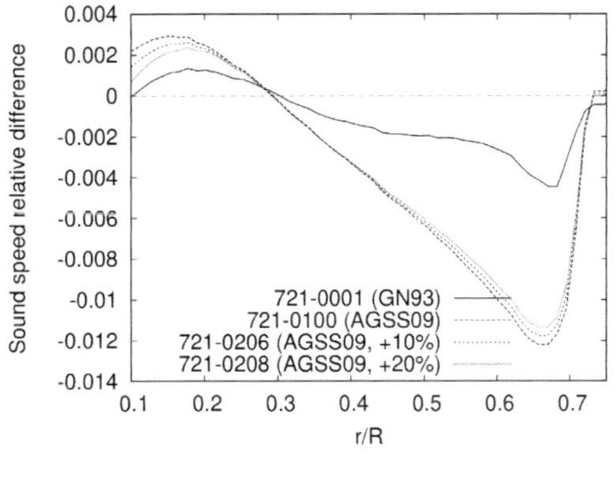

Figure 6. Sound speed profile in models with monotonic global opacity change

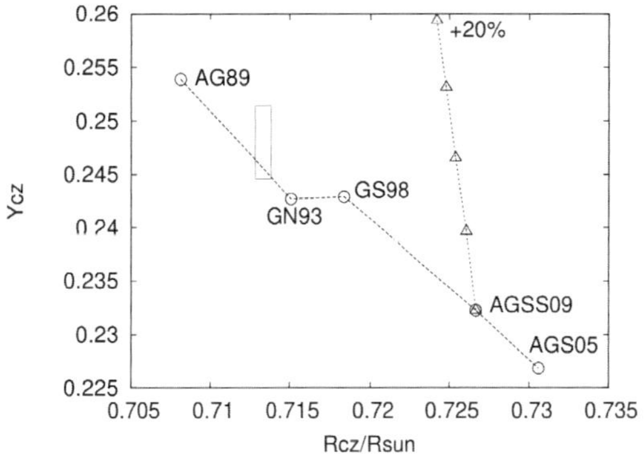

Figure 7. Models with global opacity change (triangles) on $Y - R_{cz}$ plane

3.4. Corrections localized in the core and radiative zone

To obtain model with helioseismically consistent sound speed profile we applied a set of opacity corrections localized in the core and radiative zone. Trial-and-error technique was used to search for appropriate set; result was judged by model's sound speed profile. It was found that simplest opacity change in fact is not monotonic; opacity is to be significantly decreased in central regions. Model with such correction has proper sound speed and R_{cz} but low Y abundance in the envelope. Since global opacity correction can change Y_{cz}, we can add 19% global correction to obtain helioseismically consistent model (721-0347; Fig. 10, 11).

It was also found that increasing opacity in the core makes convection zone shallower while increasing opacity below CZ makes CZ deeper. The switching point lies at $\lg T = 6.7$ in our models; figure is not included due to space constraints.

4. Conclusions

As expected, no natural solution for low-Z problem has been found. We have attempted to apply enhanced heavy element diffusion and several kinds of opacity correction. Opacity corrections are able to bring the model into agreement with helioseismic data (sound speed profile, convection zone depth and helium abundance in the envelope). The required correction amplitude is quite

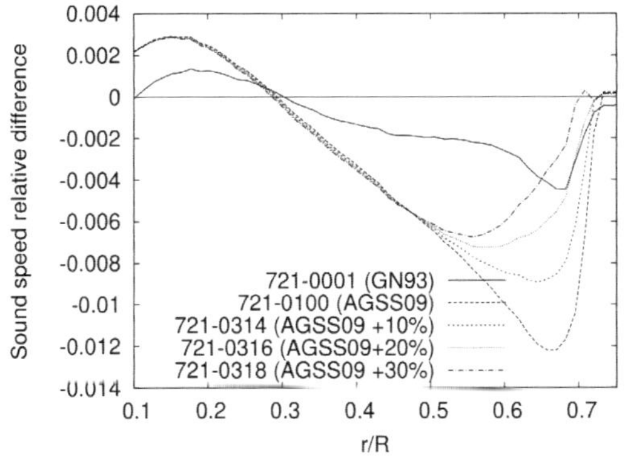

Figure 8. Sound speed in models with opacity correction localized near CZ boundary

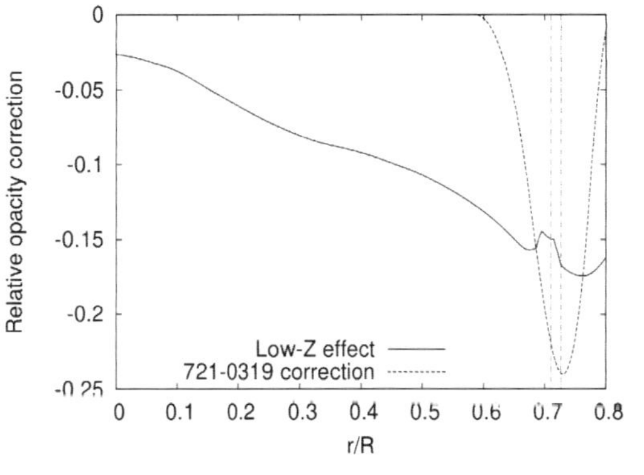

Figure 9. Opacity correction in model 721-0319

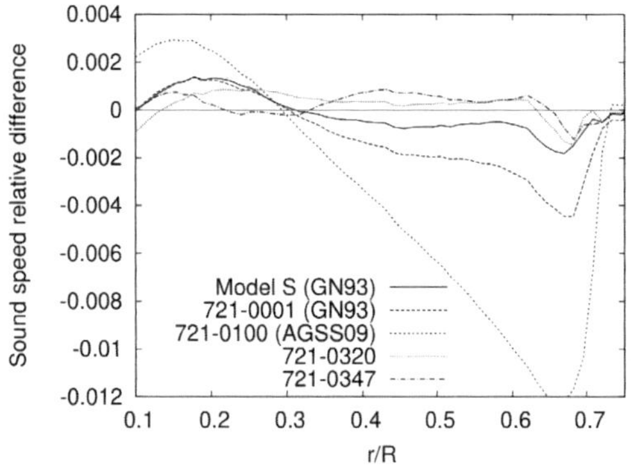

Figure 10. Sound speed in helioseismically consistent models with low-Z abundances and opacity correction (721-0320, 721-0347)

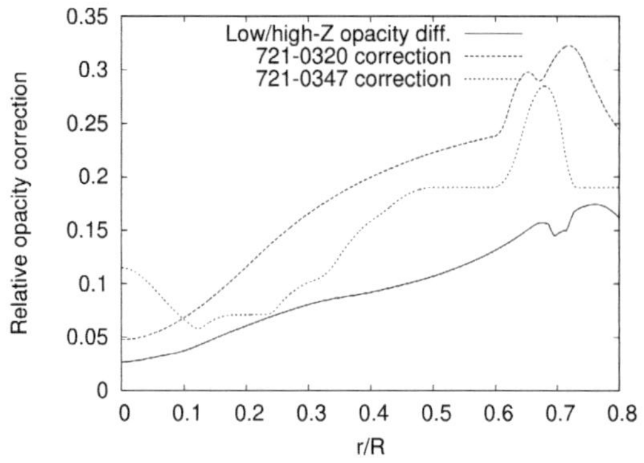

Figure 11. Opacity correction amplitude for helioseismically consistent models 721-0320, 721-0347

large and reaches 30% in the vicinity of the CZ base (Fig. 11). However our current high-Z model also has sound speed profile which significantly deviates from inversion, so high-Z model also needs opacity correction but required amplitude is approximately two times smaller than for low-Z model.

Models with opacity corrections near CZ base demonstrate that core and radiative zone in low-Z models are in disagreement with inversions and cannot be improved simply by changing CZ depth.

In author's opinion a solution to low-Z problem may be either in error of abundances and/or opacities or in existence of some mechanism which creates a gradient in chemical composition between deep layers of the convection zone and atmosphere.

Acknowledgements

This work was supported by ISTC grant 3755.

References

[1] Basu S and Antia H M 2008 *Phys. Rep.* **457** 217
[2] Christensen-Dalsgaard J et al. 1996 *Science* **272** 1286
[3] Asplund M, Grevesse N, Sauval A J, Allende Prieto C and Kiselman D 2004 *Astron. Astrophys.* **417** 751
[4] Asplund M, Grevesse N and Sauval A J 2005 *Cosmic Abundances as Records of Stellar Evolution and Nucleosynthesis* ed T G Barnes and F N Bash *ASP Conf. Ser.* **336** pp 25–38 (AGS05)
[5] Asplund M, Grevesse N, Sauval A J and Scott P 2009 *Annu. Rev. Astron. Astrophys.* **47(1)** 481 (AGSS09)
[6] Ayukov S V and Baturin V A 2010 *Astrophys. Space Sci.* **328** 23–7
[7] Basu S and Antia H M 2004 *Astrophys. J.* **606** L85–L88
[8] Montalban J, Miglio A, Noels A, Grevesse N and Di Mauro M P 2005 *Proc. SOHO 14/GONG 2004 Helio- and asteroseismology: Towards a golden future* ed D Danesy (New Haven Connecticut USA) ESA **SP-559** 574–6
[9] Bahcall J N, Basu S, Pinsonneault M and Serenelli A M 2005 *Astrophys. J.* **618** 1049–56
[10] Guzik J A, Watson L S and Cox A N 2005 *Astrophys. J.* **627** 1049–56

Are recent solar heavy element abundances consistent with helioseismology?

H. M. Antia[1] and Sarbani Basu[2]

[1] Tata Institute of Fundamental Research, Homi Bhabha Road, Mumbai 400005, India

[2] Department of Astronomy, Yale University, P. O. Box 20801, New Haven CT 06520-8101, U. S. A.

E-mail: antia@tifr.res.in, sarbani.basu@yale.edu

Abstract. During the last decade the abundances of heavy elements in the Sun have been revised downwards leading to serious discrepancy between solar models constructed using these abundances and the available seismic data. Much of these downward revision of abundances of Oxygen and other light elements was attributed to use of improved 3D solar atmospheric models. Recently, independent 3D models have been used to calculate solar abundances of these elements and calculated values are higher than the earlier estimates also obtained using 3D atmospheric models. In this work we investigate if these revised abundances are consistent with seismic data. We also investigate whether an increase in Neon abundance can help in resolving the discrepancy.

1. Introduction

During the last decade solar abundances of oxygen and other heavy elements have been revised significantly. For example, Grevesse & Sauval (1998, henceforth GS98) determined oxygen abundance to be 8.83 ± 0.06 dex, which was revised to 8.66 ± 0.05 by Asplund et al. (2005, henceforth AGS05) leading to serious discrepancy between solar models constructed using these revised abundances and the seismically inferred solar structure. Much of the differences in structure can be attributed to a mismatch between the position of the convention-zone base in these models compared to that in the Sun. Additionally, models constructed with the AGS05 abundances have a lower helium abundance in the convection zone compared with the seismic estimates of the convection-zone helium abundance of the Sun. The main cause of the revision in solar abundances was attributed to the use of improved 3D models of solar atmosphere in place of 1D models used until then. The main effect of the abundances on solar structure is through reduced opacities in the radiative interior. Considerable effort has been put into resolving this discrepancy both by modifying the solar models by changing some of the input physics like diffusion rates (see e.g., Basu & Antia 2008) as well as by independent determination of abundances. Modifications in solar models have not helped in resolving the discrepancy. Recently, Caffau et al. (2010, henceforth Caf10) have used independent 3D models to determine an oxygen abundance of 8.76 ± 0.07 which lies between the GS98 and AGS05 values, the Asplund group has revised their value to 8.69 ± 0.05 (Grevesse et al. 2010; henceforth GASS10), which is only slightly higher than their earlier, AGS05 value. While the GASS10 value is consistent with Caf10 within 1σ of uncertainties, the Caf10 value is consistent with the abundance of GS98.

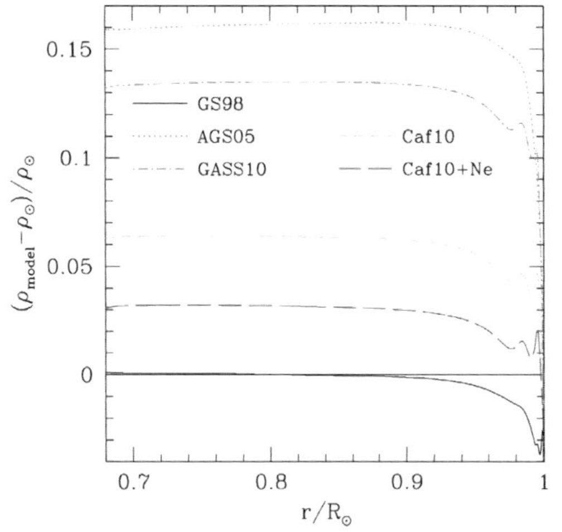

Figure 1. The relative difference in density between solar envelope models constructed with different abundances and the Sun. All models have the seismically determined value of $X = 0.739$ and position of the base of the convection zone $r_b = 0.7133R_\odot$.

Because of the revisions, the ratio of heavy element to hydrogen abundance, Z/X in the Sun has changed too: it was 0.023 for GS98, 0.0165 (AGS05) 0.0209 (Caf10) and 0.0181 (GASS10).

In this work we check whether models constructed with the latest estimates of abundances are consistent with helioseismic constraints. One of the solutions that had been proposed to alleviate the problem with the solar models constructed with AGS05 abundances was to increase the abundance of Neon since its photospheric abundance is uncertain (Antia & Basu 2005; Bahcall et al. 2005), we take a second look at this suggestion. We first look at density differences in solar envelope models with different Z/X to determine how much opacity modifications will be needed to get a model that agrees with seismic constraints, we then look at full, evolutionary solar models to look at discrepancies in all parts of the Sun and lastly, we try to see if the ionisation-zones can tell us something about which of the abundance estimates is closer to the Sun.

2. Results using envelope models

Following the approach of Antia & Basu (2005), we determine the convection-zone density differences between solar envelope models constructed with different abundances and the Sun. The envelope models were constructed with the seismically determined value of $X = 0.739$ (Basu & Antia 2004) and the position of the convection-zone base $r_b = 0.7133R_\odot$ (Basu & Antia 1997, 2004). With these constraints, the sound-speed profile in the convection zone is very similar to the seismically determined solar profile, differences in structure caused by differences in abundance show up as density differences and the results are shown in Fig. 1. As expected from earlier results, the GS98 model agrees well with the Sun, the AGS05 model does not. What we find is that while the GASS10 model fares better than AGS05, it is still very discrepant. The convection-zone density in Caf10 models is 6% larger than that in the Sun. If neon abundance is increased by a factor of 2, the density profile in the corresponding envelope model matches the seismic profile in the convection zone. If the neon abundance is increased by a factor of $\sqrt{2}$ the results are in between as shown in Fig. 1. It should be noted that all models were constructed with OPAL opacities (Iglesias & Rogers 1996) that were calculated for the respective mixtures. Since Caf10 do not give abundances of all required elements, abundances from Lodders et al. (2009) were used for the remaining elements.

By comparing differences in the density profile in the lower convection zone between the envelope models with the seismically derived solar density profile, it is possible to estimate the opacity change needed near the base of the convection zone to get an envelope model with a given

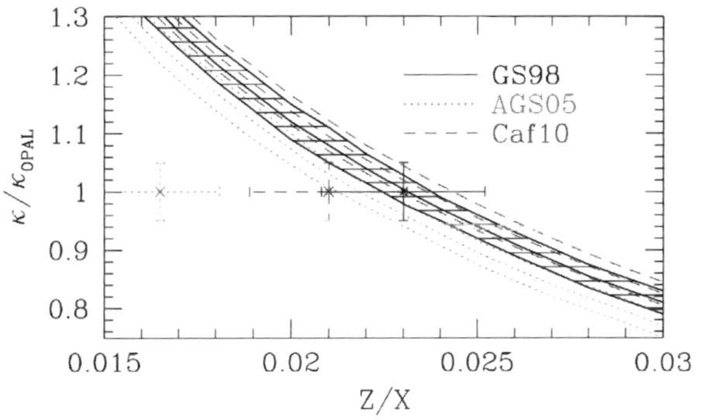

Figure 2. Opacity modification as a function of Z/X for different heavy element mixtures needed for the corresponding envelope models to satisfy the seismic density constraint in the convection zone.

value of Z/X and a given relative abundance to match the Sun. For this purpose we multiply OPAL opacities by a fixed factor and determine the factor needed to give density profile within 1.5% of the seismic value. This error includes systematic errors in inversions and solar models. The results for different mixtures are shown in Fig. 2. The actual value of Z/X of the respective mixture is marked by the points with errorbars. It is clear that GS98 abundances are consistent with seismic constraint, while the AGS05 values are well outside the constraints. The errorbars for the recent abundances of Caf10 graze the allowed region.

3. Results with full models

Since solar envelope models do not satisfy constraints in the inner parts of the Sun very well, we have also used full, evolutionary solar models to study the effect of the abundances. The models with different abundances were constructed using YREC the Yale Rotating Evolutionary Code in its non-rotating configuration (Demarque et al. 2008). We compare the sound-speed and density profiles of these models with those of the Sun in Fig. 3. As can be seen from Fig. 3, the best agreement is obtained with GS98 abundances, but the model using Caf10 abundances is also fairly close. On the other hand, models with AGS05 and GASS10 abundances are significantly worse.

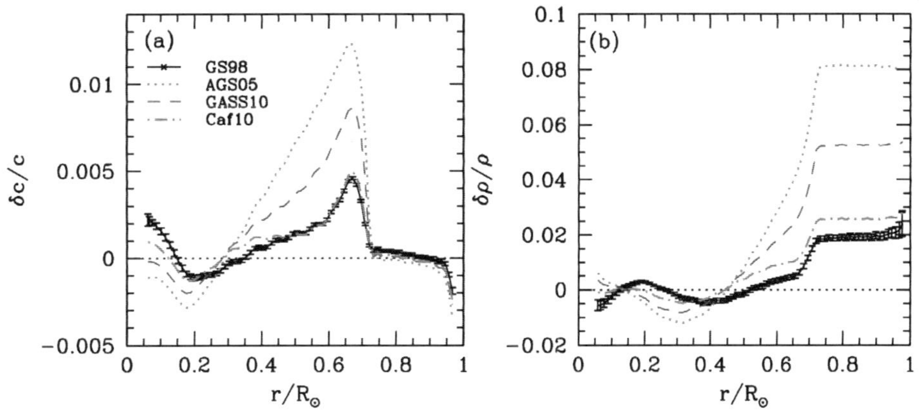

Figure 3. Relative difference in sound speed and density profile between solar models and the Sun.

All these models have been constructed using the OPAL opacities for the respective mixtures. If we construct a model using the Z/X value of Caf10 but using opacities calculated for the GS98 mixture, the resulting model is significantly worse. This can be seen in Fig. 4 as the model

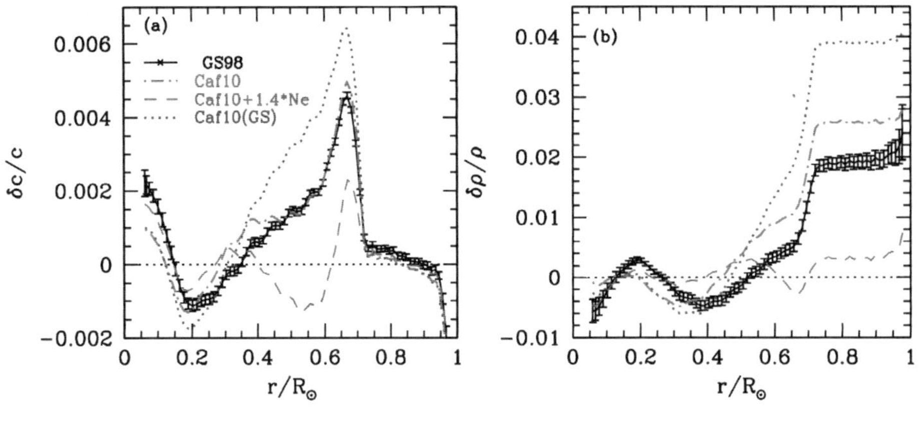

Figure 4. Relative difference in sound speed and density profile between a few other solar models and the Sun.

marked Caf10(GS). Thus it is clear that not only the total abundance, but the heavy element mixture also makes a significant difference and hence consistent opacities are a must. As in the case of envelope models, we have also constructed a neon-enhanced Caf10 model where the neon abundance is increased by a factor of $\sqrt{2}$. This model is also shown in Fig. 4 and as can be seem it is in better agreement with the Sun than even the GS98 model. All these models were constructed using OPAL opacities for the respective mixtures. Models with OP opacities (Badnell et al. 2005) will give slightly better agreement.

Unlike in envelope models, the convection-zone depth and envelope helium abundance in full models cannot be specified *a priori*. These are determined by the constraint that the models have the correct radius and luminosity at the current age of the Sun, and hence test how close to the Sun the models are and provide additional tests. Recently, Serenelli & Basu (2010) found that the initial abundance that the Sun was born with is $Y_{\rm ini} = 0.273 \pm 0.006$(random)$\pm 0.002$(systematic). In models, $Y_{\rm ini}$ is adjusted to satisfy the luminosity constraint, and thus $Y_{\rm ini}$ is another test of the models. In Table 1, we compare the position of the convection-zone base, the current surface helium abundance and the initial helium abundance of the models shown in Figures 3 and 4 with the seismically derived quantities for the Sun. It can be seen that while models using GS98 or Caf10 abundances are in reasonable agreement with the seismic values, the model with AGS05 and GASS10 abundances have values which are significantly different from seismic constraints.

4. The ionisation zones

The effect of abundances on solar models is primarily due to change in opacities. Nevertheless, the change in Equation of State (EOS) also makes a small difference, which can be noticed in the convection zone, where stratification is determined by the EOS. In the ionisation zones of various elements the adiabatic index, Γ_1 is reduced below its ideal gas value of 5/3 and this can be measured by inverting oscillation frequencies. A similar technique has been used to determine the helium abundance in the solar convection zone (see Basu & Antia 2008 for a review). In addition to Γ_1 it is also possible to use the dimensionless sound speed gradient (Gough 1984),

$$W(r) = \frac{r^2}{Gm}\frac{dc^2}{dr}, \tag{1}$$

where, G is the gravitational constant and m is the mass enclosed in a sphere of radius r. $W(r)$ has the ideal-gas value of $-2/3$ everywhere in the convection zone except at the ionisation

Table 1. The position of the base of the convection zone, r_b, and the helium abundance in the convection zone, Y_{CZ}, and the initial helium abundance, Y_{ini}, in solar models with different abundances.

Abundances	Z/X	$r_b/R_\odot$	Y_{CZ}	Y_{ini}
Observed		0.7133 ± 0.0005	0.2485 ± 0.0035	0.273 ± 0.006(rand) ± 0.002(syst)
GS98	0.0230	0.7154	0.2464	0.2768
AGS05	0.0165	0.7272	0.2296	0.2601
Caf10	0.0209	0.7166	0.2425	0.2725
GASS10	0.0181	0.7225	0.2363	0.2666
Caf10+Ne*1.4	0.0219	0.7127	0.2460	0.2756
Caf10(GS)	0.0209	0.7181	0.2395	0.2696

zones of different elements. The deviation from $-2/3$ caused by the HeII ionisation zone has been used to determine the solar helium abundance (e.g., Antia & Basu 1994). The ionisation zones of different heavy elements overlap making it almost impossible to separate out the effect of individual elements, However, the combined effect of all heavy elements is discernible in inversion results (Antia & Basu 2006). Figure 5, shows $W(r)$ for a few solar models with different abundances. Also shown are $W(r)$ values for the Sun as inferred from inversions of GONG and MDI data sets. We show solar results for all available data sets from both projects and the resulting lines combine to give bands which give an estimate of errors in the inversion results. Since the OPAL EOS is not available for different mixtures, we use the CEFF EOS (Christensen-Dalsgaard & Dappen 1992) in these calculations. As expected, the GS98 model agrees quite well with the observations, while AGS05 does not. The Caf10 model is close to observations, though there are some differences. Increasing neon abundance helps in improving the agreement of the Caf10 model near the base of the convection zone, a neon increase also helps the AGS05 model. It is unfortunate that none of the more sophisticated EOSs are available with different mixtures. Since the detailed structure of Γ_1 and $W(r)$ depend on the EOS for a given set of abundances, availability of the the better EOSs for more mixture will allow a more reliable tests for models constructed with different abundances.

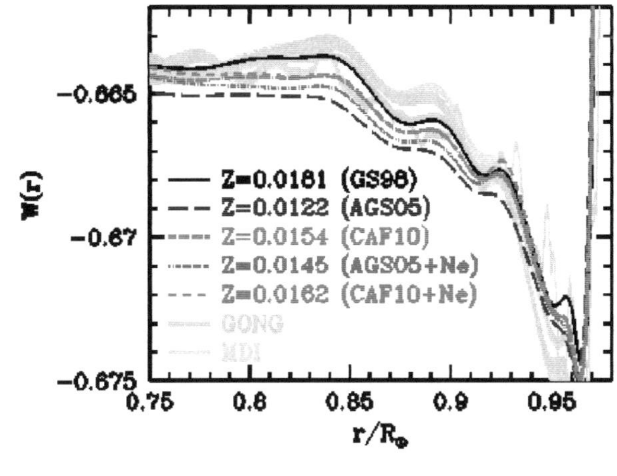

Figure 5. The dimensionless sound speed gradient, $W(r)$ in the convection zone in solar envelope models with various abundances is compared with that inferred from observed data from GONG and MDI. The model AGS05+Ne is with neon abundance increased by a factor of 2, while model Caf10+Ne is with neon abundance increased by a factor of $\sqrt{2}$.

5. Summary

We have compared the structure of models constructed using different heavy-element abundances with the seismically determined structure of the Sun. In particular, we have compared the sound-speed and density profiles, the position of the convection-zone base, the current and initial helium abundances, as well as signatures of the ionisation zones.

We find that although the Caf10 abundances are lower than the GS98 abundances, the solar models constructed with Caf10 abundances show almost as good an agreement with the Sun as do models constructed with GS98 abundances. This is a result of the very different *relative* abundances of heavy elements in the two mixtures which results in a different opacity at the same temperature and density. In fact, a model constructed with the Caf10 value of Z/X but with the relative abundances of GS98 [the model we call Caf10(GS)] fares much worse. This shows that it is extremely important to use opacities calculated for the correct mixture before constructing a solar model. A marginal increase in neon abundance by a factor of $\sqrt{2}$ can improve the agreement still further. Models with abundances from AGS05 or GASS10 are not consistent with seismic constraints even if neon abundance is increased by a factor of 2.

Acknowledgements

This work utilises data obtained by the Global Oscillation Network Group (GONG) project, managed by the National Solar Observatory, which is operated by AURA, Inc. under a cooperative agreement with the National Science Foundation. The data were acquired by instruments operated by the Big Bear Solar Observatory, High Altitude Observatory, Learmonth Solar Observatory, Udaipur Solar Observatory, Instituto de Astrofisico de Canarias, and Cerro Tololo Inter-American Observatory. This work also utilises data from the Solar Oscillations Investigation/ Michelson Doppler Imager (SOI/MDI) on the Solar and Heliospheric Observatory (SOHO). SOHO is a project of international cooperation between ESA and NASA. SB acknowledges support from NSF grant ATM 0348837.

References

[1] Antia H M and Basu S 1994 *ApJ* **426**, 801
[2] Antia H M and Basu S 2005 *ApJ* **620**, L129
[3] Antia H M and Basu S 2006 *ApJ* **644**, 1292
[4] Asplund M, Grevesse N and Sauval A J 2005 in *Cosmic Abundances as Records of Stellar Evolution and Nucleosynthesis*, eds., T. G. Barnes, F. N. Bash, ASP Conf. Ser. **336**, p. 25 (AGS05)
[5] Badnell N R, Bautista M A, Butler K, Delahaye F, Mendoza C, Palmeri P, Zeippen C J and Seaton M J 2005 *MNRAS* **360**, 458
[6] Bahcall J N, Basu S and Serenelli A M 2005 *ApJ* **631**, 1281
[7] Basu S and Antia H M 1997 *MNRAS* **287**, 189
[8] Basu S and Antia H M 2004, ApJ 606, L85
[9] Basu S and Antia H M 2008 *Phys. Rep.* **457** 217
[10] Caffau E, Ludwig H-G, Steffen M, Freytag B and Bonifacio P 2010 *Solar Phys.* (in press) (Caf10)
[11] Christensen-Dalsgaard J and Däppen W 1992 *Astron. Astrophys. Rev.* **4**, 267
[12] Delahaye F and Pinsonneault M H 2006 *ApJ* **649**, 529
[13] Demarque P, Guenther D B, Li L H, Mazumdar A and Straka C W 2008 *Astrophys. Sp. Sci* **316**, 31
[14] Gough D O 1984, *Mem. Soc. Astron. Ital.* 55, 13
[15] Grevesse N and Sauval A J 1998 *Space Sci. Rev.* **85**, 161 (GS98)
[16] Grevesse N, Asplund M, Sauval A J and Scott P. 2010 *Astrophys. Space Sci.* (in press) (GASS10)
[17] Iglesias C A and Rogers F J 1996 *ApJ* **464**, 943
[18] Lodders K, Palme H and Gail H-P 2009 *Astronomy, Astrophysics, and Cosmology, B: Solar System*, Landolt–Börnstein, New Series **VI/4**, Springer, Berlin, p. 560
[19] Serenelli A M and Basu S 2010 *ApJ* **719** 865

Theoretical and experimental activities on opacities for a good interpretation of seismic stellar probes

S Turck-Chièze[1], G Loisel[1,2], D Gilles[1], F Thais[2], S Bastiani[3], C Blancard[4], M Busquet[5], T Caillaud[4], P Cosse[4], T Blenski[2], F Delahaye[6], J E Ducret[1], G Faussurier[4], F Gilleron[4], J Guzik[7], J W Harris[8], D P Kilcrease[7], N H Magee[7], L Piau[1], J C Pain[4], M Poirier[2], Q Porcherot[4], C Reverdin[4], V Silvert[4], B Villette[4], C Zeippen[6]

[1]IRFU/ CEA/ CE Saclay, F-91190 Gif sur Yvette, [2] IRAMIS/ CEA/ CE Saclay, F-91190 Gif sur Yvette, [3]LULI, Ecole Polytechnique, CNRS, CEA, UPMC, F-91128 Palaiseau Cedex, [4]CEA DAM/DIF, F-91297 Arpajon, [5]ARTEP, Ellicott City, MD 21042, USA, [6]LERMA and LUTH, Observatoire de Paris, ENS, UPMC, UCP, CNRS, 5 Place Jules Janssen, F-92195 Meudon, [7]Theoretical Division LANL, Los Alamos NM 87545,USA, [8]AWE, Reading, Berkshire, RG7 4PR, UK

E-mail: sylvaine.turck-chieze@cea.fr

Abstract. Opacity calculations are basic ingredients of stellar modelling. They play a crucial role in the interpretation of acoustic modes detected by SoHO, COROT and KEPLER. In this review we present our activities on both theoretical and experimental sides. We show new calculations of opacity spectra and comparisons between eight groups who produce opacity spectra calculations in the domain where experiments are scheduled. Real differences are noticed with real astrophysical consequences when one extends helioseismology to cluster studies of different compositions. Two cases are considered presently: (1) the solar radiative zone and (2) the beta Cephei envelops. We describe how our experiments are performed and new preliminary results on nickel obtained in the campaign 2010 at LULI 2000 at Polytechnique.

1. Introduction

Opacity coefficients are basic elements of stellar equations like equation of state and reaction rates in the stellar zones where the radiative gradient is smaller than the adiabatic one, that means the regions where the transport of energy is dominated by the photon interaction with matter. These coefficients, $\kappa(T, \rho, X_i)$ expressed in cm^2/g, represent the interacting cross sections of photon with matter. They are calculated at each mesh of a stellar model in radius and time and depend strongly on the temperature and composition but less on the density. In stellar interiors, these interactions are considered in Local Thermodynamic Equilibrium and quasi instantaneous. This last point could appear less justified than the first one due to the stochastic displacement of photons but stays a reasonable approximation to follow the great stages of evolution which describe the radiative transport of energy. It was established several decades ago, that these cross sections correspond to the Rosseland mean values of the corresponding photon spectrum within the diffusion approximation (Cox & Giuli, 1968; Clayton 1983).

[1] J E Ducret, Present address: CELIA, Bordeaux, France

The corresponding complex spectra must contain all the different processes that the different constituants of the plasma experience and they strongly depend on the degree of ionization of each species. It is known that these spectra result from a well knowledge of atomic and plasma physics. Their production needs a dedicated huge work which has been mainly done by Los Alamos (Huebner et al. 1977) and Livermore (Iglesias, Rogers & Wilson, 1987) groups producing tables for the astrophysical uses. Then the completude of elements and a lot of corrections have been introduced in the OPAL tables of Livermore (Iglesias & Rogers 1996) following also some comparisons with first experiments. During the last decade, another team has produced OP tables and spectra for astrophysical application (Seaton 2005 and references therein). It is important to recall that the opacity spectra are not only useful to describe, through the mean Rosseland values, the transport of energy, they are also important for the estimate of the radiative acceleration along the stellar lifetime and for the prediction and interpretation of the stellar acoustic modes (Turck-Chièze et al. 2009).

Today with the success of global helioseismology (Turck-Chièze et al. 1993, Thompson et al. 1996, Turck-Chièze et al. 2001, Basu et al. 2009) and the development of asteroseismology (Aerts, Christensen-Dalsgaard & Kurtz 2010), it is the appropriate time to focus more deeply on this fundamental ingredient of stellar evolution to properly identify and interpret the present and future space missions. In this review, we describe the new activities developed by our team to give credit to the used opacity calculations.

In Turck-Chièze et al. (2009), we point out two cases which must be examined in great details: (1) the radiative zone of Sun and solar like stars, where the observed sound speed (Turck-Chièze et al. 2001, Basu et al. 2009, Turck-Chièze & Couvidat 2010) is not yet understood, (2) the envelop of β Cephei where the excitation of the oscillations is due to the opacity peak of the iron group and for which there is some well identified difficulty of interpretation (Pamyatnykh 1999). To progress on these two fields we have formed a consortium between plasma and astro physicists in order to compare calculations and to perform experiments on high energy lasers.

2. The opacity coefficients in radiative zones of Sun and solar-like stars

Figure 1 illustrates the successive role of the heavy elements in the increase of the opacity cross section of the solar radiative zone along radius (Turck-Chièze et al. 1993). In fact the central region is sufficiently hot for all the elements to be completely ionized except for the elements in Z equal or greater than iron. Then one observes the role of lighter and lighter elements from the center to the base of the convective zone. A radiative gradient greater than the adiabatic one results from the strong increase of opacity due to the partial ionization of oxygen (the third element in abundance). When this phenomenon appears, the convective transport takes over. Just below the photosphere (the zoom at the right part of the figure), helium then hydrogen dominates successively with bound-bound and bound-free contributions in addition to free-free or diffusion ones. This property of the solar opacity coefficients is known for a long time and by chance the different calculations, OPAL and OP, agree reasonably well in Rosseland mean values (Seaton & Badnell 2004) within 5%. Nevertheless, these coefficients depend strongly on the detailed composition and more specifically on O, Ne, and Fe contributions.

Figure 2 shows, on the contrary, that the agreement on elemental Rosseland values is not so good for most of the heavy elements comparing OP and OPAS calculations. Just below the base of the convective zone and down to 0.5 $R_\odot$, differences, up to 60% for some specific elements, are noticed between OP and OPAS. OPAS are new calculations of opacity dedicated to stellar applications including 21 elements and performed by a new team in CEA (Blancard, Cossé and Faussurier 2010). The noticed differences are important for the treatment of the radiative acceleration in microscopic diffusion which uses the individual spectra (see Turck-Chièze et al. 2009, for the different expressions used in stellar evolution for treating the different radiative effects).

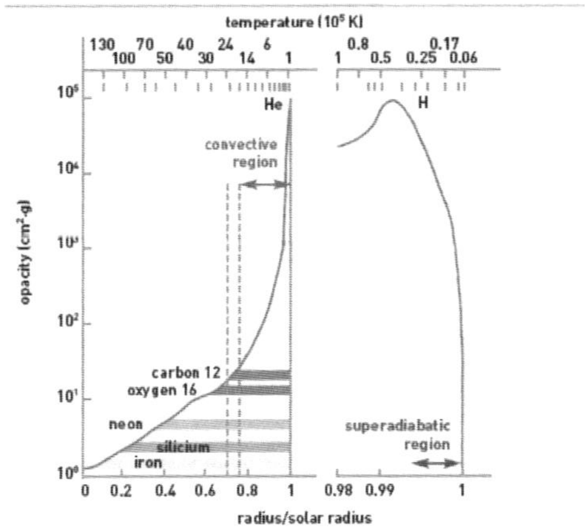

Figure 1. Main heavy elements contributors to the opacity of the radiative zone of the Sun along the solar radius. From Turck-Chièze et al. 1993.

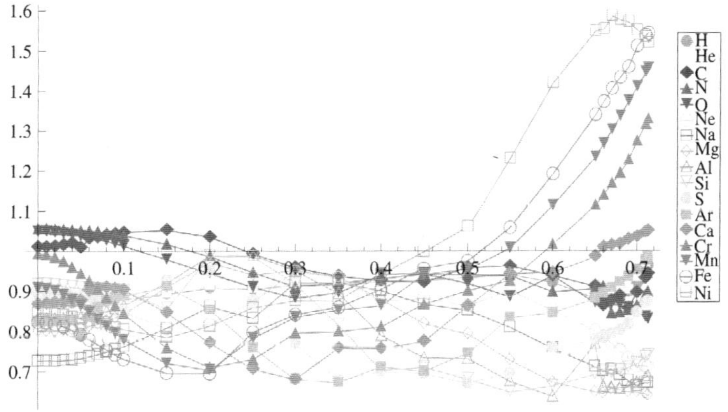

Figure 2. Difference versus solar radius between OPAS and OP Rosseland mean values for each individual contributor to the solar opacity. From Blancard, Cossé and Faussurier 2010.

The present situation illustrated by figures 1 and 2 is not totally surprising because the total Rosseland mean value integrates contribution of a lot of elements and depends strongly on the free-free process of helium, the bound-free of iron and the position of the main bound-bound lines of the partially ionized elements. On the opposite, the individual mean Rosseland value for each element strongly depends on the specific lines considered for each element, their width and the Stark effect, all strongly dependent on the approach used. It is why in addition to check the validity of the calculations used it is important to perform also experiments which reveal some aspects of the spectra. The solar radiative conditions are not yet been checked but a first experiment has been realized in the Z pulsed power facility of Sandia lab. in Albuquerque at 1.8 10^6 K and an electron density about 100 times smaller than the base of the convective zone which is in reasonable agreement with OPAL and OPAS estimate of the iron and magnesium spectra (Bailey et al. 2007, 2009).

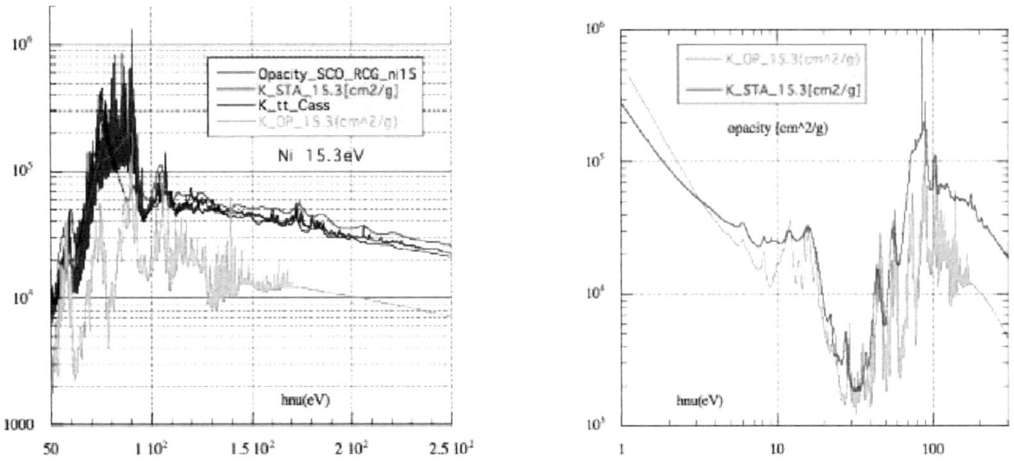

Figure 3. Theoretical opacity nickel spectra corresponding to 15.3 eV and 5.7 10^{-3} g/cm^3 and obtained from different approaches : OP (Seaton & Badnell, 2004; Seaton, 2007), Cassandra (Crowley et al. 2001), STA (Bar-Shalom, 1989) and SCO-RCG (Porcherot et al., 2010) in two wavelength ranges: the left one corresponds to the range that must be explored to calculate the mean Rosseland value, the right one corresponds to a zoom near the maximum at 62 eV of the Rosseland mean value.

3. Preliminary results on Nickel for envelops of β Cephei

The β Cephei stars ($8 < M < 12M_\odot$) pulsate by κ mechanism due to a strong peak of opacity of the iron group. Unfortunately the two available opacity tables OP and OPAL lead to strong differences in mean Rosseland values for the four contributors Cr, Mn, Fe and Ni (see Turck-Chièze 2010). This fact leads to the difficulty to choose which table is the best to use for interpreting the corresponding pulsations observations (Daszynska-Daszkiewicz & Walczak 2010, Degroote et al. 2009). So in order to better understand these differences we decided to perform an experiment on these elements and impulse some comparison between codes.

Unfortunately, it is not possible to perform an experiment at the too low densities of these envelops so we have determined equivalent conditions of plasma where the degree of ionization is rather similar. We decided also to check two or three conditions of temperature because the opacity spectra change quickly with temperature. We have shown how the iron spectrum itself can differ between the 8 calculations which are participating to the comparison (see Turck-Chièze et al., 2010 for the spectra and a rapid code description). So, one may hope a good discrimination from the experiment. Figure 3 shows different calculations for the nickel spectrum in two bands of wavelengths corresponding to the experiences we have performed, it appears clearly that OP calculations differ strongly from the other calculations for temperature around 15 eV (170 000 K). Nickel, contrary to iron, has never been measured in the wavelength range corresponding to the mean Rosseland value at such temperature. In the following, we report on the first nickel spectrum measurements done this year by our consortium.

The spectral opacity measurements have been performed at LULI 2000 with two complementary lasers (see www.luli.polytechnique.fr). A nanosecond laser delivers an energy between 30-500 J in a 500 ps duration pulse. This laser is used to irradiate a gold cavity (hohlraum) on which a foil of the considered element (here nickel) is deposited and heated. After a delay chosen to get the required density and temperature (this delay is determined by a

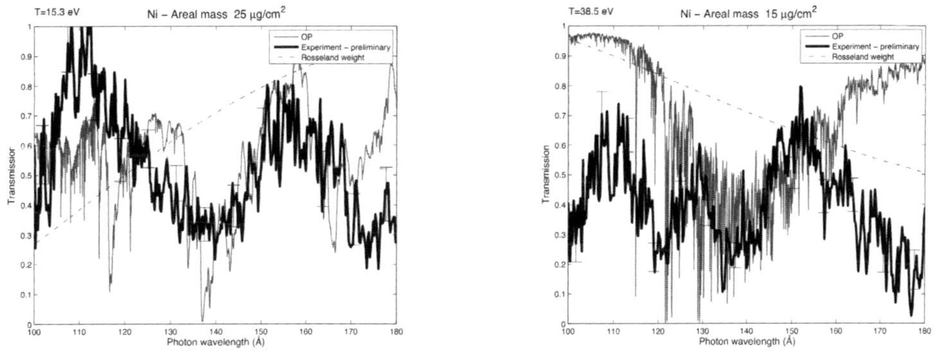

Figure 4. Preliminary result of the transmission spectra obtained for nickel (in heavy black) measured at LULI 2000 compared to the OP calculation (blue) already shown in Figure 3 for two conditions of temperature (15.3 eV and 38 eV) and respectively $\rho = 2.7$ mg/cm^3 Ne= 3.16 10^{20}cm^{-3}. Superimposed is the weight of the Rosseland mean value normalized to 1 in dotted line. The statistical error bars are shown on the experimental spectra and the final comparison will integrate the temperature gradient in the foil and a proper estimate of the mean temperature.

simulation of the geometry and heating of the experiment), a picosecond laser interacts with a backlighter foil to produce x rays in a short pulse (10-30ps) to probe the formed plasma.

The transmission spectrum of the photons: $T(\nu) = \exp(-\kappa(\nu)\rho r)$ where r is the thickness of the foil, is measured by a streak camera placed behind a specifically designed XUV-ray spectrometer. A detailed description of the experimental set up and of the experiment analysis will be published by Loisel et al. (2011). The quality of these measurements requires a sufficient spectral accuracy (about 1 eV) and a rather small emission of the cavity during the measurement of the spectrum. A detailed analysis of all the previous experiments and their required conditions can be found in Chenais-Popovics (2002), Bailey et al. (2009) and Loisel et al. (2009).

For the specific measurements that we have done on Cr, Ni, Fe, Cu and Ge, we need to form a plasma in LTE at the required conditions. This supposes first a good simulation of the whole experiment to probe the foil at the best moment to get the appropriate temperature without pollution of the gold of the cavity. Moreover, the rapid expansion of the foil during the heating is limited by placing the foil between two thin samples of a low Z material here carbon. We measure the foil temperature on one side thanks to μDMX, a 12 channels spectrometer which measures x-rays energy (Bourgade et al., 2001) and we have limited the gradient of temperature below 10% inside the foil in separating the incident ns beam in two parts and placed the foil between two cavities. The streak camera gives a time dependence of the phenomenon (resolution of about 50 ps) allowing to discriminate between backlight signal and self emission of the cavity.

Figure 4 shows the transmission spectra obtained for nickel and compared to OP calculation for two conditions of temperature around the value which corresponds to the same degree of ionisation than in β Cephei. We have also compared these spectra to the other calculations. Figure 4 does not present a perfect agreement between experiment and calculation, on the whole range observed but the same comparison disagrees more strongly with the other calculations. We have already introduced in the observed spectra a statistical error bar which increases on the limit of the range due to the relative increase of the effect of the background. Systematic error may be still present and a better knowledge of the detector systematic effect is under study. It would be interesting to see if the other spectra on iron, chromium, copper and germanium show the same preference for OP calculation and to confirm such result in a more extensive study.

This review has shown through two examples that the interpretation of the seismic observations need to use appropriate opacity coefficients to extract without ambiguity the manifestations of stellar dynamics. From the existing experimental studies one cannot exclude that the different regions of stars require more effort on the theoretical calculations and that the present ones differ essentially by their approach which is better adapted for some terms and not some others due to the methods used (statistical and detailed configuration approaches or interaction between configurations). Moreover depending on their use, different efforts have been applied. Presently, we must also be cautious because the experiments are extremely complex, different approaches are applied for the covered wavelengths (X or XUV techniques) for the machines used (Z machines or high energy lasers). Even the experimental approach exists since more than 10 years, they need certainly to be repeated and extended in wavelength ranges and elements considered face to the development of asteroseismology.

References

Aerts C, Christensen-Dalsgaard J & Kurtz D W 2010 *Asteroseismology* A&A edition, Springer
Asplund M, Grevesse N and Scott P 2009 *ARA&A* **47** 481
Bailey J et al. 2007 *Phys. Rev. Lett.* **99** 5002
Bailey J et al. 2009 *Phys. of Plasmas* **16** 058101
Bar-Shalom, A., Oreg, J., Goldstein, W. H., Schwarts, D., Zigler, Z. 1989 *Phys. Rev. A***40** 3183
Basu S et al. 2009 *ApJ* **699** 1403
Blancard C, Cossé P & Faussurier G 2010 *A&A Suppl. Series* in preparation
Bourgade J L et al. 2001 *Rev. Sci. Inst* **72** 1173
Chenais-Popovics C 2002 *Laser and Particle Beams* **20** 291
Clayton D, 1983 *Principle of stellar evolution and nucleosynthesis* Mc Graw Hill
Cox J and Giuli R 1968 *Principles of stellar structure Applications to stars* Gordon and Breach
Crowley, B.J.B et al. 2001 *JQRST* **71** 257
Daszynska-Daszkiewicz J & Walczak P 2010 *MNRAS* **403** 496
Degroote P et al. 2009 *A&A* **506** 111
Huebner W F, Merts A L, Magee N H and Argo M F 1977 *Los Alamos Sci. Lab. Rept. LA-6760 M*
Iglesias C A, Rogers F J & Wilson B G 1987 *ApJ* **322** 45
Iglesias C A & Rogers F J 1996 *ApJ* **464** 943
Loisel G et al. 2009 *HEDP* **5** 173
Loisel G et al. 2011 *MNRAS* in preparation
Porcherot Q, Gilleron F, Pain J C and Blenski T , 2010, to be published in HEDP.
Seaton M J 2005 *MNRAS* **362** 1
Seaton M J & Badnell N R 2004 *MNRAS* **354** 457
Thompson M et al. 1996 *Science* **272** 1300
Turck-Chièze S et al. 1993 *Phys Rep* **230** 59
Turck-Chièze S et al. 2001 *ApJ* **555** L69
Turck-Chièze S et al. 2009 *HEDP* **5** 132
Turck-Chièze S et al. 2010 *Astrophys & Space Science* in press

Adiabatic exponent in isentropic convective zone: a heavy elements abundance and seismic inversion

V A Baturin

Sternberg Astronomical Institute, Universitetsky pr. 13, Moscow, Russia

E-mail: vab@sai.msu.ru

Abstract. Profile of adiabatic exponent along isentropic curve is considered in context of determination of the CO and Ne contents in the solar convective zone. Tiny variations of the adiabatic exponent are caused by ionizations of elements. Isolated ionization stage leads to a narrow gap of the adiabatic exponent. Position of the gap depends on the ionization potentials and partition function of ions, which are essential parts of an equation of state. With the extended version of SAHA-S EOS we are able to study traces of elements in the adiabatic exponent profile. We give preliminary estimations on the abundance according to the inverted profile of the adiabatic exponent. Results do not reveal an indication of low-Z abundances in the adiabatic part of the convection zone.

1. Introduction

We focus on tiny variations of the adiabatic exponent Γ_1 profile. In the lower part of the solar convection zone a function $\Gamma_1\,(\lg T)$ is following to adiabatic stratification. The function $\Gamma_1\,(\lg T)_S$ appears is determined by the EOS and depends on a model only via a value of entropy of the adiabatic curve.

Adiabatic exponent is determined by several factors. First, due to almost full ionization of hydrogen and helium, Γ_1 is close to 5/3 as in the case of monatomic perfect gas. Second, Γ_1 slightly increases over 5/3 due to the Coulomb screening, which causes some sort of "Coulomb elasticity" ([1]). Third, value of Γ_1 decreases if a degree of ionization and specific capacity are increased. Hydrogen and helium may not be fully ionized, but reach limits of adiabatic ionization, and they do not contribute to lowering of Γ_1. Heavy elements gradually ionized with depth mainly contribute to Γ_1 lowering. Profiles of Γ_1 depression may be considered as an adiabatic spectrum of the element. A sum of the elements contributions gives final Γ_1 profile, what may serve for estimations of the heavy elements content. We present Γ_1 profile variations due to several factors. One of them is a value of the specific entropy which specifies an adiabatic curve. The helium content and specific contribution to Γ_1 from the heavy elements ionization, included to SAHA-S EOS, i.e. C, N, O, Ne, Si, Fe, are considered. In the conclusion we roughly estimate the heavy element content from the high-precision helioseismic inversion of Γ_1.

The results are obtained with the newly developed SAHA-S EOS ([2],[3]). The SAHA-S EOS characterized by the extended set of included elements and some details in physics (see [4]). Together with basic thermodynamic varietes, the set of the ionic distributions and some other values are available. At the moment an updated version SAHA-S2 EOS is beeing developed.

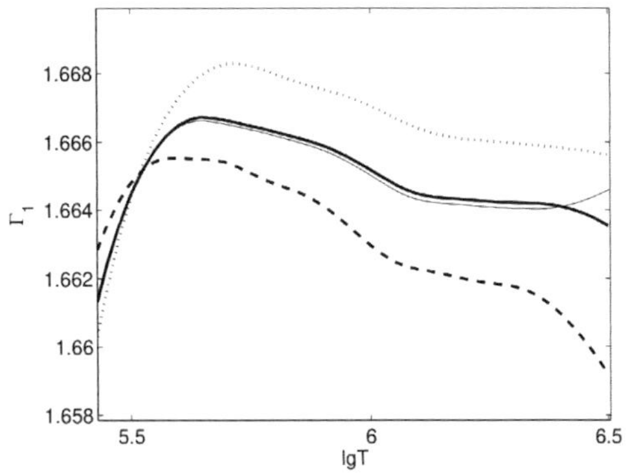

Figure 1. The value of Γ_1 along three adiabatic curves with the specific entropy $s/R_g = 25, 27, 29$ (dotted, solid, dashed curves accordingly). Thin curve shows the Γ_1 in the model S calculated with OPAL96 EOS

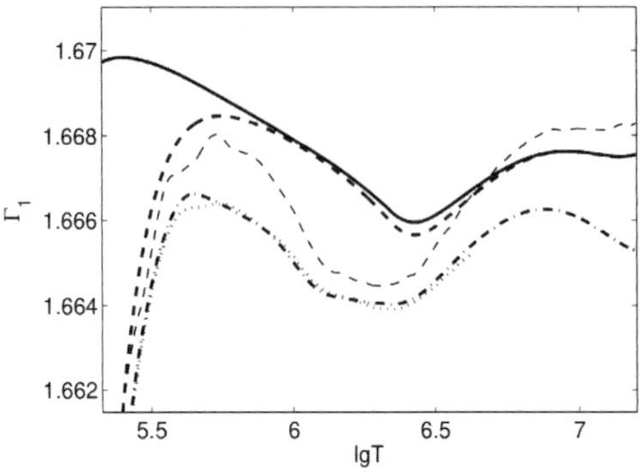

Figure 2. Γ_1 for different mixtures. Solid curve is for pure hydrogen. Dashed – for hydrogen-helium mixture with $X = 0.7$. Dot-dashed and dotted are for SAHA-S2 and SAHA-S ($X = 0.7$ and $Z = 0.02$). Thin dashed is for OPAL96 EOS.

2. Effect of the specific entropy

Actual adiabatic curve of the solar convection zone is unknown a priory. Generally, the specific entropy of the convection zone (together with the envelope helium abundance) is defined in calibration of the model to the solar radius and luminosity. The model calibration provides the convective entropy difference between solar models with similar opacities is rather small. Alternatively, the convective entropy may be found in a course of the helium ionization zone calibration (see e.g. [5]), which is essentially independent of the model calibration. Accordance of these calibrations is a necessarily condition for the adequate model of the Sun. Fig. 1 shows Γ_1 along three adiabatic curves, which are calculated for the selected values of the specific entropy - $s/R_g = 25, 27, 29$. The curves and Γ_1 have been calculated with SAHA-S EOS for the mixture with $X = 0.7$ and $Z = 0.02$. Even if the effect of entropy is rather large, the plotted variations of entropy are much larger then the model or helioseismic calibration uncertainties. The thin curve corresponds to Γ_1 in the model S [6]. The adiabatic exponent Γ_1 rises with lowering of the entropy (i.e. increase of density at the same temperature) because of higher Coulomb effect. At the same time, the ionization depression of Γ_1 (and the radiative lowering) becomes smaller with smaller entropy, so a maximum of Γ_1 may exceed of $5/3$.

3. Chemical composition effect

Fig. 2 shows effects of main components of solar mixture on general Γ_1 profile, calculated for the model set of points. The maximum of the adiabatic exponent Γ_1 is most remarkable on pure hydrogen and determined by the nonideal Coulomb "excess" of Γ_1. The depression due to helium ionization reduces the maximum noticeably and shifts it downward. The Z-ionization depression of the maximum is comparable with the effect of helium. It is interesting that the Z-depression increases with the helium content due to smaller total particle concentration. So exact value of Γ_1 depends on accurate helium content in the convection zone. Fig. 2 shows effect of EOS, calculated on the same set of points. The difference with a "classic" version of OPAL96 EOS [7] is rather remarkable. Comparison with OPAL2001 and OPAL2005 is considered in [4]. But reason of general exceess in Γ_1 of OPAL EOS over SAHA-S EOS is still unclear. Γ_1 plotted from SAHA-S2 EOS, which differs from a predecessor with newly added heavy elements (S and Mg) and revised system of exited states of ions.

4. Heavy elements contributions to the Γ_1 depression

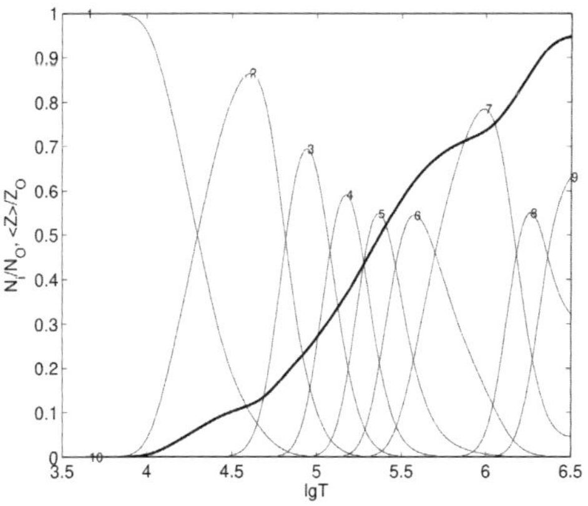

Figure 3. The distribution of the oxygen ions in solar models. The relative parts of ions marked with the degree of ionization are plotted together with the mean charge of ions

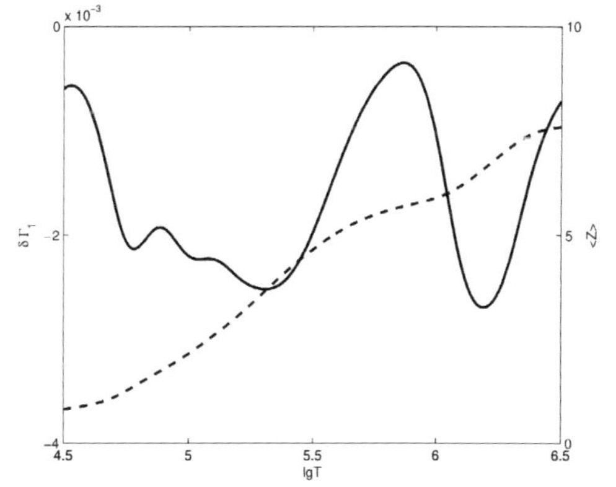

Figure 4. The depression of the adiabatic exponent due to ionization of oxygen (solid curve, left axis) and the mean charge of the oxygen ions (dashed curve, right axis)

Last considered effect is the ionization depression of adiabatic exponent induced by the heavy elements. The depression of adiabatic exponent in a region of ionization arises due to an increase of the specific capacity c_V. Heavy elements are considered as small impurities on the background of hydrogen and helium. The technique of calculations of Γ_1 -contribution in this approximation is described in [8]. We present distributions of the elements ions along model points for two examples, oxygen and neon. On Fig. 3 and Fig. 5 the relative parts of ions n_j^Z / N^Z in j-stages of ionization are plotted for oxygen and neon. The mean charge of ion divided by atomic number $\langle Z\rangle/Z_{el}$ is plotted as the monotonically increasing solid curve. With enlarge of atomic number Z_{el}, the regions of ionized ions are overlapped and an "jumping" over some ionization stage becomes possible. It can be clearly seen on example of neon and iron (see figures for the iron in [4]). The effect appears when the maximum relative part of ion is smaller then neighbors and causes a stepper rise of the mean charge. Fig. 4 and 6 give contributions to Γ_1 depressions due to ionization of oxygen and neon. Also, the mean charge $\langle Z_j\rangle$ of ions are plotted as dashed curves versus the right axis. The main purpose is searching a relationship between a course of ionization and "gaps" Γ_1 depression profiles from individual element. The C, N and O are very similar and do not show ionization "jumping" in the solar convection zone. But "gaps" of Γ_1 do exist. They are connected with ionization of the last two electrons from K-shell. The potentials of K-ionization are large and these ionizations are well separated from others. Both two K-ionizations are overlapped to one gap within the convective zone. Moreover, all three main contributions from CNO are overlapped and appear as common depression near the bottom of convection zone. Interestingly, that the neon Γ_1 -depression also demonstrate a narrow gap, although the neon does not reach K-shell ionizations inside the convection zone. The neon gap is due to ionization "jumping" and localizes at $\lg T \approx 5.6$, where coinciding with the local maximum of the Γ_1. So increase of the neon content affects the peak of Γ_1 and may lead to "camel humps" on Γ_1 profile. We can conclude, that the elements specificaly contribute to Γ_1, what allow analyse a content of a element or a group of elements.

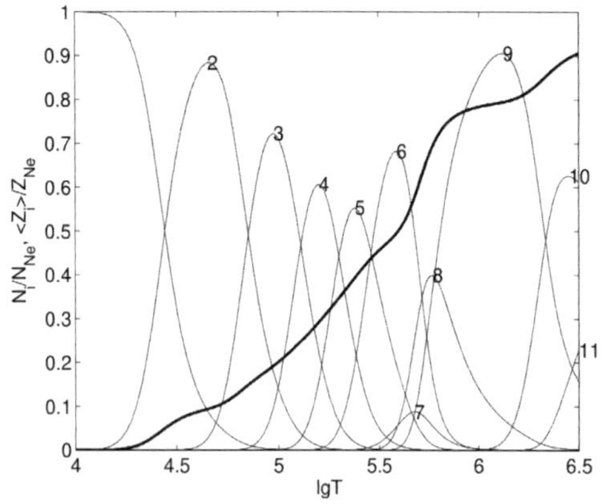

Figure 5. The same as on Fig.3 for the neon

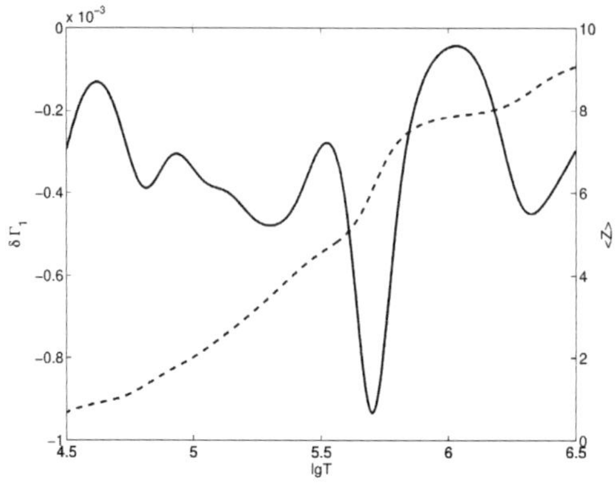

Figure 6. The same as on Fig.4 for the neon

5. Conclusion

Detailed analysis of ionization of heavy elements reveales specific variations Γ_1 from the individual elements. The gaps of lowering have specific position at temperature and are highly narrow in the case of the neon and iron. We explain them as result of ionization "jumping". After comparison of the result of helioseismic inversion of Γ_1 (see [9], and figure in [4]) and SAHA-S profiles for different abundances, we preliminary find that the neon abundance may be somewhat less than assumed in [10], but the content of CNO is hardly less then supposed in [10]. That is low-Z content [11] is not revealed in Γ_1 profile. These conclusions are preliminary and based on tiny features of Γ_1 profile. Also, it is based on the high-accuracy inversion results, which is extremely difficult problem.

6. Acknowledgment

I am indebted to the authors of SAHA-S EOS: A.N. Starostin, V.K. Gryaznov, I.L. Iosilevsky for exclusive access to the SAHA-S EOS data. I am very grateful to S.V. Vorontsov for the results of the helioseismic inversion of Γ_1. The study has been supported by ISTC project 3755.

References

[1] Baturin V A, Däppen W, Gough D O and Vorontsov S V 2000 *Mon. Not. R. Astron. Soc.* **316** 71-83

[2] Gryaznov V K, Ayukov S V, Baturin V A, Iosilevskiy I L, Starostin A N and Fortov V E 2004 *Equation-of-state and phase-transition issues in models of ordinary astrophysical matter (Leiden)* ed V Celebonovic, W Däppen and D Gough (Melville, New York: AIP conference proceedings **731**) 147

[3] Gryaznov V K, Ayukov S V, Baturin V A, Iosilevskiy I L, Starostin A N and Fortov V E 2006 *J. Phys. A, Math. General* **39** 4459

[4] Baturin V A 2010 it Astrophys. Space Sci. **328** 147-151

[5] Baturin V A 2006 *Beyond the spherical Sun (Sheffield)* ed K Fletcher and M J Thompson (Noordwijk: ESA Publications Division SP-624)

[6] Christensen-Dalsgaard J et al. 1996 *Science* **272** 1286

[7] Rogers F, Swenson F and Iglesias C 1996 *Astrophys. J* **456** 902 (OPAL96)

[8] Baturin V A 2007 *Unsolved problems in stellar physics (Cambridge)* ed R J Stancliffe, J Dewi, G Houdek, R G Martin and C A Tout (Melville, New York: AIP conference proceedings **948**) 213

[9] Vorontsov S V 2004 *Equation-of-state and phase-transition issues in models of ordinary astrophysical matter (Leiden)* ed V Celebonovic, W Däppen D Gough (Melville, New York: AIP conference proceedings **731**) 47

[10] Grevesse N and Noels A 1993 *Origin and Evolution of the Elements* ed N Prantzos, E. Vangioni-Flam and M Casse (Cambridge: Cambridge Univ. Press) pp. 14-25 (GN93)

[11] Asplund M, Grevesse N, Sauval A J and Scott P 2009 *Annu. Rev. Astron. Astrophys.* **47(1)**, 481 (AGSS09)

How to turn gravity waves into Alfvén waves and other such tricks

Maric E Newington and Paul S Cally

Centre for Stellar and Planetary Astrophysics, Monash University, Victoria, Australia 3800

E-mail: `paul.cally@monash.edu`

Abstract. Recent observations of travelling gravity waves at the base of the chromosphere suggest an interplay between gravity wave propagation and magnetic field. Our aims are: to explain the observation that gravity wave flux is suppressed in magnetic regions; to understand why we see travelling waves instead of standing waves; and to see if gravity waves can undergo mode conversion and couple to Alfvén waves in regions where the plasma beta is of order unity. We model gravity waves in a VAL C atmosphere, subject to a uniform magnetic field of various orientations, considering both adiabatic and radiatively damped propagation. Results indicate that in the presence of a magnetic field, the gravity wave can propagate as a travelling wave, with the magnetic field orientation playing a crucial role in determining the wave character. For the majority of magnetic field orientations, the gravity wave is reflected at low heights as a slow magneto-acoustic wave, explaining the observation of reduced flux in magnetic regions. In a highly inclined magnetic field, the gravity wave undergoes mode conversion to either field guided acoustic waves or Alfvén waves. The primary effect of incorporating radiative damping is a reduction in acoustic and magnetic fluxes measured at the top of the integration region. By demonstrating the mode conversion of gravity waves to Alfvén waves, this work identifies a possible pathway for energy transport from the solar surface to the upper atmosphere.

1. Introduction

Gravity waves, presumably excited by granulation, have recently been observed at multiple heights near the base of the quiet-Sun chromosphere, displaying the signature of upward (group) propagation [1]. Standard atmospheric models (*e.g.*, VAL C) suggest that these waves are confined to a gravity wave cavity in the absence of magnetic field, roughly $0 < z < 1.2$ Mm at the relevant frequencies (about 1 mHz). The observations indicate significant associated wave energy fluxes, perhaps an order of magnitude larger than co-spatial acoustic waves. However, the gravity waves appear to be substantially suppressed by magnetic fields.

These observations raise several interesting questions that we set out to answer theoretically:

(i) Why do the gravity waves avoid magnetic regions?

(ii) Why do we see them as travelling waves and not standing waves?

(iii) What happens to gravity waves as they propagate higher and inevitably encounter regions where the plasma beta is of order unity?

(iv) Can (as informally postulated in [1]) the gravity waves couple into Alfvén waves and so escape the "gravity cavity"?

Using a combination of dispersion diagrams, ray calculations, and direct numerical solution of the wave equations, [2] find some very elegant and surprising answers to these questions, which suggest that highly inclined magnetic field may form a crucial link in the chain between the solar surface and the upper atmosphere. We summarize these results here, but then extend them to briefly explore the effects of radiative loss on atmospheric gravity waves.

2. Method

In this paper we consider waves with frequency of 1 mHz and horizontal wave numbers of 2 Mm^{-1}, subject to a uniform magnetic field. The field orientation is described by its inclination from the vertical (θ) and its azimuthal angle from the plane of wave propagation (ϕ). The atmospheric quantities were given by the VAL C model up to a height of 1.6 Mm, with an isothermal layer appended above.

We refer the reader to [2] for the form of the dispersion relation, the governing equations of motion, and details of the method of their numerical solution, in the limit of adiabatic wave propagation. The equations used to describe radiatively damped wave propagation will be described in a forthcoming paper.

3. Results

3.1. In the absence of a magnetic field

Figure 1 shows the dispersion diagram relating the vertical component of the wave number k_z to the height z for the 1 mHz, $k_x = 2$ Mm^{-1} gravity wave in the absence of magnetic field. The closed curve indicates that it is trapped as a standing wave in a "gravity cavity", travelling upwards to a height of about 1.2 Mm before being reflected back downwards. Note that the lower branch of the curve represents the up-going gravity wave.

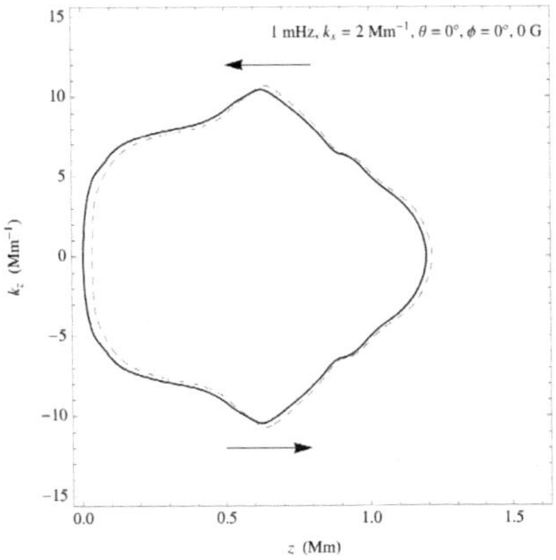

Figure 1. The dispersion diagram for the gravity wave in the absence of magnetic field. Arrows indicate the direction of energy transport. The heights at which waves reflect are evident. The full and dashed curves correspond to two different forms of the acoustic cutoff frequency: the isothermal and Deubner-Gough forms respectively (see [2] for full details). Since there is little difference, the isothermal form $\omega_c = c/2H$ will be used throughout here, where c is the sound speed and H the density scale height.

3.2. Low to moderate field inclination

Application of a magnetic field allows the gravity wave to escape the gravity wave cavity and propagate as a travelling wave. For vertical or moderately inclined field the gravity wave reflects as a down-going slow MHD wave. This explains the observations that gravity waves appear to avoid magnetic regions. Figure 2 demonstrates this behaviour with a ray diagram for the scenario of a gravity wave propagating in a 10 G vertical field.

3.3. Highly inclined field

When the magnetic field is highly inclined to the vertical (large θ values), the ramp effect, which reduces the acoustic cutoff frequency by a factor $\cos\theta$, enables the wave to penetrate

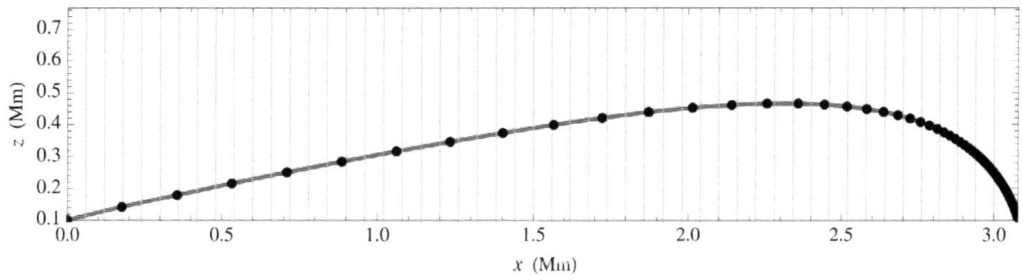

Figure 2. Ray diagram for the scenario of a gravity wave propagation in a 10 G vertical field (field lines are shown in grey). This represents the path of the wave packet in the x-z plane. The dots represent the wave packet location at one minute intervals. Note that after the reflection at a height $\sim$ 0.4 Mm, the 1 minute dots are more closely spaced, showing that the wave has converted to a slow wave.

the equipartition level (where the Alfvén and sound speeds are equal) and mode conversion to either field-aligned acoustic waves or Aflvén waves can occur. The dispersion diagram in Figure 3 illustrates a high field inclination scenario where conversion to the Alfvén wave occurs.

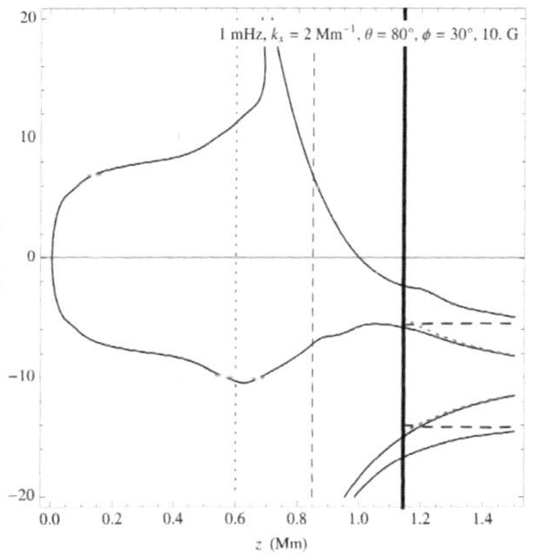

Figure 3. Dispersion diagram for a wave in a highly inclined field that penetrates the equipartition level (the vertical solid black line). Asymptotic solutions of the dispersion relation are shown – the blue dashed line is the field-aligned acoustic solution and the red dotted line is the Alfvén solution. The up-going gravity wave (bottom branch) converts to an Alfvén wave near the equipartition level.

The character of the converted wave (acoustic or magnetic) depends on the azimuthal orientation (ϕ) of the field. In cases where ϕ=0, coupling to the Alfvén wave is not possible and the gravity wave converts to a field guided acoustic wave. When ϕ is non-zero, conversion to both Alfvén and acoustic waves can occur. The orientation of the field determines the connectivity in the dispersion diagram to the field-aligned acoustic wave or the Alfén wave.

For some field orientations, (high inclination, moderate azimuthal angles) the Alfvén and field-aligned acoustic solutions are only separated by a narrow gap in the dispersion diagram. In that case, conversion to both types of wave is possible, as the energy can tunnel across the gap.

Numerical solution of the wave equation was used to measure the acoustic and magnetic fluxes at the top of the integration region. Figure 4 shows the behaviour of the fluxes as a function of the field inclination for a moderate azimuthal angle (ϕ = 30°). The dot-dashed (magnetic) and dashed (acoustic) curves represent the result of solving the adiabatic equations. Numerical solution reveals that the connectivity implied by the dispersion diagrams usually represents the dominant behaviour, even when tunnelling occurs.

3.4. Effect of radiative damping

The assumption of adiabatic wave propagation is invalid in the photosphere and low chromosphere. We attempted to incorporate the effects of radiative damping in our simulations by including a Newton cooling term in the energy equation. Our preliminary work employs a constant value for the radiative relaxation time (τ_R) up to 1.0 Mm. We have considered τ_R values from 200 s to 1 ks. Figure 4 shows the acoustic and magnetic fluxes (full and dashed curves respectively) as a function of field inclination when a 1 ks τ_R has been included in the equations. Comparison with the adiabatic results suggests that the primary effect of including Newton cooling is a reduction of flux magnitude – the smaller the damping time, the larger the flux reduction. Dispersion diagrams for damped wave propagation suggest that the effect of damping is largest in the photosphere and that the mode conversion pathways are preserved (Figure 5).

4. Conclusion

Gravity waves can couple to Alfvén waves and propagate up into the atmosphere when the magnetic field is very highly inclined. Preliminary results suggest that this coupling still occurs in the presence of radiative damping, though this is being explored further.

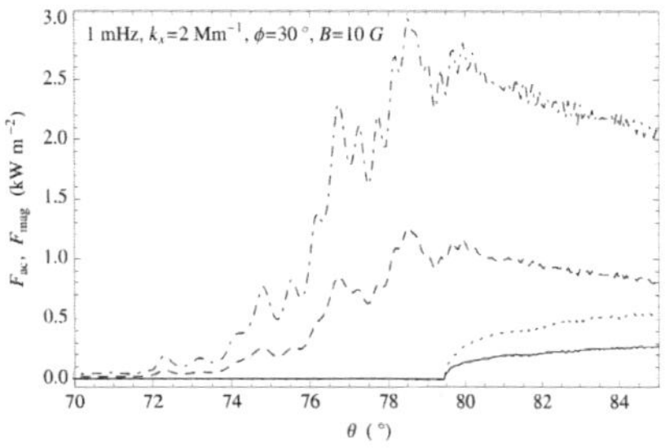

Figure 4. Magnetic and acoustic fluxes measured at the top of the model atmosphere. The dot-dashed and dashed curves represent the magnetic fluxes in the adiabatic and damped simulations respectively. The dotted and full curves are the acoustic fluxes in the adiabatic and damped simulations. The damped results shown were obtained using $\tau_R = 1$ ks.

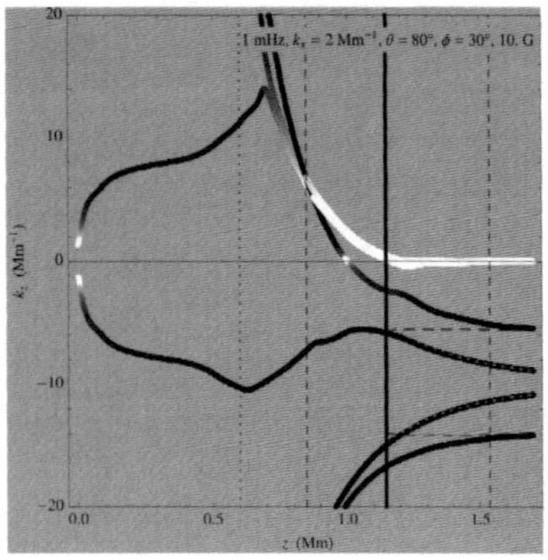

Figure 5. Dispersion diagram for the same scenario as in Figure 3, except with the gravity wave subject to weak radiative damping. The colour of the curves indicates the relative size of the imaginary part of k_z, with white sections showing where the wave is most heavily damped ($\mathrm{Im}\, k_z \geq \mathrm{Re}\, k_z$). The up-going gravity wave is most heavily damped in the photosphere. The damping does not alter the connectivity of the up-going wave to the Alfvén solution in the chromosphere.

References

[1] Straus T, Fleck B, Jefferies S M, Cauzzi G, McIntosh S W, Reardon K, Severino G, Steffen M, 2008 *ApJL* **681** L125

[2] Newington M E and Cally P S 2010 *MNRAS* **402** 386

Probing the parameter space of HD 49933: a comparison between global and local methods

O L Creevey[1,2], **M Bazot**[3,4]

[1] Instituto de Astrofísica de Canarias (IAC), E-38200 La Laguna, Tenerife, Spain
[2] Dept. de Astrofísica, Universidad de La Laguna (ULL), E-38206 La Laguna, Tenerife, Spain
[3] Centro de Astrofísica da Universidade do Porto, Rua das Estrelas, 4150-762 Porto, Portugal
[4] Institut for Fysik og Astronomi, Aarhus Universitet, DK-8000 Aarhus C., Denmark

E-mail: `orlagh@iac.es, bazot@astro.up.pt`

Abstract. We present two independent methods for studying the global stellar parameter space (mass M, age, chemical composition X_0, Z_0) of HD 49933 with seismic data. Using a local minimization and an MCMC algorithm, we obtain consistent results for the determination of the stellar properties: M = 1.1–1.2 $M_{\odot}$, Age $\sim$ 3.0 Gyr, $Z_0 \sim 0.008$. A description of the error ellipses can be defined using Singular Value Decomposition techniques, and this is validated by comparing the errors with those from the MCMC method.

1. Introduction

HD 49933 is a main sequence solar-type star that was observed by CoRoT. It is the first stellar object where solar-like oscillations were clearly detected in the photometric signal [1, 2]. There has been some controversy over the original labelling of the oscillation frequencies with their mode degrees, and in this work we present results based on the data published by [1], although we note that since this publication there has been a preference among the scientific community towards an inverted mode-labelling [2]. We present two independent methods of fitting the observational data, while placing an emphasis on defining the boundaries of the parameter space where the model of this star lies. The best-fitting models are determined using the Levenberg-Marquardt (LM) and Markov Chain Monte Carlo (MCMC) algorithms, while the uncertainties and the form of the parameter space that is constrained by the set of observations is described by both Singular Value Decomposition (SVD) and MCMC.

We use the frequencies from [1], and we compute the frequency separations to use as the observational data to model. The stellar models are computed using ASTEC [3] which include the EFF equation of state, OPAL 95 opacities together with Kurucz low-T opacities, NACRE reaction rates and overshooting, and the oscillation frequencies are calculated with ADIPLS [4].

2. Local minimization of stellar observables

Using as observational constraints the $\ell = 0$ frequency separations, $T_{\rm eff}, L_{\odot}, [M/H]$ from [1], we obtain the following set of stellar parameters for HD 49933 using LM: $M = 1.12$ $M_{\odot}$, Age = 2.9 Gyr, $Z_0 = 0.007$, $\alpha = 1.80$, and $R = 1.39$ $R_{\odot}$. $X_0 = 0.70$ was fixed, and the analysis was repeated using various initial parameter values [5]. In Fig. 1, each dot is the result of a minimization, clearly showing a sensitivity to the initial guess. This sensitivity is not uncommon for local methods, however, when we repeated our analysis using the data from [2] we found a rather stable solution of $M = 1.20$ $M_\odot$, Age $= 2.5$ Gyr, $Z = 0.0105$, $\alpha = 1.70$ and $R = 1.44$ $R_\odot$.

3. Global stellar parameter solution

Markov Chain Monte Carlo (MCMC) algorithms allows one to perform stochastic samplings of probability densities using properties of Markov Chains. This is a Bayesian methodology and we use it to estimate the posterior Probability Density Function (PDF) of the parameter(s) of our stellar model. The strength of this approach is that, once the PDF is derived, we can apply the classical tools of statistical inference to estimate the stellar parameters and associated confidence intervals — a long-standing problem in stellar physics [6, 7, 5]. This methodology was applied successfully to α Cen A [8].

We ran the MCMC algorithm using two different values of initial parameters ($M = 1.1, 1.2$ $M_\odot$, $Z_0 = 0.007, 0.010$) while including a prior on Z_0 for the first run only. Using the same observational constraints as those described in Sect. 2, we computed the marginal distributions of the global stellar properties. Using their mean values as the best-fitting parameter and considering the standard deviation of the sample as the corresponding 1-σ error bar, we obtain $M = 1.104 \pm 0.010$ $M_\odot$, Age $= 3.16 \pm 0.11$Gyr, $Z = 0.0065 \pm 0.0006$, and $\alpha = 2.002 \pm 0.080$. However, we also find that the second MCMC run hints at a possible (much less populated) alternative solution with a mass near 1.2 $M_\odot$.

4. SVD describing the error ellipses

When we obtain the best-fitting set of parameters P, then locally the models can be described as linear. With this assumption, we proceed to calculate the SVD ($D = UWV^T$) of the matrix $D = \frac{\partial O}{\partial P}\sigma^{-1}$, where O are the observables and σ the observational errors. The matrix V describes the N-dimensional (in our case $N = 4$) parameter correlations (essentially 4 4-element vectors), while W describes the magnitudes of these vectors. Fig. 2 shows the (longest) two-dimensional projections of the four-dimensional vectors, for the parameters of M, age and Z_0. These 2-D vectors are represented by the red arrows in the figures, and the black ellipses are the error-ellipses defined by these vectors.

5. Uncertainties

Fig. 2 shows the error ellipses described by SVD. The models generated from the MCMC runs are represented by the blue dots, with the size of the dots proportional to the likelihood values calculated from the MCMC (i.e. larger dots are more likely to be the true models). The error ellipses clearly encloses most of the (good) models from the MCMC method. This is strong evidence that, although we are assuming linearity for these models, the analytical approximation describes quite well the correlations that we expect to find, and the magnitudes of these correlations.

Because SVD gives a good indication of the shape of the parameter spaces, we can use it to investigate the effects of including extra observational constraints. In Fig. 3 we compare the two-dimensional error-ellipses for various parameters when using the same constraints as Fig. 2 (black ellipses), and then including the large frequency separations from the $\ell = 1$ modes (blue ellipses). The allowed parameter spaces defined by the extra set of constraints is reduced in volume by a factor of 2 for the 4 parameters, and the individual uncertainties are reduced by a significant amount for Z_0 and α.

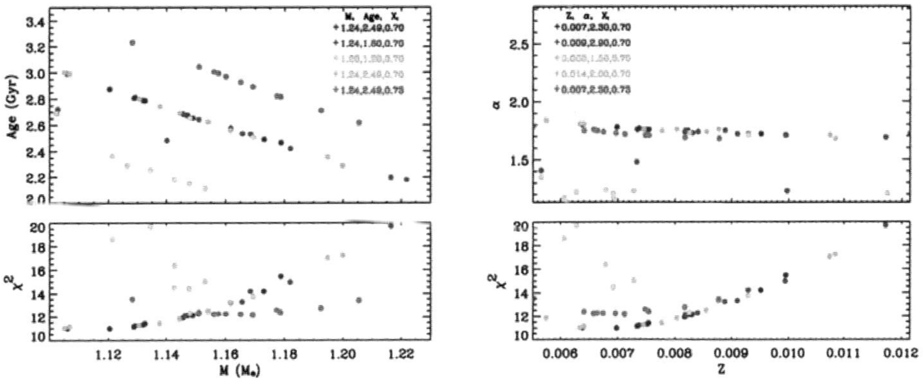

Figure 1. Levenberg-Marquardt results for the best-fitting parameters using the $\ell = 0$ large frequency separations and $T_{\rm eff}, L_\odot, [M/H]$ data from [1] as constraints on the models. The various colours indicate different initial guesses of the parameters, all having fixed $X_0 = 0.70$ except for the red dots which have fixed $X_0 = 0.73$. Each colour has a set of solutions, because we also varied the initial guess of the mass.

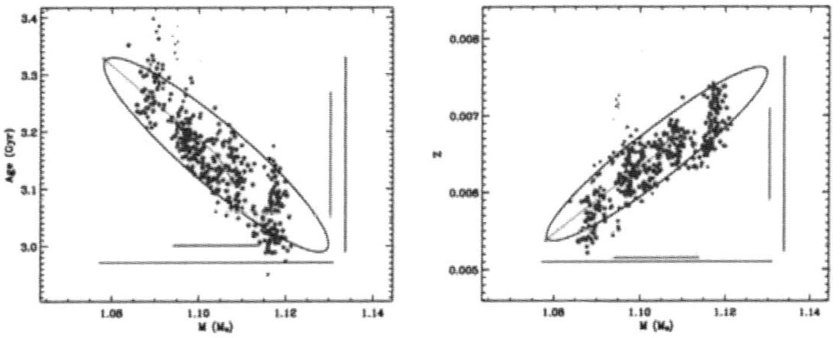

Figure 2. 2-D probability distributions for M and Age (left) and M and Z_0 (right) from the MCMC run. The size of the blue dots is proportional to the likelihood of the corresponding model. The red arrows are the two-dimensional scaled projections of M and Age as defined by SVD, while the black ellipse is the analytical error-ellipse given by these vectors.

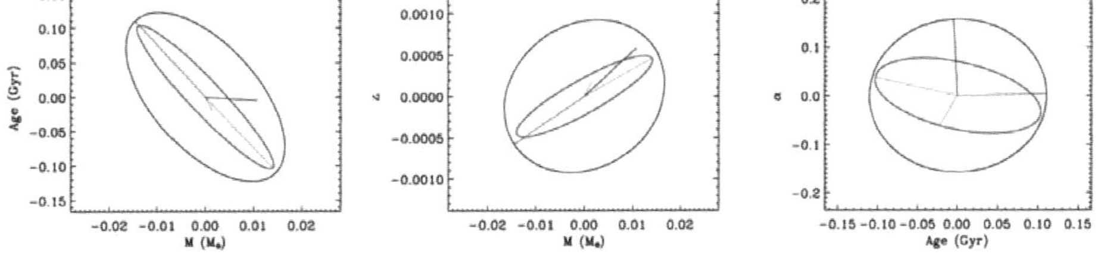

Figure 3. Error ellipses for M and Age (left), M and Z_0 (centre) and Age and α (right) as defined by SVD, assuming a set of constraints comprising $T_{\rm eff}, L_\odot$ and the large frequency separations for $\ell = 0$ only (black), and then including the $\ell = 1$ frequency separations (blue).

6. Conclusions

We have analysed the seismic data from HD 49933 and we have found that the global stellar properties using two independent methods (local and global) are in agreement. Using the mode-identification and frequencies given by [1] and using the frequency separations as the observational data to compare to the models we obtain $M = 1.12\ \mathrm{M}_\odot \pm 2\%$, Age = 2.9 Gyr $\pm$ 5%, $X_0 = 0.70$ (fixed), $Z_0 = 0.007 \pm 18\%$, $\alpha = 1.80 \pm 9\%$, and $R = 1.39\ \mathrm{R}_\odot$ from the local minimization method. This solution is supported by the results from the global method: $M = 1.1\ M_\odot$, Age = 3.16 Gyr, $Z_0 = 0.0065$ and $\alpha = 2.00$. The MCMC run also finds a possible alternative solution with $M = 1.2\ M_\odot$. A similar solution is also found from the local method while using the reversed mode-tagging [2]: $M = 1.20\ \mathrm{M}_\odot$, Age = 2.5 Gyr, $Z = 0.0105$, $\alpha = 1.70$ and $R = 1.44\ \mathrm{R}_\odot$.

We also studied the parameter space defining the correlations and uncertainties using the analytical approach of SVD. The results from the global MCMC validated the analytical description. In particular we found that including the $\ell = 1$ frequency separations should reduce the volume of parameter space by at least a factor of two in all dimensions, with the individual parameter uncertainties reducing significantly for Z_0 and α.

Acknowledgments

This research was in part supported by the European Helio- and Asteroseismology Network (HELAS), a major international collaboration funded by the European Commission's Sixth Framework Programme.

References

[1] Appourchaux T, Michel E, Auvergne M et al. 2008 *A&A* **488** 705
[2] Gruberbauer et al. 2009 *A&A* **506** 1043
[3] Christensen-Dalsgaard J 2008 *Ap&SS* **316** 113
[4] Christensen-Dalsgaard J 2008 *Ap&SS* **316** 13
[5] Creevey O L, Monteiro M J P F G, Metcalfe T S, Brown T M, Jiménez-Reyes S J and Belmonte J A 2007 *ApJ* **659** 616
[6] Jørgensen B R and Lindegren L 2005 *A&A* **436** 127
[7] Takeda G Ford E B, Sills A Rasio F A Fischer D A and Valenti J A 2007 *ApJS* **168** 297
[8] Bazot M, Bourguignon S and Christensen-Dalsgaard J 2008 *MSAI* **79** 660

Evolutionary scenarios and chemical inhomogeneities of extended horizontal branch stars

Lisa Esch, Pierre Demarque and Sarbani Basu

Yale University, New Haven, CT 06520-8101, USA

E-mail: `lisa.esch@yale.edu`

Abstract. Extended Horizontal Branch (EHB) stars are observed in many globular clusters and as field stars in the Galactic halo. They belong to old stellar populations of the halo and the old disk. Their evolutionary status is unclear, and still a current subject of debate. Current interest in these stars arise from their association with the discoveries of helium abundance inhomogeneities in the globular clusters ω Cen and NGC 2808. The origin of the inhomogeneities is not yet understood, but there are many interpretations.

In order to better understand EHB stars, we explore the evolution of standard blue Horizontal Branch (HB) models using up-to-date physics. We present several grids of post Zero Age Horizontal Branch (post-ZAHB) evolutionary models to include both canonical and non-canonical evolutionary scenarios, as well as to compare models that contain semi-convection to models without semi-convection. We follow the models to the termination of nuclear helium burning. The detailed properties of the models, including shell flashes and breathing pulses, are described.

1. Introduction

We have explored the post Zero Age Horizontal Branch (post-ZAHB) evolution of Extended Horizontal Branch (EHB) models with a canonical evolutionary history and without a prescription for semi-convection and learned that they undergo a series of shell flashes after core helium exhaustion. In this poster we want to explore the effects of different evolutionary scenarios as well as the effect of semi-convection on the post-ZAHB evolution of EHB models.

Besides the canonical evolutionary history of evolving onto the EHB after the helium flash at the tip of the Red Giant Branch (RGB), there are alternative evolutionary scenarios that can populate the EHB as well. One such alternative is the late hot flasher scenario [1]. In this scenario there is enough mass loss on the RGB that the helium flash is delayed until the star is on the white dwarf cooling curve. There are two types of mixing that can be associated with this late helium flash, deep and shallow mixing. The details of the two mixing events can be found in Lanz et al., 2004 [1]. We have mimicked these evolutionary scenarios by changing the surface abundances of our models in order to study if different evolutionary scenarios affect post-ZAHB evolution.

We have also created models that contain semi-convection. Semi-convection occurs during core helium burning when a region around the convective core becomes unstable to convection and expands, pulling surface material into the core and depositing core material onto the surface. Semi-convection is known to occur in horizontal branch stars through observations of the ratio of Asymptotic Giant Branch (AGB) stars to RGB stars [2]. Because of this fact we wanted to include semi-convection in our models in order to study what affect semi-convection has on the post-ZAHB evolution of EHB stars.

2. Models

We have created a grid of Horizontal Branch (HB) models using the Yale Rotating and Evolutionary Code (YREC) in the non-rotating configuration [3]. These models have an envelope mass of $0.001M_{\odot}$ and are evolved from the ZAHB to helium core exhaustion. Models with canonical evolution with and without semi-convection were constructed with different helium abundances $Y = 0.14, 0.24, 0.39$ and metallicities $Z = 0.0001, 0.0002, 0.01, 0.02$ and $M = 0.4887M_{\odot}$.

Models with a non-canonical evolution and without semi-convection were constructed with ZAHB models that have artificially enhanced surface abundances to be consistent with the deep and shallow mixing scenarios as described by Lanz et al. [1] with a mass $M = 0.4488M_{\odot}$.

We used the OPAL opacities for moderate to high tempratures [3], the Ferguson opacties [4] for low temperatures, the OPAL equations of state [5] and the nuclear reaction rates from the NACRE collaboration [6].

3. Results

After creating our models we discovered that non-canonical models have evolutionary tracks characterized by oscillations in the Hertzsprung Russell (HR) diagram that occur after the smooth central helium burning phase (Fig. 1).

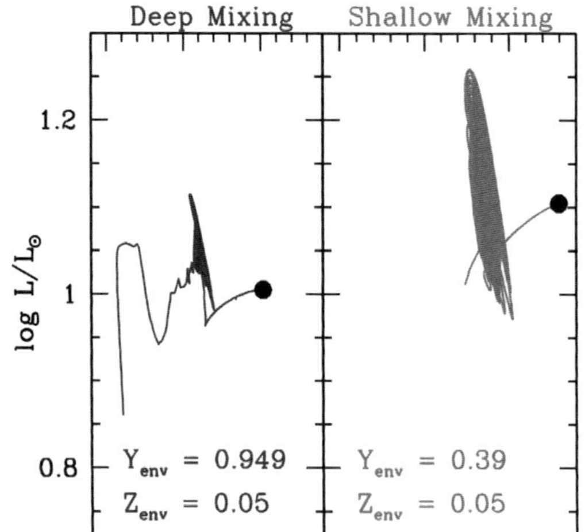

Figure 1. Evolutionary tracks in the HR diagram for sdB models who arrived on the ZAHB via the late hot flasher scenario as described in Lanz et al. (2004). The model on the left undergoes deep mixing, while the model on the right undergoes shallow mixing. The level of mixing in each models affects the surface abundance on the ZAHB (black dot). Notice how each model enters an evolutionary phase characterized by oscillations in the HR diagram following the smooth phase of core helium burning caused by thin shell flashes just as the models with canonical pre-ZAHB evolution.

These are the same type of oscillations that are exhibited by the evolutionary tracks of the canonical models as seen in Figure 2. We find no fundamental difference in horizontal branch evolution between models that have different pre-ZAHB evolutionary scenarios.

After creating the models of EHB stars that contain a prescription for semi-convection we discovered that the models with semi-convection evolve to lower temperatures and lower luminosities than the models without semi-convection, and their evolution is characterized by small loops in the HR diagram (Fig. 2).

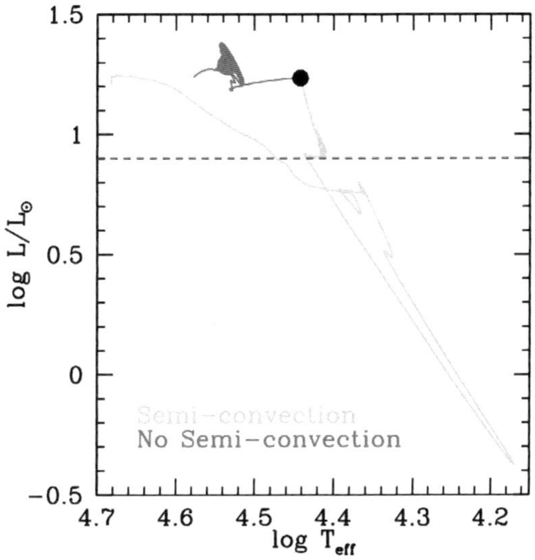

Figure 2. Evolutionary tracks of two models with Y=0.14 and Z=0.01, one that contains a prescription for semi-convection and one that does not and both undergo canonical pre-ZAHB evolution. Notice how the model with semi-convection evolves to lower temperatures and lower luminosities. The evolution of the model with semi-convection is characterized by small loops during the core helium-burning phase caused by breathing pulses. The evolution of the model without semi-convection is characterized by loops in the HR diagram following the core helium-burning phase. The semi-convection model spends most of its evolutionary lifetime above the blue line.

These loops are caused by breathing pulses, where the convective core expands into the envelope, pulling helium-rich material into the core, which increases the central helium abundance (Fig. 3).

The lower luminosity and temperature of the model with semi-convection can be explained through these breathing pulses. When the convective core expands, surface material is brought to the core, and core material, which is extremely metal-rich, is brought to the surface. The increase in the surface metal abundance increases the opacity of the star, which will lower the effective temperature and the luminosity of the star. Including a prescription for semi-convection suppresses the shell flashes that are seen in the models that do not have semi-convection because it increases the size of the carbon-oxygen core. Another important difference between models with and without semi-convection is that due to the increase in helium throughout central helium burning, the lifetime of the star is increased by a factor of 3 over the model without semi-convection.

This work was the necessary preparatory work for an asteroseismic study of EHB stars. The next step in this research is to explore the acoustic spectrum and its dependence on input parameters such as total mass, helium core mass and chemical composition.

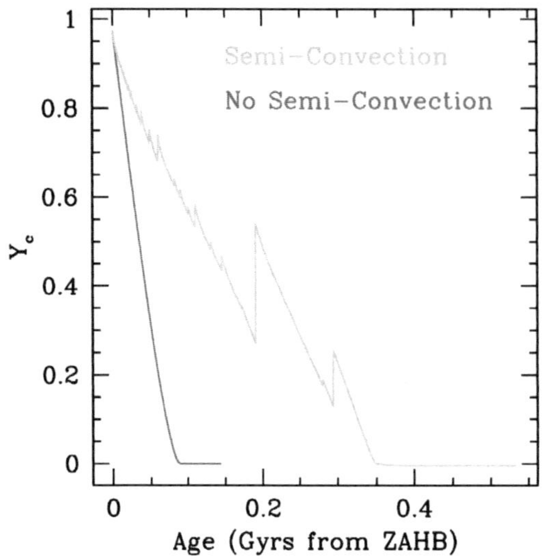

Figure 3. Central helium abundance plotted against time for a models with semi-convection(red) and models without semi-convection (black) that have Y=0.14 and Z=0.01. Notice how the addition of semi-convection adds spikes in Yc and lengthens the core-helium burning lifetime by a factor of 3 as compared to the model without semi-convection. The spikes in Yc are caused by breathing pulses that are characteristic of semi-convection. The convective core expands into the envelope, which increases Yc, adding more fuel to the core and thereby lengthening the lifetime.

References

[1] Lanz T, Brown T, Sweigart A V, Hubeny I and Landsman W B 2004 *ApJ* **602** 342

[2] Caputo F, Castellani V and Wood PR 1978 *MNRAS* **184** 377
Buzzoni A, Fusi Pecci F and Buonanno R 1983 *A&A* **128** 94
Buonanno R, Corsi C E and Fusi Pecci R 1985 *A&A* **145** 97

[3] Demarque P, Guenther D B, Li L H, Mazumdar A and Straka C W 2008 *Ap&SS* **316** 31

[4] Iglesias C A and Rogers F J 1996 *ApJ* **464** 943

[5] Ferguson J W, Alexander D R, Allard F, Barman T, Bodnarik J G, Hauschildt P H, Heffner-Wong A and Tamanai A 2005 *ApJ* **623** 585

[6] Rogers F J and Nayfonov A 2002 *ApJ* **576** 1064

[7] Angulo C, Arnould M, Rayet M, Descouvemont P, Baye D, Leclercq-Willain C, Coc A, Barhoumi S, Aguer P, Rolfs C et al. 1999 *Nu. Phy. A.* **656** 3

Magneto-acoustic waves in sunspots from observations and numerical simulations

T Felipe[1,2], E Khomenko[1,2,3], M Collados[1,2] and C Beck[1,2]

[1] Instituto de Astrofísica de Canarias, 38205, C/ Vía Láctea, s/n, La Laguna, Tenerife, Spain
[2] Departamento de Astrofísica, Universidad de La Laguna, 38205, La Laguna, Tenerife, Spain
[3] Main Astronomical Observatory, NAS, 03680, Kyiv, Ukraine

E-mail: `tobias@iac.es`

Abstract. We study the propagation of waves from the photosphere to the chromosphere of sunspots. From time series of cospatial Ca II H (including its line blends) intensity spectra and polarimetric spectra of Si I λ 1082.7 nm and He I λ 1083.0 nm we retrieve the line-of-sight velocity at several heights. The analysis of the phase difference and amplification spectra shows standing waves for frequencies below 4 mHz and propagating waves for higher frequencies, and allows us to infer the temperature and height where the lines are formed. Using these observational data, we have constructed a model of sunspot, and we have introduced the velocity measured with the photospheric Si I λ 1082.7 nm line as a driver. The numerically propagated wave pattern fits reasonably well with the observed using the lines formed at higher layers, and the simulations reproduce many of the observed features. The observed waves are slow MHD waves propagating longitudinally along field lines.

1. Introduction

The magnetic field of the sunspots affects the propagation of waves and produces the appearance of new wave modes not present in the quiet Sun. The study of waves in sunspots is interesting from several points of view. It gives information about the atmospheric structure and dynamics. Also, waves are believed to constitute one of the heating mechanisms of the chromosphere. Observations show that the properties of waves greatly vary with the region where they are observed: at the umbral photosphere they present 5 minute period with amplitudes around 100 m s^{-1}, while the chromospheric umbral oscillations have basically 3 minute period and amplitudes of a few kilometers per second.

Recent simulations have been able to reproduce many of the observed properties of waves between the photosphere and the chromosphere by means of the introduction of a photospheric pulse which drives a wave spectrum close to the solar one [1]. In this work we go a step further and we have developed numerical simulations where the real photospheric velocity pattern retrieved from observations in the Si I λ 1082.7 nm line is used to drive the simulated wave pattern. The oscillations obtained with the simulation at the high photosphere and chromosphere are compared with those measured with lines which are formed at those heights.

2. Observations

We have obtained simultaneous cospatial maps of the LOS velocity measured with photospheric lines (Si I λ 1082.7 nm and the Fe I lines at 396.54, 396.61, 396.66, 396.74 and 396.93 nm from

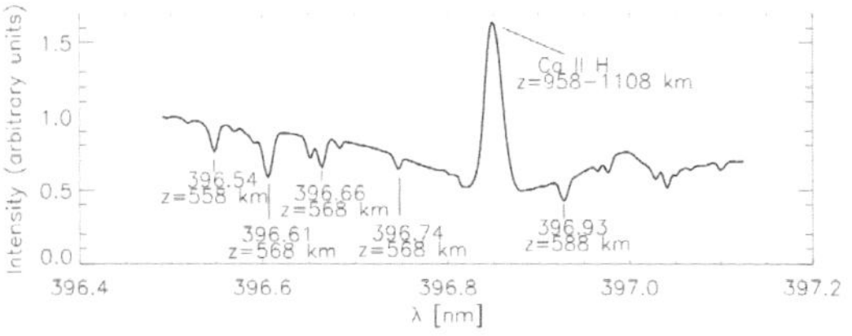

Figure 1. Spectra of the Ca II H line. The lines used and their formation height are marked.

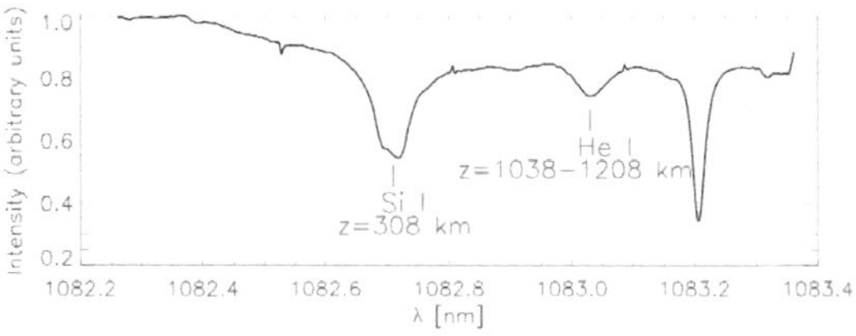

Figure 2. Spectra of the region at 1083 nm. The lines used and their formation height are marked.

the wings of Ca II H) and chromospheric lines (Ca II H core and He I λ 1083.0 nm). The phase and amplification spectra between several pairs of lines in this set of observations have been analized before in [2]. From the fit of the observations to a simple model of linear upward wave propagation we have infered temperature and formation height of the lines. Figures 1 and 2 show the observed spectral region. The relative formation height of the spectral lines is written next to each line.

3. Numerical simulations

The numerical simulations used in this work were performed with a 3D non-linear MHD code [1]. Energy losses are implemented following Newton's cooling law, accounting for the damping of the temperature fluctuations due to radiative losses with Spiegel formula for the typical cooling time [3].

We have constructed a MHS model of sunspot using observational data. The stratification of the thermodynamical magnitudes, obtained from the inversion of the umbra and the quiet Sun photospheric spectra of Si I line (using SIR [4]) are set at the axis of the sunspot and at 40 Mm far from it, respectively. These two atmospheres are smoothly merged along the radial distance which separates them [5]. The model is an axysymmetric thick flux tube with distributed currents and has no twist. In the horizontal directions the computational domain covers 8.4×5.2 Mm, with an horizontal spatial steps of $\Delta x = \Delta y = 100$ km. In the vertical direction it spans form $z = -0.6$ Mm to $z = 1$ Mm, excluding the special layer added at the boundaries to absorb the waves without reflections, with a spatial step of $\Delta z = 25$ km. The MHS model of the sunspot is pertubed around the photospheric formation height of the Si I line with the velocity measured with this line. This driver generates an oscillatory pattern identical to the observed one at the photosphere, and these waves propagate upward to higher layers.

4. Results

In order to compare the numerical simulation with the observational data, we have assigned a fixed z to the formation height of each spectral line, and we assume that the vertical velocity at that height corresponds to the velocity measured from the Doppler shift of the line. (see Figures 1 and 2) [2].

Figure 3 shows the observational and numerical velocity maps, corresponding to the chromosphere. It is the highest spectral line observed, and these waves have propagated upward about 800 km from the formation height of the Si I line in order to reach this layer. Note that in the simulated velocity map (middle panel) the velocity signal is almost zero during the first 2 minutes, due to the time spent by the slow waves to cover the distance between the driver and this height travelling at the sound speed. During this travel the period of the waves is reduced to around 3 minutes and their amplitude increases, reaching peak-to-peak values of almost 8 km s^{-1}. Bottom panel of Figure 3 shows that the oscillations develop into shocks. This

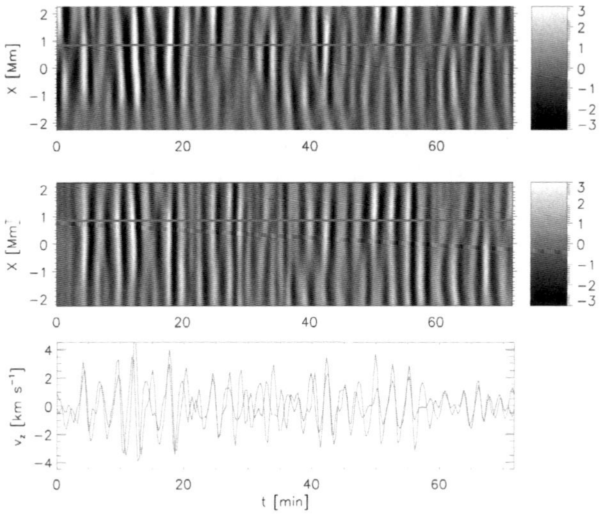

Figure 3. Velocity maps for the chromosphere inside the umbra. Top: observational (measured from the Doppler shift of the He I line); middle: numerical (vertical velocity at the formation height of the He I); bottom: temporal variations of numerical (blue) and observed (red) velocities at $x = 0.9$ Mm.

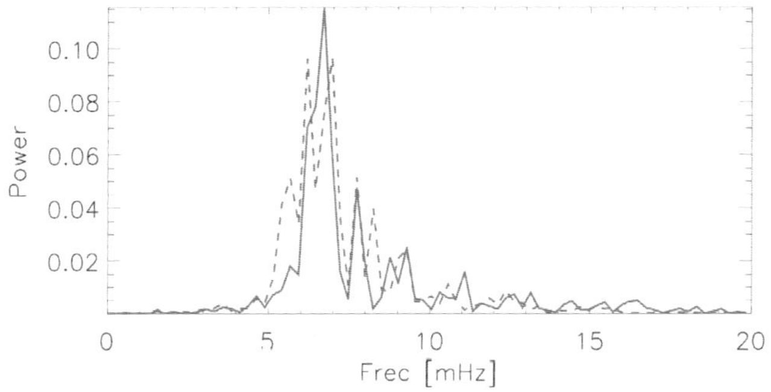

Figure 4. Power spectra of the velocity measured with the He I line (red dashed line) and the simulated at the corresponding height (blue line).

behavior is well reproduced by the numerical simulation. The simulated velocity map reproduces reasonably well the observed oscillatory pattern. Only in the temporal lapse between $t = 27$ and $t = 40$ min the simulated pattern differs significatively from the observations.

The power spectra at chromospheric heights is dominated by 3 minute oscillations in the band between 5 and 8 mHz (Figure 4). The observational power has three power peaks in this band, located at 5.5, 6 and 7 mHz. The simulated power is concentrated at a single peak between the two highest peaks of the observations. The simulated peak at 5.5 mHz is lower than the observed one. The simulations also reproduce the power peaks at 7.7 mHz and 9 mHz, and the low power at frequencies below 5 mHz. At frequencies above 13 mHz the simulated power is slightly higher than the observational one.

Phase spectra give the delay between two signals at every frequency. The phase difference between the velocity signals measured in the photospheric Si I line and the chromospheric He I line is zero for frequencies below 4 mHz (left panel of Figure 5). At these frequencies the atmosphere oscillates as a whole, *i.e.*, the waves are stationary. From 4 mHz to 7 mHz, the phase difference increases linearly with the frequency. It indicates that waves at these frequencies propagate from the photospheric layer, where the Si I line forms, to the chromospheric layer, where He I forms. Note the excellent agreement between observational and numerical data from 0 to 7 mHz, where the coherence of the observations is high. At higher frequencies the coherence of the observed phase difference is lower, but the simulated one keeps its linear increase. With regards to the amplification spectra, for frequencies above 1.5 mHz the simulated spectra reproduces properly the observed one.

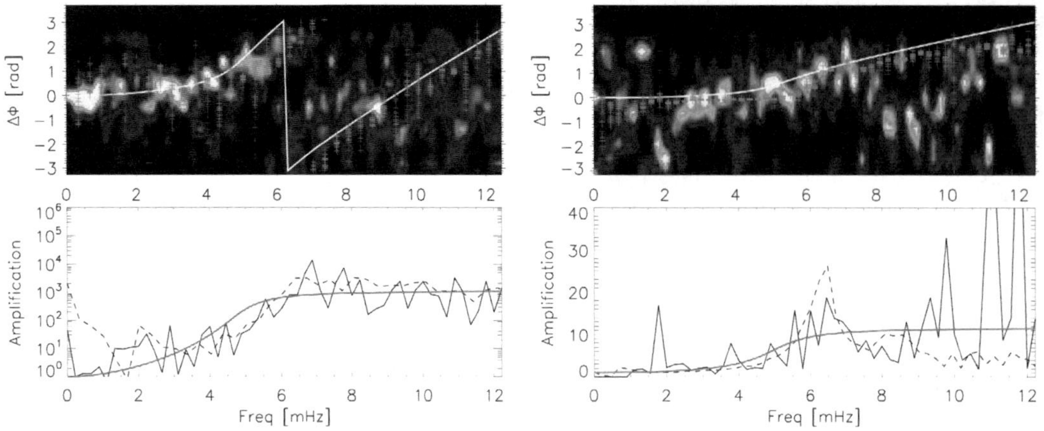

Figure 5. Left panel: Phase difference (top) and amplification spectra (bottom) between Si I and He I velocities. Right panel: Phase difference (top) and amplification spectra (bottom) between Si I and Fe I velocities. The background color in the top panels indicates histograms of the relative occurrence of a given value in the observed phase (from black -low- to yellow -high-). Green crosses are the results of the simulation for all the spatial points. The white line is the fit to a model of linear wave propagation [2]. The bottom panels show the ratio of the power at the heights of the two signals (amplification spectra) as a function of frequency for the observation (black solid line), the simulation (black dashed line) and the model (red line).

Right panel of Figure 5 shows the phase and amplification spectra between two photospheric lines, the Si I and the Fe I λ 396.93 lines. The observed phase difference has high coherence between 2 and 8 mHz, and in this frequency range the simulated phase delay fits the observational one reasonably well, showing a $\Delta\phi = 0$ for frequencies below 4 mHz and a slow increase for higher frequencies. For frequencies below 2 mHz and above 8 mHz the observed phase difference spreads out and has lower coherence. The behaviour of the simulated amplification spectra is similar to the observed one between 0 and 8 mHz, but the numerical amplification is significatively higher at the peak around 6.5 mHz.

5. Conclusions

We have studied the propagation of waves in sunspots from the photosphere to the chromosphere, using observationally driven numerical simulations. We have been able to simulate the travel of the waves through the sunspot atmosphere, with a remarkable match between observations and simulations. The numerical calculations reproduce the wave pattern (Figure 3), the power spectra (Figure 4) and the phase and amplification spectra between several pairs of lines (Figure 5), indicating that observed waves are slow longitudinal acoustic waves dampled by radiative losses.

References

[1] Felipe T, Khomenko E and Collados M 2010 *Apj* **719** 357
[2] Felipe T, Khomenko E, Collados M and Beck C 2010 *Apj* **722** 131
[3] Spiegel E A 1957 *ApJ* **126** 202
[4] Ruiz Cobo B and del Toro Iniesta J C 1992 *Apj* **398** 375
[5] Khomenko E and Collados M 2008 *Apj* **689** 1379

Diffusion segregation of heavy elements in the Sun

A B Gorshkov and V A Baturin

Sternberg astronomical institute of Moscow State University, Russia

E-mail: `gorshkov@sai.msu.ru`

Abstract. Lowering of heavy element abundances Z according to modern determinations presents a problem of the solar modeling. We estimate settling rates of several heavy elements such as C, N, O, Ne, Si and Fe from the solar convective zone during evolution according to the Bahcall-Thoul technique. Acceleration of the heavy elements settling due to partial ionization and radiation-driven slowing are taken into account. Diffusion profile in modern Sun reveals a narrow region under the convective zone where heavy elements abundance remains almost the same due to specific effect of thermal diffusion. A rate of the elements settling from the convective zone depends on the convective zone depth and on the overshooting mixing around the low convective boundary. According to our results, diffusion settling is fairly small to explain entirely a low Z value in the photosphere.

1. Introduction

In this work we: 1) estimate the rate and its variation of different mechanisms of heavy elements diffusion inside the Sun, and 2) solve diffusion equation on the evolutionary set of solar models to get chemical composition profiles with the diffusion taken into account. Rotation, magnetic field and non-radial flows are ignored.

Solar models given initially by evolutionary track 721-0001 based on SAHA-S equation of state with hydrogen and helium diffusion included (S. Ayukov, private communication) as $T(r,t)$, $P(r,t)$ etc. and did not changed in simulations. We treat the solar convective zone as a region with ultrafast diffusion (taking into account that convection rapidly equalizes chemical composition over all convective zone).

2. Problem Formulation

For component i of solar plasma (ion or electrons) we can write equations for momentum and energy conservation:

$$\nabla P_i - \rho_i \frac{F_i}{m_i} = \sum_k \left[G_{ik}^{(1)}(w_i - w_k) + \mu_{ik} G_{ik}^{(2)} \left(\frac{h_i}{m_i} - \frac{h_k}{m_k} \right) \right] \quad (1)$$

$$\frac{5}{2} P_i \frac{\nabla T}{T} = \sum_k \left[\frac{5}{2} \frac{\mu_{ik}}{m_i} G_{ik}^{(2)}(w_i - w_k) + G_{ik}^{(5)} h_i + G_{ik}^{(6)} h_k \right] \quad (2)$$

where P_i, ρ_i – partial pressure and mass density, T – temperature, external force $F_i = m_i g - q_i E$, m_i, q_i – atomic mass and charge, reduced mass $\mu_{ik} = m_i m_k/(m_i + m_k)$, $G_{ik}^{(l)}$ – "friction" coefficients describing momentum and energy exchange in particles' collisions, and, finally, w_i and h_i – diffusion velocities and partial residual heat flows correspondingly.

This hydrodynamic approach was proposed by [1] and later developed by [2], [3] and [4].

Following [3], we complement these equations with conditions of current neutrality and mass conservation (ρ_{ei} – charge density of component i):

$$\sum_k \rho_{ek} w_k = 0, \qquad \sum_k \rho_k w_k = 0. \tag{3}$$

So, for N species we have a linear system of $2N+2$ equations for $2N+2$ unknowns: N diffusion velocities w_i, N partial residual heat flows h_i, local electric field E and gravitational acceleration g. Treatment gravity acceleration as unknown allows to not exclude explicitly one equation from the overdetermined system. Note that species i can be different chemical elements as well as differently charged ions of the same element.

In every (r,t) point of models we search a solution in a form (K_0 is numerical coefficient)

$$w_i = K_0 \frac{T^{5/2}}{\rho} \left(A_P \frac{d\ln P}{dr} + A_T \frac{d\ln T}{dr} + \sum_k A_{Xk} \frac{d\ln X_k}{dr} \right), \tag{4}$$

and get coefficients A. Substituting $w_i(r,t)$ in the diffusion equation

$$\frac{\partial X_i}{\partial t} = -\frac{1}{\rho r^2} \frac{\partial (r^2 \rho w_i X_i)}{\partial r}, \tag{5}$$

and integrating (5) over t with appropriate boundary ($dX/dr(0,t) = 0$, $dX/dr(R_{Sun},t) = 0$) and initial ($X(r,0) = const$) conditions, we finally calculate composition profiles $X_i(r,t)$ for every component i of plasma.

Representation (4) of diffusion velocity makes possible to compare incomes to w_i from barodiffusion, thermodiffusion etc.

Here we present diffusion simulation of mixture $H^+ + He^{2+} + Me^{Z+} + e^-$, with initial heavy element mass-fraction $X_{Me}(r,0) = 0.02$, where Me^{Z+} denotes heavy element under consideration with charge Z (in general case $Z = Z(r,t)$).

3. Results

3.1. Effects of photodiffusion and radius-depended ionization stage

Since diffusion rates depends on the plasma interaction cross section, highly ionized ion of the element has bigger cross section and lower diffusion velocity.

Hydrogen and helium are fully ionized everywhere in the solar interiors, but heavy ions are presented in solar plasma in several ionization stages, and a diffusion velocity is higher for ions with smaller charge. We calculate ionization stages according to SAHA-S EOS tables ([5]). CNO and Ne elements reveal to be fully ionized in the convective zone while Si and Fe do not reach full ionization even in the center of the Sun.

Since heavy elements are not fully ionized in solar interiors, the effect of interaction with radiation flux is noticeable (so called photodiffusion). We take into account radiative accelerations modifying gravitational acceleration g in (1) by a factor $(1 - g_{rad\ i}/g)$, where radiative accelerations g_{rad} are calculated according to [6].

We analyzed the settling of C, N, O, Ne, Mg, Al, Si, S, Ar, Ca and Fe in $Z(r,t) = Z_{max}$ approximation, and C, N, O, Ne, Si, and Fe – with $Z = Z(r,t)$ and photodiffusion taken into account. We compare diffusion profiles for carbon and silicon (Fig. 1). Diffusional settling rates from the convective zone of six heavy elements are demonstrated on Fig. 2.

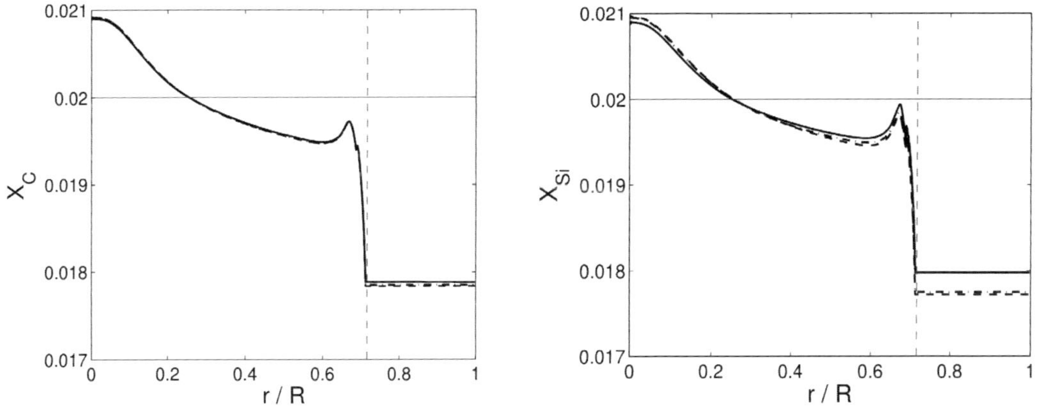

Figure 1. On the left: carbon diffusion profiles. Solid line – basic diffusion model (full ionization $Z = Z_{max}$, no photodiffusion), dashed line – a model with $Z = Z(r, t)$, and dash-dotted line – a model with $Z = Z(r, t)$ and photodiffusion. With ions charge depended on depth, elements are settling rapidly, but interaction with radiation prevent settling and reduce the effect. In the center summary effect increases carbon enhancement on 1.6 %, in convective zone – enlarges depletion on 1.4 % in comparison with basic model. On the right: silicon diffusion profiles. The summary effect of $Z = Z(r, t)$ and photodiffusion taken into account leads to additional silicon settling on 6.1 % in the center and on 11 % in the convective zone.

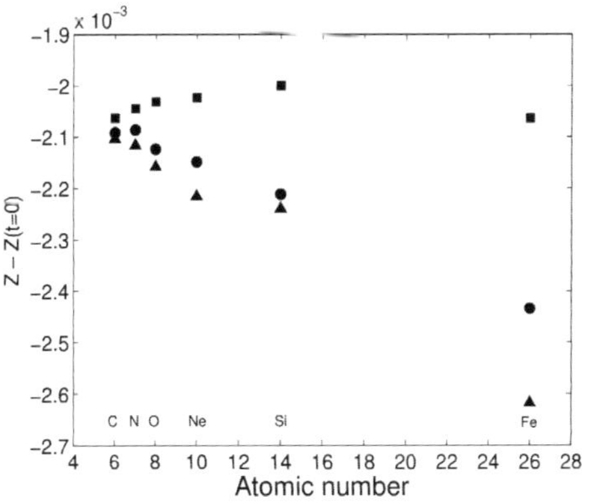

Figure 2. Effects of photodiffusion and radius-depended ionization stage. Changes in heavy elements abundances in convective zone are shown. Squares denote results for the model using full ionization approximation; triangles – for the model with radius-depended ionization stage of the heavy elements; circles – for the model with radius-depended ionization stage and radiative pressure taken into account.

3.2. Elements settling as a function of mixing zone boundary position

To test the effect of shifting of the mixing zone boundary on the rate of hydrogen enhancement in solar envelope we performed several simulations with fixed evolutionary solar model sequence and different positions of mixing zone boundary. Resulting diffusion profiles are shown on the Fig. 3. From the experiment we conclude that diffusion enchancement (for hydrogen) and diffusion settling (for heavier elements) decrease in deeper convective zone.

3.3. Temperature profile effect

From equations (4) and (5) follows that $\partial X_i/\partial t$ is proportional to the second derivative of the temperature. Temperature gradient lowering in the narrow region under the mixing zone leads to the thermodiffusion acting against the gravity. As a result, a local maximum of heavy element profiles arises (Fig. 4). For additional check we performed simulations without taking thermodiffusion into account, and the local maximum of element profile has disappeared.

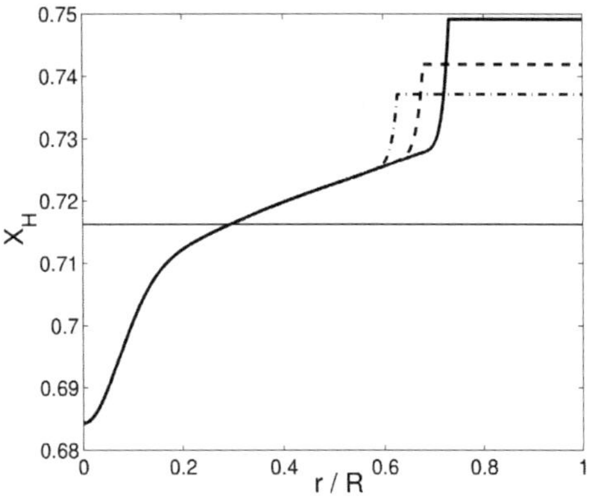

Figure 3. Hydrogen diffusion profiles for t = 4.6 Gyr in the case of the original boundary position ($0.73R$, solid line), and in cases of deeper mixing zones ($0.68R$ and $0.63R$, dashed and dot-dashed lines correspondingly). X_H values in mixing zone are 0.749, 0.742 and 0.737 correspondingly. Initial abundance equals 0.716. The effect of nuclear burning on X_H in solar core is not shown for clarity.

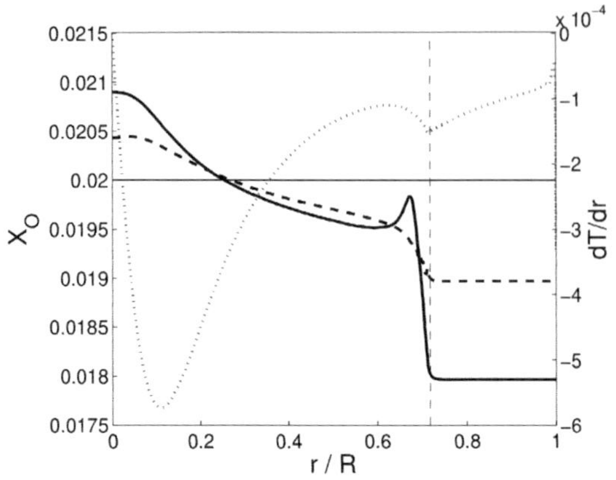

Figure 4. Effect of temperature gradient profile on element settling under the mixing zone boundary. Solid line – oxygen diffusion profile, dashed line – oxygen diffusion profile calculated without thermodiffusion, dotted line – temperature gradient profile for $t = 4.6$ Gyr.

4. Conclusions

- Taking into account effects of photodiffusion and radius-depended ionization stage provide more accurate treatment of diffusion. Superposition of these effects noticeably increase diffusion velocity for neon and heavier elements.
- According to [7] the heavy element abundance in solar atmosphere is lower by 0.006 in comparison to previous works. These figures are not directly related to our computations because we compute only differences with primordial abundances, which are generally unknown. Our results show that settling could not explain entirely the low-Z heavy element abundances in the atmosphere according to [7].
- Nevertheless, diffusional settling of elements from convective zone is sensitive to a location of the convective zone boundary – the most important value in low-Z problem consideration – and increases in shallower convective zone.

Acknowledgments

The study was supported by ISTC project 3755.

References

[1] Grad H 1949 *Communs. Pure and Appl. Math.* **2** 331

[2] Burgers J M 1969 *Flow Equations for Composite Gases* (New York: Academic)

[3] Thoul A A, Bahcall J N and Loeb A 1994 *Astrophys. J.* **421** 828

[4] Zhdanov V M 2009 *Transfer processes in multicomponent plasma* (Moscow: Physmathlit) in Russian

[5] Gryaznov V K, Ayukov S V, Baturin V A, Iosilevskiy I L, Starostin A N and Fortov V E 2004 *AIP Conf. Proc.* vol 731 p 147

[6] Seaton M J, Yu Yan, Mihalas D and Pradham A K 1994 *MNRAS* **266** 805

[7] Asplund M, Grevesse N and Sauval A J 2005 *ASP Conf. Ser.* vol 336 (eds Barnes T G and Bash F N) p 25

Numerical simulations of conversion to Alfvén waves in solar active regions

E. Khomenko[1] **and P. S. Cally**[2]

[1] Instituto de Astrofísica de Canarias, 38205, C/ Vía Láctea, s/n, Tenerife, Spain

[2] School of Mathematical Sciences, Monash University, Clayton, Victoria 3800, Australia

E-mail: `khomenko@iac.es`

Abstract. We study the coupling of magneto-acoustic waves to Alvén waves using 2.5D numerical simulations. In our experiment, a fast magnetoacoustic wave of a given frequency and wavenumber is generated below the surface. The magnetic field in the domain is assumed homogeneous and inclined. The efficiency of the conversion to Alfvén waves near the layer of equal acoustic and Alfven speeds is measured calculating their energy flux. The particular amplitude and phase relations between the oscillations of magnetic field and velocity help us to demonstrate that the waves produced after the transformation and reaching upper atmosphere are indeed Alfvén waves. We find that the conversion from fast magneto-acoustic waves to Alfvén waves is particularly important for the inclination θ and azimuth ϕ angles of the magnetic field between 55 and 65 degrees, with the maximum shifted to larger inclinations for lower frequency waves. The maximum Alfvén flux transmitted to the upper atmosphere is about 2–3 times lower than the corresponding acoustic flux.

Conversion from fast-mode high-β magneto-acoustic waves (analog of p modes) to slow-mode waves in solar active regions is relatively well studied both from analytical theories and numerical simulations (e.g., [1, 2, 3, 4, 5, 6]), see [7] for a review. In a two-dimensional situation, the transformation from fast to slow magnetoacoustic modes is demonstrated to be particularly strong for a narrow range of the magnetic field inclinations around 20–30 degrees to the vertical. However, no generalized picture exists so far for conversion from magneto-acoustic to Alfvén waves in a three-dimensional situation. Studies of this conversion were initiated by Cally & Goossens [8], who found that the conversion is most efficient for preferred magnetic field inclinations between 30 and 40 degrees, and azimuth angles between 60 and 80 degrees, and that Alfvénic fluxes transmitted to the upper atmosphere can exceed acoustic fluxes in some cases. Newington & Cally [9] studied the conversion properties of low-frequency gravity waves, showing that large magnetic field inclinations can help transmitting an important amount of the Alfvénic energy flux to the upper atmosphere.

Motivated by these recent studies, here we attack the problem by means of 2.5D numerical simulations. The purpose of our study is to calculate the efficiency of the conversion from fast-mode high-β magneto-acoustic waves to Alfvén and slow waves in the upper atmosphere for various frequencies and wavenumbers as a function of the field orientation. We limit our study to a plane parallel atmosphere permeated by a constant inclined magnetic field, to perform a meaningful comparison with the work of Cally & Goossens [8]. Numerical simulation will allow generalization to more realistic models in our future work.

We numerically solve the non-linear equations of ideal MHD assuming all vectors in three

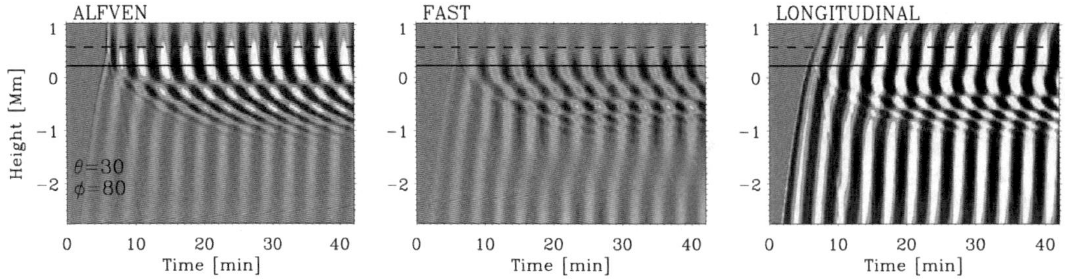

Figure 1. Time-height variations of the three projected velocity components corresponding to $\hat{\mathbf{e}}_{\rm perp}$ (Alfven wave, left), $\hat{\mathbf{e}}_{\rm tran}$ (fast wave, middle) and $\hat{\mathbf{e}}_{\rm long}$ (slow wave, right) for $\nu = 5$ mHz in a simulation with B inclined by $\theta = 30°$ and $\phi = 80°$. The solid line marks the position $c_S = v_A$, and the dashed line marks the cut-off layer $\nu = \nu_c \cos\theta$. The colour scaling is the same in all panels. The amplitudes are scaled with $\sqrt{\rho_0 v_A}$ (first two panels) $\sqrt{\rho_0 c_S}$ (last panel).

spatial directions and all derivatives in two directions (i.e. 2.5D approximation, see [10, 11]), though perturbations are kept small to approximate the linear regime. An acoustic wave of a given frequency and wave number is generated at -5 Mm below the solar surface in a standard model atmosphere permeated by a uniform inclined magnetic field. The top boundary of the simulation box is 1 Mm above the surface, and 0.8 Mm above the layer where the acoustic speed, c_S, and the Alfvén speed, v_A, are equal. We consider frequencies $\nu = 3$ and 5 mHz and wave numbers $k_X = 1.37$ Mm^{-1} and $k_Y = 0$. The simulation grid covers field inclinations θ from 0° to 80° and field azimuths ϕ from 0° to 160°. The field strength is kept at $B = 500$ G. To separate the Alfvén mode from the fast and slow magneto-acoustic modes in the magnetically dominated atmosphere we use velocity projections onto three characteristic directions:

$$\begin{aligned} \hat{\mathbf{e}}_{\rm long} &= [\cos\phi\sin\theta,\ \sin\phi\sin\theta,\ \cos\theta]; \\ \hat{\mathbf{e}}_{\rm perp} &= [-\cos\phi\sin^2\theta\sin\phi,\ 1-\sin^2\theta\sin^2\phi,\ -\cos\theta\sin\theta\sin\phi]; \\ \hat{\mathbf{e}}_{\rm trans} &= [-\cos\theta,\ 0,\ \cos\phi\sin\theta]. \end{aligned} \tag{1}$$

To measure the efficiency of conversion to Alfvén waves near and above the $c_S = v_A$ equipartition layer, we calculate acoustic and magnetic energy fluxes, averaged over time:

$$\mathbf{F_{ac}} = \langle p_1 \mathbf{v}_1 \rangle; \qquad \mathbf{F_{mag}} = \langle \mathbf{B_1} \times (\mathbf{v}_1 \times \mathbf{B_0})/\mu_0 \rangle. \tag{2}$$

Figure 1 shows an example of the projected velocities in our calculations as a function of space and time. In this representation the larger inclination of the ridges mean lower propagation speeds and vice versa. Note, that by projecting the velocities, we are able to separate the modes only in the magnetically dominated atmosphere, i.e. above the solid line in Fig. 1. The figure shows how the incident fast mode wave propagates to the equipartition layer and then splits into several components. The Alfvén wave is produced by mode conversion above 0.2 Mm (left panel) and propagates upwards with the (rapid) Alfvén speed, confirmed by almost vertical inclination of the ridges. Conversely, the essentially magnetic fast-mode low-β wave produced in the upper atmosphere (middle panel) is reflected, and its velocity variations in the upper layers vanish with height. The (acoustic) slow-mode low-β wave escapes to the upper atmosphere tunnelling over the cut-off layer due to the field inclination of $\theta = 30°$. The amplitudes of the velocity variations of the Alfvén wave are comparable to those of the slow wave.

To confirm the Alfvén nature of the transformed waves, as revealed by the projection calculations, we checked the amplitude and phase relations for all three modes reaching the upper atmosphere. For the Alfvén mode the magnetic field B_1 and velocity variations V_1 should

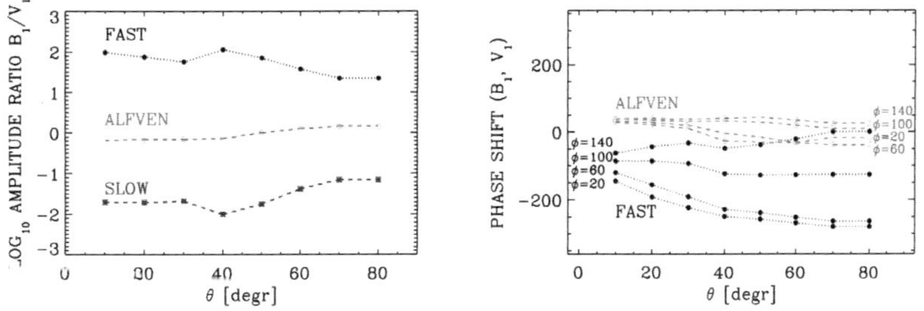

Figure 2. Left panel: Log_{10} of the ratio B_1 to $V_1/\sqrt{\mu_0\rho_0}$ for projected velocities and magnetic field variations, averaged over all ϕ, as a function of θ. Black line: fast mode ($\hat{\mathbf{e}}_{\text{tran}}$ projection); red line: Alfvén mode ($\hat{\mathbf{e}}_{\text{perp}}$); blue line: slow mode ($\hat{\mathbf{e}}_{\text{long}}$). Right panel: phase shift between the projected variations of V_1 and B_1, as a function of θ for selected ϕ. Red lines: Alfvén mode; black lines: fast mode.

be in equipartition (i.e. $B_1 = V_1/\sqrt{\mu_0\rho_0}$), and both magnitudes should oscillate in phase (see Priest [12]). Figure 2 presents the calculations of the amplitude ratio $B_1\sqrt{\mu_0\rho_0}/V_1$ and temporal phase shift between B_1 and V_1, where both velocity and magnetic field variations are projected in the corresponding characteristic direction for each mode (Eq. 1). This calculation confirms that, indeed, for all magnetic field orientations θ and ϕ, the amplitude ratio for the Alfvén mode ($\hat{\mathbf{e}}_{\text{perp}}$ projection) is around one (left panel). This is clearly not the case for the slow and fast modes. For the fast mode, the amplitude ratio is two orders of magnitude larger, and for the slow mode, it is two orders of magnitude lower than one. For the Alfvén mode the phase shifts group around zero for all ϕ, unlike the case of the fast mode (right panel). We did not calculate the phase shifts for the slow mode as the variations of the magnetic field are negligible. Thus, we conclude that the properties of the simulated Alfvén mode separated by the projection correspond to those expected for a classical Alfvén mode.

An example of the height variations of the acoustic and magnetic fluxes is given in Figure 3. The total vertical flux (dotted line) is conserved in the simulations except for the limitations caused by the finite grid resolution not resolving slow small-wavelength waves in the deep layers (see Fig. 1). Both acoustic and magnetic fluxes show strongest variations near the conversion layer and become constant above it between 0.5 and 1 Mm height. The fluxes reaching the upper atmosphere depend crucially on the orientation of the field. In this example, the acoustic flux decreases with θ whilst the magnetic flux increases with θ and becomes larger than the acoustic fluxes for $\theta = 60°$. As the fast wave is already reflected in the upper atmosphere (see Fig. 1), the magnetic flux at these heights is due to the propagating Alfvén wave.

Finally, Figure 4 gives the time averages of the vertical magnetic and acoustic fluxes at the

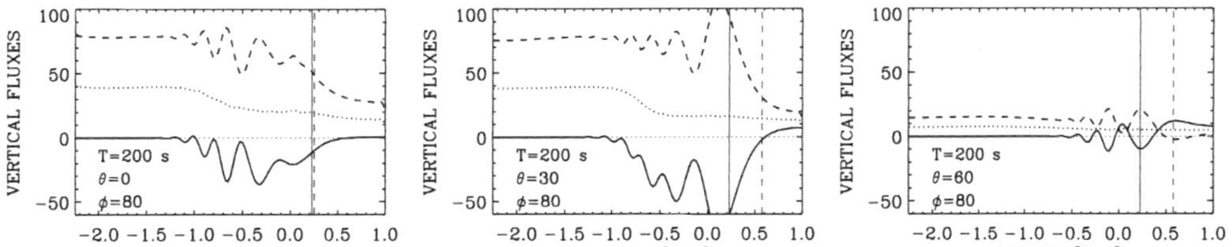

Figure 3. Examples of the height dependence of the magnetic (solid line) and acoustic (dashed line) vertical fluxes, defined by Eq. 2, for $\nu = 5$ mHz and several θ and ϕ. Solid vertical line marks the position $c_S = v_A$, dashed vertical line marks the cut-off layer $\nu = \nu_c \cos\theta$.

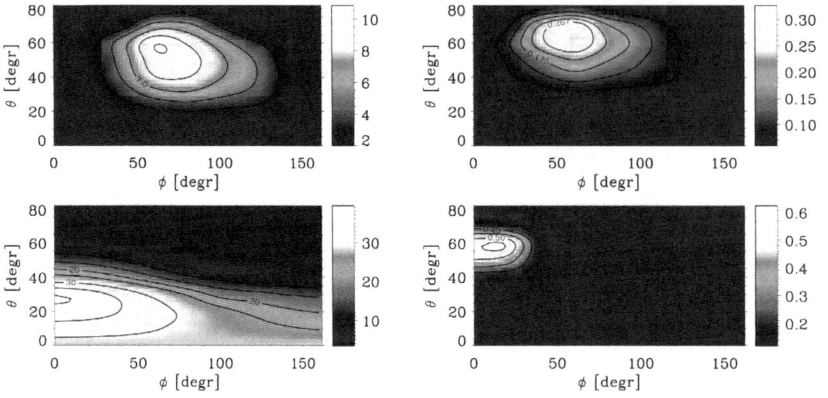

Figure 4. Vertical fluxes measured at the top of the atmosphere at 1 Mm for waves with $\nu = 5$ mHz (left panels) and 3 mHz (right panels). Upper panels give magnetic fluxes and lower panels give acoustic fluxes.

top of the atmosphere as a function of the field orientation. As proven above, the magnetic flux at 1 Mm corresponds to the Alfvén mode. At $\nu = 5$ mHz, the maximum of the magnetic flux corresponds to $\theta = 50°$ and $\phi = 65°$. This maximum is shifted to larger inclinations $\theta = 65°$ for waves with $\nu = 3$ mHz. The presence of the sharp maximum of the Alfvénic flux transmission agrees well with the conclusions made previously by Cally & Goossens [8], though the exact position of the maximum is shifted to somewhat larger inclinations. The maximum of the transmitted acoustic flux corresponds to inclinations $\theta \approx 30°$ for $\nu = 5$ mHz waves, and to $\theta \approx 55°$ for $\nu = 3$ mHz waves, again, in agreement with previous calculations [3, 8]. The absolute value of the fluxes is about 30 times lower for 3 mHz compared to 5 mHz. At some angles the Afvén magnetic flux transmitted to the upper atmosphere is larger than the acoustic flux. However, at angles corresponding to the maximum of the transmission, the Alfvén flux is 2-3 times lower than the corresponding acoustic flux.

It is important to realize that quantitatively simulating mode transformation numerically is a challenge, as any numerical inaccuracies are amplified in such second-order quantities as wave energy fluxes. The tests presented in this paper prove the robustness of our numerical procedure and offer an effective way to separate the Alfvén from magneto-acoustic modes in numerical simulations. This will allow us in future to study the coupling between magneto-acoustic and Alfvén waves in more realistic situations resembling complex solar magnetic structures.

References

[1] Zhugzhda Y D and Dzhalilov N S 1982 *Astron. Astrophys.* **112** 16

[2] Cally P S and Bogdan T J 1997 *Apj* **486** L67

[3] Schunker H and Cally P S 2006 *MNRAS* **372** 551

[4] Cally P S 2006 *Phil. Trans. R. Soc. A* **364** 333

[5] Khomenko E, Kosovichev A, Collados M, Parchevsky K and Olshevsky V 2009 *Apj* **694** 411

[6] Felipe T, Khomenko E and Collados M 2010 *Apj* **719** 357

[7] Khomenko E 2009 *ASP Conf. Series* vol 416 ed. M Dikpati, T Arentoft, I González Hernández, C Lindsey, & F Hill p. 31

[8] Cally P S and Goossens M 2008 *Solar Phys.* **251** 251

[9] Newington M E and Cally P S 2010 *MNRAS* **402** 386

[10] Khomenko E and Collados M 2006 *Apj* **653** 739

[11] Khomenko E, Centeno R, Collados M and Trujillo Bueno J 2008 *Apj* **676** L85

[12] Priest E R 1984 *Solar magneto-hydrodynamics* Geophysics and Astrophysics Monographs, Dordrecht: Reidel

The solar core as never seen before

A. Eff-Darwich[1,2], **S.G. Korzennik**[3]

[1] Instituto de Astrofísica de Canarias (IAC), E-38200 La Laguna, Tenerife, Spain
[2] Dept. de Edafología y Geología, Universidad de La Laguna (ULL), E-38206 La Laguna, Tenerife, Spain
[3] Harvard-Smithsonian Center for Astrophysics, 60 Garden St. Cambridge, MA,02138, USA

E-mail: `adarwich@iac.es`, `skorzennik@cfa.harvard.edu`

Abstract. One of the main drawbacks in the analysis of the dynamics of the solar core comes from the lack of consistent data sets that cover the low and intermediate degree range ($\ell = 1, 200$). It is usually necessary to merge data obtained from different instruments and/or fitting methodologies and hence one introduces undesired systematic errors. In contrast, we present the results of analyzing MDI rotational splittings derived by a single fitting methodology applied to 4608-, 2304-, etc..., down to 182-day long time series. The direct comparison of these data sets and the analysis of the numerical inversion results have allowed us to constrain the dynamics of the solar core and to establish the accuracy of these data as a function of the length of the time-series.

1. Introduction

Ground-based helioseismic observations, *e.g.*, GONG [5], and space-based ones, *e.g.*, MDI [10] or GOLF [4], have allowed us to derive a good description of the dynamics of the solar interior [12], [2], [6]. Such helioseismic inferences have confirmed that the differential rotation observed at the surface persists throughout the convection zone. The outer radiative zone ($0.3 < r/R_\odot < 0.7$) appears to rotate approximately as a solid body at an almost constant rate (≈ 430 nHz), whereas the innermost core ($0.19 < r/R_\odot < 0.3$) rotates slightly faster than the rest of the radiative region.

Most of these studies have used averaged MDI, GONG or GOLF frequencies to produce a high precision data set to infer time-averaged properties of the solar interior. The resulting mode set is reduced to the modes fitted for all, or at least most, epochs. The consequences of this restriction is particularly dramatic for the low degree and the low frequency modes - *i.e.* the modes that carry significant information about the rotation of the inner radiative zone [3]. This is why the time-averaging should be carried out by fitting the corresponding (long) time series instead. Fitting epochs of various lengths leads to averaging these epochs at the power spectrum level. This allows fitting spectra with higher signal to noise and thus extracting the most significant information, and leading to values consistent among the different sets.

We have used frequency sets resulting from fitting MDI time-series of various lengths in order to study the effect of the length of the time series on the estimation of the solar rotation in the inner radiative region. Specifically, rotational splitting sets were derived from fitting 4608-, 2304-, 1456-, 728-, 360- and 182-day long time series. An iterative inversion methodology [3] was applied to the data sets to infer the rotation rate in the solar interior.

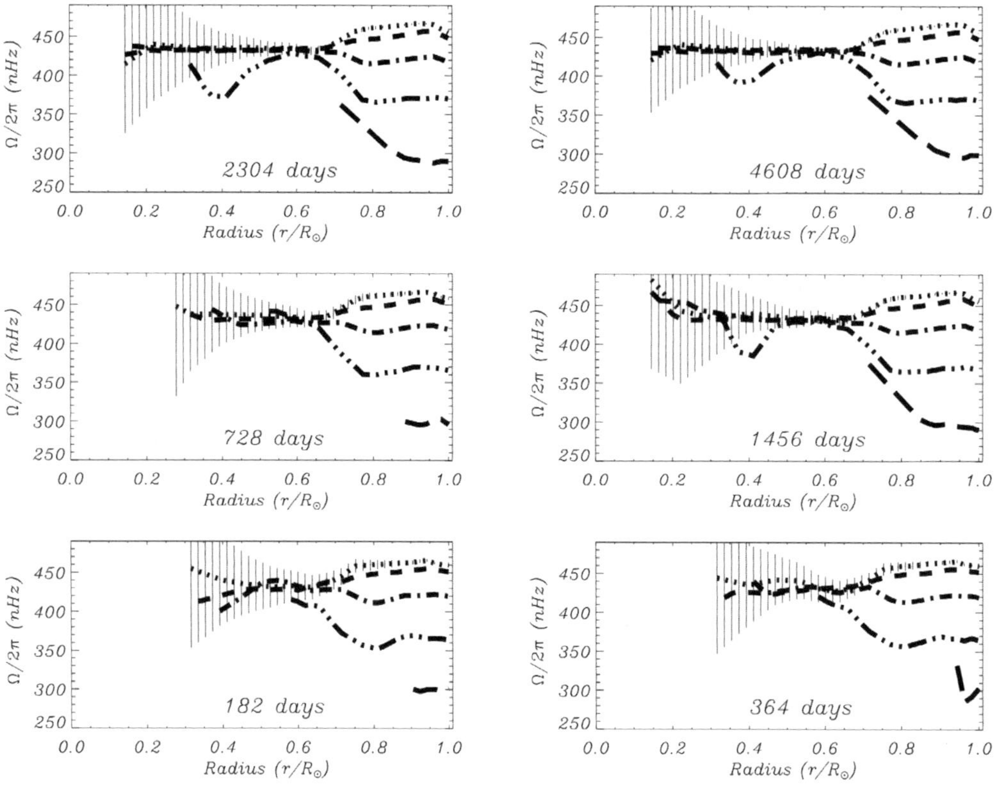

Figure 1. Inversions of the MDI data sets, plotted as a function of radius at several co-latitudes. The dotted, dashed, dotted-dashed, triple-dotted-dashed and long-dashed lines correspond to the rotation rate for the equator, and at latitudes of 20°, 40°, 60° and 80°, respectively.

2. Results and conclusions

The average rotational profiles obtained from the inversions of the 4608-, 2304-, 1456-, 728-, 360- and 182-day long sets are shown in Fig. 1. The longer the time-series associated with the data, the deeper is the significance of the inversions. Moreover the latitudinal extent of the inversions also increase with the length of the time-series, resulting in better estimations of the rotational rate at high latitudes.

The rotational rates obtained from 4608- and 2304-day long time series might presently represent the best estimates of the dynamical conditions of the solar inner radiative region, including the core. The sun seems to rotate as a rigid solid in the entire radiative region from the equator up to approximately a co-latitude of 50 degrees. A striking feature shows up at approximately $0.4R_\odot$ and 60 degrees in latitude, namely a "dip" that is clearly visible not only in the inversions of 4608-, 2304-day long sets, but also in those associated to the 1456-day long time series. The rotational rate down to approximately $0.18R_\odot$ is nearly constant within error bars, not exceeding in any case 600 nHz. The poor estimations of the solar rotational rate in the inner radiative region using shorter time-series (360- and 182-day long) also happens using the GONG pipeline data [2], that still uses 108-day long time series and a methodology developed back in 1990 [1]. The same applies to MDI observations, since it is using 72-day long time series and an algorithm devised in 1992 [11]. Moreover both GONG and MDI pipelines contain systematic errors [8], [9].

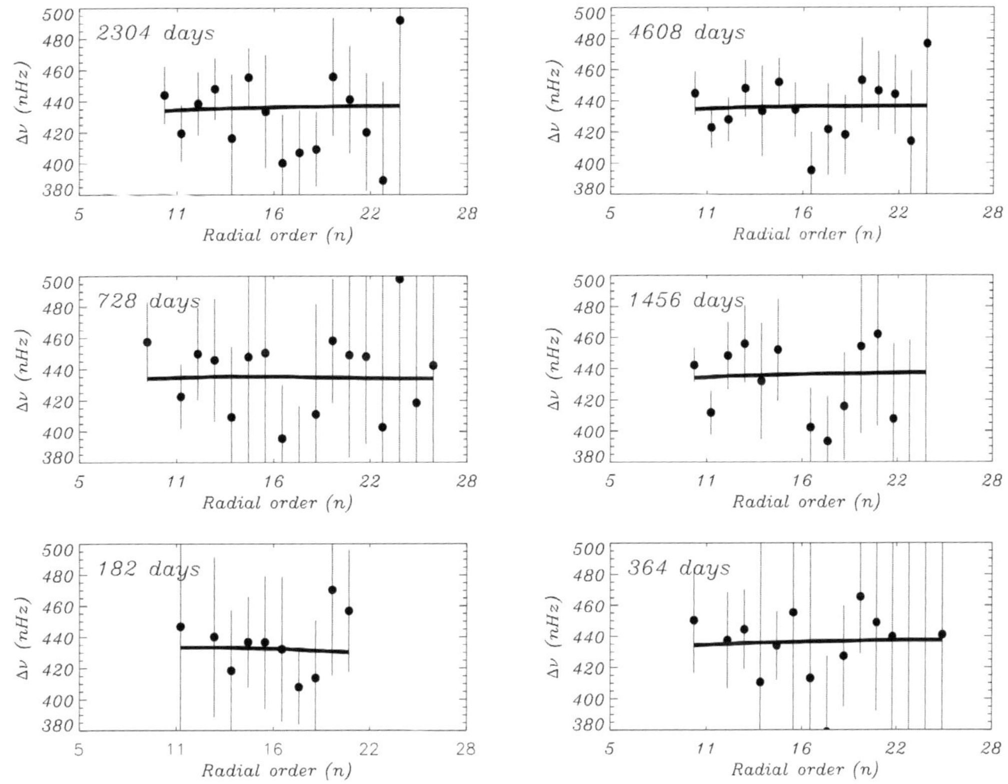

Figure 2. Observational sectoral frequency splittings as a function of radial order for $\ell = 1$ modes. The different panels correspond to the average splittings obtained from the time-series defined in this work (see text for details). The solid lines correspond to the theoretical splittings obtained from the rotational profiles of Fig. 1, under the supposition the radiative region rotates rigidly at 433 nHz.

However, the inversion results are still far from representing the actual variations of frequency splittings at low degrees (Figures 2 and 3). Longer time series mean lower uncertainties, revealing that the observational splittings (specially at larger frequencies) do not match the theoretical splittings obtained from the inverted rotational rates. This is an intriguing result that could be just an artifact of the fitting methodology or the observational time-series; although it could be an unknown solar dynamical feature that ought to be explored.

Acknowledgments
The Solar Oscillations Investigation-Michelson Doppler Imager project on SOHO is supported by NASA grant NAS5-3077 at Stanford University. SOHO is a project of international cooperation between ESA and NASA. This work utilizes data obtained by the Global Oscillation Network Group (GONG) Program, managed by the National Solar Observatory, which is operated by AURA, Inc. under a cooperative agreement with the National Science Foundation. This work was in part funded by NASA grant NNX09AB15G, the Spanish grant AYA2004-04462 and the Canarian grant SolSubC200801000243.

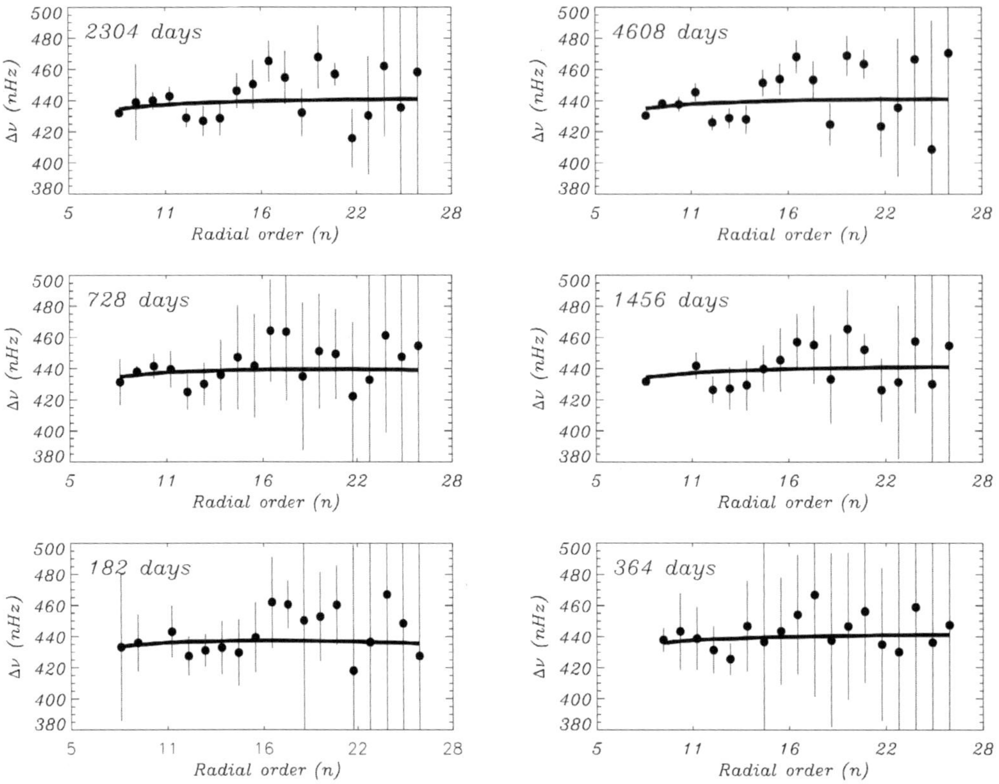

Figure 3. As in Fig. 2, but for $\ell = 2$ sectoral modes.

References

[1] Anderson, E. R., Duvall, T. L.,Jr., & Jefferies, S. M. 1990, *ApJ*, **364**, 699

[2] Eff-Darwich, A., Korzennik, S. G., Jiménez-Reyes, S. J., & García, R. A. 2008, *ApJ*, **679**, 1636

[3] Eff-Darwich, A., Korzennik, S. G., & García, R. A. 2009, *AN*, in press

[4] Gabriel, A. H., et al. 1995, Sol. Phys, **162**, 61

[5] Harvey, J. W., et al. 1996, Science, **272**, 1284

[6] Howe, R. 2009, Living Reviews in Solar Physics, **6**, 1

[7] Korzennik, S. G., Rabello-Soares, M. C., & Schou, J. 2004, *ApJ*, **602**, 481

[8] Korzennik, S. G. 2005, *ApJ*, **626**, 585

[9] Korzennik, S. G. 2008, Astronomische Nachrichten, **329**, 453

[10] Scherrer, P. H., et al. 1995,Sol. Phys. , **162**, 129

[11] Schou, J. 1992, Ph.D. Thesis,

[12] Thompson, M. J., Christensen-Dalsgaard, J., Miesch, M. S., & Toomre, J. 2003, Ann. Rev. Astron. Astrophys., **41**, 599

The dynamics of the radiative zone of the Sun

S Turck-Chièze[1], S Couvidat[2], V Duez[3], S Mathis[1], J Marques[4], A Palacios[5], L Piau[1]
[1]IRFU/ CEA/ CE Saclay, Gif sur Yvette, France [2] HEPL, Stanford, USA
[3]Argelander-Institut fur Astronomie, Universitat Bonn, Germany [4] Observatoire de Meudon, Meudon, France [5]GRAAL Montpellier, France

E-mail: sylvaine.turck-chieze@cea.fr

Abstract. Helioseismology puts strong constraints on the internal sound speed and on the rotation profile in the radiative zone. Young stars of solar type are more active and faster rotators than the Sun. So we begin to build models which include different rotation histories and compare the results with all the solar observations. The profiles of the rotation we get have interesting consequence for the introduction of magnetic field in the radiative zone. We discuss also the impact of mass loss deduced from measured flux of young stars. We deduce from these comparisons some quantitative effect of the dynamical processes (rotation, magnetic field and mass loss) of these early stages on the present sound speed and density. We show finally how we can improve our present knowledge of the radiative zone with PICARD and GOLFNG.

1. Time evolution of the internal rotation induced by its transport of momentum

The Sun is rotating and the internal seismic rotation pushes us to introduce the effect of the time evolution of the rotation in solar model. Turck-Chièze et al., 2010a have followed three rotation histories: a very low initial rotation, an academic case where the initial value is so low that the Sun naturally slows down, and two higher initial rotations which lead to 20 and 50 km/s at the arrival on the main sequence plus some braking to reach the present surface value of 2.2 km/s. Their time internal evolution are followed, modified by the transport of momentum induced in the radiative zone by both advection and diffusion terms.

We notice that a radial differential rotation is present in all the models we have computed and that its radial profile is mainly established during the contraction phase which corresponds to the first million years (see Figure 7 of Turck-Chièze et al. 2010a and tables 4 and 5). It seems that a gradient stays present in the core of the Sun today if one believes the detection of gravity modes with the GOLF instrument aboard SoHO (Turck-Chièze et al. 2004, Garcia et al. 2007, Mathur et al.2008). Of course in our computations, the order of magnitude of the gradient depends strongly on the presence or not of braking. Our conclusions on the transport of momentum by rotation alone are the following:

1) This transport during the main sequence appears extremely small. The meridional circulation in the radiative zone is smaller by about 10 orders of magnitude in comparison with the observed convective meridional circulation velocity measured at 99 % $R_{\odot}$.

2) The sound speed and density profiles are not largely modified by rotation alone but the difference with obervation slightly increased (Figure1 left).

3) Although the combined effect of meridional circulation and shear-induced turbulent associated to rotation is small, this study leads to radial rotation profiles that can be directly

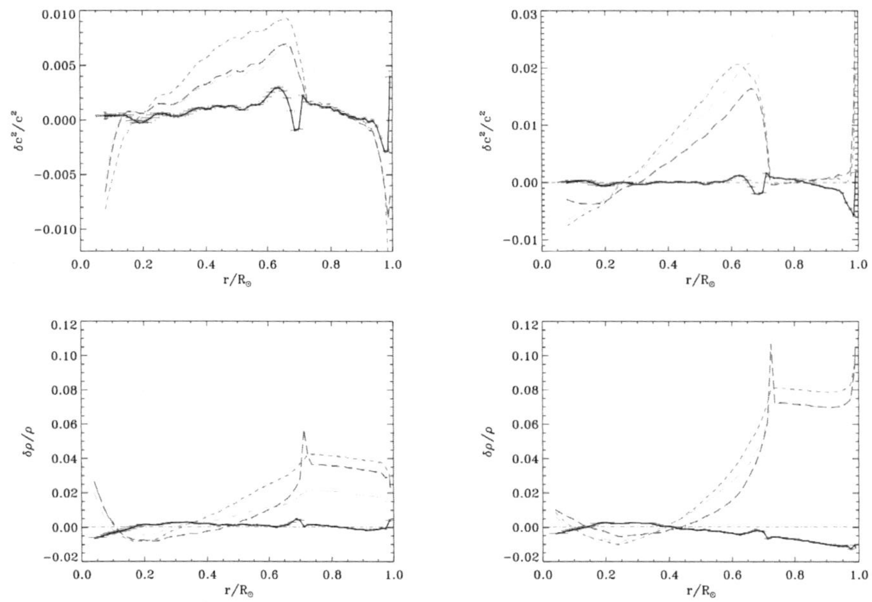

Figure 1. Left: Radial squared sound speed and density differences between observations and models. Seismic model is in black with error bars coming from the seismic data. The standard GN composition model, model A (slow rotation) and model B (large rotation) are respectively in green (dotted dashed line), blue (large dashed line) and red (small dashed line). Right: Idem: now the standard model calculated with the most recent Asplund composition, model with turbulence in the tachocline and model with 0.33 $M_\odot$ are respectively in red (dashed line), green (dot dashed line) and blue (large dashed line). From Turck-Chièze et al. 2010a,b.

compared to the seismically observed one. The comparison with observations sustains the idea that the Sun was not a rapid rotator (Figure 2 left). The low initial rotation (the academic case) model helps us to quantify the structural consequences of rotation.

4) We believe that the radial rotation profile during the contraction phase could be reduced by the transport of momentum by magnetic field in the core and that the diffusion of this field can flatten the profile in the rest of the radiative zone. A first tentative has been performed by Eggenberger et al. (2007), but the stability of that field, if one believes that it is still present, supposes probably a mixture of poloidal and toroidal components.

It is why we have begun to introduce a magnetic configuration which may exist in the solar-type radiative regions and developed the modified structural equations. Probably, these fields cannot exceed several MG. If it was not the case, their impact on the solar shape would have been clearly established already. Consequently, the β parameter that gives the ratio between the total pressure and the magnetic pressure stays large in the whole radiative zone and the microscopic quantities including the sound speed must not be modified significantly by the presence of such a field. In Duez et al., 2010, we have calculated the solar quadrupole moment which results from such a field. So the order of magnitude of the variation of the different terms at the arrival on the main sequence is small (Duez, Mathis & Turck-Chièze, 2010). But the description of the early stage is more complex and justifies to look deeper on young stars.

2. The strong activity of the young solar like stars

Young solar-like stars are generally more active and are rotating faster than the present Sun. The review of Gudel (2007) gives some specific properties of young analogs of the Sun. Their observation confirms the need to introduce magnetic field in stellar modelling to derive an appropriate comparison with all the present observables.

The specific observation of young stellar rotations shows different temporal rotation profiles during the contraction phase and the PMS (Figures 1 and 3 of Bouvier 2008a). If a typical braking law $dJ/dt \propto \Omega^3$ is observed between α Persei, Pleiades, Hyades and the Sun, other observations of young stars show that this law is not universal and that it exists slow and fast young rotators. This difference may be explained by the coupling timescale of core and envelope (Bouvier 2008b).

The magnetized stellar winds are efficient to brake the magnetically active stars. Their efficiency is about a factor 1000 greater than any other phenomenon. So, even their effect in the radiative zone appears small for the present Sun, magnetic field of the young Sun is the central ingredient which governs the rotational evolution of solar-like stars and probably already in the very early stage. We have not yet results for this phase because one needs to know how to introduce a potential growing of the field probably through a dynamo process. We begin to introduce another effect connected to UV loss observed in young analogs, in modelling a mass loss in the early stage directly deduced from the study of a lot of young stars.

3. Young Sun mass loss and its impact on luminosity and its present sound speed

In standard model studies, the lost of luminosity by X and UV due to the strong activity of the young Sun is ignored. For the early stage, these phenomena are measured in young stars (Ribas, 2009). Moreover we now know that the present dynamo, which produces the Hale cycle, leads to luminosity loss 10^6 greater than was supposed by the standard model along the last decade (about some 10^{-3} instead of 10^{-9}). So it is clear that one needs to introduce the lost of luminosity along age produced by different phenomena.

To progress on this direction, we have used the expressions given by Ribas to estimate some properties of the Sun in the first stage, knowing that there is difficulty to separate periods of accretion and mass loss and the duration of each of them. Our first model leads to a more luminous Sun during the formation of planets phase which can help to resolve partly the solar paradox. Figure 1 right illustrates the impact of mass loss on the sound speed and density. One confirms that contrary to the effect of transport of momentum by rotation (figure 1 left), the introduction of mass loss in the early stage could reduce the discrepancy that we observe with the observed sound speed and density at the present time.

4. The need for new observational constraints

The whole story of the solar activity supposes to introduce one by one all the different processes which maintain this activity up to now. To check its relevance, all the observations we may gather on young cluster stars or on the present Sun are useful. It is why it will be particularly useful to predict how internal magnetic field and rotation profile impact on the shape of the present Sun and how this shape evolves at small and large time scales in using the measurements of SODISM/PICARD satellite (Thuillier, Dewitte and Schmutz 2007). We show, in Figure 2 right, how different profiles of rotation (a flat profile outside the core and an increase below R_C) could influence the solar quadrupole moment. We find that they are of the same order than the effect of a deep field. Up to now, the aspheric shape of the Sun is not yet understood, the balloon shows values of oblateness of 4.3 to 10. 10^{-6} increasing with the cycle (Emilio et al., 2007) and we can hope to determine the different components of this shape with PICARD.

In addition to SODISM, a better description of the solar core after the SoHO measurements is of fundamental importance to know precisely the density, temperature and rotation profiles. It

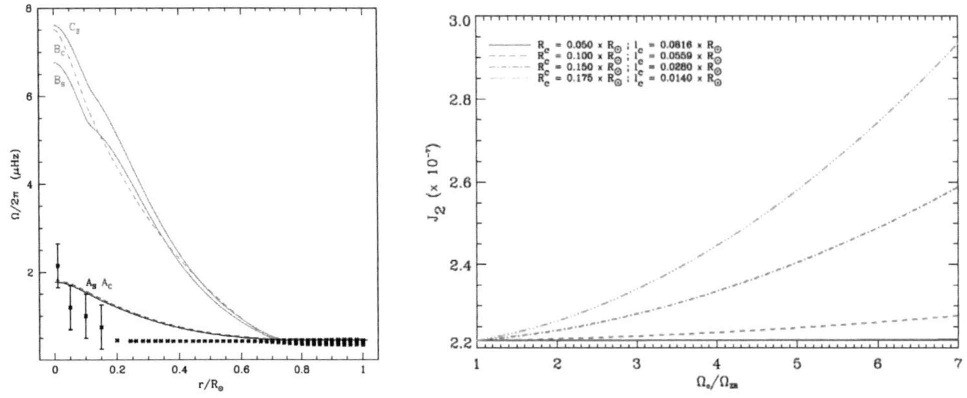

Figure 2. Left: Comparison between the solar internal profile predicted by different models (A for slow initial rotation, B and C for fast initial rotation) and that deduced from helioseismology. The data down to r/R$_\odot$ = 0.2 are deduced from the acoustic mode splittings determined by GOLF, MDI and GONG instruments (Couvidat et al. 2003b, Eff-Darwich, 2008). The data in the core mimic the information extracted from the gravity mode study. From Turck-Chièze et al. 2010a). Right: Quadrupole moment for different profiles of rotation versus the ratio central rotation over surface rotation. From Duez, Turck-Chièze, Mathis 2011.

is for such objective, that we have prepared the successor of GOLF/SoHO, through the building of the GOLF-NG prototype for which we have now demonstrated all the improved performances and solved all the technical difficulties . All the tests done have successfully shown the ability of this instrument to measure properly the time variation of the Doppler velocity on at least 6 heights in the atmosphere in order to eliminate part of the granulation noise and to measure quicker the small amplitude of useful gravity modes (Turck-Chièze et al. 2006, 2008, 2011).

References

Bouvier J 2008a in *Stellar Magnetism*, EDP science, p 199
Bouvier J 2008b *A&A* **489** L53
Couvidat S, Turck-Chièze S & Kosovichev A 2003a *ApJ* **599** 1434
Couvidat S, et al. 2003b *ApJ* **597** L77
Duez V, Mathis S & Turck-Chièze S 2010 *MNRAS* **402** 271
Duez V, Turck-Chièze S & Mathis S 2011 *A&A* in preparation
Emilio M. et al. 2007 *Astrophys. J.* 660L 161
Garcia R A, Turck-Chièze S et al. 2007 *Science* **316** 1537
Garcia R A et al. 2008 *Astron Notes* **329** 476
Eff Darwich A et al. 2008 *ApJ* **679** 1636
Eggenberger P, Maeder A & Meynet G 2005 *A&A* **440** L9
Gudel M 2007 *Living Review Solar Phys* **4** 3
Mathis S & Zahn JP 2004, *A&A*, **425** 229
Mathur S, Eff-Darwich A, Garcia R A& Turck-Chize S. 2008 *A&A* **484** 517
Thuillier G, Dewitte S & Schmutz W 2007 *Adv. Space Res.* **26** 1792
Turck-Chièze S. et al. 2001 *ApJ* **555** L69
Turck-Chièze S et al. 2004a *Phys Rev L* **93** 211102
Turck-Chièze et al. 2004b *ApJ* **604** 555
Turck-Chièze S. et al. 2006 *Adv. Space. Res.* **38** 1812
Turck-Chièze S. et al. 2008 *Astron Notes* **329** 521
Turck-Chièze S. et al. 2010a *ApJ* **715** 1539
Turck-Chièze S. et al. 2010b *ApJ lett* submitted
Turck-Chièze S. et al. 2011 *Experimental Astronomy* in preparation

Unveiling stellar magnetic activity using CoRoT seismic observations

Savita Mathur[1], Rafael A. García[2], David Salabert[3,4], Jérôme Ballot[5], Clara Régulo[3,4], Travis S. Metcalfe[1], and Annie Baglin[6]

[1] High Altitude Observatory, 3080 Center Green Drive, Boulder, CO, 80301, USA
[2] Laboratoire AIM, CEA/DSM-CNRS-Université Paris Diderot; CEA, IRFU, SAp, F-91191, Gif-sur-Yvette, France
[3] Instituto de Astrofísica de Canarias (IAC), E-38200 La Laguna, Tenerife, Spain
[4] Dept. de Astrofísica, Universidad de La Laguna (ULL), E-38206 La Laguna, Tenerife, Spain
[5] Laboratoire dAstrophysique de Toulouse-Tarbes, Université de Toulouse, CNRS, F-31400, Toulouse, France
[6] LESIA, Universit Pierre et Marie Curie, Université Denis Diderot, Observatoire de Paris, 92195 Meudon Cedex, France

E-mail: savita@ucar.edu

Abstract. It is well known that in the Sun, the frequencies and amplitudes of acoustic modes vary throughout the solar cycle. Indeed, while the magnetic activity goes towards its maximum, the frequencies of the modes increase and their amplitudes decrease. We have analyzed data from the CoRoT mission on a few stars that exhibit solar-like oscillations. The study of HD49933 (observed during 60 days and 137 days spanning a total of 400 days) showed a modulation of the maximum amplitude per radial mode and the frequency shifts of the modes, showing magnetic activity in this rapidly rotating star. Moreover, both properties vary in an anticorrelated way and the data allowed us to establish a lower limit for the activity cycle period of 120 days. Measurements in Ca H and K lines confirmed that this star is in the "active stars" category. We will also discuss the results obtained for other targets such as HD181420 and HD49835 for which we have investigated a similar behavior.

1. Introduction

Many improvements have been made on the mechanisms behind the solar activity cycle, which can be seen in the simulations of the solar dynamo [e. g. 1]. However, the last solar minimum that was unexpectedly long emphasized the fact that our knowledge is not completely accurate and that classical proxies do not say the same thing as seismic parameters [2, 3, 4, 5, 6]. The recent discovery of a 2-year modulation in the seismic observables [7] raises more questions to the understanding of such cycles in the Sun.

One way to better understand the solar cycle is to study magnetic activity cycles in other stars. Baliunas et al. [8] showed that stellar dynamos depend on the evolutionary time scale of the star. There is already an empirical law that has been drawn out thanks to the observations of a few tens of stars [9, 10]. For cool stars like the Sun and with an $\alpha\Omega$ dynamo, a longer rotation period implies a longer cycle period. But this law depends utterly on the characteristics of the convective zone, which can be well determined thanks to asteroseismology. We know that the activity of the Sun interacts with the acoustic modes and it might also have an impact on the high-frequency p modes [11, 12].

The last decade, asteroseismology made a big jump. Thanks to the CoRoT mission, several solar-like stars could be observed continuously during a few months with a good signal-to-noise ratio (SNR) [e.g. 13, 14, 15] or with a much lower one [16] allowing us to measure individual low-degree p modes in some of the former cases. Here we show how asteroseismology can allow us to detect a magnetic activity cycle in the Sun and other stars.

2. Methodology

For this analysis, we studied the temporal variations of the power spectrum, starspots number, frequency shifts of the p modes, and maximum amplitude per radial mode.

2.1. Wavelet analysis

We obtain an estimate of the surface rotation of the star with the wavelets techniques [17,18]. It was shown in [12, 19] that this tool is very powerful for helio- and asteroseismology. In particular, it distinguishes between the fundamental period and the first harmonic of the rotation period as in the classical power spectrum, we have this ambiguity sometimes. The wavelet power spectrum also shows the temporal evolution of the magnetic activity of the star.

2.2. Proxy of the "starspots number"

We can also notice that the time series have the fingerprints of the starspots and the bright plages appearing and moving across the stellar disk. By calculating the standard deviation of small subseries of the data, we are able to follow the evolution of the global coverage of the starspots on the stellar surface [20]. This is a rough "proxy" for the magnetic activity cycle of a star as we cannot resolve the stellar disk.

2.3. Calculating the frequency shifts of the p modes

We also studied the variation of the frequency shift of the global modes or the position of the p-mode hump. We have used two different techniques to obtain these frequency shifts.

The first one consists of studying the global envelope of the p-mode hump. We compute the power spectrum density of subseries and subtract the background, modeled with three components (Harvey-law model, a power law, and white noise) and six free parameters (ignoring the p-mode region). The cross-correlation function [21] is computed in the region of the p modes. These cross-correlation functions look like Gaussians. To determine a possible shift in this cross-correlation function, we use a frequency range of $\pm$ 7 μHz to estimate the third order moment of this function, which measures the asymmetry. Then, starting with the lag given by the asymmetry, we fit a Gaussian function using a window of $\pm$ 7 μHz. The maximum of this fitted Gaussian is used as the position of the cross-correlation peak.

We can also measure the shift of individual modes. Here, we fit the modes of each subseries with a standard likelihood maximization function [22]. We fit the modes with a Lorentzian profile over frequency windows containing the l=0, 1, and 2 modes. Then we compute the difference of frequency compared to a reference. Finally, the frequency shift is the average of the differences over the frequency range studied.

2.4. Estimation of the maximum amplitude per radial mode

We use the A2Z pipeline [19] to calculate the rms maximum amplitude per radial mode, A_{max} and its variation with time. We subtract the background fit and smooth the power spectrum density over $2\times\Delta\nu$, where $\Delta\nu$ is the mean large separation of the modes of the star. We fit the envelope of the modes with a Gaussian and convert the maximum power to bolometric amplitude per radial mode [23].

We have successfully applied our methodology to the Sun [20] using 10 years of data from the VIRGO (Variability of solar IRradiance and Gravity Oscillations [24]) instrument aboard the SoHO spacecraft, taking the average of the three independent SPM (Sun PhotoMeters) channels (red, blue, and green).

3. Analysis of the star HD49933

We analysed the data obtained by the CoRoT (Convection Rotation and planetary Transits) satellite for the star HD49933, which is an F5V star of 1.2 $M_{\odot}$ and 1.3 $R_{\odot}$. Around 50 acoustic modes have been identified thanks to the 60+137 days of CoRoT observations [25,26].

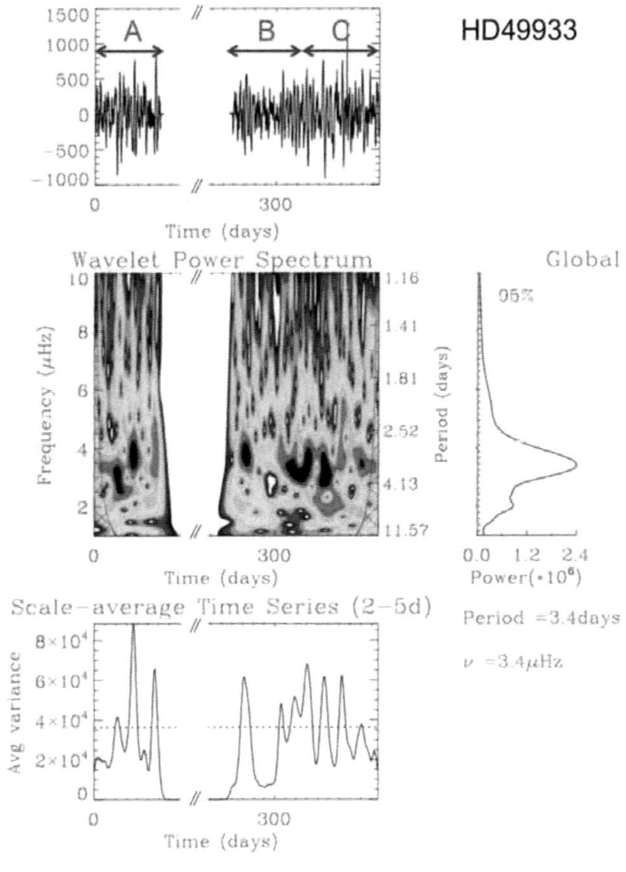

Figure 1. Wavelet analysis of HD49933. Top panel: time series. Middle panel: wavelet power spectrum (left) and global wavelet power spectrum (right). Bottom panel: scale average variance between a period of 2 and 5 days.

Fig. 1 (top) shows the light curve of HD49933 observed during 60 days of the CoRoT initial run in 2007 and again during 137 days in 2008. It presents clear signatures of active regions. A surface rotation is confirmed at 3.5 d with some differential rotation between 2 and 5 days (middle panel of Fig. 2). The collapsogram between 2 and 5 days (bottom panel of Fig. 2) shows an increase in the average signal in the first run, A, while the first part of the long run, B, is a more quiet period. Finally, C is a more active period.

We computed the starspots proxy as described in Section 2.3 leading to Fig. 2 (2nd panel from the top). The minimum of this proxy occurs at around 300 d.

We analyzed the parameters of the acoustic modes. We took subseries of 30 days shifted every 15 days. We calculated the cross-correlation of the p-mode hump (between 1460 and 2100 μHz) of each subseries with a reference one. The latter is taken during the minimum of activity. It gives the red triangles of Fig. 2 (3rd panel). The frequency shifts of the individual modes were calculated in the range 1460-2070 μHz. We removed the outliers above 5 σ. The frequency shifts obtained are the black curve in Fig. 2 (3rd panel). After subtracting the background, we fitted the envelope of the modes with a Gaussian function to obtain the rms maximum amplitude per radial mode (Fig. 2, bottom panel).

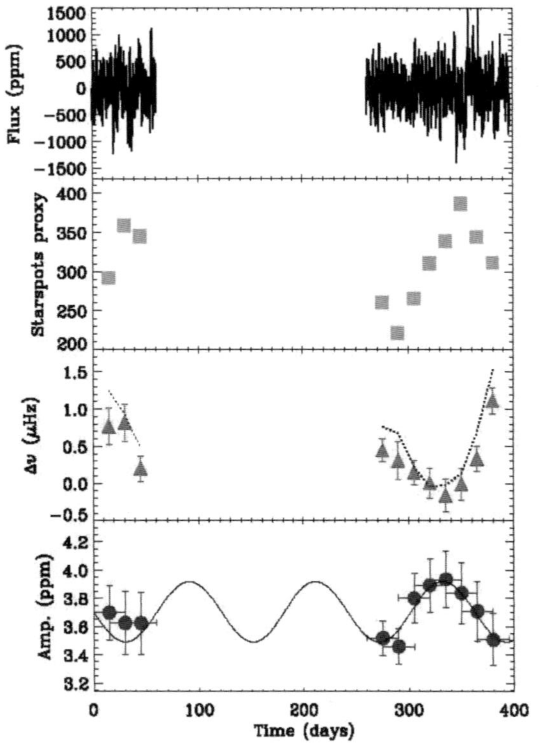

Figure 2. Top: lightcurve of HD49933 obtained with CoRoT. Second panel: temporal evolution of the starspots proxy obtained as explained in Sect. 2.1. Third panel: temporal evolution of the frequency shifts ($\Delta\nu$) using the cross-correlation (red triangles) and the individual mode fitting (dotted line). Bottom: maximum amplitude per radial mode versus time. A tentative fit with a sinus wave is overplotted giving a lower limit for the cycle period of 120 days (see text for details).

We clearly see the anticorrelation in the temporal evolution between the amplitude variation and the frequency shifts variation, confirming the existence of a magnetic activity cycle, which also agrees with the wavelet analysis. We fitted the whole set of data with a sinus wave. The best fit is obtained for a period of 120 days, which is a lower limit of the cycle period (the sinusoide is overplotted to guide the eye). Indeed, we do not have a full coverage of the cycle and we know from the Sun that the periods of maximum activity can have a "plateau" region that sometimes has some structure such as a double maximum. Because our fit is obtained around the minimum activity we should be cautious about the real length of the cycle and even about the fact that is a regular cycle.

Besides, we see a small time lag between the minimum of activity in the starspots proxy while in the asteroseismic data, the minimum comes 30 days later. This could be explained with the inclination angle from which we observe the star as it influences other seismic parameters as the rotational splitting [27, 28]. Some simulations show that depending on the angle of inclination, the minimum (or maximum) of activity is shifted in time and with more or less strength when we look for the signature of the starspots [29], This is a consequence of the interaction between the active longitudes, the migration of the spots towards the equator when the cycle evolves and the inclination of the star.

Some results were obtained thanks to observations of the Calcium H and K on April 13, 2010, showing that this is an active star with a Mount Wilson S-index of 0.3 [20] confirming previous conclusions [30]. Additional Ca H and K observations are scheduled for this fall.

4. Other CoRoT targets

We also applied the same methodology to other stars observed by CoRoT. Figs. 3 and 4 show the temporal variations of the amplitude and the frequency shifts for HD181420 [13] and HD49385 [14]. We know that the first star has a rotation period of 2.6 d, while the rotation period of HD49385 is still uncertain, maybe around 10 days. For HD181420, there is a small hint of anticorrelation between the two parameters with a maximum for the amplitude around 60 d. But we cannot confirm it because of the large error bars, specially for the frequency shifts. For HD49385, we do not see any anticorrelation, which could be due to the fact that for such a rotation period, the cycle might be much longer than the observation length.

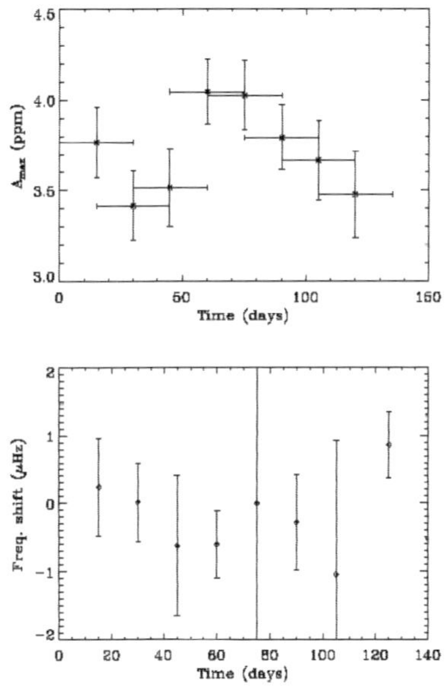

Figure 3. Amplitude and frequency shift variation for HD181420.

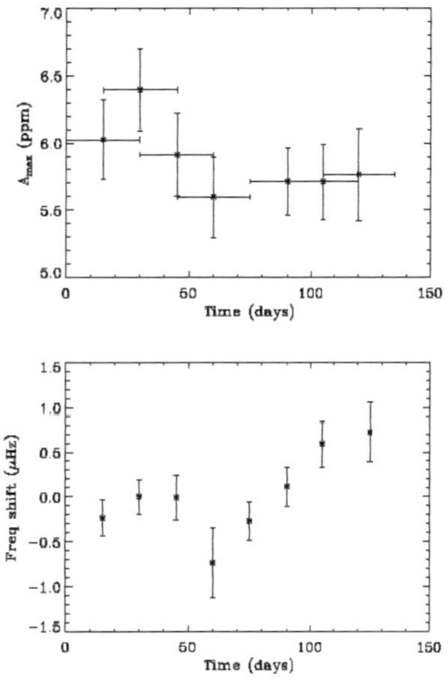

Figure 4. Amplitude and frequency shift variation for HD49385.

5. Conclusions

These analyses show that HD49933 has a magnetic activity and we obtain a lower limit on the cycle period of 120 days. Actually, such a short cycle might not be so uncommon as observations in Ca HK of HD17051 show a cycle period of 1.6 yr [31]. The long-term observations of Kepler will allow us to apply this methodology on more stars and give new constraints for stellar dynamo models. Thanks to the accuracy we can have on the convective zone parameters and on the modes parameters, it would enable us to better constrain the simulations. The fact that such short cycles can exist tells us that we should be able to observe these magnetic activity cycles during the 3.5 years of observations by the Kepler mission that is observing solar-like oscillating stars [32, 33, 34, 35, 36, 37].

Acknowledgments

NCAR is supported by the National Science Fundation. This work has been partially supported by the Spanish National Research Plan (grant PNAyA2007-62650) and by the Programme National de Physique Stellaire at CEA Saclay. The CoRoT space mission, launched on December 27th 2006, has been developed and is operated by CNES, with the contribution of Austria, Belgium, Brazil, ESA (RSSD and Science Program), Germany and Spain.

References

[1] Brun A S, Miesch A and Toomre J 2004 *Astroph. J.* **614** 1073
[2] Broomhall A M, Chaplin W J, Davies G R et al. 2009 *Month. Not. Roy. Astron, Soc.* **396** 100
[3] González Hernández I, Howe R, Komm R and Hill F 2010 *Astroph. J.* **713** L16
[4] Hathaway D H and Rightmire L 2010 *Science* **327** 1350
[5] Howe R, Christensen-Dalsgaard J, Hill F, Komm R, Schou J and Thompson M J 2009 *Astroph. J.* **701** L87
[6] Salabert D, García R A, Pallé P L and Jiménez-Reyes S J 2009 *Astron. & Astroph.* **504** L1
[7] Fletcher S T, Broomhall A M, Salabert D et al. 2010 *Astroph. J.* **718** L19
[8] Baliunas S L, Donahue R A, Soon W H et al. 1995 *Astroph. J.* **438** 269
[9] Ossendrijver A J H 1997 *Astron. & Astroph.* **323** 151
[10] Jouve L, Brun A S and Browning B 2010 *Astron. & Astroph.* **509** 32
[11] Karoff C and Kjeldsen H 2008 *Astroph. J.* 678 **73**
[12] Kumar B, Mathur S, García R A and Venkatakrishnan P 2010 *Astroph. J.* **711** L12
[13] Barban C, Deheuvels S, Baudin F et al. 2009 *Astron. & Astroph.* **506** 51
[14] Deheuvels S, Brunt H, Michel E et al. 2010 *Astron. & Astroph.* **515** 87
[15] García R A, Régulo C, Samadi R et al. 2009 *Astron. & Astroph.* **506** 41
[16] Mathur S, García R A, Catala C et al. 2010 *Astron. & Astroph.* **518** A53
[17] Torrence C and Compo G P 1998 *Bull. of the Amer. Met. Soc.* **79** 61
[18] Liu Y, Liang X S and Weisberg R H 2007 *J. Atmos. & Ocean. Tech.* **24** **(12)** 2093
[19] Mathur S, García R A, Régulo C et al. 2010 *Astron. & Astroph.* **511** 46
[20] García R A, Mathur S, Salabert D, Ballot J, Régulo C, Metcalfe T S and Baglin A 2010 *Science* **329** 1032
[21] Pallé P L, Régulo C and Roca Cortès T 1989 *Astron. & Astroph.* **224** 253
[22] Salabert D, Chaplin W J, Elsworth Y, New R and Verner G A 2007 *Astron. & Astroph.* **463** 1181
[23] Michel E, Samadi R, Baudin F et al. 2009 *Astron. & Astroph.* **495** 979
[24] Frohlich C et al. 1995 *Solar Physics* **162** 101
[25] Appourchaux T, Michel E, Auvergne M. et al. 2008 *Astron. & Astroph.* **488** 705
[26] Benomar O, Baudin F, Campante T L et al. 2009 *Astron. & Astroph.* **507** L13
[27] Gizon L and Solanki S K 2003 *Astroph. J.* **589** 1009
[28] Ballot J, García R A and Lambert P 2006 *Month. Not. Roy. Astron, Soc.* **369** 1281
[29] Vázquez Ramió H, Mathur S, C Régulo C and R A García 2010 *J. Phys.: Conf. Series* **This volume**
[30] Mosser B, Bouchy F, Catala C et al. 2005 *Astron. & Astroph.* **431** L13
[31] Metcalfe T S, Basu S, Henry T J et al. 2010 *Astroph. J.* **723** L213
[32] Bedding T R, Huber D, Stello D et al. *Astroph. J.* **713** L176
[33] Chaplin W J, Appourchaux T, Elsworth Y et al. 2010 *Astroph. J.* **713** L169
[34] Stello D, Basu S, Brunt H et al. 2010 *Astroph. J.* **713** L182
[35] Hekker S, Elsworth Y, De Ridder J et al. 2010 *Astron. & Astroph.* **In press**
[36] Huber D, Bedding T R, Stello D et al. 2010 *Astroph. J.* **723** 1607
[37] Kallinger T, Mosser B, Hekker S et al. 2010 *Astron. & Astroph.* **522** A1

New insights on the solar core

R.A. García[1], D. Salabert[2,3], J. Ballot[4], A. Eff-Darwich[2,5], R. Garrido[6], A. Jiménez[2,3], S. Mathis,[1], S. Mathur[7], A. Moya[8], P.L. Pallé[2,3], C. Régulo[2,3], K. Sato[1], J.C. Suárez[6] and S. Turck-Chièze[1]

[1] Laboratoire AIM, CEA/DSM-CNRS, Université Paris 7 Diderot, IRFU/SAp, Centre de Saclay, 91191, Gif-sur-Yvette, France
[2] Instituto de Astrofísica de Canarias (IAC), E-38200 La Laguna, Tenerife, Spain
[3] Dept. de Astrofísica, Universidad de La Laguna (ULL), E-38206 La Laguna, Tenerife, Spain
[4] Laboratoire d'Astrophysique Toulouse-Tarbes - OMP, 14 avenue Edouard Belin, 31400 Toulouse, France
[5] Dept. de Geología, Universidad de La Laguna (ULL), E-38206 La Laguna, Tenerife, Spain
[6] Instituto de Astrofísica de Andalucía (CSIC), Apartado 3004, 18080 Granada, Spain
[7] High Altitude Observatory, 3080 Center Green Drive, Boulder, CO, 80302, USA
[8] Centro de Astrobiología (INTA-CSIC), Madrid, Spain

E-mail: `rgarcia@cea.fr`

Abstract. Since the detection of the asymptotic properties of the dipole gravity modes in the Sun, the quest to find individual gravity modes has continued. An extensive and deeper analysis of 14 years of continuous GOLF/SoHO observational data, unveils the presence of a pattern of peaks that could be interpreted as individual dipole gravity modes in the frequency range between 60 and 140 microHz, with amplitudes compatible with the latest theoretical predictions. By collapsing the power spectrum we have obtained a quite constant splitting for these patterns in comparison to regions where no g modes were expected. Moreover, the same technique applied to simultaneous VIRGO/SoHO data unveils some common signals between the power spectra of both instruments. Thus, we are able to identify and characterize individual g modes with their central frequencies, amplitudes and splittings allowing to do seismic inversions of the rotation profile inside the solar core. These results open a new ligh t on the physics and dynamics of the solar deep core.

1. Introduction

Little progress has been done during the last few years on the structure [1,2,3] and Dynamics [4,5,6,7] of the solar core even after the detection of the asymptotic spacing of the dipole gravity modes [8,9] Indeed no general consensus has been obtained for the detection of individual g modes yet [10]. This is because the increasing convective background level towards the lower frequencies (e.g. [11]), combined with very small amplitudes of those modes (several mm/s in the case of g modes, [12]) are the limiting factors for their detections. In the case of the low-degree, low-frequency p modes, accurate measurements are hardly obtained below 1 mHz [13,14,15].

This situation might change in the near future when we have a few years of data coming from the new instrumentation available, such as PICARD [16], and the very promising HMI and AIA aboard SDO [17] or the new projected instrumentation (e.g. GOLF-NG [18]).

Fortunately, time goes by and the signal-to-noise ratio of the Global Oscillations at Low Frequency (GOLF) instrument [19] and the Sun Photometers (SPM) on the Variability of

IRradiance and Global Oscillations [20] aboard the Solar and Heliospheric Observatory (SoHO) mission increases. In this work we uncover the presence of peaks in the power spectral density (PSD) that could be individual ℓ=1 modes. These peaks are regularly spaced in period in the positions determined by the asymptotic periodicity measured by [8] Moreover, these peaks present a regular pattern in frequency that could be the signature of the rotational splitting. Thus this work study potential candidates that might merit further investigation, a step forward of what it has been done during the last few years (e.g. [21,22,23]).

2. Observations and Data analysis

In this work, we have used two different sets of data starting on April 11, 1996. The first one contains 4472 days and the second 5163 days finishing on May 30, 2010. The GOLF signal has been calibrated into velocity [24] following the methods described in [25,26]. We have worked with a single – full resolution – power spectrum even knowing that GOLF has been observing in two different configurations (blue and red) with a different sensitivity to the visible solar disk [27,28]. Because the analysis we are going to do is very sensitive to the amplitude of the peaks, we have computed a 5 times zero-padded power spectrum to reduce the problem of the discretization in frequency (see the discussion concerning this problem in [29]). We have also used time series from the SPM/VIRGO package on the same time span.

To uncover the peaks in the PSD responsible for the measurement of the periodic signal found in GOLF by [8] and interpreted as the spacing of the dipole gravity modes, we follow the same procedure usually done in low signal-to-noise ration (SNR) targets in asteroseismology. Thanks to the observations of solar-like stars done by CoRoT [30] and Kepler [31] and some ground based campaigns [32] we now know how to tackle the problem and measure individual acoustic (p) modes in such stars [33,34,35,36] without ambiguity. Even when the SNR of these modes is very small, we are still able to determine the large separation even without seeing the individual p modes in the power spectrum (e.g. [37,38,39,40,41,42,43]). In such cases, the only thing that is generally done is to perform a more or less heavy smooth of the power spectrum to unveil the p-mode hump in the region in which the frequency spacing is found.

3. Results

As said previously, we have computed the 5 times zero-padded Power Spectral Density (PSD) of the GOLF time series, which is represented in Figure 1 (top) as function of the period. Then, we took the predicted frequencies of the dipole gravity modes ℓ=1, m=0 components from the Saclay seismic model [44,2] and we overplotted them to guide the eye of the reader (vertical red dotted lines). We have also checked that in this frequency range of [60-140] μHz, the differences between the frequencies of this model and the ones of the model S [45] and the model M1 from Nice [46] are less than 0.05 μHz –which is inside the width of the vertical lines–. As the frequencies of the modes can be dependent of the pulsation code used to compute them, we have also verified how the predictions of the ℓ=1 in the same frequency range vary with the parameters of the pulsation codes: the differences are negligible [47].

The dashed and dot-dashed vertical red lines are the m=$\pm$1 assuming a core rotating 5 times faster than the rest of the radiative region (2 μHz). This value is compatible with the range inferred by [8] for the core rotation rate and it matches the highest peaks of the PSD. Then, as the SNR is quite low, we smoothed the PSD with a boxcar function of 41 nHz (see Figure 1, bottom). We can see that, in most of the cases, the high-amplitude peaks are located around the $m = \pm 1$ components. It is important to notice that similar results are obtained using different sizes of the boxcar filter.

To quantitatively measure these g-mode candidates, we used the rotation-corrected, m-averaged spectrum technique, the so-called collapsograms. This method was originally developed for imaged instruments (see [15] for a detailed explanation of the method) to observe low

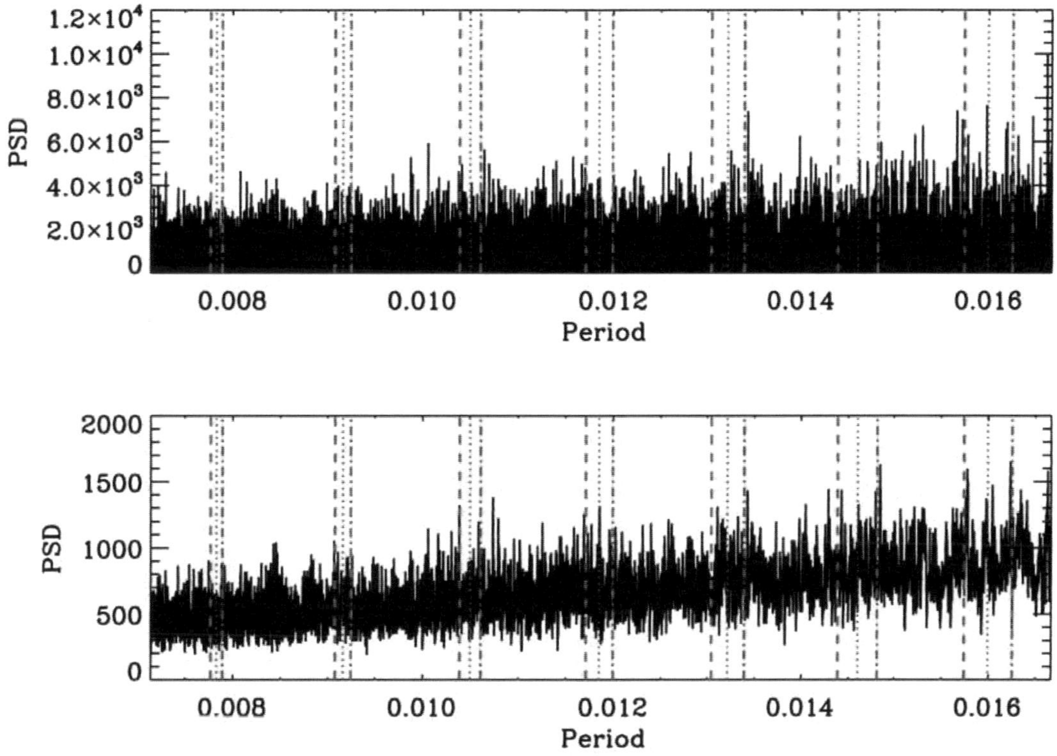

Figure 1. PSD in $(m/s)^2/Hz$ of the zero padded series (top) and the one after applying a smooth by a boxcar of 41 nHz (bottom), as a function of the period expressed in Mega seconds (inverse of μHz) in the region [60-140] μHz. The vertical dotted lines are the ℓ=1 predicted frequencies. The dashed and dot-dashed lines represent the positions of the $m = \pm 1$ components assuming a rotational splittings of 2 μHz.

SNR, low-frequency solar p modes. An m-averaged spectrum corresponds to the average of the individual m components, thus reducing the non-coherent noise. Before averaging, each m spectrum of a given mode (n, l) is shifted by a frequency that compensates for the effect of rotation and structural effects on the frequencies. The m-averaged spectrum concentrates for a given multiplet (n, l) all the m components, considerably improving the SNR of the resulting spectrum, which is then fitted using a Lorentzian profile. The shift coefficients are chosen to maximize the likelihood of the m-averaged spectra. Figure 2 shows a sequence of four snapshots of the procedure, for four different shifts (379.08, 836.73, 934.90, and 957.45 nHz respectively from left to right and top to bottom), around the theoretical frequency of the g mode ℓ=1, $n = -6$. The central frequency of the mode is given by the frequency of the fitted Lorentzian profile, while the displacement of the m-components provides a measurement of the rotational splitting.

In five out of the seven dipole gravity modes in the region [60-140] μHz we were able to obtain a stable splittings (around 850 to 950 nHz) around a central frequency laying close to the

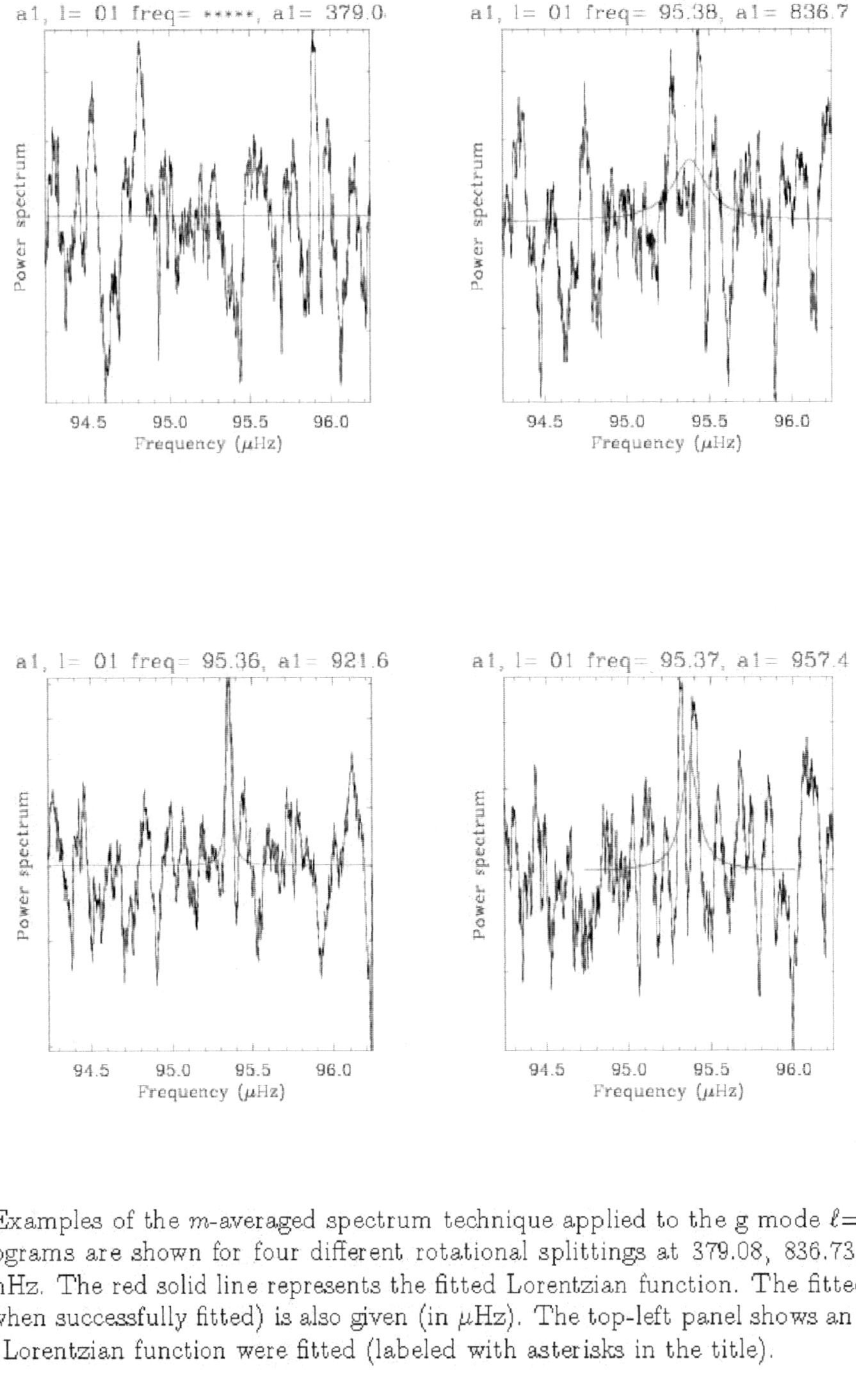

Figure 2. Examples of the m-averaged spectrum technique applied to the g mode ℓ=1, n=-6. The collapsograms are shown for four different rotational splittings at 379.08, 836.73, 934.90, and 957.45 nHz. The red solid line represents the fitted Lorentzian function. The fitted central frequency (when successfully fitted) is also given (in μHz). The top-left panel shows an example in which no Lorentzian function were fitted (labeled with asterisks in the title).

predicted frequencies. The modes we found with this method were the ℓ=1, n=[-4,-6,-8,-9,-10].

The same work has been performed using the VIRGO/SPM datasets. Although the PSD is noisier than the GOLF one, in some cases we obtained a reliable Lorentzian fitting giving the same quantitative results in the central frequency of the mode and the splittings compared to GOLF.

4. Dynamics of the solar core

Once we have characterized the central frequency and the rotational splittings of these 5 dipole gravity modes, we can include them in an inversion of the internal rotation of the Sun. To do so, we use the p-mode frequencies and splittings extracted from 4608 days of MDI data [48] for the modes $\ell \geq 3$ and the ones from GOLF, for the modes $\ell \leq 2$. The method used for the inversion is a modified Regularized Least Squared methodology [49]. Figure 3 (left) shows the inversion of the internal rotation without including the five g-mode candidates, while on the right we show the resultant inversion including those g-mode candidates.

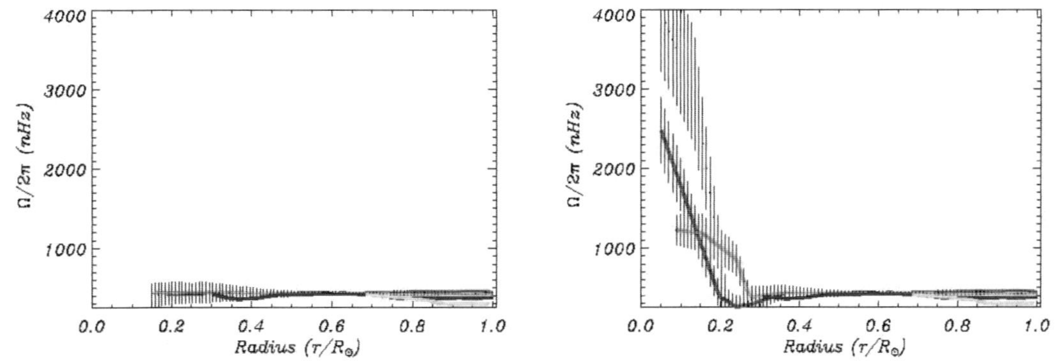

Figure 3. Inversions of the internal rotation rate of the Sun using 4608 days of MDI and GOLF data. On the left containing only p modes, on the right we have added the five candidate g modes of this work (with 40nHz error bars). The colors corresponds to the inversions at different latitudes: The equator (black), 20° (red), 40° (green), 60° (blue) and 80° (light blue).

Firstly, we notice that the inversions worked fine even considering that we have put a small error bar of 40nHz in the rotational splittings of the g-mode candidate. Some differences are found at different latitudes which can be a consequence of the small sensitivity we have in the core due to the use of only a few dipole modes. Finally, the average rotation rate obtained is compatible with the average rotation rate previously inferred using the asymptotic periodicity. However, more work is needed to understand the sensitivity of the error bars in the inversion.

5. Conclusions

In this paper we have been able to identify the individual peaks generating the periodic signal found by [8] and interpreted as the asymptotic periodicity of the dipole gravity modes. The analysis of the collapsograms gave five possible detections in the frequency range between 60 and 140 μHz with a rotational splittings in the range 850 to 950 nHz. When these candidate modes are used in the inversion of the rotational profile, we see that the rotation in the deep core reaches values close to 4000 nHz. However, more work should be done to better characterize these g-mode candidates as well as the determination of the error bars that are very important for the inversions.

Acknowledgments

SoHO is a space mission of international cooperation between ESA and NASA. DS acknowledges the support of the grant PNAyA 2007-62650 from the Spanish National Research Plan, J.B.

the support through the ANR SIROCO, and R.A.G and S.T. thanks the support form the CNES and the French Stellar Physics National Plan (PNPS). NCAR is supported by the National Science Foundation.

References

[1] Basu S, Chaplin W J, Elsworth Y, New R, Serenelli A M 2009 *Astrophys. J.* **699** 1403
[2] Mathur S, Turck-Chièze S, Couvidat S, García R A 2007 *Astrophys. J.* **668** 594
[3] García R A, Mathur S, Ballot J 2008 Solar Phys **251** 135
[4] Mathis S, Zahn J-P 2005 *Astron. & Astroph.* **440** 653
[5] Mathur S, Eff-Darwich A, García R A, Turck-Chiéze S 2008 *Astron. & Astroph.* **484** 517
[6] García R A, Mathur S, Ballot J et al. 2008 Solar Phys **251** 119
[7] Turck-Chieèze S, Palacios A, Marques J P, Nghiem P A P 2010 *Astrophys. J.* **715** 1539
[8] García R A, Turck-Chièze S, Jiménez Reyes S J et al. 2007 *Science* **316** 1591
[9] García R A , Jiménez A, Mathur S et al. 2008 *Astron. N.* **329** 476
[10] Appourchaux T, Belkacem K, Broomhall A.M. et al. 2010 *ARA&A* **18** 197
[11] Lefebvre S, García R A, Jimeénez-Reyes S J, Turck-Chièze S, Mathur S 2008 *Astron. & Astroph.* **490** 1143
[12] Belkacem K, Samadi R, Goupil M J et al. 2009 *Astron. & Astroph.* **494** 191
[13] Bertello L, Varadi F, Ulrich R K et al. 2000 *Astrophys. J.* **537** L143
[14] Broomhall A-M, Chaplin W J, Davies G R et al. 2009 *MNRAS* **396** L100
[15] Salabert D, Leibacher J, Appourchaux T, Hill F 2009 *Astrophys. J.* **696** 653
[16] Thuillier G, Dewitte S, Schmutz W 2006 *in 36th COSPAR Scientific Assembly* **170**
[17] Howe R 2010 *Proc. of GONG 2010/SOHO XXIV* In press
[18] Turck-Chièze S, Carton P-H, Ballot J., et al. 2006 *Adv. Space Res.* **38** 1812
[19] Gabriel, A H, Grec G, Charra J et al. 1995 *Solar Phys.* **162** 61
[20] Frohlich C, Romero J, Roth H et al. 1995 *Solar Phys.* **162** 101
[21] Turck-Chièze S, García R A, Couvidat S et al. 2004 *Astrophys. J.* **604** 455
[22] Jiménez A., García R. A. 2009 *Astrophys. J. Supp.* **184** 288
[23] Broomhall A-M, Chaplin W J, Elsworth Y et al. 2010 *MNRAS* **406** 767
[24] Pallé P L, Régulo C, Roca Cortés T et al. 1999 *Astron. & Astroph.* **341** 625
[25] García R A, Turck-Chièze S, Boumier P et al. 2005 *Astron. & Astroph.* **442** 385
[26] Ulrich R K, García R A, Robillot J M et al. 2000 *Astron. & Astroph.* **364** 799
[27] García R A, Roca-Cortés T, Régulo C 1998 *Astron. & Astroph. Suppl.* **128** 389
[28] Henney C J , Ulrich R K, Bertello L et al. 1999 *Astron. & Astroph.* **348** 627
[29] Gabriel A H, Baudin, F, Boumier P et al. 2002 *Astron. & Astroph.* **390** 1119
[30] Baglin A, Auvergne M, Barge P et al. 2006, *in ESA Special Publication* **1306** 33
[31] Borucki W, Koch D, Batalha N et al. 2009 *in IAU Symposium* **253** 289
[32] Arentoft T, Kjeldsen H, Bedding T et al. 2008 *Astrophys. J.* **687** 1180
[33] Appourchaux T, Michel E, Auvergne M. et al. 2008 *Astron. & Astroph.* **488** 705
[34] Benomar O, Baudin F, Campante T L et al. 2009 *Astron. & Astroph.* **507** L13
[35] Deheuvels S, Bruntt H, Michel E et al. 2010 *Astron. & Astroph.* **515** A87
[36] Chaplin W J, Appourchaux T, Elsworth Y et al. 2010 *Astrophys. J.* **713** L169
[37] Mosser B, Michel E, Appourchaux T et al. 2009 *Astron. & Astroph.* **506** 33
[38] García R A, Régulo C, Samadi R et al. 2009 *Astron. & Astroph.* **506** 41
[39] Campante T. L., Karoff C., Chaplin W. J. et al. 2010 *MNRAS* **1125**
[40] Hekker S., Broomhall A., Chaplin W. J. et al. 2010 *MNRAS***402** 2049
[41] Huber D., Stello D., Bedding T. R. et al. 2009, *Comm. in Astero.***160** 74
[42] Mathur S, García R A, Régulo C et al. 2010 *Astron. & Astroph.* **511** A46
[43] Mathur S, García R A, Catala C et al. 2010 *Astron. & Astroph.* **518** A53
[44] Couvidat S, Turck-Chiéze S, Kosovichev A G 2003 *Astrophys. J.* **599** 1434
[45] Christensen-Dalsgaard J, Dappen W, Ajukov S V et al. 1996 *Science* **272** 1286
[46] Provost J, Berthomieu G, Morel P 2000 *Astron. & Astroph.* **353** 775
[47] Moya A, Mathur S, García R A 2010 *Sol. Phys.* In press
[48] Korzennik S 2010 *J.P. Conf.* This volume
[49] Eff-Darwich A, Korzennik S G, Jiménez-Reyes S J, García R A 2008 *Astrophys. J.* **679** 1636

Amplitudes of solar p modes: *modelling of the eddy time-correlation function*

K. Belkacem[1,2], R. Samadi[2], M.J. Goupil[2]

[1] Institut d'Astrophysique et de Géophysique, Université de Liège, Allée du 6 Août 17-B 4000 Liège, Belgium.

[2] LESIA, UMR8109, Université Pierre et Marie Curie, Université Denis Diderot, Obs. de Paris, 92195 Meudon Cedex, France.

E-mail: Kevin.Belkacem@ulg.ac.be

Abstract. Modelling amplitudes of stochastically excited oscillations in stars is a powerful tool for understanding the properties of the convective zones. For instance, it gives us information on the way turbulent eddies are temporally correlated in a very large Reynolds number regime. We discuss the way the time correlation between eddies is modelled and we present recent theoretical developments as well as observational results. Eventually, we discuss the physical underlying meaning of the results by introducing the Ornstein-Uhlenbeck process, which is a sub-class of a Gaussian Markov process.

1. Introduction

Amplitudes of solar-like oscillations result from a balance between excitation and damping and crucially depend on the way the eddies are temporally correlated (see the review of [1]). Hence, the improvement of our understanding and modeling of the temporal correlation of turbulent eddies, hereafter denoted in the Fourier domain as $\chi_k(\omega)$, is fundamental to infer turbulent properties in stellar convection zones.

Most of the theoretical formulations of mode excitation explicitly or implicitly assume a Gaussian functional form for $\chi_k(\omega)$ ([2–7]). However, 3D hydrodynamical simulations of the outer layers of the Sun show that, at the length-scales close to that of the energy bearing eddies (around 1 Mm), χ_k is a Lorentzian function [8, 9]. As pointed-out by [7], a Lorentzian χ_k is also a result predicted for the largest, most-energetic eddies by the time-dependent mixing-length formulation derived by [10]. However, [7], [1], and [11] found that a Lorentzian χ_k, when used with a mixing-length description of the whole convection zone, results in a severe over-estimation for the low-frequency modes.

In a recent work, [12] introduced a cut-off frequency in the modelling of the eddy-time correlation function to account for the effect of short-time scales. Indeed, under the sweeping approximation, which consists in assuming that the temporal correlation of the eddies, (in the inertial subrange) is dominated by the advection by energy-bearing eddies, the shape of the temporal correlation function of eddies is no longer Lorentzian.

Hence, in this paper we discuss the modelling of the eddy-time correlation function in both the limit of large and short time scales. This modelling is then validated using observational constraints from amplitudes of solar-like oscillations. Eventually, the underlying physical meaning of the observed shape of the temporal correlation function is discussed by mean of a simplified stochastic process.

2. Modelling the Eulerian eddy time-correlation function

The formalism we used to compute excitation rates of radial modes was developed by [13] (see also [14]). For a radial mode of frequency $\omega_0 = 2\pi\,\nu_0$, the excitation rate (or equivalently, the energy injection rate), P, mostly arises from the Reynolds stresses and can be written as (see Eq. (21) of [15])

$$P(\omega_0) = \frac{\pi^3}{2I} \int_0^M \left[\rho_0 \left(\frac{16}{15} \right) \left(\frac{\partial \xi_r}{\partial r} \right)^2 \int_0^{+\infty} \mathcal{S}_k \, \mathrm{d}k \right] \mathrm{d}m \tag{1}$$

$$\mathcal{S}_k = \frac{E^2(k)}{k^2} \int_{-\infty}^{+\infty} \chi_k(\omega + \omega_0) \, \chi_k(\omega) \, \mathrm{d}\omega \tag{2}$$

where ξ_r is the radial component of the fluid displacement eigenfunction ($\vec{\xi}$), m is the local mass, ρ_0 the mean density, ω_0 the mode angular frequency, I the mode inertia, $\mathcal{S}_k$ the source function, $E(k)$ the spatial kinetic energy spectrum, χ_k the eddy-time correlation function, and k the wave-number.

2.1. Some preliminary definitions

We now turn to a rigourous definition of the eddy-time correlation function (χ_k). For a turbulent fluid, one defines the Eulerian eddy time-correlation function as

$$\langle \vec{u}(\vec{x} + \vec{r}, t + \tau) \cdot \vec{u}(\vec{x}, t) \rangle = \int \mathcal{E}(\vec{k}, t, \tau) \, e^{\mathrm{i}\vec{k}\cdot\vec{x}} \, \mathrm{d}^3\vec{k} \,, \tag{3}$$

where $\vec{u}$ is the Eulerian turbulent velocity field, $\vec{x}$ and t the space and time position of the fluid element, $\vec{k}$ the wave number vector, τ the time-correlation length, and $\vec{r}$ the space-correlation length. The function $\mathcal{E}$ in the RHS of Eq. (3) represents the time-correlation function associated with an eddy of wave-number $\vec{k}$.

We assume an isotropic and stationary turbulence, accordingly $\mathcal{E}$ is only a function of k and τ. The quantity $\mathcal{E}(k, \tau)$ is related to the turbulent energy spectrum according to

$$\mathcal{E}(k, \tau) = \frac{E(k, \tau)}{2\pi k^2} \,. \tag{4}$$

where $E(k, \tau)$ is the turbulent kinetic energy spectrum whose temporal Fourier transform is

$$E(k, \omega) \equiv \frac{1}{2\pi} \int_{-\infty}^{+\infty} E(k, \tau) \, e^{\mathrm{i}\omega\tau} \, \mathrm{d}\tau \tag{5}$$

where ω is the eddy frequency, and $E(k, \omega)$ is written as follows [13, 16]

$$E(k, \omega) = E(k) \, \chi_k(\omega) \quad \text{with} \quad \int_{-\infty}^{+\infty} \chi_k(\omega) \, \mathrm{d}\omega = 1 \tag{6}$$

where $\chi_k(\omega)$ is the frequency component of $E(k, \omega)$. In other words, $\chi_k(\omega)$ represents - in the frequency domain - the temporal correlation between eddies of wave-number k.

2.2. Long-time scales

A modelling of χ_k in the stellar context from the full non-linear hydrodynamic equations is a very difficult task, and no such a model has been proposed so far. This is particularly the case for large times, which we defines by the times of the order or greater than the integral time-scale that is the width at half maximum of χ_k. Hence, we adopt an alternative approach that is to postulate the functional form of χ_k and then to confront it to observations and/or 3D hydrodynamical numerical simulations.

Several functional have then been tested in the past and as mentioned in Sect. 1, theoretical and observational evidence show that $\chi_k(\omega)$ is Lorentzian, *i.e.*

$$\chi_k(\omega) = \frac{1}{\pi\omega_k} \frac{1}{1+(\omega/\omega_k)^2} \quad \text{with} \quad \omega_k = k\, u_k \quad \text{and} \quad u_k^2 = \int_k^{2k} E(k)\, \mathrm{d}k \tag{7}$$

where ω_k is by definition the width at half maximum of $\chi_k(\omega)$. In the framework of [13]'s formalism, this latter quantity is evaluated as:

$$\omega_k = k\, u_k \quad \text{with} \quad u_k^2 = \int_k^{2k} E(k)\, \mathrm{d}k \tag{8}$$

where $E(k)$ is defined by Eq. (6). Note that Eq. (7) corresponds to an exponential in the time domain.

Such a modelling of the temporal correlation of eddies has successfully been tested in the solar and stellar context (see [1] for a detailed discussion). However, when coupled with a standard 1D model of the solar convection zone, it leads to an important overestimation of the solar mode amplitudes. More precisely, a substantial fraction of the energy injected to those modes comes from the deep layers of the solar convective region. As a result, [7] and [1] then suggested that most contributing eddies situated deep in the Sun have a χ_k rather Gaussian than Lorentzian since, at fixed frequency, a Gaussian χ_k decreases more rapidly with depth. Nevertheless, this conclusion has recently been questioning by [12], which demonstrate that Eq. (7) is no longer valid for short-time scales.

2.3. Short-time scales

To infer some properties of the eddy-time correlation function at short-time scales, the function $\mathcal{E}(k,t,\tau)$ appearing in Eq. (3) is expanded for short time scales. In the inertial sub-range it gives (see [17] for a detailed derivation)

$$\mathcal{E}(k,\tau) = \mathcal{E}(k,\tau=0)\left(1 - \alpha_k|\tau| - \frac{1}{2}\left(\omega_E \tau\right)^2 + \dots\right) \tag{9}$$

where the characteristic frequency α_k can be estimated by the eddy turn-over frequency ω_k (see [12] for details), and the second characteristic frequency, $\omega_E(k)$, is the curvature of the correlation function near the origin, and defined by

$$\omega_E = k\, u_0\,. \tag{10}$$

This expression is obtained under the sweeping assumption, which consists in assuming that the velocity field $\vec{u}(k)$ associated with an eddy of wave-number $\vec{k}$ lying in the inertial-subrange (*i.e.* large $\vec{k}$ compared to $\vec{k}_0$) is advected by the energy-bearing eddies with velocity $\vec{u}_0$ (*i.e.* of wave-number $\vec{k}_0$).

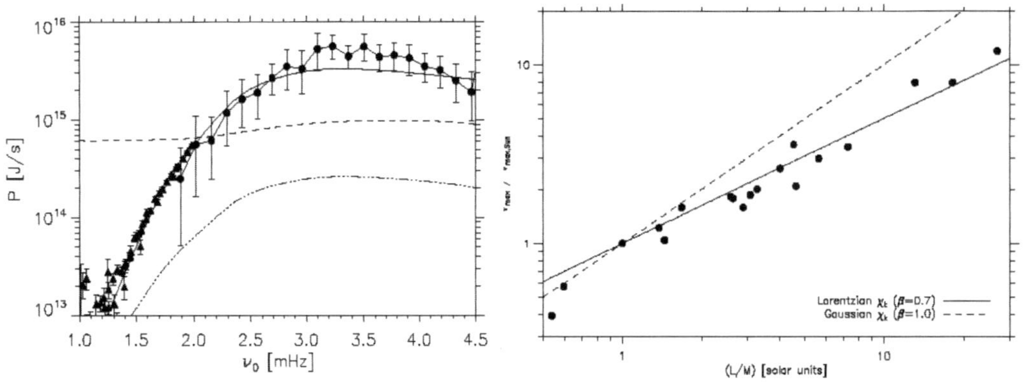

Figure 1. Left panel: Solar p-mode excitation rates as a function of the frequency ν. The dots correspond to the observational data obtained by the GONG network, as derived by [18], and the triangles corresponds to observational data obtained by the GONG network as derived by [19] for $\ell = 0$ to $\ell = 35$. The dashed line corresponds to the computation of the excitation rates using with a Lorentzien description of χ_k *without* any cut-off frequency, while the dashed triple dot line corresponds to the same computation except that the Lorentzian description of χ_k down to the cut-off frequency ω_E. The solid line corresponds to the computation of mode excitation rates using 3D numerical computation to described turbulent convection properties and a Lorentzian χ_k together *with* a cut-off frequency at $\omega = \omega_E$. Note that no scaling is applied to those calculations. **Right panel:** Ratio between ($V_{\rm max}$) the maximum of the mode velocity relative to the observed solar value ($V_{\odot,max} = 25.2cm/s$ for $\ell = 1$ modes, see Kjeldsen et al. 2008). Filled dots correspond to the stars for which solar-like oscillations have been detected in Doppler velocity (see [20]. The lines correspond to the power laws obtained from the predicted scaling laws for P_{max} and estimated values of the damping rates η_{max} (see [1] for details).

To go further, we note that the zeroth- and first-order terms in Eq. (9) are consistent with an exponential decrease of width α_k (*i.e.* a Lorentzian in the frequency domain of width ω_k, Eq. (7)) for small τ. In contrast, the zeroth-order term together with the second order term in Eq. (9) are consistent with a Gaussian behavior of width τ_E. In turn, [12] have shown that for $\omega > \omega_E$ the second order term dominates over the first order one in Eq. (9), at all length-scales. Hence, for frequencies near the micro-scale frequency ($\omega > \omega_E$), the eddy-time correlation function behaves as a Gaussian function ($e^{-(\omega/\omega_E)^2}$) rather than a Lorentzian function, resulting in a sharp decrease with ω. Hence, the contributions for $\omega > \omega_E$ are negligible.

3. Observational evidences for a Lorentzian eddy-time correlation function

3.1. *The solar p-mode energy injection rates*

When the frequency range of χ_k is extended toward infinity, computation of P fails to reproduce the observations, in particular the low-frequency shape (see Fig. 1). It is in agreement with the results of [7] and [1], and results in an over-estimation of the excitation rates at low frequency. In contrast, by assuming that the time-dynamic of eddies in the Eulerian point of view is dominated by the sweeping, the Eulerian time micro-scale arises as a cut-off frequency (see Sect. 2.3). Hence, $\chi_k(\omega)$ is modeled following Eq. (7) for $\omega < \omega_E$ and $\chi_k(\omega) = 0$ elsewhere. Using such a procedure to model $\chi_k(\omega)$ permits us to reproduce the low-frequency ($\nu < 3$mHz) shape of the mode excitation rates as observed by the GONG network (see Fig. 1). This is explained as follows; for large scale eddies near k_0^{-1}, situated deep in the convective region, the cut-off frequency ω_E is close to ω_k. As a consequence, the frequency range over which χ_k is integrated in Eq. (2) is limited, resulting in lower injection rates into the modes. This results then re-conciliates theoretical and observational evidences that the frequency dependence of the eddy-time correlation must be Lorentzian in the whole solar convective region down to the cut-off frequency ω_E.

Note that a modelling of the excitation rates, using a Gaussian for χ_k and a 3D numerical simulation to model convection, leads to an underestimation of the excitation rates, as shown by [12].

3.2. Scaling law on the maximum mode energy injection rate for solar-type pulsators

A similar conclusion has been reached by [21] for main-sequence and red giant stars. Indeed, [22] have shown that the maximum mode surface velocity scales as a function of the ratio of luminosity to the mass (L/M). Later, [21] have shown that the exponent β is sensitve to the adopted model for computing χ_k, *i.e.* $\beta = 0.7$ for a Lorentzian and $\beta = 1$ for a Gaussian. As shown in Fig. 1, the best agreement with the observations is found when a Lorentzian χ_k is assumed. In contrast, assuming a Gaussian χ_k results in a larger exponent β and do not permit to reproduce the observations. Consequently, we conclude that the eddy-time correlation function is a Lorenztian functional (or equivalently an exponential functional in the time domain) for all solar-like pulsators and in the upper-layers of convective regions. Note that the case of deep layers in convective regions is still to be investigated, but 3D numerical results from [9] also suggest χ_k is Lorentzian.

4. On the interpretation of the Lorentzian shape of χ_k

A modelling of the Eulerian time correlation function from the full non-linear hydrodynamic equations is a difficult task, making the interpretation of our results non-trivial. However, one can use a classical approach in turbulence, which consists in modelling the turbulence by a stochastic model (see the review [23]). Similarly to the Brownian motion, one can define a fluid elements and express the corresponding Langevin equation. Naturally, a Lagrangian approach is more adapted however, our modelling of mode amplitudes deals with Eulerian velocities. To overcome this difficulty we first introduced a relation between Eulerian and Lagrangian time correlation function, first introduced by [24].

4.1. The Corrsin's conjecture

The problem of relating Eulerian and Lagrangian correlation is an outstanding problem in turbulence, since the measurement of Lagrangian quantities is experimentally difficult. A general expression can be written such as

$$\langle w(t+\tau)w(t)\rangle = \int \mathrm{d}^3\vec{r}\, \langle v(\vec{x}+\vec{r}, t+\tau)v(\vec{x},t)\psi(\vec{x},t)\rangle \tag{11}$$

where w is the Lagrangian velocity and u the Eulerian velocity, and ψ is the probability distribution function of the fluid particule displacement at time t. In this framework, [24] proposed a simplified relation based on the assumption that for large times (τ), ψ and u are statistically weakly dependent. Then by using a Gaussian distribution for ψ, [24] demonstrates that one can rewritte Eq. (11) in the limit of large scales[1]

$$\langle w(t+\tau)w(t)\rangle \approx \langle v(\vec{x}, t+\tau)v(\vec{x},t)\rangle \tag{12}$$

[1] This assumption is valid for our purposes since modes are mainly excited by large scales.

This approximated relation has been extensively discussed and has been shown to be rather accurate for turbulent flows without helicity (e.g. [25, 26]).

4.2. *The Ornstein-Uhlenbeck Process*

For long time-scales, given the relation Eq. (12), we now turn to discuss some aspect of Lagrangian stochastic models. In the limit of a stationary, homogeneous turbulence, a stochastic model can be adopted that follows the Langevin equation (written in term of stochastic differential equation)

$$\mathrm{d}w(t) = -w(t)\mathrm{d}t/T + (2u'^2/T)\zeta(t)\mathrm{d}t \tag{13}$$

where T is the Lagrangian integral time-scale, u'^2 is the variance of $w(t)$, and ζ a random fluctuating term. To go further, lets us assume that ζ is a Gaussian random variable with zero mean and which is rapidly varying, *i.e.* $<\zeta(t)\zeta(t')>=\delta(t-t')$.

Consequently, the Langevin equation Eq. (13) describes a continuous Gaussian Markov process in time (it is a Wiener process or equivalently a Brownian process). More precisely, since we are interested to describe a stationary process, $w(t)$ is a stationary, Gaussian, Markov process know as the Ornstein-Uhlenbeck Process (see [27] for a more detailed definition). In other terms it corresponds to a noisy relaxation process. A general property of such a process is that its time correlation function is (e.g., [23, 27])

$$\frac{\langle w(t+\tau)w(t)\rangle}{u'^2} = e^{-|\tau|/T} \tag{14}$$

which, in the time Fourier space, corresponds to a Lorentzian function.

A more rigourous investigation has been proposed by [28], who proposed a stochastic model for turbulent dissipation. The author then proposed an expression for the time correlation function as a function of the Reynolds number

$$\frac{\langle w(t+\tau)w(t)\rangle}{u'^2} = \left[e^{-|\tau|/T} - Re^{-1/2}e^{-Re^{-1/2}|\tau|/T}(1-Re^{-1/2})^{-1}\right] \tag{15}$$

which tends to Eq. (14) for $Re \to +\infty$.

Hence, by used of the Corrsin's conjecture one can conclude that our results, obtained using mode amplitudes (*i.e.* χ_k is Lorentzian in the time Fourier domain), suggests that turbulent eddies in convective region of solar-type stars follow such a noisy relaxation process.

5. Conclusion

We have shown that the modelling of mode amplitudes gives us access to some properties of turbulent convection in stars. More precisely, we detailed the way the time correlation between eddies is actually modelled and validated using observational data. The conclusion is that the eddy-time correlation function follows a Lorentzian shape (in the time Fourier domain) down to a cut-off frequency, in convective region of solar-type stars.

We then discussed this shape and we emphasised that a Lorentzian function - or equivalently an exponential function in the time domain - is a rather common result for noisy relaxation processes. Our discussion also shows that maybe the more interesting will be to investigate the departure from a Lorentzian shape of χ_k, and its physical interpretation. To this end, we stress that more accurate observations for the Sun of solar-like pulsators is mandatory.

6. References

[1] Samadi R 2009 *Stochastic excitation of acoustic modes in stars, ArXiv e-prints (0912.0817)* (*Preprint* 0912.0817)

[2] Goldreich P and Keeley D A 1977 *ApJ* **212** 243–251

[3] Dolginov A Z and Muslimov A G 1984 *Ap&SS* **98** 15–36

[4] Goldreich P, Murray N and Kumar P 1994 *ApJ* **424** 466–479

[5] Balmforth N J 1992 *MNRAS* **255** 639–+

[6] Samadi R, Goupil M and Lebreton Y 2001 *A&A* **370** 147–156

[7] Chaplin W J, Houdek G, Elsworth Y, Gough D O, Isaak G R and New R 2005 *MNRAS* **360** 859–868

[8] Samadi R, Nordlund Å, Stein R F, Goupil M J and Roxburgh I 2003 *A&A* **404** 1129–1137

[9] Belkacem K, Samadi R, Goupil M J, Dupret M A, Brun A S and Baudin F 2009 *A&A* **494** 191–204 (*Preprint* 0810.0602)

[10] Gough D O 1977 *ApJ* **214** 196–213

[11] Houdek G 2009 *Ap&SS* 252–+ (*Preprint* 0911.4909)

[12] Belkacem K, Samadi R, Goupil M, Baudin F, Salabert D and Appourchaux T 2010 *submitted to A&A*

[13] Samadi R and Goupil M J 2001 *A&A* **370** 136–146

[14] Samadi R, Goupil M J, Alecian E, Baudin F, Georgobiani D, Trampedach R, Stein R and Nordlund Å 2005 *J. Astrophys. Atr.* **26** 171–+

[15] Belkacem K, Samadi R, Goupil M J and Dupret M A 2008 *A&A* **478** 163–174 (*Preprint* arXiv:0710.1039)

[16] Stein R F 1967 *Solar Physics* **2** 385–432

[17] Kaneda Y 1993 *Physics of Fluids* **5** 2835–2845

[18] Baudin F, Samadi R, Goupil M J, Appourchaux T, Barban C, Boumier P, Chaplin W J and Gouttebroze P 2005 *A&A* **433** 349–356

[19] Salabert D, Leibacher J, Appourchaux T and Hill F 2009 *ApJ* **696** 653–667 (*Preprint* 0902.2561)

[20] Bedding T R and Kjeldsen H 2007 Observing Solar-like Oscillations *Unsolved Problems in Stellar Physics: A Conference in Honor of Douglas Gough* (*American Institute of Physics Conference Series* vol 948) ed R J Stancliffe, G Houdek, R G Martin, & C A Tout pp 117–124 (*Preprint* 0802.0531)

[21] Samadi R, Georgobiani D, Trampedach R, Goupil M J, Stein R F and Nordlund Å 2007 *A&A* **463** 297–308 (*Preprint* arXiv:astro-ph/0611762)

[22] Houdek G, Balmforth N J, Christensen-Dalsgaard J and Gough D O 1999 *A&A* **351** 582–596 (*Preprint* arXiv:astro-ph/9909107)

[23] Pope S B 1994 *Annual Review of Fluid Mechanics* **26** 23–63

[24] Corrsin S 1959 *Advances in Geophysics* **6** 161–+

[25] Kraichnan R H 1977 *Journal of Fluid Mechanics* **83** 349–374

[26] Ott S and Mann J 2005 *New Journal of Physics* **7** 142–+

[27] Gardiner C W 1986 *Applied Optics* **25** 3145–+

[28] Sawford B L 1991 *Physics of Fluids* **3** 1577–1586

Investigating the correlation between high-frequency global oscillations and solar flares

S Chakraborty, M C Rabello-Soares, R S Bogart and T Bai
Stanford University, Stanford, CA 94305-4085, USA
E-mail: deepc@sun.stanford.edu

Abstract. Global oscillations have been observed in the high-frequency spectrum in velocity and intensity measurements above the acoustic cutoff frequency of the solar atmosphere. In a recent work, strong correlation at high frequencies, determined between helioseismic spectra and soft x-ray flux has been cited as evidence that solar flares may be one of the mechanisms whereby these high-frequency oscillations are excited. We investigate the possible dependencies of the correlation coefficient, computed for helioseismic and soft x-ray data, on the pre-processing and data analysis, and the instruments used for data acquisition.

1. Introduction

Ever since Wolff (1972) suggested the theoretical possibility of solar flares stimulating free oscillations in the Sun, several attempts have been made to establish this using observational data (cf. Karoff 2008; Kumar et al. 2010). Recently, Karoff & Kjeldsen (2008) claimed to have provided evidence supporting Wolff's idea by showing a high correlation between the high-frequency helioseismic power spectra and the solar x-ray flux (1–8 Å). More recently, Kumar et al. (2010) also presented evidence of increase in the energy of high-frequency global acoustic oscillations during the occurrence of major solar flares.

In this work we have reproduced the results of Karoff et al. (2008) by calculating frequency-time plots for acoustic data obtained from the blue, green and red channels of the Sun PhotoMeter (SPM), and then correlating the power spectra with soft (1–8 Å) x-ray flux obtained from the Solar X-ray Imager (SXI) on board the Geostationary Operational Environmental Satellites (GOES). SPM is part of the Variability of solar IRradiance and Gravity Oscillations (VIRGO) experiment on the Solar and Heliospheric Observatory (SOHO) spacecraft. However, we do not get the same results when correlating the (ℓ=0) spherical harmonic time series from the Global Oscillation Network Group (GONG) instruments with the GOES soft x-ray flux data.

2. Frequency-time and correlation plots

We generally follow the procedure described by Karoff et al. (2008) and use the least-squares spectra method (Lomb 1976) to calculate the power spectra of substrings of the helioseismic time series. For comparison purposes, we also calculate the power spectra using FFT and the results are shown in the correlation plots in figure 2. We calculate the power spectra of time strings of various lengths to bring out any dependencies in the correlation coefficient on the length of the time strings. The power spectra of these time series are normalized by the duty cycle of the time strings, and then stacked vertically (1 Jan 1996–31 Dec 2005) to produce frequency-time (f-t) diagrams (figure 1).

We correlate both the unsmoothed and smoothed f-t diagrams with x-ray flux time series, in order to determine the effect of smoothing on the results. The unsmoothed f-t diagrams are correlated with the daily averaged x-ray flux data. Roughly following Karoff et al. (2008), smoothing is done with a 2-d Gaussian filter with widths 30 μHz in frequency and 20 days in time. The x-ray flux data is smoothed by a triangular running mean of 20 days to ensure compatibility in temporal resolution with the f-t diagram.

The Pearson correlation coefficient (Rice 2007) is computed for the power spectra "time series" at each frequency bin, and the GOES x-ray flux (1–8 Å); this yields the correlation coefficient as a function of frequency (figure 2). All time strings with zero variance, and with duty cycles less than 70% for GONG and less than 90% for SPM are not included in the calculation of the correlation coefficient. These rejected data are indicated as horizontal dark blue strips of various widths in the f-t diagrams.

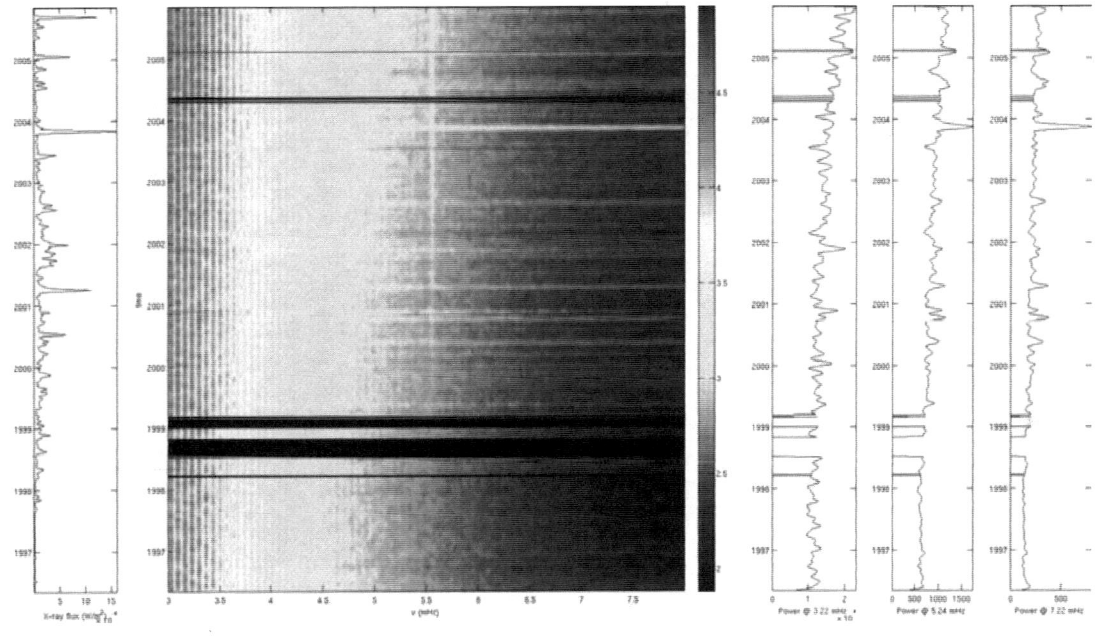

Figure 1: (From left to right) Daily averaged x-ray flux (smoothed by a running triangular mean 20 days in width); frequency-time (f-t) diagram (smoothed by a Gaussian filter 20 days wide in time and 30 μHz wide in frequency) for 8-day time strings of SPM level-1.8 blue channel disk integrated intensity (ordinate: 1996–2005; abscissa: 3–8 mHz); slices of f-t diagram at 3.22 mHz, 5.24 mHz & 7.22 mHz. Dark blue horizontal strips in f-t diagram indicate rejected data.

3. Pre-processing and data analysis

3.1. SPM level-1.8 disk integrated intensity time series

The SPM level-1.8 time series (blue, green and red channels) are available at `ftp://ftp.pmodwrc.ch/pub/data/irradiance/virgo/SSI/`. On analyzing these time series we were able to reproduce the results in Karoff et al. (2008). For the blue channel we get correlation coefficient values of ~0.4 at higher frequencies (figure 2b). FFT and the Lomb method yield results that are different, especially in the higher frequencies (figures 2a & b). From figures 2b & c we can see that smoothing before correlating does significantly raise the correlation coefficient as was suggested in Chakraborty et al. (2010). We also find that the correlation coefficient increases with the length of the time string.

3.2. GONG Doppler velocity spherical harmonic (ℓ=0) time series

The GONG Doppler velocity spherical harmonic (ℓ=0) time series was obtained from `ftp://gong2.nso.edu/TSERIES/`. In this case the correlation coefficient values reach ~0.2 at higher frequencies (figure 2e). FFT and the Lomb method yield practically indistinguishable results (figures 2d & e). Here too, the correlation coefficient increases due to smoothing, and with the length of the time string (figures 2e & f). It also seems that the p-mode power spectra (around 3 mHz) are anti-correlated with the soft x-ray flux data (figures 2e & f). This p-mode frequency anti-correlation is also seen in the SPM level-1.8 correlation plots (figures 2b & c). This merits further study.

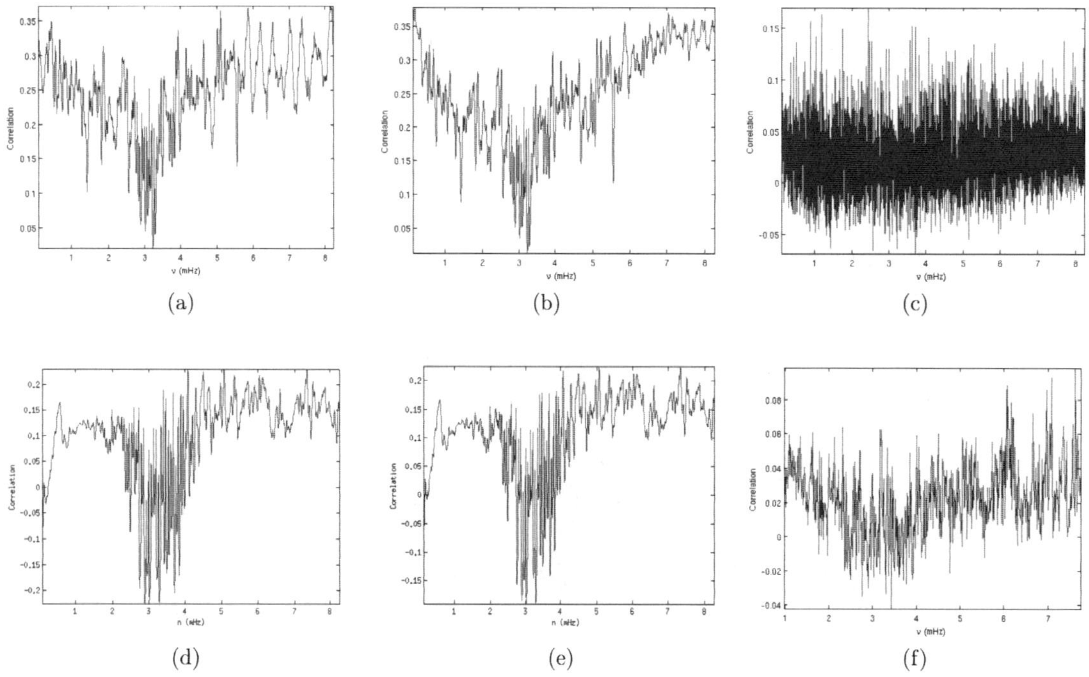

Figure 2: Pearson correlation coefficient as a function of frequency for: SPM level-1.8 blue channel (a) 8-day time string, FFT, smoothed; (b) 8-day time string, Lomb, smoothed (cf. figure 1); (c) 8-day time string, Lomb, unsmoothed; GONG (ℓ=0) (d) 8-day time string, FFT, smoothed; (e) 8-day time string, Lomb, smoothed; (f) 1-day time string, Lomb, unsmoothed.

4. Discussion and future work

We studied the correlation coefficient as a function of frequency by analyzing data from different instruments. FFT and the Lomb method yield results that are different for SPM (cf. figures 2a & b), but virtually indistinguishable for GONG (cf. figures 2d & e); the reason for this is not understood and needs investigation. SPM and GONG show very different values for the correlation coefficient at all frequencies (cf. figures 2b & e). Both SPM and GONG data show some anti-correlation between helioseismic and soft x-ray flux data in the p-mode region.

Removal of long-period oscillations from the time series using a running mean does not have any significant effect on the correlation coefficient at higher frequencies. Outlier removal does influence the correlation coefficient: the higher the standard deviation used in outlier removal, the higher the correlation in the higher frequencies. We have investigated the possibility of legitimate signals being accidentally removed in an effort to meet statistical requirements. Calculating f-t diagrams using longer time strings and/or smoothing the f-t diagram and the x-ray flux before correlating them increases the correlation. As the correlation coefficient varies with smoothing and the length of the time series, it is not possible to accurately interpret the values depicted in figure 2. At this point we have yet to reach any firm conclusions; further analysis is required.

Time series from more instruments, such as, the Michelson Doppler Imager (MDI), Luminosity Oscillations Imager (LOI/VIRGO) and the Global Oscillations at Low Frequencies (GOLF) instrument aboard the SOHO spacecraft, and eventually the Helioseismic and Magnetic Imager (HMI) on board the Solar Dynamics Observatory (SDO), need to be analyzed. The raw data from these instruments need to be carefully pre-processed (filtering, outlier removal, etc.) so as not to introduce systematic errors in subsequent analyses. B. Kumar (personal communication) suggested correlating the helioseismic data with *hard* x-ray flux data as it is associated with the possibly more relevant impulsive phase of a solar flare. The analysis needs to be carried out rigorously, and the correlation coefficient needs to be suitably normalized with respect to the appropriate parameters, e.g., dimensions of the smoothing window, to give us more confidence in the correlation found. Analytical methods such as the postflare/preflare power ratio distribution described in Chakraborty et al. (2010) to determine the statistical significance of acoustic power perturbation with respect to flare activity on the Sun may be used as an independent method of verifying the results presented in this paper.

Acknowledgments

We are grateful for the discussions with C. Karoff, and for the time series provided by him that enabled us to reproduce the results of Karoff et al. (2008).

This work utilizes data obtained by the GONG program, managed by the National Solar Observatory (NSO), which is operated by AURA, Inc. under a cooperative agreement with the National Science Foundation (NSF). The data was acquired by instruments operated by the Big Bear Solar Observatory, High Altitude Observatory, Learmonth Solar Observatory, Udaipur Solar Observatory, Instituto de Astrofísica de Canarias, and Cerro Tololo Interamerican Observatory. This work also uses data on board the SOHO which is a mission of international cooperation between the European Space Agency (ESA) and the National Aeronautics and Space Administration (NASA). The National Oceanic and Atmospheric Administration (NOAA) operates GOES. This work is part of the MDI project and is supported by NASA grant NAG5-8878 to Stanford University.

References

Chakraborty S, Bai T, Bogart R S and Rabello-Soares M C 2010 On the correlation between high-frequency global oscillations and solar flares (To be published) *Proc. 4th HELAS Int. Conf. on Seismological Challenges for Stellar Structure* (Arrecife de Lanzarote, Spain)

Karoff C 2009 *Proc. GONG 2008/SOHO XXI Conf. on Solar-Stellar Dynamos as revealed by Helio- and Asteroseismology* (Boulder: ASP Conf. Ser. 416) p 233

Karoff C and Kjeldsen H 2008 *Astrophys. J.* **678** L73

Kumar B, Mathur S, García R A and Venkatakrishnan P 2010 *Astrophys. J.* **711** L12

Lomb N R 1976 *Astrophys. Space Sci.* **39** 447

Rice J A 2007 *Mathematical Statistics and Data Analysis* 3rd ed (Duxbury Press) p 406

Wolff C L 1972 *Astrophys J.* **176** 833

The acoustic low-degree modes of the Sun measured with 14 years of continuous GOLF & VIRGO measurements

R A García[1], D. Salabert[2,3], J. Ballot[4], K. Sato[1], S. Mathur[5], A. Jiménez[2,3]

[1] Laboratoire AIM, CEA/DSM-CNRS-Université Paris Diderot; CEA, IRFU, SAp, F-91191, Gif-sur-Yvette, France

[2] Instituto de Astrofísica de Canarias, E-38200 La Laguna, Tenerife, Spain

[3] Dept. de Astrofísica, Universidad de La Laguna, E-38206 La Laguna, Tenerife, Spain

[4] Laboratoire d'Astrophysique Toulouse-Tarbes - OMP, 14 avenue Edouard Belin, 31400 Toulouse, France

[5] High Altitude Observatory, 3080 Center Green Drive, Boulder, CO, 80302, USA

E-mail: `rgarcia@cea.fr`, `salabert@iac.es`, `jballot@ast.obs-mip.fr`, `kumiko.sato@cea.fr`, `savita@ucar.edu`, `ajm@iac.es`

Abstract. The helioseismic Global Oscillation at Low Frequency (GOLF) and the Variability of solar Irradiance and Gravity Oscillations (VIRGO) instruments onboard SoHO, have been observing the Sun continuously for the last 14 years. In this preliminary work, we characterize the acoustic modes over the entire p-mode range in both, Doppler velocity and luminosity, with a special care for the low-frequency modes taking advantage of the stability and the high duty cycle of space observations.

1. Introduction

Helio- and astero-seismology are the only tools available to pierce inside the stars and provide stratified information of the structure and dynamics of their, otherwise, hidden interiors. This information is embedded in the resonant modes stochastically excited –in the case of solar-like stars– by turbulent motions taking place in the convective layers. In particular, the acoustic (p) modes with the largest horizontal scales –lowest angular degree ℓ– are the only ones that can be measured in Sun-as-a star instruments such as GOLF [1] and VIRGO [2] onboard the Solar and Heliospheric Observatory (SoHO) spacecraft. These global modes are very important since they can propagate through the deepest layers of the stars and penetrate inside their cores, giving for example information on the internal structure [e.g. 3], the internal rotation [e.g. 4], and the magnetic activity cycles, either in the Sun [e.g. 5] or in other stars [6]. In this work, we analyze observations collected with these two instruments to characterize the p-mode properties.

2. Data analysis and results

We analyzed nearly 14 years of data collected by the GOLF and VIRGO instruments: 5163 days of GOLF velocity time series [7,8] from April 11, 1996 to May 30, 2010 with a duty cycle, dc=95.4 %; and 5154 days of intensity data from the three VIRGO Sun photometers (SPM) at

402, 500, and 862 nm from April 11, 1996 to May 21, 210 (dc = 95.2 %). GOLF is a resonant scattering spectrophotometer that measures the line-of-sight integrated Doppler velocity of the Sun. It has been designed to be very stable at low frequency, allowing, for example, to indirectly detect gravity modes in the Sun [9,10]. On the other hand, the overall noise level of VIRGO/SPM is higher than in GOLF: the signal-to-noise ratio (SNR) being smaller, the frequency range on which we can extract reliable estimates of p-mode characteristics is reduced.

To characterize the p modes, we computed the power spectrum density (PSD) of the entire time series in order to maximize the frequency resolution (~ 2.24 nHz). Therefore, the obtained linewidths of the modes could be slightly overestimated because of the shift of the modes during the solar activity cycle [e.g. 5]. The acoustic modes were described using asymmetric Lorentzian profiles, extracted using a classical maximum likelihood method. The fits were performed by a multi-step iterative method [11,12]. Figures 1 and 2 show the p-mode parameters as a function of frequency extracted from GOLF and VIRGO/SPM respectively between 1000 to 4000 μHz, and up to 5000 μHz when possible (the splitting is left constant (400 nHz) for modes above 3.5 mHz). For instance, preliminary results of the linewidths extracted from GOLF are given up to 5000 μHz. Also, due to the smaller $l = 3$ visibility in the VIRGO/SPM data, the linewidth and acoustic power of the $l = 1$ mode could be fitted up to 5000 μHz, the mode blending at high frequency being the limiting factor in radial velocity measurements. Reliable estimates below 1800 μHz could not be properly extracted in VIRGO/SPM data because the smaller SNR at low frequency in intensity measurements compared to radial velocity measurements.

We also computed the average maximum amplitude per radial mode of the Sun (Fig 2 bottom right) for the three VIRGO/SPM channels, as it is commonly done in asteroseismology [e.g. 13,14]. The maximum amplitudes were corrected by the instrumental response function using the values given by [15] for the different channels.

Some new developments will be addressed in the near future: for instance, we plan to improve the analysis for the low-SNR, low-frequency modes by using specific techniques, i.e., the collapsograms [16]. We will also provide the oscillation parameters corrected from the solar activity by fitting simultaneously the parameters and their variation with activity [17].

Acknowledgments

SoHO is a space mission of international cooperation between ESA and NASA. DS acknowledges the support the Spanish National Research Plan grant PNAyA 2007-62650, J.B. the ANR SIROCO, and R.A.G the CNES grant. NCAR is supported by the National Science Foundation.

References

[1] Gabriel, A H, Grec G, Charra J et al. 1995 *Solar Phys.* **162** 61
[2] Frohlich C, Romero J, Roth H et al. 1995 *Solar Phys.* **162** 101
[3] Basu S, ChaplinW J Elsworth Y, New R, Serenelli A M 2009 *Astrophys. J.* **699** 1403
[4] Mathur S, Eff-Darwich A, García R A, Turck-Chiéze S 2008 *Astron. & Astroph.* **484** 517
[5] Jiménez Reyes S J, García R A Jiménez A, Chaplin W J 2003 *Astrophys. J.* **595** 446
[6] García R A, Mathur S, Salabert D et al. 2010 *Science* **329** 1032
[7] Ulrich R K, García R A, Robillot J M et al. 2000 *Astron. & Astroph.* **364** 799
[8] García R A, Turck-Chièze S, Boumier P et al. 2005 *Astron. & Astroph.* **442** 385
[9] García R A, Turck-Chièze S, Jiménez Reyes S J et al. 2007 *Science* **316** 1591
[10] García R A , Jiménez A, Mathur S et al. 2008 *Astron. N.* **329** 476
[11] Gelly B, Lazrek M, Grec G, et al. 2002 *Astron. & Astrophys.* **394** 285
[12] Salabert D, Chaplin W J, Elsworth Y, New R, Verner G, 2007 *Astron. & Astroph.* **463** 1181
[13] Mathur S, García R A, Catala C et al. 2010 *Astron. & Astroph.* **518** A53
[14] Mathur S, García R A, Régulo C et al. 2010 *Astron. & Astroph.* **511** 46
[15] Michel E., Baglin A, Auvergne M et al. 2008 *Science* **322** 558
[16] Salabert D, Leibacher J, Appourchaux T, Hill F 2009 *Astrophys. J.* **696** 653
[17] García R A, Mathur S, Ballot J et al. 2008 Solar Phys, **251** 119
[18] Mathur S, Turck-Chièze S, Couvidat S, García R A 2007 *Astrophys. J.* **668** 594

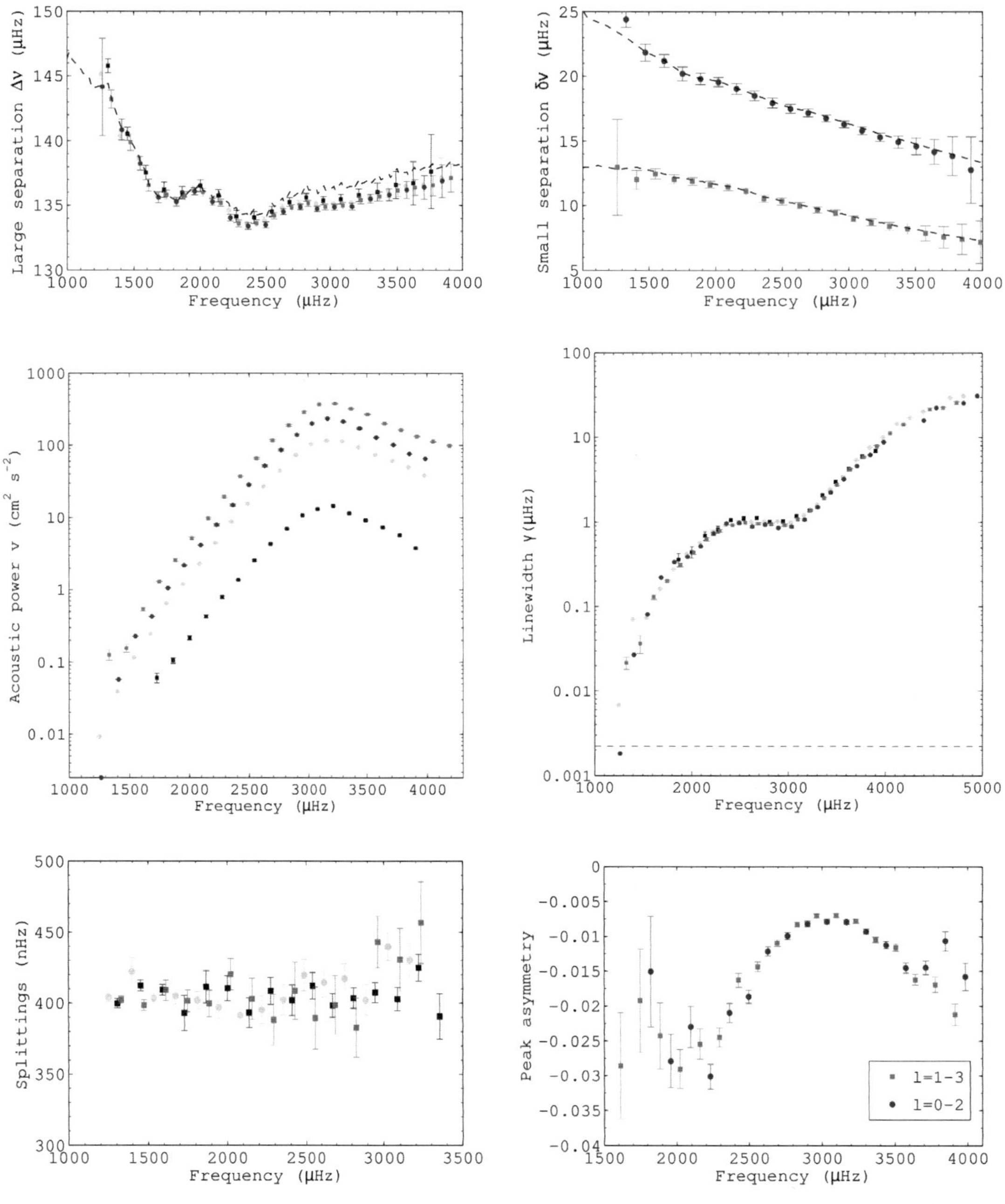

Figure 1. Top left: large separation as a function of frequency calculated from the fitted GOLF frequencies (same legend than in VIRGO/SPM plots). The formal errors were multiplied by a factor 10. The dashed lines correspond to the theoretical values using the Saclay seismic model [18]. Top right: small separation ($l = 0 - 2$ modes in red, $l = 1 - 3$ modes in blue). The formal errors were multiplied by a factor 10. Middle left: Full amplitudes (in units of ppm^2). Middle right: Linewidths. The horizontal dashed line corresponds to the frequency resolution. Bottom left: splittings. Above 3500 μHz, the splittings were fixed to 400 nHz. Bottom right: Asymmetry. The increase below 2000 μHz could not be real due to the reduction in the SNR and the fewer number of points defining the profile.

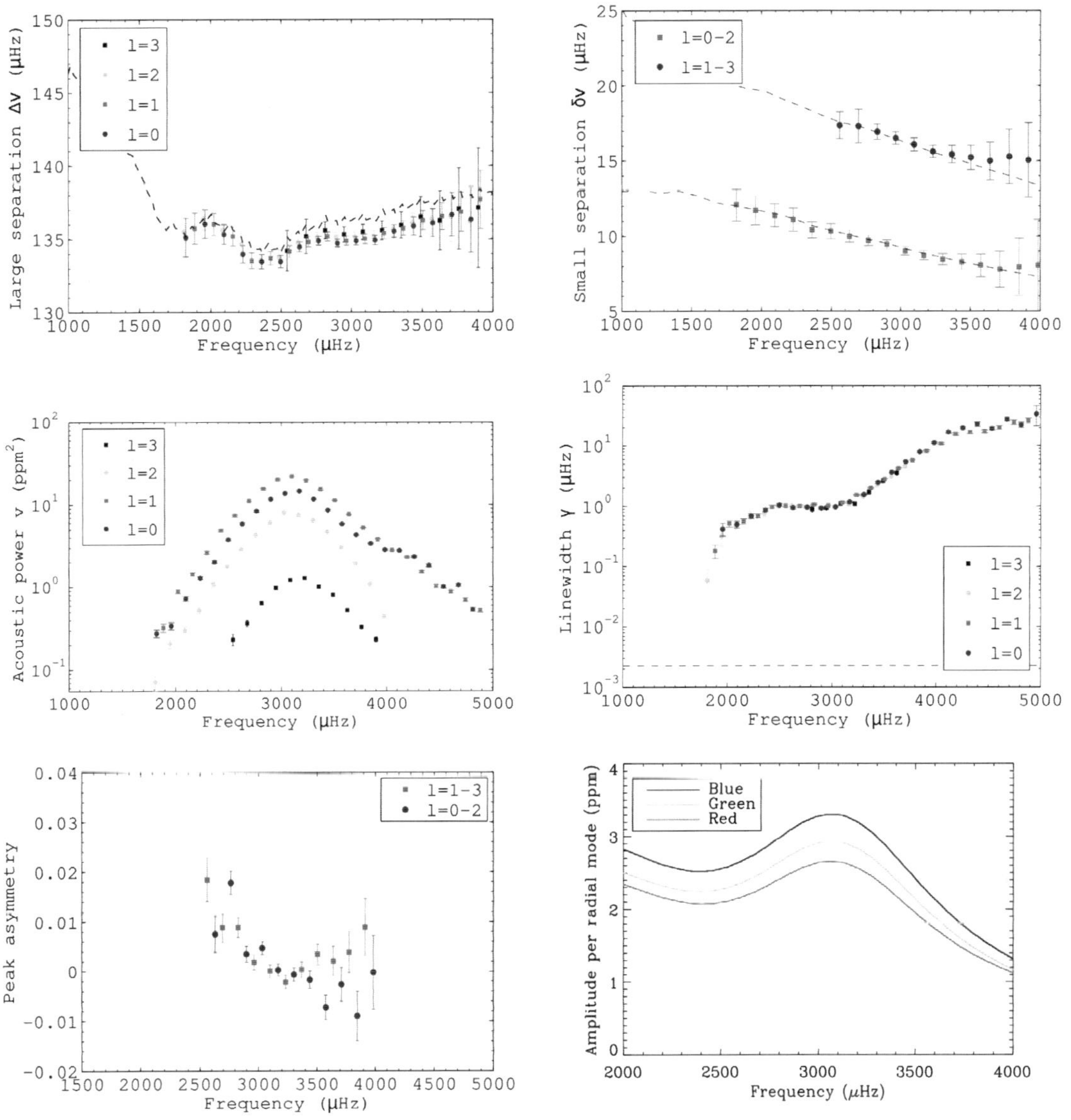

Figure 2. Top left: large separation as a function of frequency calculated from the fitted VIRGO/SPM frequencies. Top right: small separation. Middle left: Full amplitudes (in units of ppm^2). Middle right: Linewidths. Bottom left: Asymmetry. Bottom right: average maximum rms amplitudes per radial mode for the three VIRGO/SPM channels. Due to the small $l = 3$ visibility in the VIRGO/SPM data, these modes do not perturb the $l = 1$ and the linewidth and acoustic power of the $l = 1$ modes could be fitted up to 5000 μHz.

Low frequency signal in the GOLF measurements

G Grec, J Provost and C Renaud

Université de Nice-Sophia Antipolis; CNRS UMR 6202 Cassiopée; OCA. France

E-mail: grec@obs-nice.fr

Abstract. This paper shows the results obtained using a revisited method to normalize the velocity evaluation extracted from the measurements, for roughly 14 years of GOLF data.
For the search of g modes, we calculate the low frequency power spectrum of the signal with 2 different approaches:
• The classical calculation of the power spectrum of the velocity signal.
• An alternative calculation, extracting first the variations along the time of the p-mode frequencies, then calculating the power spectrum of those frequency modulation [4].
Both spectra are compared to the g-mode frequency spectrum calculated for a solar model.
Several observed frequencies are in close agreement with the calculated g modes. A careful statistical analysis of this result should now follow.

1. Introduction

The solar observatory SoHO carries several instruments designed for helioseismology. Now a long period has been devoted, in space as at ground, to detect solar g modes. The results remain today controversial, even the detectability of such oscillations is not proved [1]. Nevertheless, the low frequency part of the solar spectrum is very rich in sharp lines and some of then seem to be related to very stable features, not to noise fringes due to the solar convection noise. In addition this convection noise produces a low frequency increase of the energy. On other hand, the knowledge of the p-mode spectrum and of the solar model allow to calculate the probable spectrum of the g modes of the rotating Sun. This paper shows how the observations and the calculations can fit together. A second part is a study of the p-mode frequency variations, with the assumption of a cross modulation allowing to detect g modes.

2. The power spectrum analysis

The new procedure used here for the velocity calibration [2] produces an almost constant amplitude signal envelope over the whole duration of the observations (Fig.1). Compared to our former procedure, the conversion factor of the signal to the surface solar velocity is more effectively corrected from the variations related to the orbital parameters: the non linearity related to the orbital velocity, the temperature shift of the instrument related to the solar distance, a faint non-thermal drift with the time. The power spectrum of this velocity is then calculated with the 2.5 nHz frequency resolution corresponding to the total duration of GOLF observations. The low-frequency part of this spectrum is shown Fig.2.

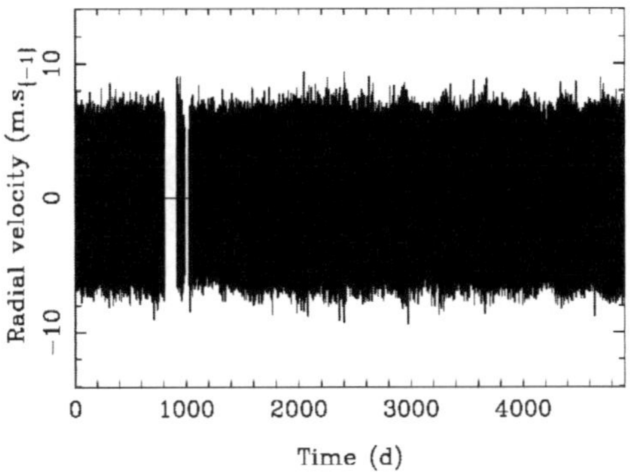

Figure 1. 14 years of velocity signal from GOLF, starting Apr. 10, 1995).

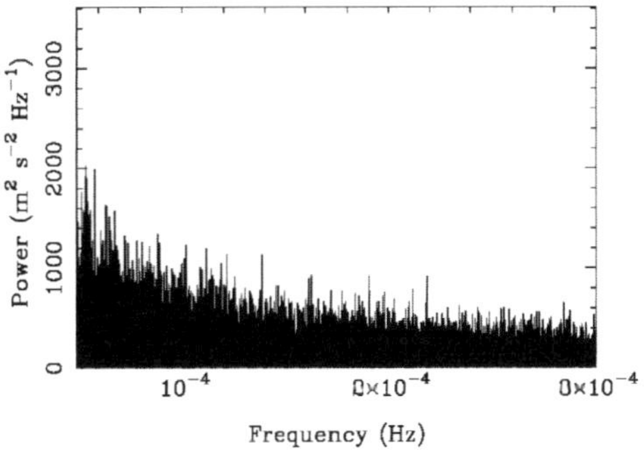

Figure 2. The velocity power spectrum in the frequency domain searched for g modes.

2.1. The lines in the 217 μHz to 221 μHz domain

In the low frequency range of the power spectrum, the new procedure used may result in a different detection of a peculiar spectral line. Obviously the power spectrum of a time sequence depends on the variation of the weighting of the signal related to a variable sensitivity along the time.

Fig. 3 shows the spectral density from 217 μHz to 222 μHz: The frequency value measured for the highest line is 218.585 μHz, the line at 220.7 μHz previously identified [5] is buried in the surrounding noise in the spectrum calculated here for 14 y of GOLF observations. A close line at 218.314 μHz is found in [3].

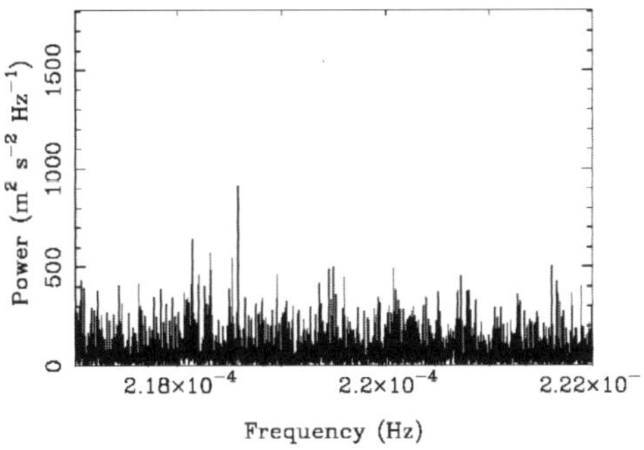

Figure 3. Part of the power spectrum: the 217 μHz to 222 μHz interval.

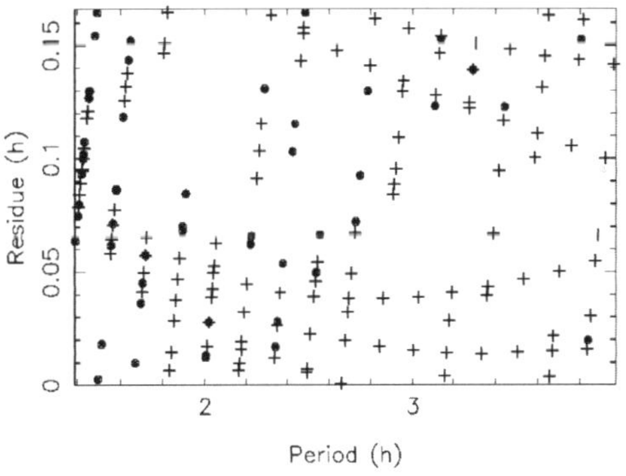

Figure 4. The periods detected in the velocity power spectrum (•) and calculated for the observable g modes for $l \leq 3$ (+).

2.2. Search for a calculated g-mode pattern

The solar model is calculated using the CESAM code [6] and calibrated in age and luminosity. The rotation, as deduced from the inversion of helioseismic data, varies with depth and latitude in the convective zone. The rotation frequency is constant $\Omega = 433$ nHz for the radiative zone and the solar core [7]. The calculations are made into inertial coordinates and translated into synodical values.

In order to make a simple detection of the higher peaks in the power spectrum, the well-known increasing trend in the low-frequency domain is cleared thanks to a normalization by a running mean filtering - assuming the common increase of signal and background is due to an increasing energy source. This results in sharp lines in a flat noisy background. The observed power spectrum is then reduced to the small number of lines having the highest amplitude in the considered frequency domain. All detected lines are sharp lines, either long phase-coherent oscillation or noise.

The comparison of the frequencies detected with the calculated g-mode spectrum has to be done globally for all g-mode in the band pass available for the observations and with a very high spectral resolution. We use a *spectre echelle* calculated for the periods (Fig.4). Unlike the p modes, the spacing of successive modes depends on the degree l. For a given period P we calculate the residual number $R = P - kP_l$ where $kP_l < P < (k+1)P_l$ and the folding factor P_l reflects the periodical behaviour of the asymptotic series of g modes. For a given degree l, $P_l = P_0 \ / \ \sqrt{l(l+1)}$ where the characteristic period P_0 depends of an integral function over the solar interior of the buoyancy force. In addition, a given mode is separated into several components by the solar rotation depending on the tesseral order m. This almost constant splitting in a frequency scale results in an increasing spacing of components as the period increases. We select the periods of geometrically visibles modes ($l + m$ even) for which the geometrical averaging over the solar disk is minimum. Fig.4 shows the periods calculated for the degrees $l = 1$, $l = 2$, $l = 3$, a mode is shown as a point at coordinates (P,R), we use a folding factor P_1 adjusted to display as clearly as possible the series of points for $l = 1$.

If the actual value of P_l determines the pattern associated to the periods for a given value of the degree l, it does not change the relative display of 2 close periods. The same diagram can display the calculated series for different l, together with the periods detected in the observations. On this display, the coincidence of the model with the observations are clearly visible. Roughly 50% of the highest lines in local domain of the power spectrum shown Fig.4 coincide with a calculated mode or are very close to it.

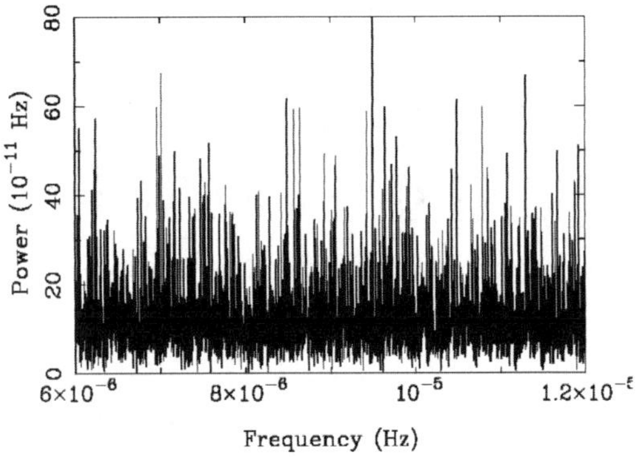

Figure 5. Averaged cross spectrum of the p-mode frequency variations measured for the 14 y of observation.

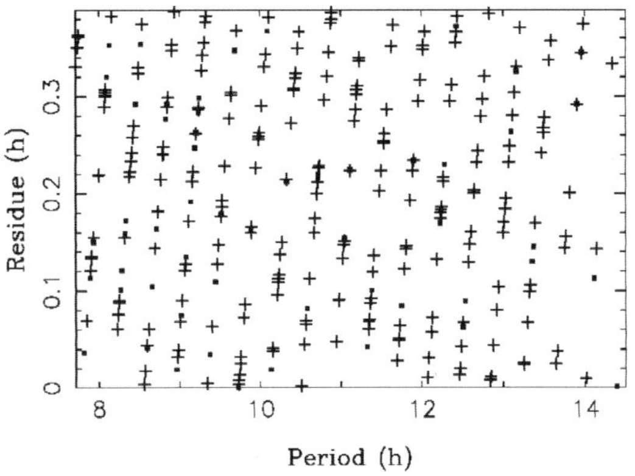

Figure 6. Diagram of periods detected in the p-mode frequency modulation (•) and of the periods calculated for splitted g modes of degree $l = 1, 2, 3$ (+).

3. Power spectrum of the p-mode frequency variations

A p-mode frequency is related of the physical condition of the corresponding resonant cavity. This frequency is related on the sound speed which depends on temperature and then them perturbations, as g-mode oscillations may induce. The idea is then to study the frequency modulation of the acoustic modes and compare the lines detected in the power spectrum to a theoretical spectrum of g modes. We use the method described in the previous paper [4] with a new set of numerical parameters, in order to improve precision and spectral resolution.

The frequency of the p modes for degree $l = 1$ and radial order $n = 11$ to $n = 25$ is measured on short time intervals, using a correlation of the power spectrum with an averaged power spectrum used as constant reference. The limits of n is fixed by the amplitude of mode, the width of the spectral lines and their dependency on the solar activity. In order to obtain a mean noiseless spectrum, we calculate the averaged cross-spectrum of the time functions giving the frequency of each mode.

Due to intrinsic limitations of the observed band pass, the analysis address the asymptotic domain of the g-mode spectrum. The comparison with the observations is made as in Sec.2.2. Presently, the longest observed periods (above 14 h) have no calculated counterpart, due to the resolution of the radial grid of the solar model. A preliminary question arrises: which g modes are observable? Let assume as a first approach that the geometry of the observable g modes ($l + m$ even) is independent on the physical observable (luminosity, radial velocity, p-mode frequency variations). Again, the display is optimized for a given value of l, for which the asymptotic series are visible (Fig.6). The periods are displayed for modes of degree $l = 1$ to $l = 3$ up to the longest calculated values. The rotational splitting produces here a complex pattern. Close coincidences of the model with the observations are detected. A statistical study should follow in order to evaluate the probability of a real g-mode detection.

4. Conclusion

On the long term observations analysed here, the previously-called 220.7 μHz line, mainly visible in the intensity variations [5], is not detected in the velocity spectrum.

Even if a small number of detected lines corresponds to calculated modes, it may confirm the validity of the numerical model, including the rigid rotation below the convective zone.

Several detected lines in the power spectrum of the velocity calculated with the new calibration process are in good agreement with the model, mainly for $l = 3$. The statistical significance of this detection remains to be estimated.

The power spectrum of the frequency modulation of the p modes contains several long period g-mode of the calculated series, either for degree $l = 1$, $l = 2$, $l = 3$. Again the statistical significance must be estimated.

5. Bibliographie

[1] T. Appourchaux, K. Belkacem, A.-M. Broomhall, W. J. Chaplin, D. O. Gough, G. Houdek, J. Provost, F. Baudin, P. Boumier, Y. Elsworth, R. A. García, B. N. Andersen, W. Finsterle, C. Fröhlich, A. Gabriel, G. Grec, A. Jiménez, A. Kosovichev, T. Sekii, T. Toutain, and S. Turck-Chièze. The quest for the solar g modes. *A&A Rev.*, 18:197–277, February 2010.

[2] Grec. G. GOLF: Calculation of a calibrated velocity signal. Paper in preparation.

[3] A. H. Gabriel, F. Baudin, P. Boumier, R. A. García, S. Turck-Chièze, T. Appourchaux, L. Bertello, G. Berthomieu, J. Charra, D. O. Gough, P. L. Pallé, J. Provost, C. Renaud, J.-M. Robillot, T. Roca Cortés, S. Thiery, and R. K. Ulrich. A search for solar g modes in the GOLF data. *A&A*, 390:1119–1131, August 2002.

[4] G. Grec and C. Renaud. Variations of the low-degree solar p-mode frequencies for 10 years of GOLF observations. *Astronomische Nachrichten*, 329:490–+, 2008.

[5] A. Jiménez and R. A. García. On the Solar Origin of the Signal at 220.7 μHz: A Possible Component of a g-Mode? *ApJS*, 184:288–297, October 2009.

[6] P. Morel. CESAM: A code for stellar evolution calculations. *A&AS*, 124:597–614, September 1997.

[7] J. Provost, G. Berthomieu, and P. Morel. Low-frequency p- and g-mode solar oscillations. *A&A*, 353:775–785, January 2000.

The acoustic cutoff frequency of the Sun and the solar cycle

A Jiménez[1,2], **R A García**[3], **P L Pallé**[1,2]

[1] Instituto de Astrofísica de Canarias, E-38200 La Laguna, Tenerife, Spain
[2] Departamento de Astrofísica, Universidad de La Laguna, E-38206 La Laguna, Tenerife, Spain
[3] Laboratoire AIM, CEA/DSM-CNRS, Université Paris 7 Diderot, IRFU/SAp, Centre de Saclay, F-91191 Gif-sur-Yvette, France

E-mail: ajm@iac.es, rgarcia@cea.fr, plp@iac.es

Abstract. The acoustic cutoff frequency –highest frequency for acoustic solar eigenmodes– is an important parameter of the solar atmosphere because it determines the upper boundary of the p mode cavities. At frequencies beyond this value, acoustic disturbances are no longer trapped waves but traveling waves. Interferences amongst them, originate higher frequency peaks -the pseudomodes- in the solar acoustic spectrum. Using data from GOLF and VIRGO instruments aboard *SoHO* spacecraft, we determine the acoustic cutoff frequency using the coherence function between both sets of data, velocity and intensity. By using data gathered by these instruments during the whole lifetime of the mission (1996 till present), it is found a variation of this parameter with the solar magnetic activity cycle.

1. Introduction

The acoustic cutoff frecuency $\nu_{\rm ac}$ is an important parameter of the solar atmosphere, given (in isothermal conditions) by $\nu_{\rm ac}$= c/2Hp, Hp being the pressure scale height and c the sound speed. Theoretically, its value is of $\sim$ 5000 μHz while from observations, a higher value has been obtained [1]. The resonant acoustic waves (p modes) are trapped in resonant cavities with lower boundary ranging from the center of the Sun up to the surface (depending of their angular degree and frequency) and their upper boundary at the surface where $\nu = \nu_{\rm ac}$. Only acoustic oscillations with frequencies $\nu < \nu_{\rm ac}$ are trapped in the resonant cavities beneath the photosphere. For $\nu > \nu_{\rm ac}$, acoustic disturbances are no longer trapped and propagate as traveling waves through the chromosphere to the base of the corona. These high-frequency peaks, hereafter, pseudomodes, show a clear structure beyond $\nu_{\rm ac}$ [2, 3, 4, 5, 6, 7, 8]. In the acoustic spectrum of the Sun, the frequency separation between consecutive modes of the same degree $\Delta\nu_{n,l}=\nu_{n,l} - \nu_{n-1,l}$ increases slightly between $\sim$ 5000-5500 μHz. This variation is due to the transition that takes place from trapped to traveling behaviour. If this frequency separation ($\Delta\nu_{n,l}$) increases around $\nu_{\rm ac}$, all the peaks with $\nu > \nu_{\rm ac}$ will be shifted respect to the peaks with $\nu < \nu_{\rm ac}$. We try to find at which frequency this shift takes place being this value a good measurement of the acoustic cutoff frequency. Using the *SoHO* data for the whole mission we study a possible relation of this value with the phase of solar activity cycle. It is also important to remember that an enhancement has been observed in the pseudomodes region during some magnetic events such as flares [9,10].

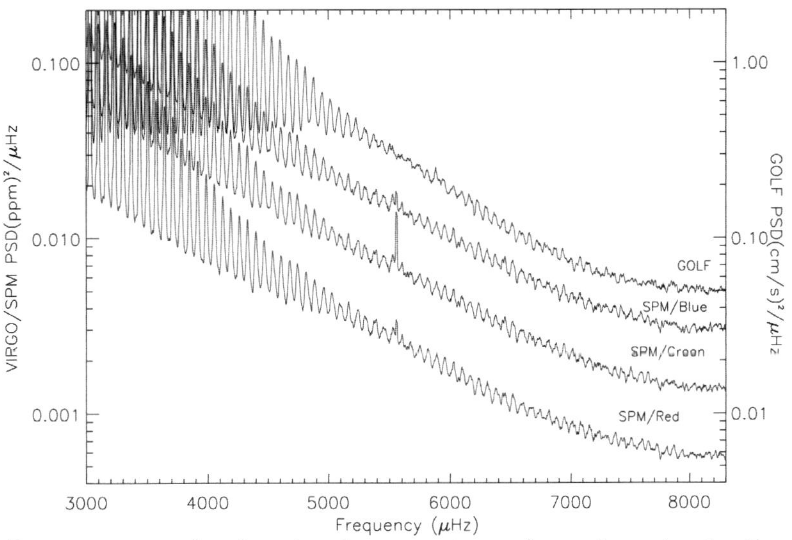

Figure 1. Power spectra showing the clear structure of pseudomodes for the different signals considered in this analysis. Bottom to top: The three intensity channels of VIRGO/SPM (Red, Green and Blue) and the velocity GOLF spectra.

2. Data sets

We use VIRGO/SPM(three colors) [11], intensity data and GOLF velocity data [12,13]. Time series of 800 consecutive days (shifted 50 days respect to the previous one) are constructed, giving a total of 85 time series from April 1996 to May 2010. Each of these 800 days time series are then sub-divided in 4-day time series. Their corresponding power spectra and coherence function are computed and the results averaged. A sample of the resulting power spectra are shown in Figure 1, where the pseudomodes signal is clearly visible.

In the present analysis, the coherence function between intensity and velocity is used instead of the use of the power spectra in order to avoid electronic contamination in the intensity data at around 5550 μHz and because the coherence function contains the coherence signals of both datasets.

3. Analysis

The Intensity-Velocity (I-V) coherence function (black curve, figure 2a) is fitted to an exponentially modulated sine wave (blue curve) between 3000 μHz and 5500 μHz to take into account the decreasing amplitude of p modes and the background. In addition, between 5000 μHz and 6500 μHz, only a single sine wave function is fitted (pseudomodes amplitudes do not vary significantly with frequency).

Both fitted curves, p modes (blue one) and pseudomodes (red one) are extended to the whole frequency range from 3500 to 6500 μHz, their maxima computed and its position plotted in Figure 2a: Black filled circles, maxima of the coherence function; Blue filled circles, Maxima of the exponentially modulated sine wave fitted to the p mode range between 3500 and 5500 μHz; Red filled circles, maxima of the sine wave fitted to the pseudomodes range between 5000 and 6500 μHz.

The frequency difference between all consecutive maxima of the coherence and of the fitted curves are computed and plotted in Figure 2b: blue circles for p modes and red squares for pseudomodes. The blue circles are close to zero in the p mode range because the fitted exponentially modulated sine wave, blue curve in figure 2a, models very well the p modes behaviour. But when the pseudomodes region begins to be considered, the difference between the maxima increases because the frequency separation of pseudomodes is different to the one of p modes. The red squares are close to zero in the pseudomode region because the fitted

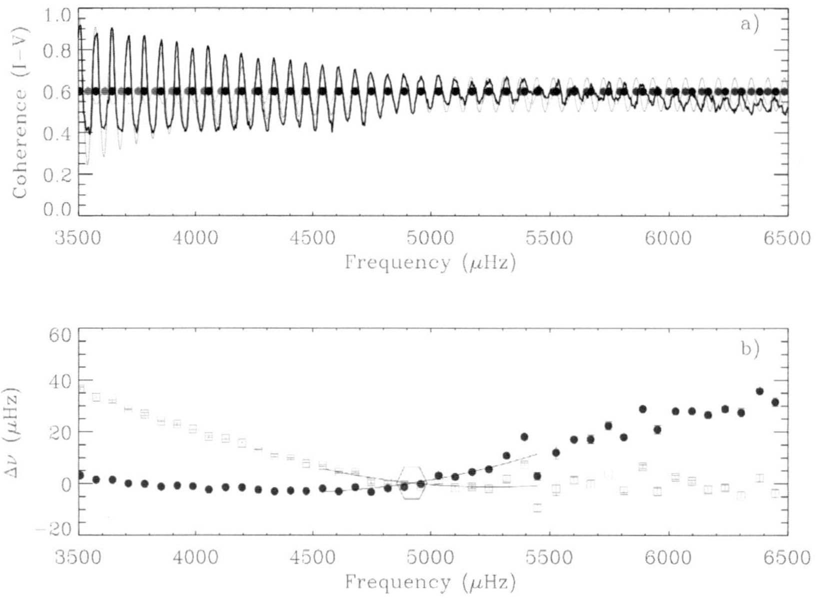

Figure 2. Coherence function and ν_{ac} determination. a) Black line: coherence function. Blue line: Exponentially decay sine wave fitted to the coherence function between 3500 and 5500 μHz (end of the p mode range) and plotted extended up to 6500 μHz. Blue and black filled circles are the maxima of the fitted sine wave and coherence function respectively. Note how the coherence shifts to higher frequencies from ~5000 μHz onwards. Red line: Cosine fitted to the frequency range between 5000 and 6500 μHz (pseudomodes range) and plotted extended to 3500 μHz . Red and black filled circles are the maxima of the fitted cosine and the coherence function. b) Blue circles: frequency differences between the maxima of the coherence function and the fitted (3500-5500 μHz) exponentially modulated sine wave. Red squares: frequency differences between the maxima of the coherence function and the fitted cosine (5000-6500 μHz). Two parabolas segments are fitted around the region where both curves cross, to obtain the best estimation of ν_{ac}

sine wave, red curve in figure 2a, reproduce nicely the pseudomodes behavior. But in the p mode range the differences increase because the frequency separation between p modes and pseudomodes is different.

Looking around 5000 μHz, blue points are always below the red ones for lower frequencies and always above for higher frequencies. The frequency interval between the transition of blue points from below to above the red points (pointed out with a hexagon) is the frequency interval of the acoustic cutoff value. These two limits are the lower and upper limits of ν_{ac}. A more accurate determination of ν_{ac} is performed by fitting two parabolas (one for red points and one for blue points) in the interval 4500μ -5500 μHz and computing the crossing frequency point of them. This is considered as the best feasible determination of ν_{ac}.

4. Results

The results obtained from the analysis of the coherence function for each one of the VIRGO/SPM channels, are the same within error, so the values of ν_{ac} have been averaged and shown in Figure 3 (left) with their associated errors. The corresponding Solar Radio Flux values for the same period are shown on the right plot. A clear positive correlation between both parameters is clearly noticeable and deserves further investigation and evaluation.

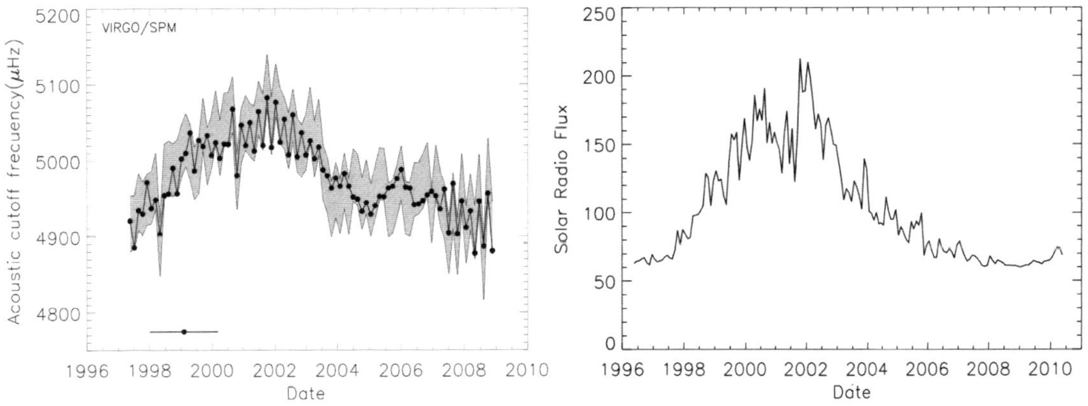

Figure 3. Acoustic cutoff frequency as a function of time as calculated from the data set (left panel) and the measured Solar Radio Flux (right panel), showsing a high correlation with the acoustic cutoff frequency determinations.

5. Conclusions

The analysis of the coherence function between SoHO-Intensity (VIRGO)/SPM) and -Velocity (GOLF) data has allowed a precise determination of the acoustic cutoff frequency of the solar atmosphere. Its variation along the continuos 14-year of the mission, shows a significant variation with the Solar activity cycle and positively correlated. This result is in agreement with previous determination of the solar acoustics radius of the Sun [14] found to be negatively correlated with activity. This work is still in progress.

Acknowledgments

The GOLF and VIRGO instruments onboard SoHO are a cooperative effort of many individuals, to whom we are indebted. SoHO is a project of international collaboration between ESA and NASA. The 10.7-cm radio flux was obtained from the National Geophysical Data Center. AJ and PLP acknowledges the support of the grant PNAyA2007-62650 from the Spanish National Research Plan, and RAG the support of the CNES/GOLF grant at the SAp, CEA-Saclay.

References

[1] Fossat E. 1992 *Astron. & Astroph.* **266** 532.
[2] Jefferies S. M. et al. 1988, in Seismology of the Sun and Sun-like Stars ed. E.Rolfe (Noordwijk: ESA SP-286) 279
[3] Libbrecht K. G. 1988 *Astroph. J.* **334** 51
[4] Duvall T. L., Harvey J. W., Jefferies S. M. and Pomerantz M. A. 1991 *Astroph. J.* **373** 308
[5] García R. A. et al. 1998 *Astroph. J.* **504** L51
[6] Chaplin W. J. et al. 2002, in Local and Global Helioseismology: The Present and Future, ed. H. Sawaya-Lacoste (Noordwijk: ESA SP- 517) 247
[7] Jiménez A., Jiménez-Reyes S.J. and García R.A. 2005 *Astroph. J.* **623** 1215
[8] Jiménez A. 2006 *Astroph. J.* **646** 1398
[9] Karoff C. and Kjeldsen H. 2008 *Astroph. J.* **678** L73
[10] Kumar B., Mathur S., García R. A. and Venkatakrishnan P. 2010 *Astroph. J.* **711** L12
[11] Frohlich C et al. 1995 *Solar Physics* **162** 101
[12] Gabriel, A. H., et al. 1995 *Solar Phys.* **162** 61
[13] García R. A. et al. 2005 *Astron. & Astroph.* **442** 385
[14] Kholikov S. and Hill F 2008 *Solar Phys* **251** 157

Meridional-Flow Measurements from 15 Years of GONG Spherical-Harmonic Time Series

S. Kholikov, I. González Hernández, F. Hill and J. Leibacher

National Solar Observatory, 950 N. Cherry Ave., Tucson, AZ, 85719, USA

E-mail: `kholikov@nso.edu`

Abstract. We present results of meridional-flow measurements for 1995 – 2009, using travel-time differences from velocity images reconstructed using GONG spherical harmonic (SH) coefficients after applying phase-velocity and low-m filters. This filtering technique increases the signal-to-noise ratio and thus extends travel-time measurements to relatively high latitudes and deep into the convection zone. Preliminary analyses shows a strong one-year periodicity presumably due to solar pole misalignment and B_0-angle artifacts, which makes it difficult to see underlying temporal variations. Removing a simple one-year-period sine wave fit reveals long-term temporal variations of the flow on top of this yearly periodicity. High-latitude measurements are affected more stronger by foreshortening and B_0-angle artifacts. We analyze different B_0-angle intervals separately, so in each hemisphere better high-latitude visibility comes six months apart. This approach suggests why at high latitudes travel-time measurements of meridional flow shows a tendency to change sign instead of continuing towards the poles.

1. Introduction

Meridional circulation transports surface plasma and magnetic field from low latitudes toward the solar poles, and as such plays an important role in the solar activity cycle [4]. According to many models, the speed of the meridional flow determines the length of near-future magnetic cycles, and therefore temporal variations of the meridional flow may play an important role in, or provide an important diagnostic of, variations of the activity cycle. Since the deep meridional flow is much weaker than solar rotation, the torsional oscillation, or active-region associated flows, measurement of temporal changes requires very precise, long-term observations, improving analysis technique and understanding systematic errors. During the last several years meridional circulation has been studied by different helioseismic techniques [1-3], [5-11].

Long term observations by MDI, GONG and TON instruments have provided detailed measurements of meridional flow and its temporal variations. In most helioseismic analyses using ring-diagram analysis, meridional-flow measurements have been studied for subsurface layers (not deeper than 30 Mm). Using SH time series allow us to extend our measurements much deeper, including the tachocline region.

2. Data and analysis technique

We use daily SH coefficients time series for $\ell = 0 - 200$ obtained by the GONG project during the 1995 – 2009 time period. Time series are organized as series of individual (ℓ,m) coefficients for 24 hour intervals. In this study, we have selected 3658 daily series with duty cycle higher than 80%.

We apply time-distance technique to measure North – South travel-time differences on velocity images that have been reconstructed. Since we are interested in travel times for acoustic waves propagated in the North – South direction, isolating such waves increases the signal-to-noise. We call this procedure "low-m filtering". In other words, we keep only North-South component of the velocity signal and remove all the rest. Only 15% of the low-m coefficients of any ℓ are used to reconstruct the velocity images. In addition we also apply phase-velocity filtering to isolate waves propagating along the same ray path and having the same lower turning points in

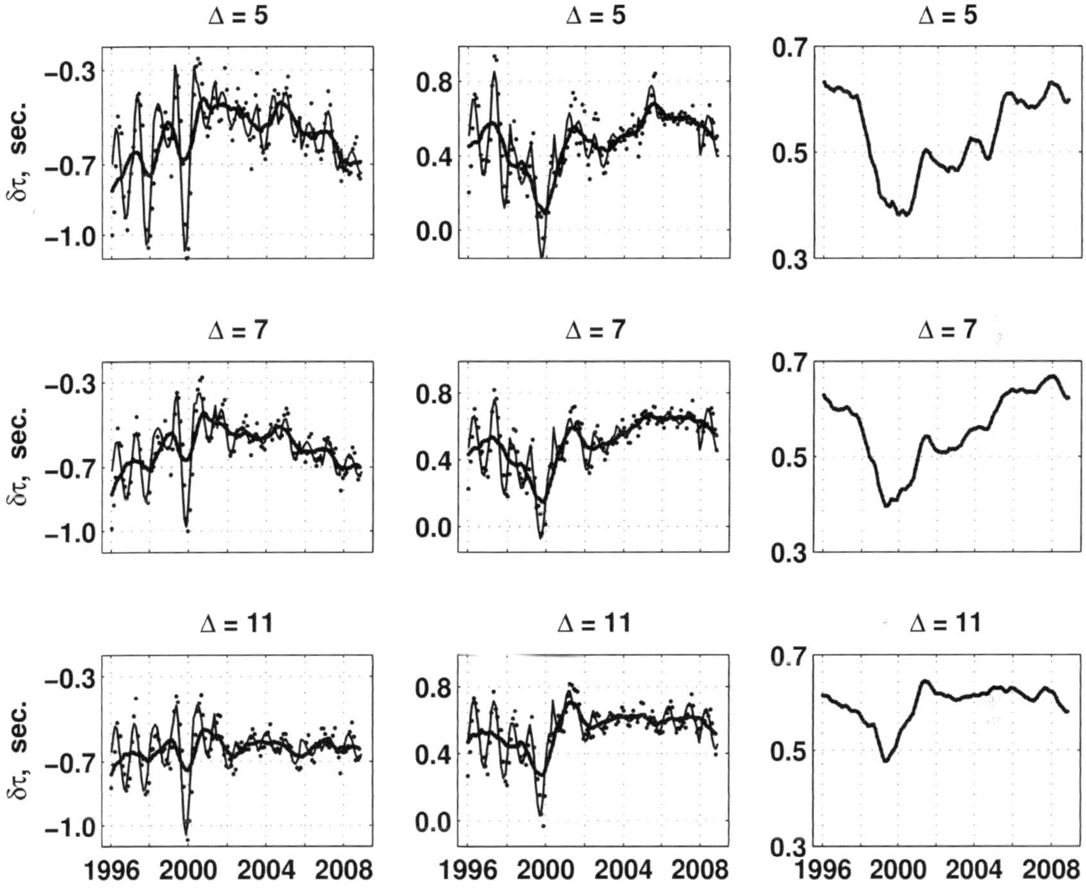

Figure 1. Travel-time differences at θ=20° for four angular separations (Δ): raw (dots), fit (thin solid line) and filtered to remove the annual (B-angle) artifact (thick line): left column for southern and middle column for northern hemispheres. Right column shows symmetrized temporal differences in the two hemispheres and running one-year sine fit of yearly intervals. A consistent decrease is seen at solar maximum, which recovers more slowly closer to the surface.

the Sun. In order to measure travel-time shifts due to meridional flow for different depths, we choose different phase speeds in our filtering process. Since standard GONG SH time series are limited to $\ell \leq 200$, our measurements cover $0.97\,R_\odot$ and deeper layers.

3. Results and discussion

We found that travel times show a strong one-year periodic signal presumably due to the B_0-angle and foreshortening, and possibly a misalignment of the solar pole [6]. Measuring the temporal variations of the meridional flow is difficult in the presence of such a strong periodic signal. At high latitudes, this artifact changes the flow direction, which can easily be misinterpreted as a second-cell structure of meridional flow [8]. To extract the changes in time, we symmetrized the temporal differences in the two hemispheres and performed a running one-year sine fit of yearly intervals. The filtered and symmetrized signal in Figure 1 shows a solar-cycle like trend. For larger separations, we do not see any systematic variations, which would be expected if these systematics are related to surface activity. Often meridional flows measured by the travel-time

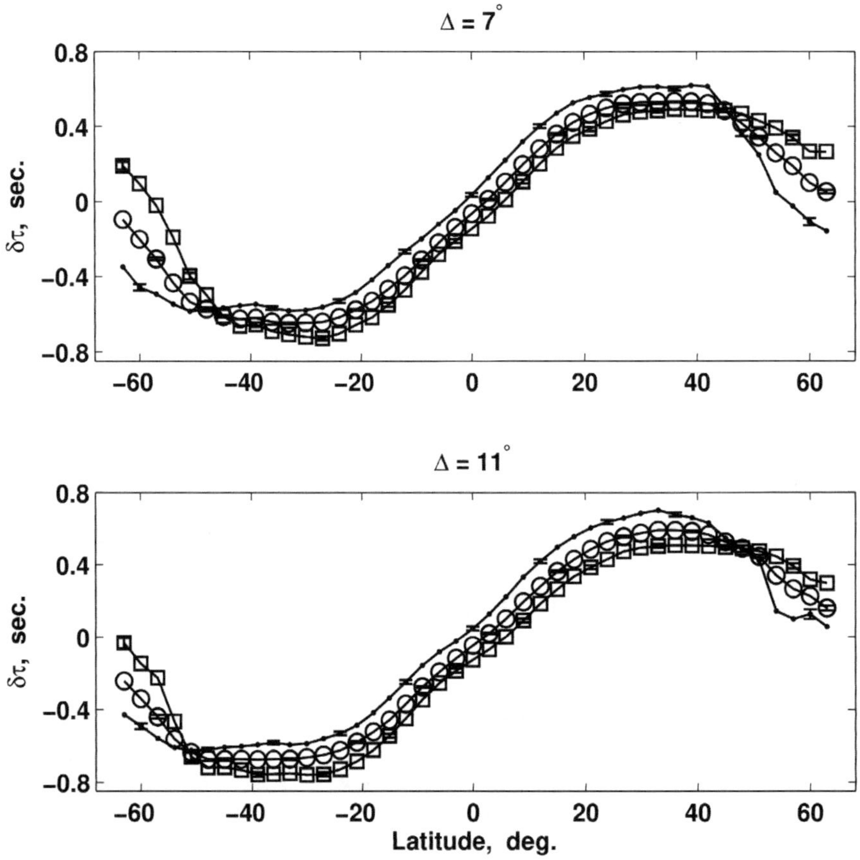

Figure 2. Travel-time difference for two angular separations: the entire time period (open circles), the 10% of the time with the South Pole most exposed (dots) and the 10% of the time with the North Pole most exposed (squares).

technique show decreasing magnitude at high latitudes starting at about $30° - 40°$, which can lead to a scenario with a reversal of the poleward flow before reaching the poles. We analyzed averaged time differences during the two maximum B_0-angle time periods. In Figure 2 we show the average time differences for positive and negative tilt periods in comparison to the average signal for the entire time period. There is obvious evidence for the decrease of this effect in favorable tilt periods. To see the depth dependence of the time differences, we have averaged measurements over the whole time period. In Figure 3, measurements with different lower turning points are presented. With increasing depth, the magnitude of the travel-time differences increases until the tachocline region and then decreases. Due to the averaging for 15 years, the typical measurement error bars are less than 0.02 seconds. To get the actual flow

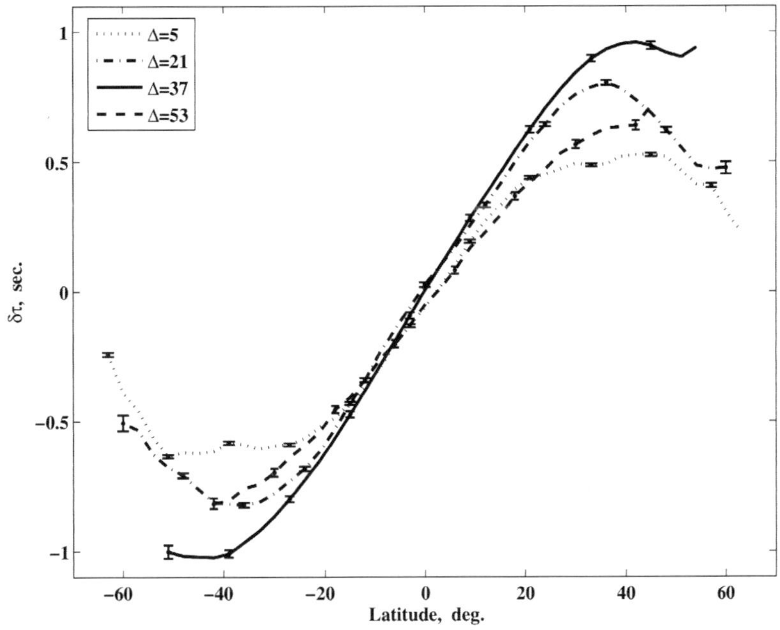

Figure 3. Travel-time differences for four angular separations. The magnitude of the travel-time differences increases for separations with lower turning point up to $0.76R_\odot$ and then decreases. Separation distances are proportional to lower turning points of the waves. These particular separation distances correspond to $0.97R_\odot$, $0.87R_\odot$, $0.76R_\odot$ and $0.67R_\odot$ respectively. $\Delta=45°$ corresponds to the tachocline region ($\approx 0.71R_\odot$).

speed and depth profile of the flow, we need to use inversion techniques. Efforts on this task are in progress.

Acknowledgments

This work utilized data obtained by the Global Oscillation Network Group (GONG) program, managed by the National Solar Observatory, which is operated by AURA, Inc. under a cooperative agreement with the National Science Foundation. The data were acquired by instruments operated by the Big Bear Solar Observatory, High Altitude Observatory, Learmonth Solar Observatory, Udaipur Solar Observatory, Instituto de Astrofísica de Canarias, and Cerro Tololo Interamerican Observatory.

References

[1] Basu, S and Antia, H M: 2003, *Astrophys. J.*, **585**, 553.
[2] Braun, D C and Fan, Y: 1998, *Astrophys. J.*, **508**, L105.
[3] Chou, D-Y and Ladenkov, O: 2005, *Astrophys. J.*, **630**, 1206.
[4] Dikpati, M and Gilman, P A: 2001, *Astrophys. J.*, **559**, 428.
[5] Duvall, T L Jr.: 1979, *Solar Phys.*, **63**, 3.
[6] Giles, P M, Duvall, T L Jr., Scherrer, P H, and Bogart, R S: 1997, *Nature*, **390**, 52.
[7] Gizon, L: 2004, *Solar Phys.*, **224**, 217.
[8] González Hernández, I, Komm, R, Hill, F, Howe, R, Corbard, T, and Haber, DA: 2006, *Astrophys. J.*, **638**, 576
[9] Haber, D A, Hindman, B W, Toomre, J, Bogart, R S, Thompson, M J, and Hill, F: 2000, *Solar Phys.*, **192**, 335.
[10] Komm, R W, Howard, R F and Harvey, J W: 1993, *Solar Phys.*, **147**, 207.
[11] Schou, J and Bogart, R S: 1998, *Astrophys. J. Lett.*, **504**, L131.

Mode visibilities in radial velocity and intensity Sun-as-a-star helioseismic measurements

D Salabert[1,2], J Ballot[3] and R A García[4]

[1] Instituto de Astrofísica de Canarias, E-38200 La Laguna, Tenerife, Spain
[2] Departamento de Astrofísica, Universidad de La Laguna, E-38206 La Laguna, Tenerife, Spain
[3] Laboratoire d'Astrophysique de Toulouse-Tarbes, Université de Toulouse, CNRS, F-31400 Toulouse, France
[4] Laboratoire AIM, CEA/DSM-CNRS, Université Paris 7 Diderot, IRFU/SAp, Centre de Saclay, F-91191 Gif-sur-Yvette, France

E-mail: `salabert@iac.es, jerome.ballot@ast.obs-mip.fr, rgarcia@cea.fr`

Abstract. We analyze more than 5000 days of Sun-as-a-star radial velocity GOLF and intensity VIRGO observations to measure the visibilities of the $l = 0$, 1, 2, and 3 modes and the m-amplitude ratios of the $l = 2$ and 3 modes in the solar acoustic spectrum. We provide observational values that we compare to theoretical predictions.

1. Introduction

In Sun-as-a-star helioseismology, it is common practice to fix the amplitude ratios between the m-components of the $l = 2$ and 3 multiplets (the so-called m-amplitude ratio) during the peak-fitting procedure when estimating the p-mode characteristics (e.g., Salabert et al. [1]), while the amplitudes of the $l = 1$, 2, and 3 modes relative to the $l = 0$ modes are left free (the so-called mode visibility). However, in asteroseismology, the mode visibilities are fixed to theoretical values due to lower signal-to-noise ratio (SNR) and shorter time series, and the m-amplitude ratios are expressed as a function of the inclination of the rotation axis only (Appourchaux et al. [2]; García et al. [3]). In both cases, they are supposed to not depend on the star magnetic activity. However, in the near future, this situation could change when stellar activity cycles will be measured in asteroseismic targets (e.g., García et al. [4]). After several years of observations collected by the Kepler mission, the SNR will be high enough to measure these parameters in a wide range of solar-like stars in the HR diagram at different evolution stages (Bedding et al. [5]; Chaplin et al. [6]). Moreover, simultaneous observations from SONG (Doppler velocity) and Kepler (intensity) will be extremely useful to better understand stellar atmospheres. Although, variations with the height in the solar atmosphere at which the measurements are obtained have been observed in the intensity VIRGO/SPM data at the beginning of the SoHO mission (Fröhlich et al. [7]), it has never been verified if these values change in the radial velocity GOLF measurements between the blue- and red-wing observing periods.

2. Observations and analysis

We used observations collected by the space-based instruments Global Oscillations at Low Frequency (GOLF) and Variability of Solar Irradiance and Gravity Oscillations (VIRGO)

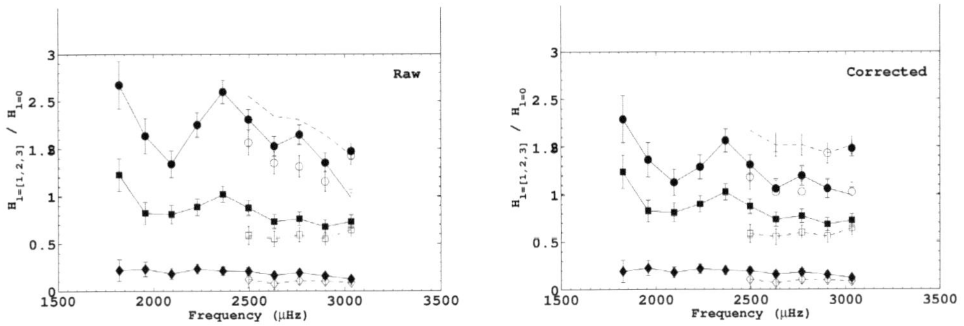

Figure 1. Raw (left) and corrected (right) mode visibilities of $l = 1$ (○), $l = 2$ (□), and $l = 3$ (◇) relative to $l = 0$ as a function of frequency in GOLF (——) and VIRGO (- - - -) observations.

onboard the *Solar and Heliospheric Observatory* (SoHO) spacecraft. GOLF (Gabriel et al. [8]) measures the Doppler velocity at different heights in the solar atmosphere depending on the wing – Blue or Red – of the sodium doublet – D1 and D2 – in which the observations were performed (García et al. [9]). VIRGO (Fröhlich et al. [10]) is composed of three Sun photometers (SPM) at 402 nm (blue), 500 nm (green) and 862 nm (red). A total of 5021 days of GOLF and VIRGO observations starting on 1996 April 11 and ending on 2010 January 8 were analyzed, with respective duty cycles of 95.4% and 94.7%. The power spectra of the time series were fitted to extract the mode parameters (Salabert et al. [11]) using a standard likelihood maximization function (power spectrum with a χ^2 with 2 d.o.f. statistics). Each mode component was parameterized using an asymmetric Lorentzian profile. Since SoHO observes the Sun equatorwards, only the $l + |m|$ even components are visible in Sun-as-a-star observations of GOLF and VIRGO. In order to obtain observational estimates of the m-amplitude ratios, the $m = \pm 2$ and $m = 0$ components of the $l = 2$ multiplet, and the $m = \pm 3$ and $m = \pm 1$ components of the $l = 3$ multiplet were fitted using independent amplitudes, assuming that components with opposite azimuthal order m have the same amplitudes ($H_{l,n,-m} = H_{l,n,+m}$). Note that the blue and red periods of GOLF were also analyzed separately, as well as the mean power spectrum of the three VIRGO SPMs.

3. Mode visibilities and m-amplitude ratios

The amplitude of a given multiplet (l, n) is defined as the sum of the amplitudes of its m-components, as $H_{l,n} = \sum_{m=-l}^{m=+l} H_{l,n,m}$. Then, the visibilities of the $l = 1$, 2, and 3 modes relative to the $l = 0$ mode are respectively defined as the ratios $H_{l=1,n}/H_{l=0,n}$, $H_{l=2,n-1}/H_{l=0,n}$, and $H_{l=3,n-1}$. The left panel of Fig. 1 shows these visibilities for both radial velocity (GOLF) and intensity (VIRGO) measurements as a function of frequency. These raw mode visibilities present a variation with frequency – especially for the $l = 1$ mode – that is due to the large variation of the mode amplitudes with frequency, even over half a large frequency separation. Thus, the visibilities are biased and in order to correct them we interpolated (using a spline interpolation) the amplitudes of the $l = 1$, 2, and 3 modes to the frequencies of the $l = 0$ mode (right panel of Fig. 1). Figure 2 shows the mode visibilities averaged over frequency for both GOLF and VIRGO observations as a function of l (see Table 1). The amplitude ratios between the m-components of the $l = 2$ and $l = 3$ multiplets, defined as $H_{l=2,m=0}/H_{l=2,m=\pm 2}$ and $H_{l=3,m=\pm 1}/H_{l=3,m=\pm 3}$ respectively, are represented on Fig. 3 in the case of the radial velocity GOLF measurements and are also reported in Table 2.

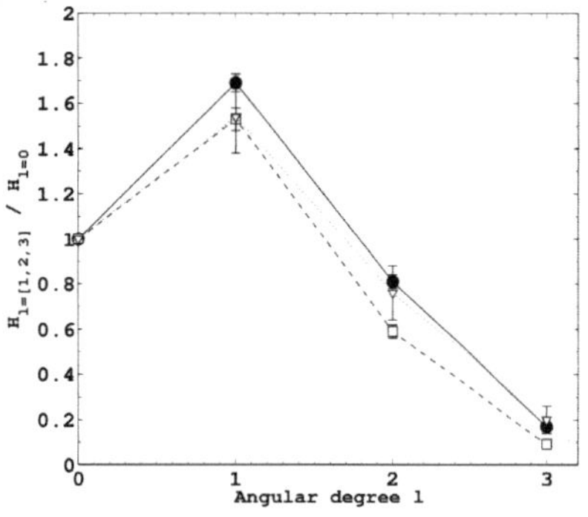

Figure 2. Mode visibilities as a function of angular degree l in GOLF (——, ●) and VIRGO (- - - -, □) measurements. For comparison, the mode visibilities of the CoRoT target HD49385 measured by Deheuvels et al. [12] are also represented (······, ▽).

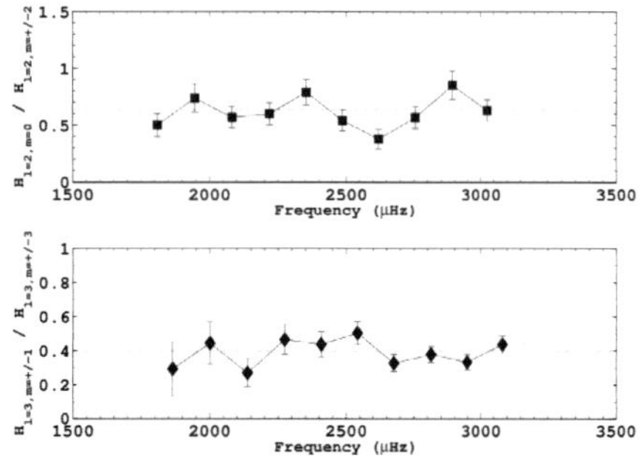

Figure 3. m-amplitude ratios of the $l = 2$ (top) and $l = 3$ (bottom) modes as a function of frequency in the GOLF measurements.

Table 1. Mode visibilities in radial velocity GOLF and intensity VIRGO measurements.

Mode visibility Radial velocity	GOLF	GOLF Blue wing	GOLF Red wing	
$H_{l=1}/H_{l=0}$	1.69±0.04	1.60±0.05	1.85±0.06	
$H_{l=2}/H_{l=0}$	0.81±0.03	0.74±0.04	0.98±0.05	
$H_{l=3}/H_{l=0}$	0.17±0.01	0.14±0.02	0.28±0.03	
Mode visibility Intensity	VIRGO	VIRGO Blue	VIRGO Green	VIRGO Red
$H_{l=1}/H_{l=0}$	1.53±0.05	1.55±0.05	1.52±0.05	1.39±0.05
$H_{l=2}/H_{l=0}$	0.59±0.03	0.63±0.03	0.57±0.03	0.42±0.03
$H_{l=3}/H_{l=0}$	0.09±0.02	0.10±0.02	0.09±0.02	0.05±0.02

Table 2. m-amplitude ratios in radial velocity GOLF and intensity VIRGO measurements.

m-amplitude ratio	GOLF	VIRGO
$H_{l=2,m=0}/H_{l=2,m=\pm2}$	0.63±0.03	0.75±0.06
$H_{l=3,m=\pm1}/H_{l=3,m=\pm3}$	0.40±0.02	0.63±0.06

Table 3. Modeled visibilities and m-amplitude ratios in intensity VIRGO and radial velocity GOLF measurements.

Mode visibility & m-amplitude ratio	VIRGO	GOLF Blue wing	GOLF Red wing
$H_{l=1}/H_{l=0}$	1.51	1.84	1.86
$H_{l=2}/H_{l=0}$	0.53	1.09	1.14
$H_{l=3}/H_{l=0}$	0.025	0.27	0.31
$H_{l=2,m=0}/H_{l=2,m=\pm 2}$	0.67	0.59	0.58
$H_{l=3,m=\pm 1}/H_{l=3,m=\pm 3}$	0.60	0.43	0.40

4. Models

When the contribution of a solar-disk element to the total flux depends only on its distance to the limb, the mode visibility and the m-amplitude ratio are decoupled (e.g. Gizon & Solanki [13]; Ballot et al. [14]). For VIRGO observations, this is verified since the contribution depends mainly on the limb-darkening. However, for GOLF, this is no more the case and we have performed complete computation taking into account the instrumental response, which differ for the blue and red wings. Results of these computations are listed in Table 3 . The limb-darkening law of Neckel & Labs [15] has been used. Indeed, visibility values for GOLF vary with frequency by a few percents due to the horizontal motions of modes that increase at low frequency. In general, these predictions agree with the observations. There is nevertheless some shortcomings: (i) even if the trend is correct, the difference between blue and red wings for GOLF is larger than expected; (ii) the visibility of the $l = 3$ modes in VIRGO are sensitively higher than expected. That could be explained by stronger effects of limb-darkening.

Acknowledgments

The authors want to thank Catherine Renaud and Antonio Jiménez for the calibration and preparation of the GOLF and VIRGO datasets. The GOLF and VIRGO instruments onboard SoHO are a cooperative effort of many individuals, to whom we are indebted. SoHO is a project of international collaboration between ESA and NASA. DS acknowledges the support of the grant PNAyA2007-62650 from the Spanish National Research Plan. This work has been partially supported by the CNES/GOLF grant at the SAp/CEA-Saclay.

References

[1] Salabert D, Fossat E, Gelly B, Kholikov S, Grec G, Lazrek M and Schmider F X 2004 *A&A* **413** 1135
[2] Appourchaux T et al. 2008 *A&A* **488** 705
[3] García R A et al. 2009 *A&A* **506** 41
[4] García R A, Mathur S, Salabert D, Ballot J, Régulo C, Metcalfe T and Baglin A 2010 *Science* **329** 1032
[5] Bedding T R et al. 2010 *ApJ* **713** L176
[6] Chaplin W J et al. 2010 *ApJ* **713** L169
[7] Fröhlich C et al. 1997 *Solar Physics* **170** 1
[8] Gabriel A H et al. 1995 *Solar Physics* **162** 61
[9] García R A et al. 2005 *A&A* **442** 385
[10] Fröhlich C et al. 1995 *Solar Physics* **162** 101
[11] Salabert D, Chaplin W J, Elsworth Y, New R and Verner G A 2007 *A&A* **463** 1181
[12] Deheuvels S et al. 2010 *A&A* **515** A87
[13] Gizon L and Solanki S K 2003 *ApJ* **589** 1009
[14] Ballot J, García R.A., Lambert P 2006 *MNRAS* **369** 1281
[15] Neckel H and Labs D 1994 *Solar Physics* **153** 91

Variation in the frequency separations with activity and impact on stellar parameter determination

O L Creevey[1,2], **D Salabert**[1,2], **R A García**[3]

[1] Instituto de Astrofísica de Canarias (IAC), E-38200 La Laguna, Tenerife, Spain
[2] Dept. de Astrofísica, Universidad de La Laguna (ULL), E-38206 La Laguna, Tenerife, Spain
[3] Laboratoire AIM, CEA/DSM-CNRS, Université Paris 7 Diderot, IRFU/SAp, Centre de Saclay, F-91191 Gif-sur-Yvette, France

E-mail: orlagh@iac.es, salabert@iac.es, rgarcia@cea.fr

Abstract. Frequency separations used to infer global properties of stars through asteroseismology can change depending on the strength and at what epoch of the stellar cycle the p-mode frequencies are measured. In the Sun these variations have been seen, even though the Sun is a low-activity star. In this paper, we discuss these variations and their impact on the determination of the stellar parameters (radius, mass and age) for the Sun. Using the data from maximum and minimum activity, we fitted an age for the Sun that differs on average by 0.2 Gyr: slightly older during minimum activity. The fitted radius is also lower by about 0.5% for the solar effective temperature during minimum.

1. Introduction

The p-mode oscillation frequencies vary with the activity cycle of the Sun [1] and other solar-like stars [2]. However, as observed in the Sun, these temporal variations present differences in amplitude, phase, and frequency among individual modes. Thus the frequency separations computed will be observed to vary with time. Comparing these frequency separations with those from stellar models will lead to differences in the fitted global parameters (radius, mass, age) of the star. Generally, when the frequencies are measured we do not know the phase of the activity cycle, unless the star can be measured continuously for long periods of time. We should then take into account when inferring the stellar parameters, that these values may depend on when the data was obtained.

2. Analysis of time-series observations and determination of frequencies

We analyzed 5202 days of velocity observations collected by the space-based instrument GOLF onboard *SoHO* spacecraft, covering more than 14 years between 1996 and 2010 with an overall duty cycle of 95.4% [3]. This dataset was split into non-independent, contiguous 365-day subseries with 91.25-day overlap and their associated power spectra fitted to extract the mode parameters [4] using a standard likelihood maximization function (power spectrum with a χ^2 with 2 d.o.f. statistic). The formal uncertainties in each parameter were then derived from the inverse Hessian matrix.

The individual frequency separations measured with 365-day time series were obtained with a precision of about 0.07 μHz. Figure 1 (left panel) shows the temporal variations of the individual

$l = 0$ and $l = 2$ mode frequencies averaged between 2200 and 3300 μHz (the reference values being taken as the average over 1996–1997). It is clear that the individual Sun-as-a-star p-mode frequencies have different temporal variations [1], which are consistent between radial velocity (GOLF) and intensity VIRGO observations [5]. The averaged large ($\Delta\nu$) and small ($\delta\nu$) frequency separations also show significant temporal variations with solar activity. For example, Figure 1 (right panel) shows that the small frequency separation $\delta\nu_{02}$ varies from peak-to-peak by about $0.2 \pm 0.02\mu$Hz over the solar cycle, which is presumably consistent with being signatures from surface effects. For a broader perspective, it is very important to note that the Sun is considered a low-activity star, and many solar-type stars could exhibit much larger variations with activity cycle.

3. Solar global parameters

3.1. *Fitting strategy*

In order to determine the global model parameters of the Sun P, we match the observational data O to the observables M from stellar models [6, 7] which are characterized by P. P comprises the mass M, age A, initial hydrogen (or helium) mass fraction X_0, initial heavy element mass fraction Z_0, and the mixing-length parameter α (from the standard mixing-length theory of [8]). The parameters that best describe the data are obtained by minimizing a χ^2 function:

$$\chi^2 = \sum_{i=1}^{M} \left(\frac{O_i - M_i}{\sigma_i} \right)^2, \qquad (1)$$

where there are $i = 1, 2, ..., M$ independent observations and σ is the observational error. We use the Levenberg-Marquardt algorithm (LM) to minimize Eq. 1 [9, 10, 11, 12, 13].

The global objective of this work is to investigate the effect of the shifted frequency values on the determination of P for any star with solar-like oscillations. So we must define a coherent and consistent method to fit the observational data to obtain the set of best-fitting parameters not only for the Sun, but for other stars too. Our strategy is to first fit the non-seismic and the average seismic quantities to obtain P to a reasonable range, and then using P as initial guesses, proceed to use the individual frequency separations (large and small) to obtain the final set of best-fitting parameters.

LM is a local minimization method and can therefore be sensitive to initial conditions. To ensure that we obtain a global minimum we search for the best parameters by initializing the minimization with different parameter values. We set the initial M as 1.0 $M_\odot$, and the initial A varies between 4 and 6 Gyr in steps of 0.5 Gyr. Additionally, we hold the parameter X_0 fixed during each minimization, and instead repeat the process for three values of X_0: 0.69, 0.71, 0.73. This gives in total 15 minimizations using the input data at maximum activity and again at minimum activity.

3.2. *Results*

From the 15 best-fitted P, we selected those sets where $\chi^2 \leq 29.1$ (4 fitted parameters, 19 data points to give 14 degrees of freedom at the 1% confidence level) and where the fitted $T_{\rm eff}$ fell to within 1σ of 5777 K ($T_{\rm eff}$ was not a constraint in the second part of the fitting strategy). These sets are represented in Fig. 2, where we show the values of age versus mass on the left panel, and radius versus effective temperature on the right. The dark/lighter filled circles are the results for fitting during maximum/minimum activity.

Fig. 2 left panel shows an offset in the fitted P between minimum and maximum activity. To quantify this offset, we fitted a linear function to each group of results. The difference between the fitted ages at opposite activity phases is 0.2 Gyr. Alternatively if we take 4.8 Gyr as the solar age, the mass is fitted as a 1.0 $M_\odot$ star at minimum, and 1.03 $M_\odot$ at maximum.

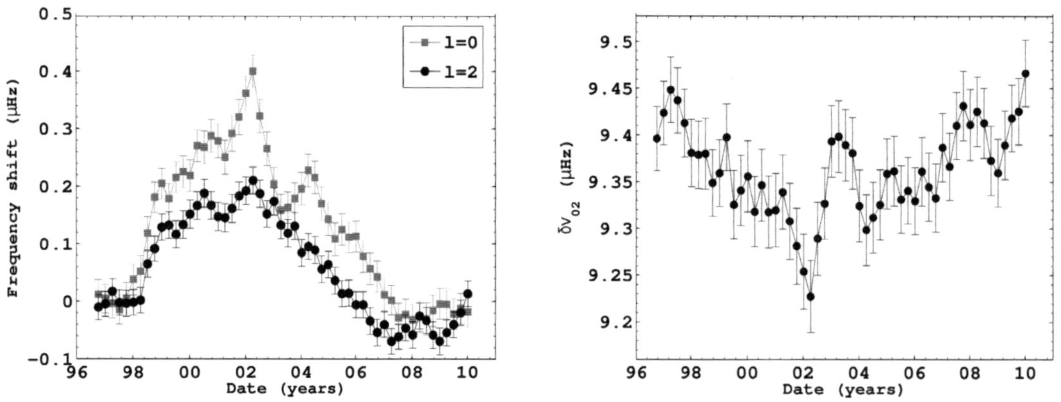

Figure 1. Left: Frequency shifts (μHz) of the $l = 0$ (●) and $l = 2$ (■) modes measured from GOLF data. Right: Temporal variations of the averaged small frequency separation $\delta\nu_{02}$ (μHz).

The right panel of Fig. 2 shows the fitted R and $T_{\rm eff}$ for these same models (neither were input constraints). We find that at minimum activity, the Sun's $T_{\rm eff}$ is about 50 K cooler for a given R, or about 0.5% smaller in R for a given $T_{\rm eff}$.

4. Conclusions

Measuring the oscillation frequencies during different phases of a stellar activity cycle will lead to different values of the individual frequencies. Thus, estimates of the large and small frequency separations will be different depending on the observing period. Using 365-day subseries of GOLF data, we determined the individual frequency separations to about 0.07 μHz.

We used stellar models to determine the best-fitting parameters using the observations taken at minimum and maximum activity. We found that the age of the star is on average 0.2 Gyr older using the values from minimum activity, or for a given age, the mass is about 2% larger using the data at maximum activity. We also found a small discrepancy in the fitted radius and effective temperature. At minimum activity, the Sun is about 50 K cooler for a given fitted radius, or the radius is about 0.5% smaller for a given effective temperature.

Although we still need to study whether these differences in fitted values are detectable given the expected uncertainties, we note that [14] quote an error in the age of the planet-hosting star HAT-P-7 of 0.26 Gyr, of the order of this detected change. The Sun, however, is considered a low-activity star, and much stronger activity cycles on other solar-type stars have already been detected [2]. With long time-series data such as those from CoRoT or Kepler, we will not only detect stellar activity cycles, but we will obtain very high precision on the seismic data, that the uncertainties in the fitted parameters will be smaller than the fitted changes in these parameters.

Acknowledgments

The authors want to thank Catherine Renaud for the calibration and preparation of the GOLF dataset. The GOLF instrument onboard SoHO is a cooperative effort of many individuals, to whom we are indebted. SoHO is a project of international collaboration between ESA and NASA.This research was in part supported by the European Helio- and Asteroseismology Network (HELAS), a major international collaboration funded by the European Commission's Sixth Framework Programme. DS acknowledges the support of the grant PNAyA2007-62650 from the Spanish National Research Plan. This work has been partially supported by the CNES/GOLF grant at the SAp CEA-Saclay.

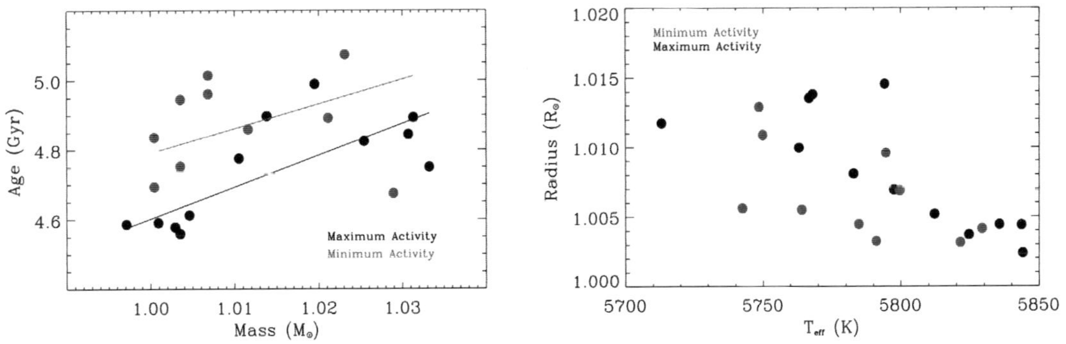

Figure 2. Left: Fitted age versus mass using the individual large and small frequency separations calculated at maximum (black) and minimum (grey) activity. Each point corresponds to a minimization. The lines are linear fits to the results. Right: The fitted radius and effective temperature for the same minimizations.

References

[1] Salabert D, García R A, Pallé P L and Jiménez-Reyes S J 2009 *A&A* **504** L1
[2] García R A et al. 2010 *Science* **329** 1032
[3] García R A et al. 2005 *A&A* **442** 385
[4] Salabert D, Chaplin W J, Elsworth Y, New R and Verner G A 2007 *A&A* **463** 1181
[5] Salabert D, García R A, Pallé P L and Jiménez A 2010 *J. Phys.: Conf. Series* These proceedings
[6] Christensen-Dalsgaard J 2008a *Ap&SS* **316** 113
[7] Christensen-Dalsgaard J 2008b *Ap&SS* **316** 13
[8] Böhm-Vitense E 1958 *ZA* **46** 108B
[9] Creevey et al. 2007 *ApJ* **659** 616
[10] Metcalfe T, Creevey O L and Christensen-Dalsgaard J 2009 ApJ **699** 373
[11] Stello et al. 2009 *ApJ* **700** 1589
[12] Chaplin et al. 2010 *ApJL* **713** 169
[13] Mathur et al. 2010 *A&A* **511** 46
[14] Christensen-Dalsgaard J et al. 2010 *ApJ* **713** 164

Angular-degree dependence of p-mode frequencies during solar cycle 23

S. C. Tripathy[1], K. Jain[1], D. Salabert[2,3], R. A. Garcia[4], F. Hill[1] and J. W. Leibacher[1]

[1] National Solar Observatory, Tucson, AZ 85719, USA
[2] Instituto de Astrofísica de Canarias, E-38200 La Laguna, Tenerife, Spain
[3] Departmento de Astrofísica, Universidad de La Laguna, E-38205 La Laguna, Tenerife, Spain
[4] Laboratoire AIM, CEA/DSM-CNRS, Université Paris 7 Diderot, IRFU/SAp, Centre de Saclay, F-91191 Gif-sur-Yvette, France

E-mail: stripathy@nso.edu

Abstract. We analyze simultaneous helioseismic observations collected by the ground- and space-based instruments during solar cycle 23 by computing oscillation frequencies for low- and intermediate-degree p-modes on a time scale of 36 days. We find that the frequency shifts corresponding to different angular degree, ℓ, indicate different epochs for the onset of the solar cycle 24. The analysis also indicates the presence of double minima between cycles 23 and 24 for some range of ℓ values.

1. Introduction

The frequencies of solar oscillations vary with the solar activity cycle and the variations are used to infer the solar-cycle related changes inside the Sun. The extended minimum between cycles 23 and 24 has proved to be rather unusual. For the first time, the global mode frequency measurements showed an anti-correlation with the activity proxies during the extended minimum [1]. Further, oscillation modes with low angular degrees, particularly $\ell = 0$ and $\ell = 2$, suggest the onset of the cycle 24 occured in the last quarter of 2007 [1, 2] while $\ell = 1$ modes from Global Oscillations at Low frequency (GOLF) and Birmingham Solar Oscillations Network (BISON), and intermediate degree modes from Global Oscillation Network Group (GONG) and Michelson Doppler Imager (MDI) onboard Solar and Heliospheric Observatory (SOHO), show the onset to be in early 2009 [3]. Therefore, it appears that the onset of the new cycle varies with the mode degree.

Here we present an updated analysis of the GOLF and GONG observations by computing oscillation frequencies with a cadence of 36 days with a 18-day overlap. The GOLF power spectrum, starting on 1996 May 1 and ending on 2010 January 1, is fitted with a Lorentzian profile. The series with a filling factor less than 70% were not used in this analysis resulting in a total of 269 time series for the GOLF data. The details of the fitting procedure can be found in [5]. The GONG data covers the period from 1996 May 1 to 2009 October 31 and consists of 273 overlapping data sets. The intermediate degree GONG frequencies were calculated using the standard GONG pipeline where the individual resonant peaks in the multi-tapered power spectrum are fitted with a Lorentzian profile [4]. For comparison and validating, we also use

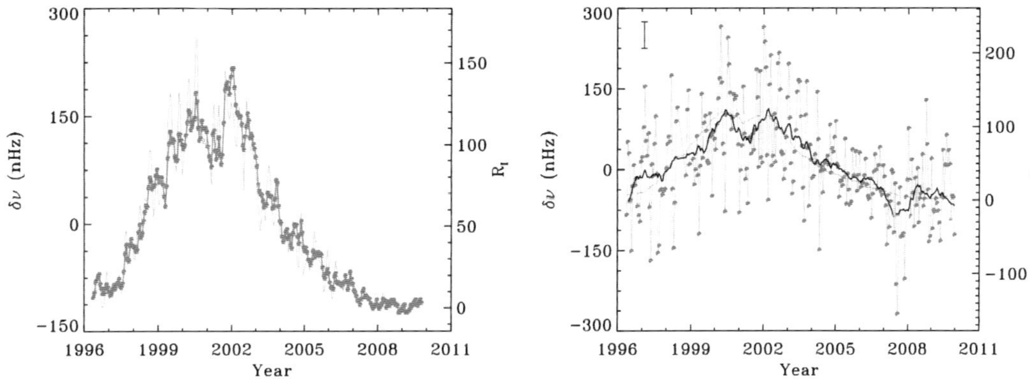

Figure 1. Temporal evolution of the frequency shifts (symbols) calculated from GONG (left panel) and GOLF (right panel) 36-day time series. The red/thin line in each panel represents the International sunspot number, R_I. The black/thick line in the right panel depicts the smoothed frequency shifts obtained from the boxcar average over 21 points. The vertical line on the right panel indicate the error bar associated with the GOLF frequency measurements.

the integrated GONG time series to compute the low degree modes ℓ = 0, 1, and 2 [6]. The temporal variations of the p-mode frequencies, i.e. the frequency shifts ($\delta\nu$), were calculated by comparing each fitted frequency with a reference frequency in the frequency range of ν = 2000 and 3300 μHz.

2. Results

2.1. Variation over solar cycle

The temporal evolution of the frequency shifts over the complete solar cycle 23 for both GONG and GOLF is shown in Figure 1. The GONG shifts are calculated from the common modes between ℓ = 0 and 100 and comprise about 160 modes that are present in all time samples. The symbols when joined by a line (to guide the eye) clearly demonstrate the variation with the solar activity cycle shown by the red/thin line representing the International sunspot number, R_I. In comparison, the GOLF shifts which are calculated from ℓ = 0 – 2 modes appear to be noisier due to averaging over a fewer modes as compared to the GONG. However the solar cycle pattern is clearly visible when the points are smoothed by taking a boxcar average over 21 points (black/thick line).

3. Extended Minimum Phase

3.1. Sun-as-a-star observation

In order to investigate the extended solar minimum phase, we present the frequency shifts between 2007 and 2010 corresponding to individual low degree modes in Figure 2. Modes of different degree suggest different onsets of the cycle 24. However, as discussed earlier, the frequency shifts appear to be noisy and within the error estimates, so we compute a boxcar average over 21 points (shown by thick solid line). This curve confirms the result of [1], where ℓ = 0 and 2 modes show the onset to be late 2007 while ℓ = 1 shows the onset to be second quarter of 2009, generally agreeing with the observations of the surface activity indicators. Similar results are also inferred from the integrated GONG data (lower panels), ℓ of 0 and 2 shows the onset to be earlier than the ℓ = 1 mode. As described in [1, 3], the zonal and sectoral modes are sensitive

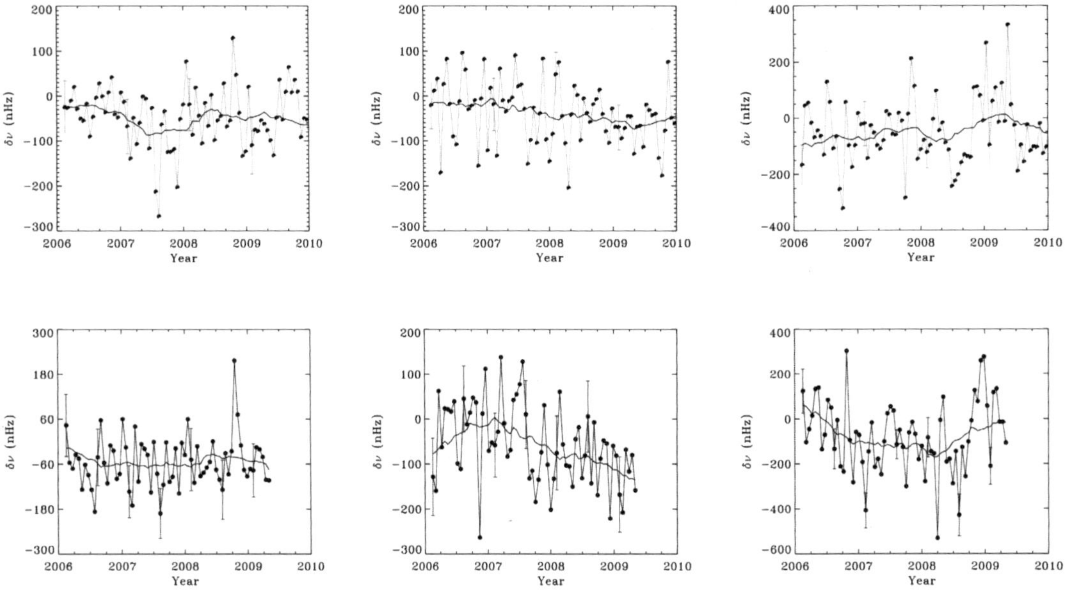

Figure 2. Temporal evolution of the frequency shifts corresponding to (left) $\ell = 0$, (middle) $\ell = 1$, and (right) $\ell = 2$ modes obtained from GOLF (upper panels) and integrated GONG (lower panels) 36 day time series. The heavy solid line is the smoothed shifts obtained from a boxcar average over 21 points.

to the different latitudes, the modes that are sensitive to high latitudes of the Sun indicate an early onset of the cycle.

3.2. Resolved observations

We also analyze the frequency shifts as a function of the degree of the mode using the intermediate-degree mode frequencies calculated from the standard 36 day GONG time series. These shifts for different ℓ ranges are presented in Figure 3. Panels (a-b) suggest a minimum around early 2008, panel (c) shows the presence of double minima, one in late 2007 and the other around early 2009, while other panels show a minimum around early 2009.

4. Conclusions

We have analyzed integrated helioseismic observations collected by the space-based GOLF and ground based GONG instruments and investigated the angular degree dependence of the *p*-mode frequencies calculated on a time-scale of 36 days. We find that the temporal variations associated with the individual low degree modes are noisy while a running mean over 21 points equivalent to frequencies calculated from a period of 378 days confirm the findings of [1]; $\ell = 0$ and $\ell = 2$ mode frequencies show an upturn from the end of 2007 while the variation of $\ell = 1$ mode frequencies agree with those of the activity proxies. The frequencies obtained from the resolved GONG observations suggest that the onset of the new cycle is a function of the angular degree of the mode. We finally note that the presence of a 2-year periodicity in the Sun-as-a-star frequency shifts [7, 8] may affect the conclusions about the onset period of the new solar cycle.

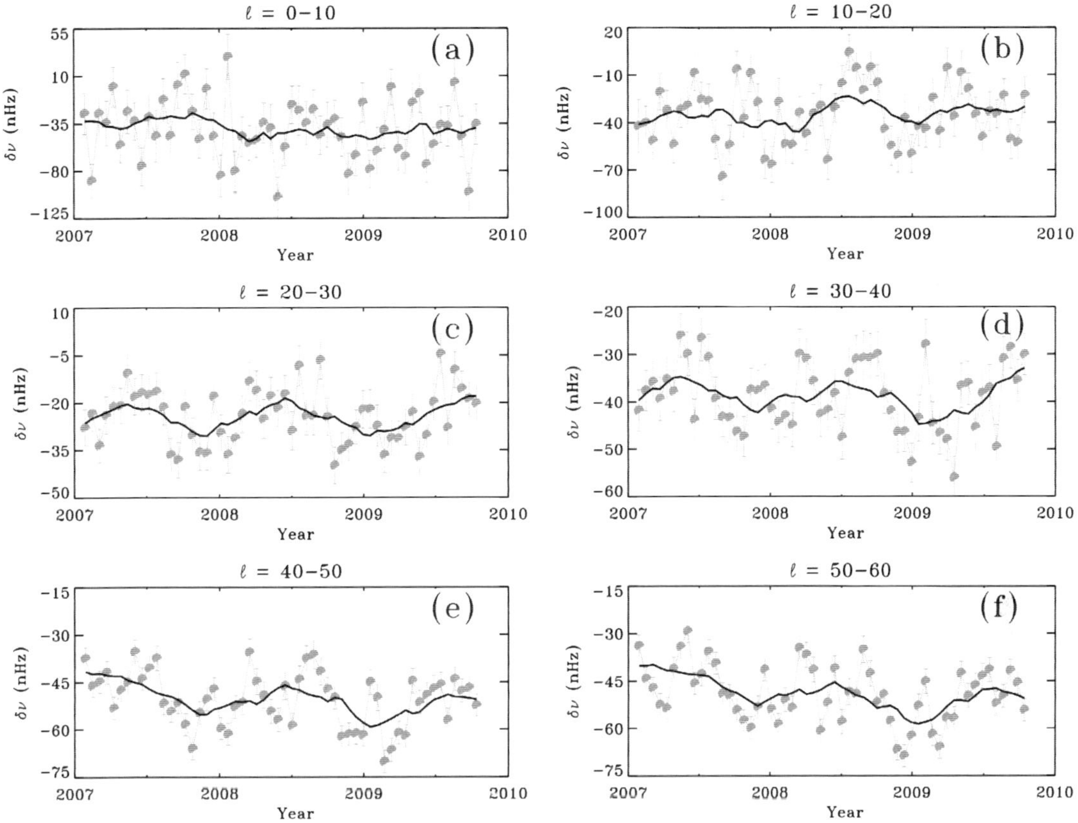

Figure 3. Temporal evolution of the GONG frequency shifts for different ℓ ranges as marked on the top of each panel. The position of seismic minimum is different in different plots.

Acknowledgments

This work utilizes data obtained by the Global Oscillation Network Group (GONG) project, managed by the National Solar Observatory, which is operated by AURA, Inc. under a cooperative agreement with the National Science Foundation. The data were acquired by instruments operated by the Big Bear Solar Observatory, High Altitude Observatory, Learmonth Solar Observatory, Udaipur Solar Observatory, Instituto de Astrofísico de Canarias, and Cerro Tololo Interamerican Observatory. The GOLF instrument onboard SOHO is a cooperative effort of many individuals. SOHO is a mission of international cooperation between ESA and NASA. This work was partially supported by NASA grant NNG 08EI54I to National Solar Observatory and CNES/GOLF grant at the SAp/CEA-Saclay.

References

[1] Salabert D, Garcia R A, Palle P L and Jimenez-Reyes S J 2009 *A&A* **504** L1-4

[2] Fletcher S T, New R, Broomhall A-M, Chaplin W J and Elsworth Y 2009 *ASP Conf. Ser.* **428** 43-49

[3] Tripathy S C, Jain K, Hill F and Leibacher J W 2010 *Astrophys. J.* **711** L84-8

[4] Tripathy S C, Jain K, Hill F and Leibacher J W 2007 *Solar Phys.* **243** 105-20

[5] Garcia R A, Turck-Chieze S and Boumier P et al. 2005 *A&A* **442** 385-95

[6] Salabert D, Leibacher J W, Appourchaux T and Hill F 2009 *Astrophys. J.* **696** 653-67

[7] Broomhall A-M, Chaplin W J, Elsworth Y, Fletcher S T and New R 2009 *Astrophys. J.* **700** L162-65

[8] Fletcher S T, Broomhall A-M, Salabert D, Basu S, Chaplin W J, Elsworth Y, Garcia R and New R 2010 *Astrophys. J.* **718** L19-23

Effect of line-of-sight inclinations on the observation of solar activity cycle: Lessons for CoRoT & Kepler

H. Vázquez Ramió[1,2], S. Mathur[3], C Régulo[1,2], R A García[4]

[1] Instituto de Astrofísica de Canarias (IAC), E-38200 La Laguna, Tenerife, Spain

[2] Dept. de Astrofísica, Universidad de La Laguna (ULL), E-38206 La Laguna, Tenerife, Spain

[3] High Altitude Observatory, 3080 Center Green Drive, Boulder, CO, 80302, USA

[4] Laboratoire AIM, CEA/DSM-CNRS-Université Paris Diderot; CEA, IRFU, SAp, F-91191, Gif-sur-Yvette, France

E-mail: `hvr@iac.es, savita@ucar.edu, crr@iac.es, rgarcia@cea.fr`

Abstract. CoRoT and Kepler missions are collecting data of solar-like oscillating stars of unprecedented quality. Moreover, thanks to the length of the time series, we are able to study their seismic variability. In this work we use numerical simulations based on the last 3 solar cycles to analyze the light curves as a function of the line-of-sight inclination angle. These preliminary results showed that the direct observation of the light curve can induce some bias in the position of the maximum of the cycle.

1. Introduction

The structure of the stars can be perturbed by their magnetic fields. The interaction of convection, rotation and magnetic fields induces the development of dynamos that can be observed as cyclic modulations in several parameters. Several surveys done on magnetically sensitive absorption lines (Ca H and K) have shown a high number of stars in our vicinity showing such cycles [e.g. 1]. In the case of the Sun, the cyclic variation of the sunspots are well known, as well as for many other parameters. In particular, frequencies and amplitudes of the acoustic modes propagating inside the Sun react to the magnetic field and show an 11 year modulation [e.g. 2]. Long and high-quality asteroseismic data are now being collected on solar-like stars thanks to the new satellites such as CoRoT and Kepler [e.g. 3, 4, 5, 6, 7, 8] or the ground-based campaigns [e.g. 9]. The recent discovery of an activity cycle in HD49933 [10] –using seismic analysis of the two observed runs performed by CoRoT [11, 12]– opens such studies to asteroseismology. However, a shift in the position of the maximum of the cycle was found between the inferences obtained from seismology indicators and from the light curve.

In the present work we will check the effect of the inclination angle of the star on the light curve as we know that it has an important effect on other parameters like the visibility of the modes [13, 14]. The visibility of the spots is utterly related to the active latitudes of the star and to the spots migration process. Thus, at a specific moment of observation, depending on the inclination angle, we can see or not these spots on the visible disk.

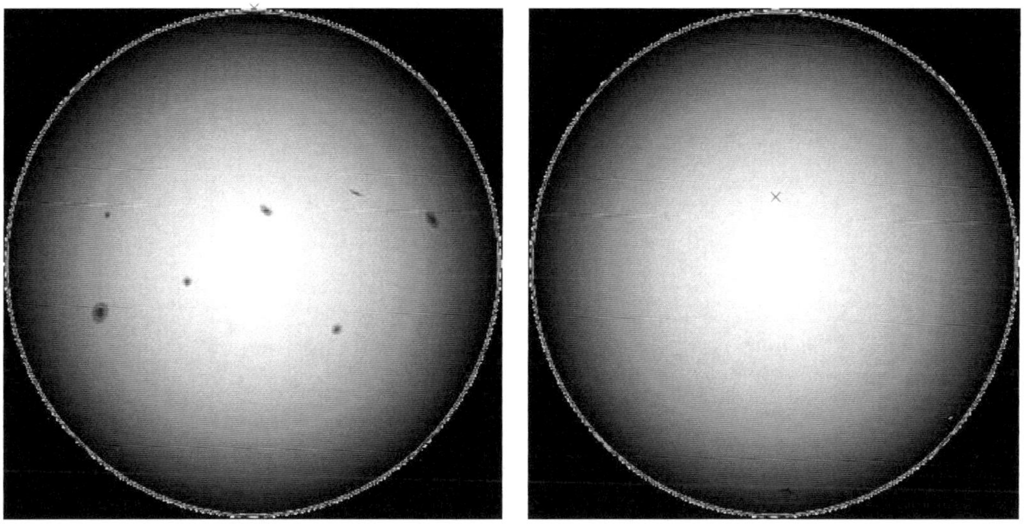

Figure 1. Examples of the numerical simulation of the same day with an inclination angle of 90 and 15 degrees, left and right panels respectively. The solar pole is indicated by a cross in each image.

2. Numerical Simulation

To explore the influence of the line-of-sight inclination angle coupled with the active latitudes of the star during the evolution of the cycle, we perform a numerical simulation of a Sun in which we change the inclination angle from 90° (the real Sun) till 0° by increments of 5°. The model was designed to simulate different solar photospheric patterns such as granulation, supergranulation and sunspots [15]. The spots areas and their positions on the solar disk have been taken from the Solar Geophysical Data Centre (*http://www.ngdc.noaa.gov/*). Every sunspot is considered to be circular with an area equivalent to the total of the reported sunspot group. Data from 1980 to 2004 were employed, covering two minima and three maxima of the solar activity cycle. Some samples of the numerical simulation are shown in Fig.1: from left to right angles 90° and 15°.

3. Analyses and Results

To analyze the light curves we use two methods: the first one consists of applying wavelet tools [16] to obtain the evolution of the sunspots with time [17]. We also build a sunspots proxy by computing the standard deviation of the light curves in subseries of 365 days, 50% overlapping (following [10]).

3.1. Observations at 90°: The Sun

In Figure 2 (Left) we can see the synthetic light curve (top) based on the last three solar cycles observed with an observation line-of-sight inclination of 90 degrees. The global wavelet power spectrum (middle panel) presents clear signatures of the 11-year activity cycle. A surface rotation is confirmed by doing a collapsogram in the temporal direction (middle right panel) obtaining a periodicity of 25 days. The second harmonic of the rotation rate is also visible between 0.90 and 0.95 μHz (between 12 and 13 days). In both cases the peak is above the limit of 90% confidence level. The collapsogram in the frequency direction –corresponding to periods in a range between 20 and 30 days (bottom panel)– shows a similar behavior than the light curve, which means that this is another way to compute a starspots proxy.

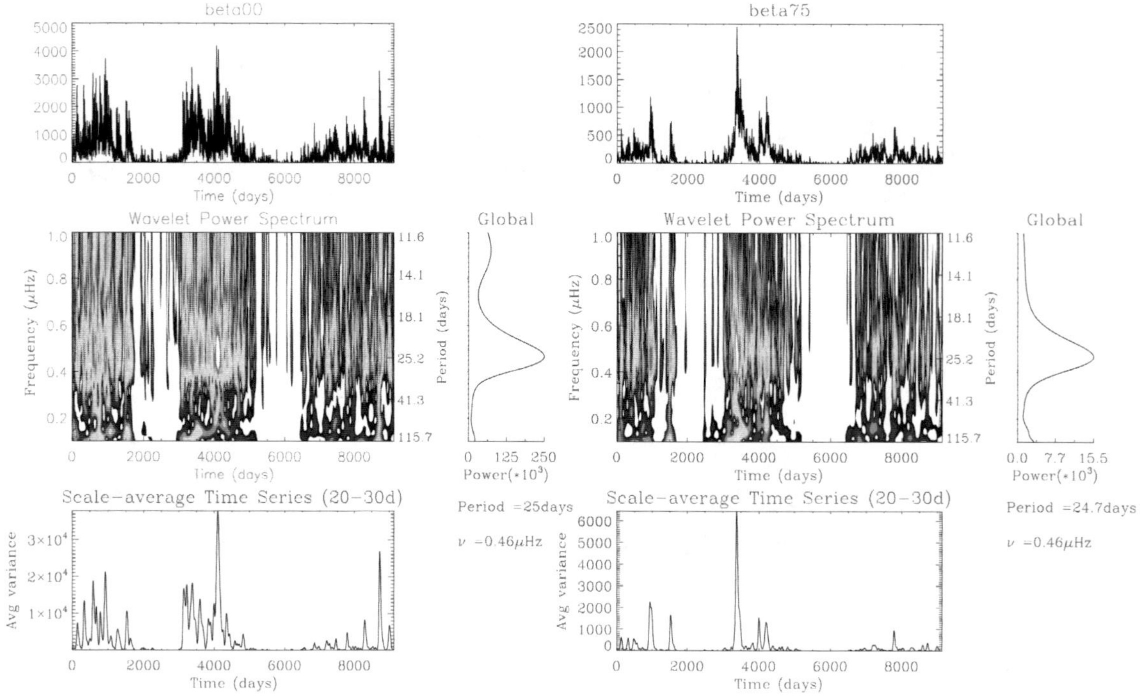

Figure 2. Analysis of the light curves from simulations with inclination angles 90 and 15 degrees (left and right respectively) using wavelet techniques. In the top panels the absolute value of the light curves obtained with both inclinations are shown in ppm. The middle panels are the global wavelet power spectrum. The bottom panels are the collapsograms of the wavelet power spectrum along in the vertical axis. The right small panels are the collapsograms in the time domain.

The magnetic activity proxy (Fig. 3) –built from the standard deviation of the light curve– is similar to the collapsogram of the vertical axis of the wavelets periodogram (see Fig. 2 bottom panels) and shows clearly the presence of the maximum activity as well as the periods of minima. The ascending and descending phases of the cycle are well determined thanks to the high signal-to-noise ratio.

3.2. Observations at 15°: similar to HD49933

When the angle is small –as in the case of HD49933 (estimated by [12] to be $17^{o}{}^{+7}_{-9}$)– the shape of the activity cycle from the analyses of the light curve is different. The inferred rotation periodicity is still correct but there is no signature of the second harmonic of the rotation. Moreover, the maximum of the activity cycles is shifted compared to the observations at 90 degrees (See Fig 2 and 3 right panels) and is lower in amplitude than in the previous case. Finally, the ascending and descending phases are not so well defined.

4. Conclusions

In the present work we have developed a numerical simulation of the variations induced in the light curves due to starspots crossing the visible disk of a star. To build such a simulation we have used real solar data measured after more than 30 years of solar observations. We have shown that the activity cycle inferred from the light curve is heavily influenced by the inclination

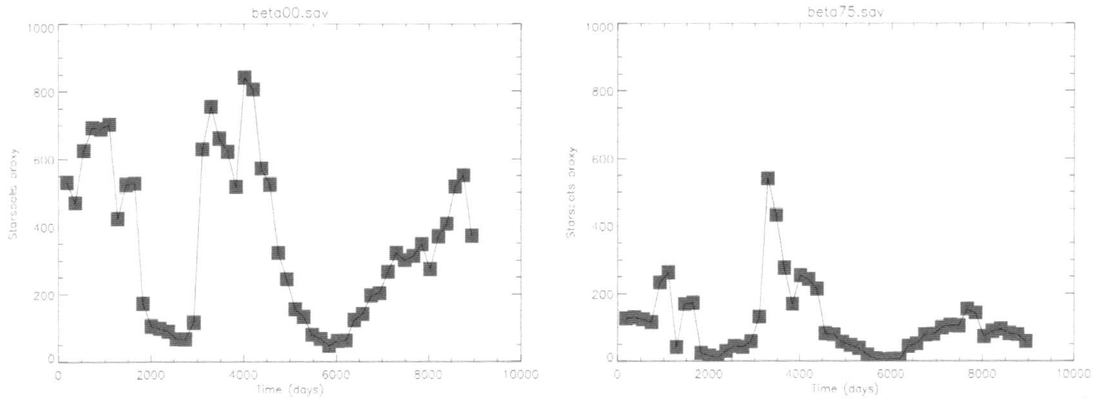

Figure 3. Starspot proxy of the simulation at 90 (left) and 15 degrees (right).

angle under which the star is observed. Beside, the maximum of the cycle changes as well as the ascending and descending phases of the cycle.

Acknowledgments

The CoRoT (Convection, Rotation and planetary Transits) space mission has been developed and is operated by CNES, with the contribution of Austria, Belgium, Brazil, ESA (RSSD and Science Program), Germany and Spain. This work has been partially supported by the CNES/GOLF grant at the SAp CEA-Saclay and by the French PNPS program.

References

[1] Baliunas S L, Donahue R A, Soon W H et al. 1995 *ApJ* **438** 269
[2] Charbonneau P 2005 *Living Reviews in Solar Physics* **2**
[3] Barban C, Deheuvels S, Baudin F et al. 2009 *A&A* **506** 51
[4] García R A, Régulo C, Samadi R et al. 2009 *A&A* **506** 41
[5] Bedding T R, Huber D, Stello D et al. 2010 *ApJ* **713** L176
[6] Chaplin W J, Appourchaux T, Elsworth Y et al. 2010 *ApJ* **713** L169
[7] Deheuvels S, Brunt H, Michel E et al. 2010 *A&A* **515** A87
[8] Mathur S, García R A, Catala C et al. 2010 *Astron. & Astroph.* **518** A53
[9] Arentoft T, Kjeldsen H, Bedding T R et al. 2008 *ApJ* **687** 1180
[10] García R A, Mathur S, Salabert D, et al. 2010 *Science* **329** 1032
[11] Appourchaux T, Michel E, Auvergne M. et al. 2008 *A&A* **488** 705
[12] Benomar O, Baudin F, Campante T L et al. 2009, *A&A* **507** L13
[13] Gizon L and Solanki S K 2003 *ApJ* **589** 1009
[14] Ballot J, García R A and Lambert P 2006 *A&A* **369** 1281
[15] Vázquez Ramió H, Régulo C and Roca Cortés T 2006 *Proceedings ESA-SP* **624** 69
[16] Torrence C and Compo G P 1998 *Bull. of the Amer. Met. Soc.* **79** 61
[17] Mathur S, García R A, Régulo C et al. 2010 *Astron. & Astroph.* **511** 46

Localized averaging kernels for probing the deep meridional flow with data from GONG, MDI and HMI

H.-P. Doerr and M. Roth

Kiepenheuer-Institut für Sonnenphysik, Schöneckstraße 6, 79104 Freiburg, Germany

E-mail: `doerr@kis.uni-freiburg.de`

Abstract. The Fourier-Legendre decomposition (FLD) method to study wave absorption in sunspots is a helioseismic technique which is also suited for the measurement of the meridional flow. Because the FLD method is sensible to low-degree oscillation modes, it bears the potential to be used to probe the average meridional flow in much deeper layers as it is currently possible with other methods. In order to rate the suitability of the available data for inversions of the deep flow, we compare localized averaging kernels as computed with data from the Global Oscillation Network Group (GONG), the Michelson Doppler Imager (MDI) aboard the SOHO spacecraft and also with preliminary data from the Helioseismic and Magnetic Imager (HMI) aboard the SDO spacecraft.

1. Introduction

The solar meridional circulation has a mainly poleward directed surface flow of about 15 m/s (eg. [1–3]) and plays a key role in the theoretical description of the dynamics of the differential rotation and the overall solar convection zone [4–6]. Moreover the motion is important for some type of solar dynamo models, where the revolution time-scale of the motion might define the length of the the solar cycle [7–10]. In contrast to the surface measurements, knowledge about the structure of the meridional flow in the solar interior is marginal at best. Local helioseismology techniques as ring-diagram analysis and time-distance helioseismology have investigated the flow down to a depth of 15–20 Mm below the solar surface [11–13]. In these layers the measurements show a mainly poleward, highly variable flow with amplitudes up to 40 $\mathrm{ms^{-1}}$.

The Fourier-Legendre decomposition (FLD) originally developed by [14] for studying wave absorption by sunspots can also find application for determining the subsurface structure of the meridional flow [15]. First comparisons of the meridional flow measurements obtained by ring-diagram analysis and the FLD [16] in the solar subsurface layers seem to be in qualitatively good agreement. In order to decide whether the method is also capable to probe layers below 20 Mm depth we study in this contribution the localization of inversion kernels as they are obtained from a SOLA (Subtractive Optimally Localized Averages) inversion method. The input data that forms the basis of our analysis were recorded by the instruments of GONG (Global Oscillation Network Group), SOHO/MDI (Michelson Doppler Imager aboard the Solar and Heliospheric Observatory; [17]) and SDO/HMI (Helioseismic and Magnetic Imager aboard the Solar Dynamics Observatory).

2. Methods

The time dependent, two-dimensional oscillation signal

$$\delta V(\theta,\, \phi,\, t) = \sum_{l m \nu} \left[A_{lm\nu} X_l^m(\theta) + B_{lm\nu} \left(X_l^m\right)^* (\theta) \right] \mathrm{e}^{i(m\phi + 2\pi\nu t)}, \tag{1}$$

on the solar surface can be decomposed into the mode amplitudes $A_{lm\nu}$ and $B_{lm\nu}$ of two wavefields traveling in poleward and equatorward direction respectively. Here θ denotes the co-latitude and ϕ the longitude on the solar surface, ν is the temporal frequency, l the harmonic degree and m the azimuthal order. The basis functions X_l^m and it's complex conjugate X_l^{m*} are superpositions of the associated Legendre functions P_l^m and Q_l^m:

$$X_l^m(\theta) = N_l^m \left[P_l^m(\cos\theta) - \frac{2i}{\pi} Q_l^m(\cos\theta) \right], \tag{2}$$

where N_l^m is a normalization factor. A meridional flow will result in a slight frequency shift $\delta\nu$ between the poleward and equatorward components. This frequency shift can be measured by fitting Lorentzian profiles to the single peaks in the power spectra obtained for each solar hemisphere according to [18]. The guess frequencies for the fits are obtained from the standard solar 'Model S' [19]. According to [15, 20] this frequency shift is a result of the advection due to the meridional flow

$$\Delta\nu_{nl} = \frac{l}{\pi R_\odot} \int \bar{U}_{\mathrm{mer}}(r) K_{nl}(r)\, \mathrm{d}r\ , \tag{3}$$

where $\bar{U}_{\mathrm{mer}}(r)$ is the averaged meridional flow over the observed region of interest as a function of the position r inside the Sun. The sensitivity kernel $K_{nl}(r)$ is the energy density of a given mode which is also a function of the position r inside the Sun. We then employ the SOLA inversion techniqe [21] to construct localized average inversions kernels at a given depth r.

2.1. Data

We use the GONG merged full-disk Dopplergrams from January 2006. There are no GONG Dopplergrams available for 2010 yet and the GONG duty cycle was comparably high for this month (40704 out of 44640 Dopplergrams available). This data set also was used previously for FLD-based flow measurements [16]. The GONG Dopplergrams have a resolution of 1 k×1 k and the cadence is 60 seconds.

In the case of MDI we use the level 1.8 full-disk Dopplergrams from May 2010. Up to now this is the best data set from MDI that overlaps with HMI, however it has not a particularly high duty cylce and only 31175 out of 44640 Dopplergrams are available. The MDI Dopplergrams have a resolution of 1 k×1 k and the cadence is 60 seconds also.

The final calibration of the HMI data products is not yet published so we use the preliminary data of May 2010 from the `hmi_test.v_45s` dataseries [22]. The HMI full-disk Dopplergrams come with a 4 k×4 k resolution at a cadence of only 45 seconds. At this point we must emphasize that the preliminary status of the HMI data series does not allow us to draw any conclusions about the quality of the final calibration.

Data is processed in chunks of one hour. To end up with only the net surface oscillation signal the mean Dopplergram of each chunk is removed from each Dopplergram in the chunk. The Dopplergrams are then interpolated to an equidistant heliographic grid and the region of interest is cut out. In order to increase the SNR for low-degree modes we use a rather large region of interest of w×h = 112×56 degrees centered on the central meridian and solar latitudes of ±38 degrees for the northern and southern hemispheres.

The Dopplergrams prepared in such a way are then feed into the Fourier-Legendre decompositon procedure as described above. The result of the Fourier-Legendre decomposition are time series of the coefficients $A_{lm\nu}$ and $B_{lm\nu}$ from Equation (1) for both hemispheres. From these time series of the equatorward and poleward traveling waves the power spectra are computed and averaged for modes with identical harmonic degree l and azimuthal order of $m = -25, \ldots, 25$.

3. Results

The theoretical positions of the power-ridges and the frequencies fitted on the three data sets are shown in Figure 1. Only fits with an error of $\leq 1.5\,\mu\mathrm{Hz}$ are regarded. For GONG we obtain 4067 successful frequency fits, for MDI 3746, and for HMI 4707.

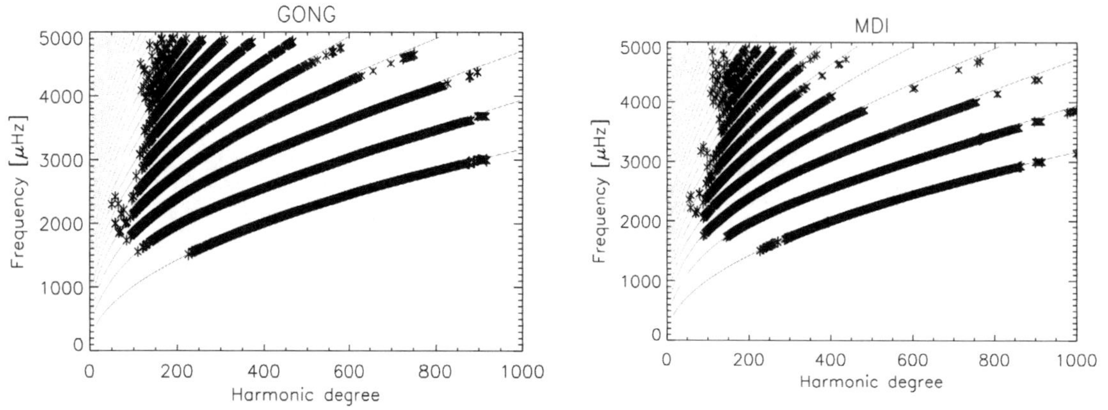

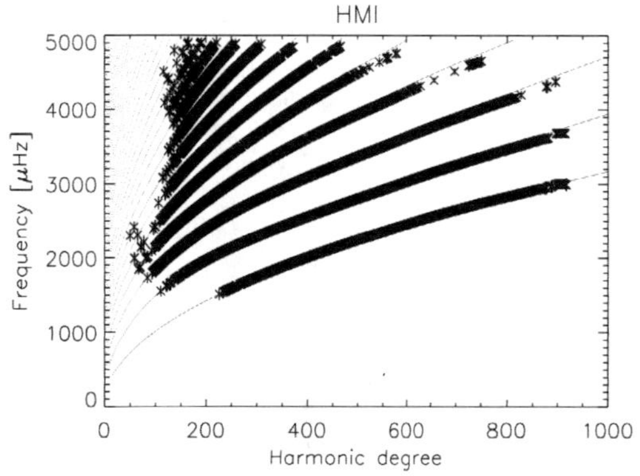

Figure 1. Mode frequencies (asterisk) obtained from one month long time series of Dopplergrams from GONG (upper left), MDI (upper right) and HMI (left). For better visibility the error bars are enlarged by a factor of 50. The dotted lines mark the positions of the f and p-mode ridges according to Model 'S'.

For the construction of the averaging kernels we use oscillation eigenfunctions which were obtained with the ADIPAK code [23]. The target kernels have Gaussian shapes with a width that is proportional to the sound speed c at the target location r_0 [24].

Figure 2 shows averaging kernels obtained for a target depth of $r_0 = 20\,\mathrm{Mm}$ and $r_0 = 70\,\mathrm{Mm}$. The kernel units are given by the normalization

$$\int_0^{R_\odot} \mathcal{K}(r_0, r)\, \mathrm{d}r = 1 \qquad (4)$$

of the kernels, where $\mathcal{K}(r_0, r)$ is the averaging kernel at target depth r_0 as a function of r.

The input parameters and results of the average kernels are shown in Table 1. For a target depth of 20 Mm the resulting average kernels for all three data sets are located at 20 Mm with a FWHM of about 20 Mm. The kernels from the different data sets are virtually identical. At

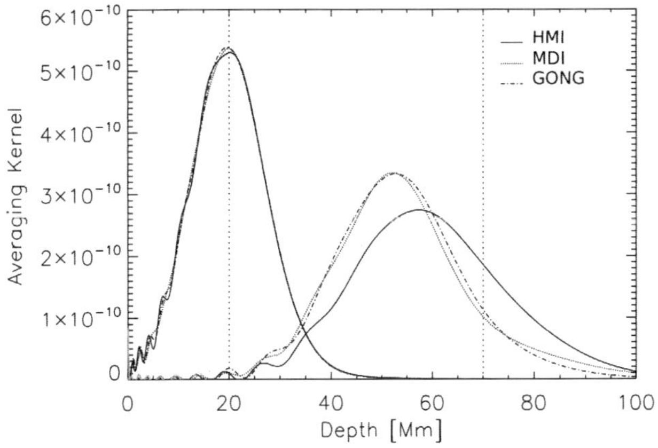

Figure 2. Averaging kernels obtained with a SOLA method for the modes displayed in Figure 1. The kernels constructed for HMI, MDI, and GONG are plotted for two target depths at $r_0 = 20$ Mm and $r_0 = 70$ Mm (indicated by the vertical dotted lines).

a target depth of 70 Mm the resulting average kernels for GONG and MDI are located at about 53 Mm with a FWHM of about 33 Mm while the one for HMI is a bit broader and located at a depth of 60 Mm.

Table 1. Input parameters and results for the averaging kernels. All numbers are in Mm.

Data set	Target depth	Kernel depth	Target width	Kernel width
GONG	20	19.8	20.9	19.9
MDI	20	19.8	20.9	20.3
HMI	20	19.8	20.9	19.9
GONG	70	53.3	43.8	32.7
MDI	70	53.3	43.8	33.8
HMI	70	60.0	43.8	39.2

4. Discussion & Conclusion

We have constructed localized averaging kernels for probing the meridional circulation within the upper third of the convection zone. The eigenmodes used for the construction of the kernels were obtained by Fourier-Legendre decomposition of one month long series of Dopplergrams from the GONG, MDI, and HMI instruments.

The method allows to obtain well localized kernels at depths down to 60 Mm. The kernels from the three data sets are very similar at a depth of about 20 Mm, while the kernels from HMI tend to give slightly better results in greater depth. Given the much better spatial resolution of the HMI instrument and its better technical specifications in general this seems not surprising. However the inverted flow velocities do not only depend on the shape of the used averaging kernels but also on the actually measured frequency shifts. Even with identical kernels the inverted flow velocities could differ a lot between the datasets and it will be interesting to see how HMI compares to GONG and MDI once the data has undergone the final calibration.

Concerning the Fourier-Legendre decomposition based on large region of interests, we conclude that this method promises to probe greater depths in the convection zone for the meridional flow once longer time series are used for the inversion. This will increase the SNR and also allow for a more precise frequency determination which is currently limiting the depth that can be reached with the method. In addition it has to be studied for which harmonic degree the frequency shifts can be used for inverting for the meridional circulation [25].

Acknowledgments

M.R. thanks Doug Braun for useful discussions. The authors thank Hannah Schunker for useful information on mode fitting. This work was supported by the European Helio- and Asteroseismology Network (HELAS) which was funded as Coordination Action under the European Commission's Sixth Framework Programme (FP6). This work utilizes data obtained by the Global Oscillation Network Group (GONG) program, managed by the National Solar Observatory, which is operated by AURA, Inc. under a cooperative agreement with the National Science Foundation. The data were acquired by instruments operated by the Big Bear Solar Observatory, High Altitude Observatory, Learmonth Solar Observatory, Udaipur Solar Observatory, Instituto de Astrofísica de Canarias, and Cerro Tololo Interamerican Observatory. Data used in this publication is provided by the SOHO/MDI consortium. SOHO is a project of international cooperation between ESA and NASA.

References

[1] Duvall Jr T L 1979 *Solar Phys.* **63** 3–15

[2] Hathaway D H 1996 *Astrophys. J.* **460** 1027–1033

[3] Komm R W, Howard R F and Harvey J W 1993 *Solar Phys.* **147** 207–223

[4] Bjerknes V 1926 *Astrophys. J.* **64** 93–121

[5] Rüdiger G 1980 *Geophysical and Astrophysical Fluid Dynamics* **16** 239–261

[6] Küker M and Stix M 2001 *Astron. Astrophys.* **366** 668–675

[7] Choudhuri A R, Schüssler M and Dikpati M 1995 *Astron. Astrophys.* **303** L29–32

[8] Dikpati M and Charbonneau P 1999 *Astrophys. J.* **518** 508–520

[9] Nandy D and Choudhuri A R 2002 *Science* **296** 1671–1673

[10] Rempel M 2006 *Astrophys. J.* **637** 1135–1142 (*Preprint* `arXiv:astro-ph/0610133`)

[11] Haber D A, Hindman B W, Toomre J, Bogart R S, Larsen R M and Hill F 2002 *Astrophys. J.* **570** 855–864

[12] Zhao J and Kosovichev A G 2004 *Astrophys. J.* **603** 776–784

[13] Zaatri A, Komm R, González Hernández I, Howe R and Corbard T 2006 *Solar Phys.* **236** 227–244

[14] Braun D C, Duvall Jr T L and Labonte B J 1987 *Astrophys. J. Lett.* **319** L27–L31

[15] Braun D C and Fan Y 1998 *ApJl* **508** L105–L108

[16] Doerr H P, Roth M, Zaatri A, Krieger L and Thompson M J 2010 *Astr. Nachr.(in press)*

[17] Scherrer P H, Bogart R S and Bush e a 1995 *Solar Phys.* **162** 129–188

[18] Anderson E R, Duvall T L and Jefferies S M 1990 *Astrophys. J.* **364** 699–705

[19] Christensen-Dalsgaard J, Däppen W, Ajukov S V, Anderson E R, Antia H M, Basu S, Baturin V A, Berthomieu G, Chaboyer B, Chitre S M, Cox A N, Demarque P, Donatowicz J, Dziembowski W A, Gabriel M, Gough D O, Guenther D B, Guzik J A, Harvey J W, Hill F, Houdek G, Iglesias C A, Kosovichev A G, Leibacher J W, Morel P, Proffitt C R, Provost J, Reiter J, Rhodes E J, Rogers F J, Roxburgh I W, Thompson M J and Ulrich R K 1996 *Science* **272** 1286–1292

[20] Gough D O and Toomre J 1983 *Solar Phys.* **82** 401–410

[21] Pijpers F P and Thompson M J 1994 *Astron. Astrophys.* **281** 231–240

[22] Schou J, Scherrer P H and Bush R I 2010 *in prep.*

[23] Christensen-Dalsgaard J 2008 *Astrophys. Spa. Sci.* **316** 113–120 (*Preprint* `0710.3106`)

[24] Thompson M J 1993 *GONG 1992. Seismic Investigation of the Sun and Stars* (*Astronomical Society of the Pacific Conference Series* vol 42) ed T M Brown pp 141–+

[25] Gough D and Hindman B W 2010 *Astrophys. J.* **714** 960–970 (*Preprint* `0911.2013`)

Low-degree helioseismology with AIA

R Howe,[1] F Hill,[1] R Komm,[1] A-M Broomhall,[2] W J Chaplin[2] and Y Elsworth[2]

[1]National Solar Observatory, 950 N. Cherry Ave., Tucson, AZ 85719, USA
[2]School of Physics and Astronomy, University of Birmingham, Edgbaston, Birmingham B15 2TT, UK

E-mail: `rhowe@noao.edu`

Abstract. We form unresolved-sun time series from the 1600 and 1700 Angstrom images produced by the Atmospheric Imaging Assembly, and find a clean low-degree p-mode spectrum at each wavelength. The time series and spectra are compared with Doppler velocity and continuum intensity time series from the Helioseismic and Magnetic Imager and velocity series from the Birmingham Solar Oscillation Network. The UV data have a slight phase shift with respect to the velocity, and show more sensitivity to high-frequency and less to low-frequency modes. Unlike the HMI (visible) continuum observations, the UV spectra show little or no granulation noise at low frequencies and thus potentially allow more low-frequency modes to be recovered. These results suggest that asteroseismology at near-UV wavelengths should be very feasible and even an improvement on visible-wavelength intensity measurements, at least in low-activity stars.

1. Introduction

The Atmospheric Imaging Assembly [AIA] aboard the Solar Dynamics Observatory [SDO] is best known for spectacular coronal imagery. However, AIA also observes at two near-UV wavelength bands, 1600 Å (C IV) and 1700 Å (continuum), which are formed in the upper photosphere and are sensitive to the five-minute oscillation signal. (Sadly, high-cadence observations in the 4500 Å visible continuum band were discontinued after the first two months of observations due to an instrumental flaw.) As a test of the usefulness of the near-UV bands for helioseismology, we use the "sun as a star" time series formed by extracting the mean value of each image; this allows us to study the solar oscillation signature without handling large volumes of image data. For comparison, we form similar time series for the velocity and continuum observables from the Helioseismic and Magnetic Imager [HMI], which uses the 6173 Å Fe I line. We also compare the observations with a sample of data from the Birmingham Solar-Oscillations Network [BiSON], which observes the unresolved Sun in the potassium D line at 7690 Å.

2. Data and analysis

For AIA, we consider the "internal" time series aia.lev1c_nrt, aia.lev1d_nrt, and aia.lev1e_nrt, which together cover the period from 2010 April 08 to the present. The export time series aia_test.lev1 is somewhat better calibrated but starts on 2010 June 10; the down-sampled export series aia_test.synoptic and aia_test.synoptic2, starting in 2010 May, have slightly too long a cadence for satisfactory helioseismic analysis. To form the time series, we extract from the

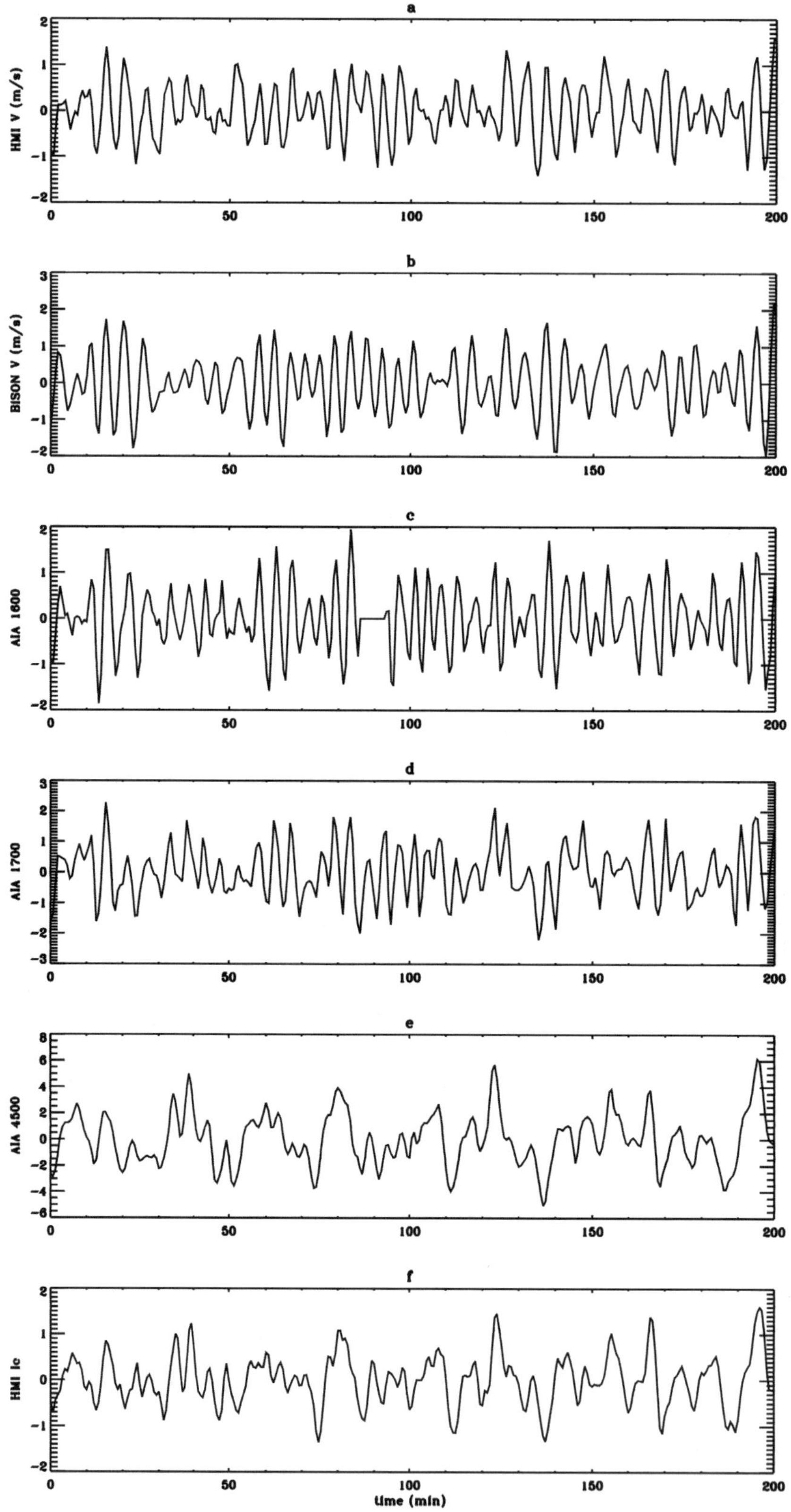

Figure 1. Contemporaneous residuals for (a) HMI V, (b) BiSON V, (c) AIA 1600, (d) AIA 1700, (e) AIA 4500, (f) HMI Ic. The photometric data have been scaled to match the amplitude of the velocity signal.

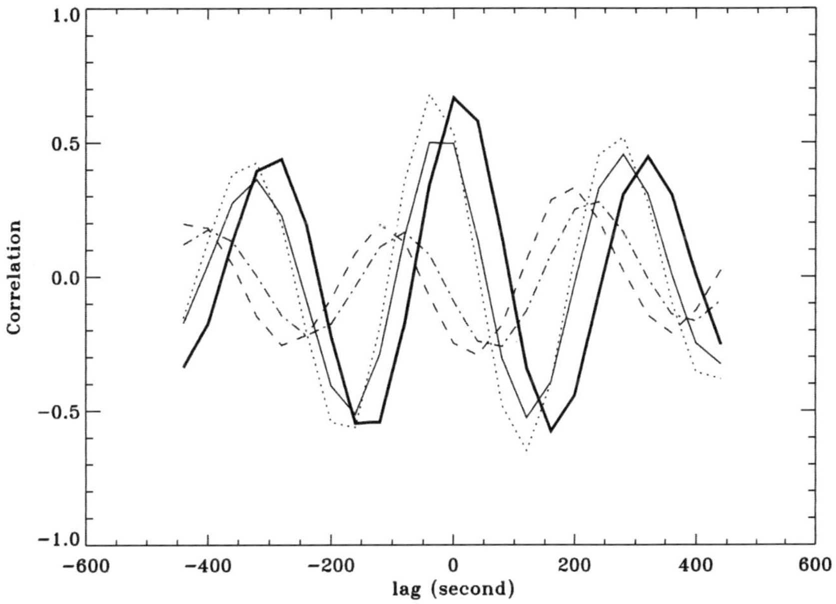

Figure 2. Cross-correlation functions with HMI velocity for AIA 1600 (dotted), AIA 1700 (solid), HMI Ic (dashed), AIA 4500 (dot-dashed), and BiSON velocity (thick solid line) over the whole of May 2010.

database the `T_OBS`, `DATAMEAN` and `INT_TIME` keywords. The `DATAMEAN` values are normalized by the `INT_TIME` values, detrended by subtracting a moving mean over 15 minutes, and interpolated to a uniform 40-second cadence. Finally, spikes of $\geq 5\sigma$ are rejected, and data in observing gaps are set to zero.

The HMI data analyzed here are based on the hmi_test.V_45s [HMI V] series for velocity and the hmi_test.Ic_45s series [HMI Ic] for continuum intensity. These series are more or less continuous from 2010 May 1. The preliminary calibration of the HMI velocity is imperfect, resulting in a daily variation in the time series. This was removed (or at least greatly reduced) by subtracting a quadratic fit to the `OBS_VR` keyword rather than by simply subtracting the observer velocity. These data have a regular 45-second cadence, but were interpolated to 40s intervals for comparison with the BiSON data

The BiSON data examined for this paper consist of 31 days starting 2010 May 1. The time series has a fill of 75%, a cadence of 40 s, and was created using a moving mean of 23 points (15.3 mins).

3. Results

3.1. Residuals and phase shifts

Figure 1 shows short segments of each time series, with the photometric series scaled to match the HMI velocity. The two visible-continuum series show more low-frequency variation than the UV and velocity series, but the five-minute oscillations are visible in all the series.

Figure 2 shows cross-correlation functions for HMI V with each of the other time series, calculated over the whole of May 2010. The visible continuum series appear to lag the HMI V by approximately 5 time samples, or 200 s, for HMI Ic and by 6 samples, or 240 s, for AIA 4500. The AIA 1600 and 1700 time series, on the other hand, lead HMI V by about one time sample,

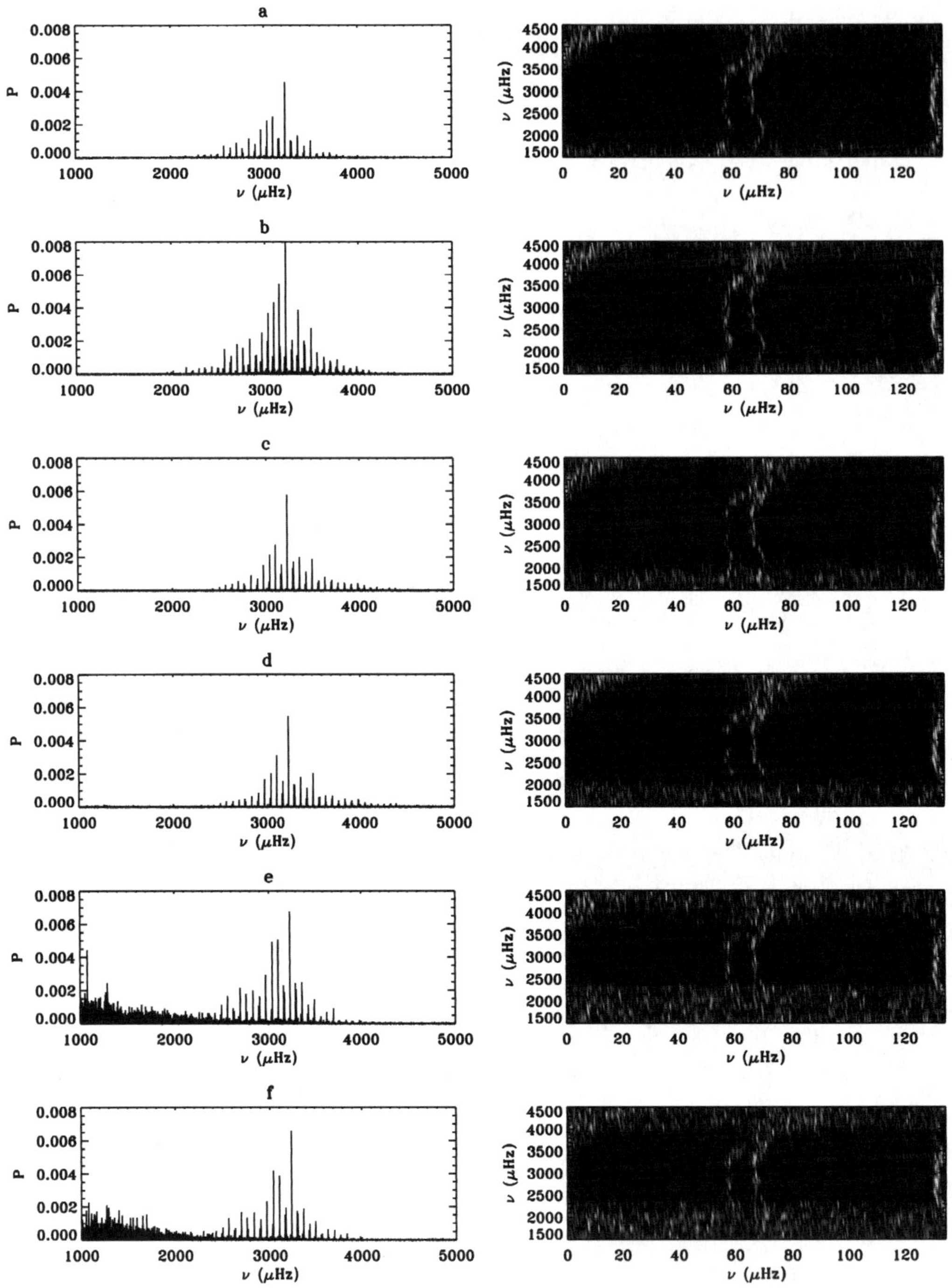

Figure 3. "Sun as a star" power spectra (left) and normalized echelle power maps (right) for May 2010, for (a) HMI V, (b) BiSON V, (c) AIA 1600, (d) AIA 1700, (e) AIA 4500, (f) HMI Ic. The power spectra are shown in arbitrary units.

or 40 s. The fact that the timing differences between AIA 4500 and HMI Ic and between AIA 1600/1700 and HMI V are in different senses gives us some confidence that they do not arise simply from an indexing error. A phase difference between continuum intensity and velocity measurements is expected; the phase difference has elsewhere been observed to be a function of frequency, and not the constant 90 deg predicted by a simple model of the atmosphere [1].

3.2. Sample spectra
Figure 3 shows sample power spectra and normalized echelle diagrams for the three AIA, two HMI and one BiSON data sets. In all cases, the five-minute p-mode spectrum is clearly visible. The AIA 4500 Å and HMI continuum also show a substantial contribution from granulation below 2000 μHz.

3.3. Comparison of power spectra
Figure 4 shows smoothed power spectra for HMI V, AIA 1600, and BiSON for May, superimposed on the same linear/log axes. The other spectra have been scaled to match the HMI V spectrum in the 2500 – 3500 μHz range. The noise levels in the AIA and BiSON spectra at 2000 μHz are very similar, while the HMI V spectrum has lower noise by a factor of two or three. The improvement over BiSON may be partly due to the better duty cycle. The velocity spectra have stronger signal and lower noise in the modes between about 1500 and 2000 μHz; on the other hand, while the noise is also slightly higher, the AIA 1600 spectrum appears to have more distinct signal in the high-frequency range, between 4500 and 5000 μHz. In these series the AIA 1700 has higher low-frequency noise than AIA 1700, but as these are still preliminary data we do not emphasize this finding.

4. Discussion
The 1600 and 1700 Å bands of AIA show promise for use in helioseismology, with a clean p mode signal and little or no granulation noise. This augurs well for the prospect of using ultraviolet photometry in seismology of sun-like stars, as is planned, for example, for the Stellar Imager mission [2]. However, it should be noted that these bands are sensitive to flare activity and may be less useful for seismology in stars more active than the current Sun.

The high signal-to-noise of the HMI velocity signal is also encouraging.

We have been able to estimate the phase shift between the p-mode signals in AIA near-ultraviolet intensity and the HMI velocity and intensity, and future work may allow a detailed comparison of the peak profiles and parameters that will be useful for the study of the behavior of the modes in the photosphere.

In the near future, we hope to analyze longer data series, allowing for the estimation of mode parameters. It would also be interesting to analyze the UV data using the methods of global and local resolved-sun helioseismology; simultaneous measurements from lines in the upper and lower photosphere using the MOTH instrument [3], for example, have already demonstrated that multi-wavelength measurements can yield interesting results. Longer time series will also allow us to study the phase differences in more detail.

Acknowledgments
SDO data courtesy SDO (NASA) and the AIA and HMI consortia. BiSON is operated by the High Resolution Optical Spectroscopy group of the School of Physics and Astronomy at the University of Birmingham, UK, in collaboration with Sheffield Hallam University, UK. It is funded by the Science and Technology Facilities Council (STFC).

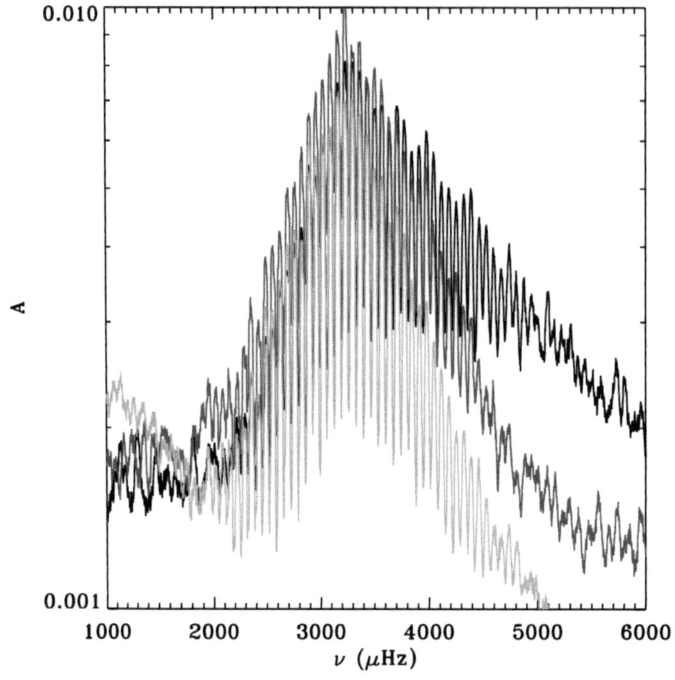

Figure 4. Power spectra, smoothed over 101 frequency bins and plotted on a logarithmic scale, for AIA 1600 (black), BiSON V (dark gray) and HMI V (light gray).

References

[1] Oliviero M, Severino G, Straus T, Jefferies S M and Appourchaux T 1999 *Astrophys.J.* **516** L45–8

[2] Carpenter K G, Schrijver C J, Karovska M and The Si Vision Mission Team 2009 *Astronomical Society of the Pacific Conference Series* (*Astronomical Society of the Pacific Conference Series* vol 412) ed D G Luttermoser, B J Smith, & R E Stencel pp 91–101

[3] Finsterle W, Jefferies S M, Cacciani A, Rapex P, Giebink C, Knox A and Dimartino V 2004 *Solar Phys.* **220** 317–31

High frequency power in HMI ring diagrams

Charles S. Baldner[1], **Richard S. Bogart**[2], **and Sarbani Basu**[1]

[1] Astronomy Department, Yale University, P.O.Box 208101, New Haven, CT, 06530-8101
[2] Hansen Experimental Physics Laboratory, Stanford University, Palo Alto, CA, 94305

E-mail: `charles.baldner@yale.edu`

Abstract. Coherent power above the photospheric acoustic cut-off frequency (at $\nu \sim 5.3$mHz) is known to exist in solar helioseismic observations. The arrival of high quality velocity data with high temporal and spatial resolution from the Helioseismic & Magnetic Imager (HMI) on the Solar Dynamics Observatory (SDO) spacecraft allows the high frequency regime to be explored with greater precision than previously. We analyse mode frequencies, widths, and amplitudes from HMI ring diagrams at high frequencies, and compare them to earlier results.

1. Introduction

Solar normal modes of oscillation with frequencies below the acoustic cut-off frequency ($\nu \sim 5.2$mHz) are trapped in the solar interior and are used in helioseismology to infer the interior structure and dynamics of the Sun. Modes with higher frequencies are not trapped and penetrate the chromosphere. High frequency coherent power has nevertheless been observed in the Sun [1, 2, 3, 4].

The high frequency p-modes have been found by [1, 4] to have fairly constant line widths above the cutoff frequency. The ridges have been found to obey a dispersion relation derived for completely trapped modes [1]. The authors [4] found evidence for an avoided crossing at frequencies slightly higher than the acoustic cutoff.

In this work, we use ring diagrams from the science commissioning observations of the Helioseismic and Magnetic Imager (HMI) on board the Solar Dynamics Observatory (SDO) to study temporal and spatial frequency power in the solar oscillation spectrum.

2. Method

The ring diagrams used here are constructed from $16^\circ \times 16^\circ$ degree line-of-sight Dopplergram data taken with the HMI instrument. These rings are standard data products from the HMI local helioseismology pipeline. The data are tracked for 1728 minutes at a 45 second cadence. The regions are centred at the equator, and cross disk centre at the mid-time of the tracking interval. The data are projected to a rectilinear grid using Postel's projection and appodized to a circular region 15.36° in diameter. The ring diagram itself is the three dimensional Fourier transform of this data cube. For this work, 40 individual ring diagrams were averaged together to improve as much as possible the signal to noise.

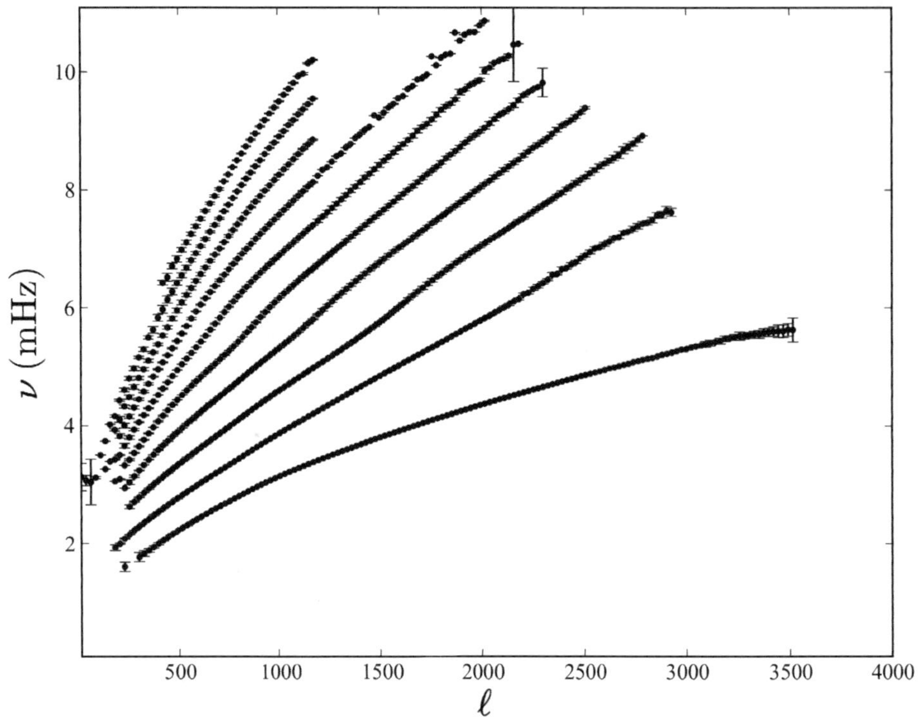

Figure 1. The fitted frequencies and uncertainties to the averaged HMI power spectrum. The fitted ridges are the f-mode and the first 9 p-modes.

3. Results

In order to fit the spectra, we azimuthally average the ring. The ridges are fit with a symmetric Lorentzian at constant wavenumber k:

$$P(\nu) = \frac{A\Gamma}{(\nu - \nu_0)^2 + \Gamma^2} + B, \tag{1}$$

where A is the amplitude, Γ is the width of the ridge, ν_0 is the central frequency, and B is the background power. Each ridge is fit independently. The fits minimise the squares of the differences between the data and the model profile using a Levenberg-Marquardt algorithm.

An $\ell - \nu$ diagram of the fitted ridges is shown in Figure 1. The ridges are not symmetric, particularly at low temporal frequency, and the frequencies we measure here are underestimated somewhat. At higher frequencies, however, particularly above the acoustic cutoff, the ridges become more or less symmetric and our estimates of the frequencies are therefore reasonable. Figure 1 shows fits to the f-mode ridge and the first 9 p-mode ridges. We succeed in fitting some ridges up close to the temporal Nyquist frequency.

Figure 2 shows the amplitude A and the width Γ. The amplitudes peak at approximately 3.5mHz, as expected, and fall off for higher frequencies. The slope of the amplitudes as a function of frequency above the acoustic cutoff is shallower than the slope below the cutoff. The widths of the f-mode ridge are substantially different than the p-mode ridge widths. The p-mode widths

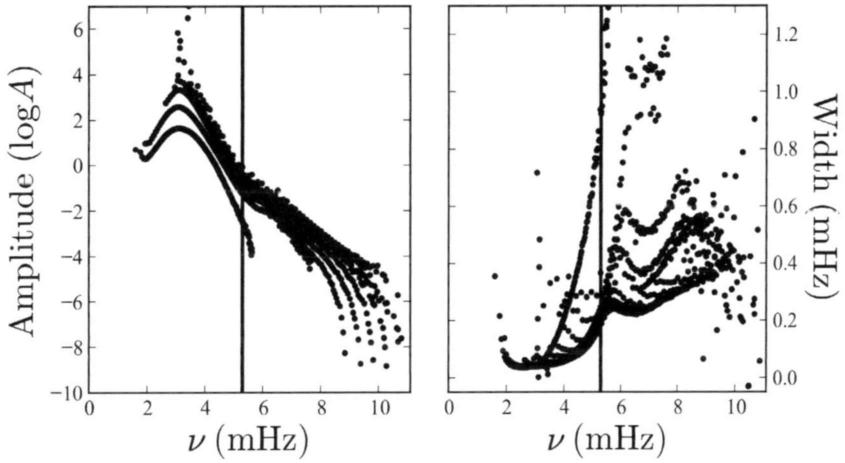

Figure 2. Amplitudes and widths for the fitted modes. The log of the amplitudes are shown in the top panel, and the widths are shown in the bottom panel. The solid black line shows the approximate location of the acoustic cutoff frequency.

behave differently below the acoustic cutoff frequency than above. The widths are larger above the acoustic cutoff.

To see clearly the avoided crossing above the acoustic cutoff frequency, we plot the difference of the nth ridge and $n-1$ ridge in Figure 3. The bump above the acoustic cutoff shows up in all p-mode ridges and is a signal of the avoided crossing.

4. Conclusions

We have fit high temporal and spacial frequency p-mode ridges from HMI ring diagram data. We are able to fit the first 9 p-mode ridges, as well as the f-mode. We find that the p-modes clearly extend well above the acoustic cutoff frequency at $\nu \sim 5.3$mHz.

Earlier work has found that mode widths become roughly constant as a function of frequency above $\nu \sim 5.3$mHz [1]. In this work, we find that widths as a function of frequency do become substantially flatter above $\nu \sim 5.3$mHz, but they are not constant. We find a peak in width just above the cutoff, and a slight negative slope or flat slope up to approximately $\nu \sim 7$mHz. Widths then increase with frequency. This is not inconsistent with [1], since we probe higher frequencies than that work.

Following the work of [4], we examine the frequency differences $\nu_n - \nu_{n-1}$. As they did, we find a bump between 5 and 6 mHz, which could be evidence of an avoided crossing above the acoustic cutoff frequency.

This work was performed with the HMI data that was available. As more data becomes available, better signal-to-noise can be achieved. Further more, the effects of solar activity on high frequency power can be explored. Further theoretical work to understand what information is contained in the high frequency power is, however, the most necessary component of further research.

Acknowledgments

CB is supported by a NASA Earth & Space Sciences Fellowship NNX08AY41H. SB is supported by NASA grant NNG06GD139 and NSF Career grant ATM-0348837. The Helioseismic and

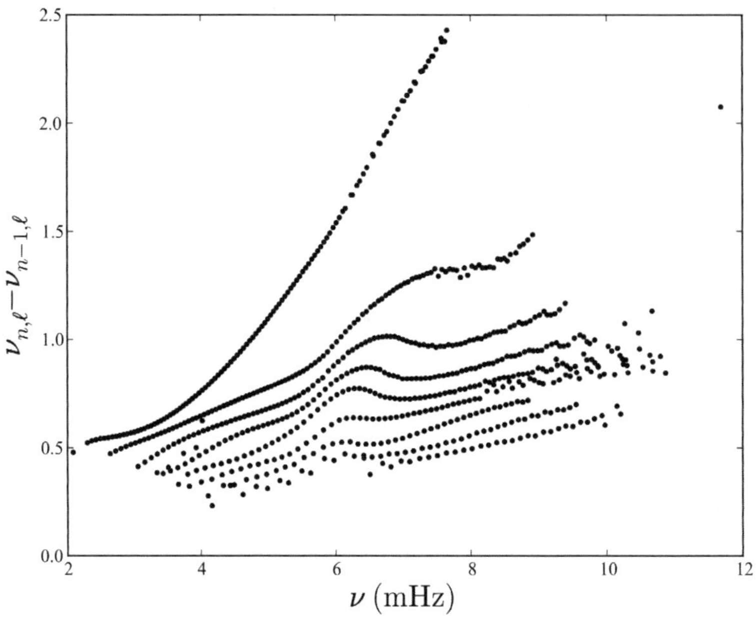

Figure 3. The frequency differences of adjacent modes with the same degree. Above the cutoff frequency, there is a distinct bump in all p-modes – evidence for an avoided crossing. The magnitude of the bump decreases with increasing n.

Magnetic Imager (HMI) project on SDO is supported by NASA grant NAS5-02139 to Stanford University.

References

[1] Duvall, T. L., Jr., Harvey, J. W., Jefferies, S. M., & Pomerantz, M. A. 1991, ApJ, 373, 308

[2] Fernandes, D. N., Scherrer, P. H., Tarbell, T. D., & Title, A. M. 1992, ApJ, 392, 736

[3] Kneer, F., & von Uexkull, M. 1993, A&A, 274, 584

[4] Antia, H. M., & Basu, S. 1999, ApJ, 519, 400

Comparison of HMI Dopplergrams with GONG and MDI data

R Howe,[1] K Jain,[1] F Hill,[1] R Komm,[1] I González Hernández[1] and R Bogart[2]

[1]National Solar Observatory, 950 N. Cherry Ave., Tucson AZ 85719, USA
[2]Hansen Experimental Physics Laboratory, Stanford University, Stanford, CA 94305, USA

E-mail: rhowe@noao.edu

Abstract. We compare sample Dopplergrams from the Helioseismic and Magnetic Imager, the Michelson Doppler Imager and the Global Oscillation Network Group. Each instrument has a distinct static velocity patterm across the disk; once this has been subtracted and the images interpolated to a common grid, the agreement is satisfactory.

1. Introduction

A useful check of the calibration of the velocity images from the Helioseismic and Magnetic Imager [HMI] can be made by comparing them with those taken simultaneously from other instruments. Here we compare sample Dopplergrams from HMI, the Michelson Doppler Imager [MDI] [1] and the Global Oscillation Network Group [GONG] [2].

2. Data

The data chosen for analysis are from observations on April 27, 2010. The HMI data are 4096×4096 Dopplergrams from the hmi_test.V_45s series, taken at 45-second cadence. The GONG images are single-site Dopplergrams, registered (but not accurately aligned) on an 839×839 pixel grid and obtained at 60-second cadence. For MDI there are a small number of 1024×1024 Doppler images available for this day.

Figure 1 shows a sample image from each instrument, to the same pixel scale. All images used in this analysis are cropped at 930 arcsec from disk center to avoid limb noise.

3. Analysis

Each instrument has a large-scale, static background pattern in the Doppler velocity map, due to instrumental distortions, uncorrected observer motion effects and the mostly-horizontal nature of supergranular motions. The first step in the comparison is to remove this background. We average the images from approximately ten days of observations and subtract the average map from the individual observations. Figure 2 shows the background image for each instrument after east-west detrending. The background image used for GONG was based on merged images from September 2009, because the more recent data have not yet been processed in sufficient quantity.

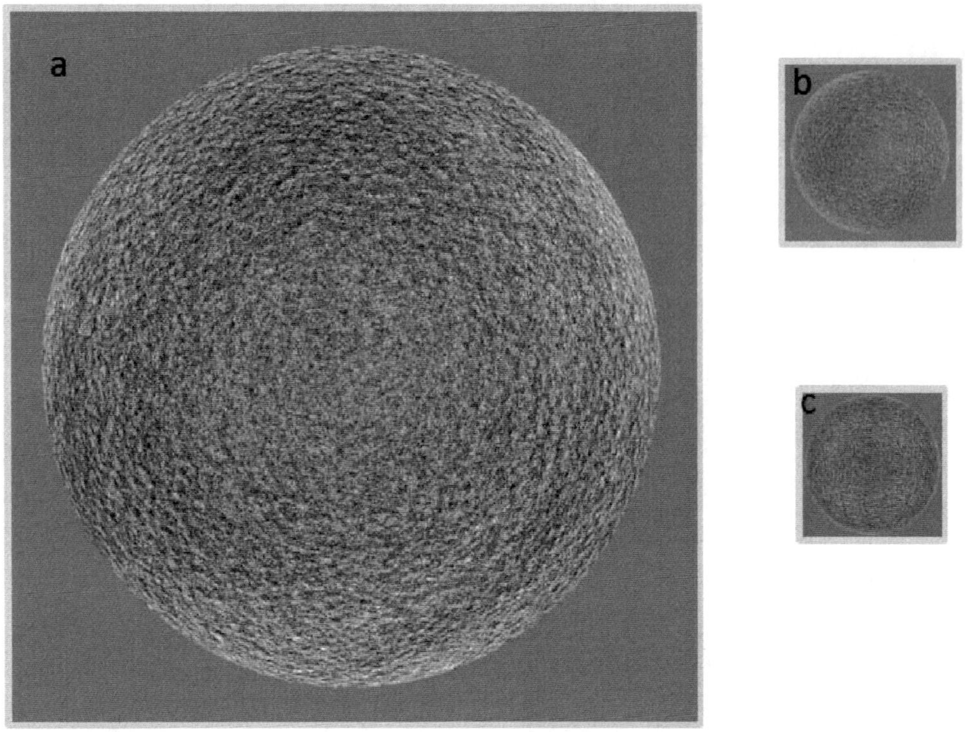

Figure 1. Sample velocity images from HMI (a), MDI (b) and GONG (c), shown to the same pixel-per-inch scale to highlight the different resolutions of the instruments. A linear east-west trend has been removed from each image.

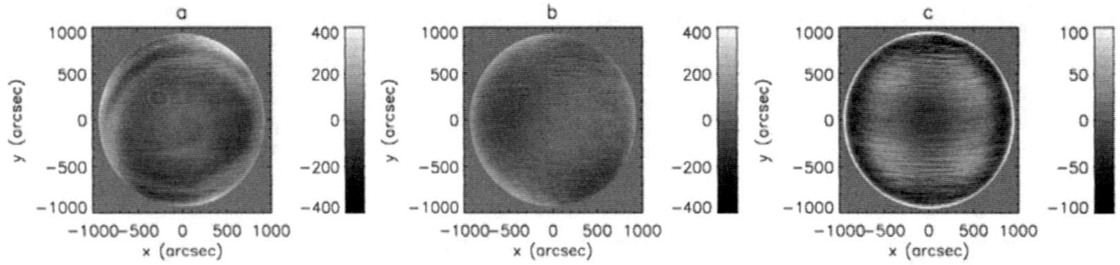

Figure 2. Background velocity pattern for (a) HMI, (b) MDI, (c) GONG. Note that the grayscale for GONG is more sensitive by a factor of 4. The grayscales are in meters per second.

Next, the images are rescaled to a 1000×1000 pixel grid with two-arcsecond pixels, and any necessary small adjustments in angle and position are applied to obtain the best correlation. To better match the resolution of the different instruments, the interpolated images are smoothed with a two-dimensional 7-pixel running mean. Figure 3 compares horizontal and vertical cuts across each image after this processing has been applied, and Figure 4 shows scatter plots of MDI and GONG against HMI. The correlation coefficients are 0.81 and 0.89 respectively. The linear regression is not strictly appropriate because the data have scatter along both axes, but in the absence of error estimates it is the best that can be done.

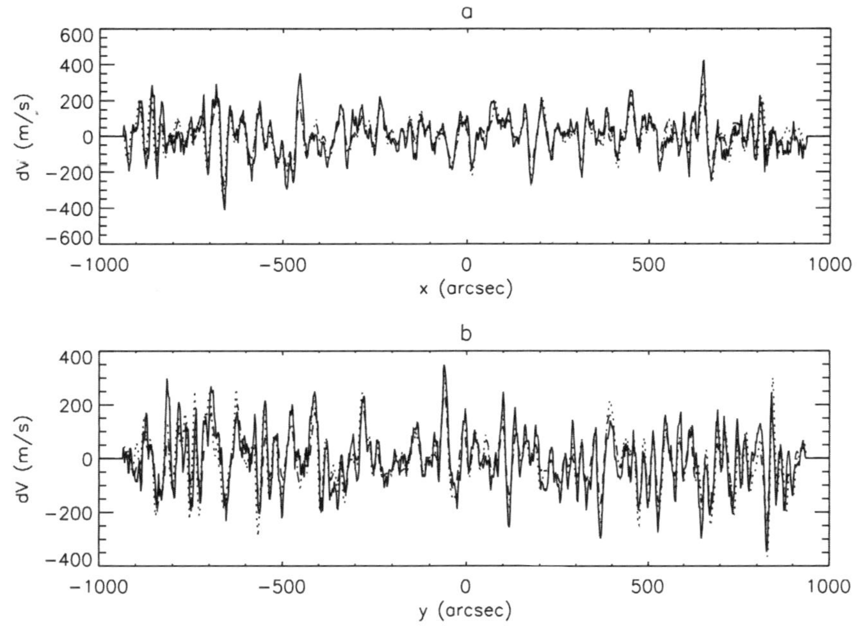

Figure 3. Slices at $y = 0$ (a) and $x = 0$ (b) across detrended, interpolated, registered and smoothed velocity maps for HMI (solid line), MDI (dotted line) and GONG (dashed line).

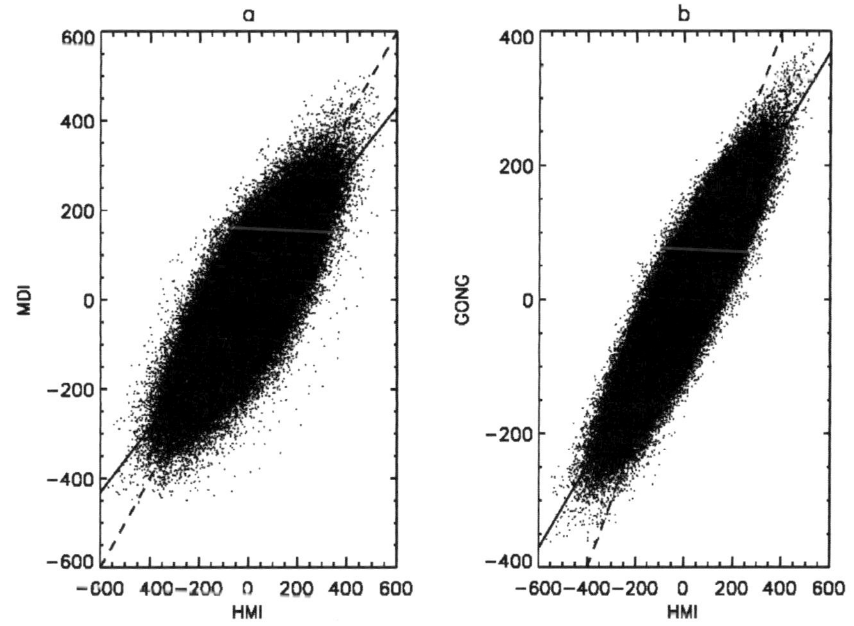

Figure 4. Scatter plots of detrended, interpolated, registered and smoothed velocity maps for (a) MDI and (b) GONG versus HMI. The solid lines represent the best least-squares fit and the dashed line the $x = y$ line.

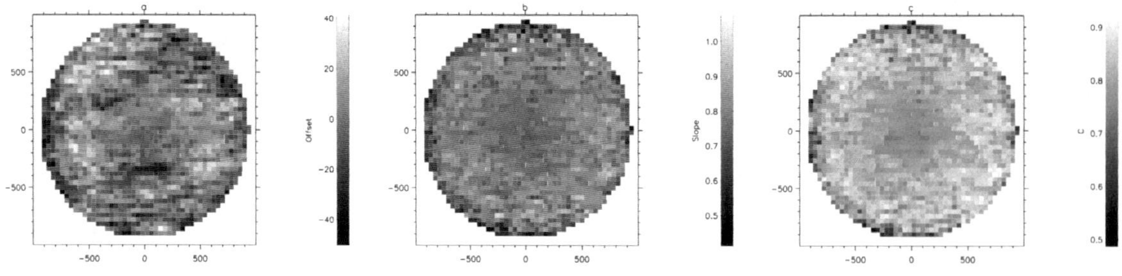

Figure 5. Offset (a), slope (b) and correlation coefficient (c) for 20-arcsec square patches as a function of disk position, averaged over 30 image pairs, for GONG (El Teide) compared with HMI.

We take about thirty images from the 2010-04-27 site day for the El Teide GONG sites and compare them with the corresponding HMI images, following the same methodology described above. After registration, we calculate the slope, offset, and correlation coefficient for fits between GONG and HMI images in sub-images 20 arcsec square, as shown in Figure 5. Even after the background subtraction, there is still some structure in the offset values across the disk, probably because of the different "fingerprints" of the GONG site images relative to the merged-image mean that was subtracted. The radial structure in the slope and correlation plots can be explained by the radial variation in the strength of the supergranulation signal. For all the images, the average GONG – HMI correlation is around 0.8 to 0.9.

4. Discussion

GONG, HMI, and MDI are seeing the same Sun, though they have different patterns of systematic velocity offset across the disk. There is clearly still work to be done on the calibration of the HMI velocity images. After detrending and smoothing, the correlation between HMI and GONG images is around 80 – 90% throughout the day, which seems an acceptable level given the differences in the instruments (including the use of a different spectral line) and the slight offsets in observation time arising from the different cadences.

Acknowledgments

This work utilizes data obtained by the Global Oscillation Network Group (GONG) Program, managed by the National Solar Observatory, which is operated by AURA, Inc. under a cooperative agreement with the National Science Foundation. The data were acquired by instruments operated by the Big Bear Solar Observatory, High Altitude Observatory, Learmonth Solar Observatory, Udaipur Solar Observatory, Instituto de Astrofísica de Canarias, and Cerro Tololo Interamerican Observatory. SOHO is a project of international collaboration between ESA and NASA. HMI data courtesy of SDO (NASA) and the HMI consortium.

References

[1] Scherrer P H, Bogart R S, Bush R I, Hoeksema J T, Kosovichev A G, Schou J, Rosenberg W, Springer L, Tarbell T D, Title A, Wolfson C J, Zayer I and MDI Engineering Team 1995 *Solar Phys.* **162** 129–88

[2] Harvey J W, Hill F, Hubbard R, Kennedy J R, Leibacher J W, Pintar J A, Gilman P A, Noyes R W, Title A M, Toomre J, Ulrich R K, Bhatnagar A, Kennewell J A, Marquette W, Patrón J, Saá O and Yasukawa E 1996 *Science* **272** 1284–6

First Global Rotation Inversions of HMI Data

R Howe,[1] T P Larson,[2] J Schou,[2] F Hill,[1] R Komm,[1] J Christensen-Dalsgaard[3,4] and M J Thompson[4,5]

[1] National Solar Observatory, 950 N. Cherry Ave., Tucson AZ 85719, USA
[2] Hansen Experimental Physics Laboratory, Stanford University, Stanford, CA 94305, USA
[3] Department of Physics and Astronomy, Aarhus University, DK-8000 Aarhus C, Denmark
[4] High Altitude Observatory, National Center for Atmospheric Research, P.O. Box 3000, Boulder, CO 80307-3000, USA
[5] School of Mathematics & Statistics, University of Sheffield, S3 7RH, U.K.

E-mail: `rhowe@noao.edu`

Abstract. We present the first 2-dimensional global rotational inversions of medium-degree *p*-mode data from the Helioseismic and Magnetic Imager, and compare the results with inversions of Michelson Doppler Imager data for the same time period. The inferred rotation profiles show good agreement between the two instruments.

1. Introduction

The Helioseismic and Magnetic Imager [HMI] aboard the Solar Dynamics Observatory [SDO] commenced observations in late March, 2010, and a continuous sequence of Dopplergrams at 45s cadence is available starting May 1. The instrument will eventually supersede the Michelson Doppler Imager [MDI] aboard SOHO [1], which will cease operations before the end of 2010.

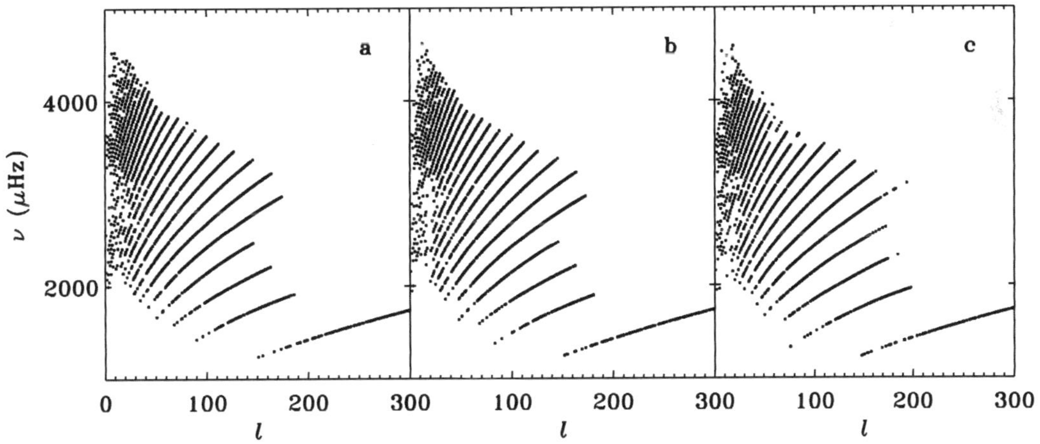

Figure 1. Coverage of fitted modes in the $l - \nu$ plane for HMI (a), MDI FD (b) and MDI Vw (c).

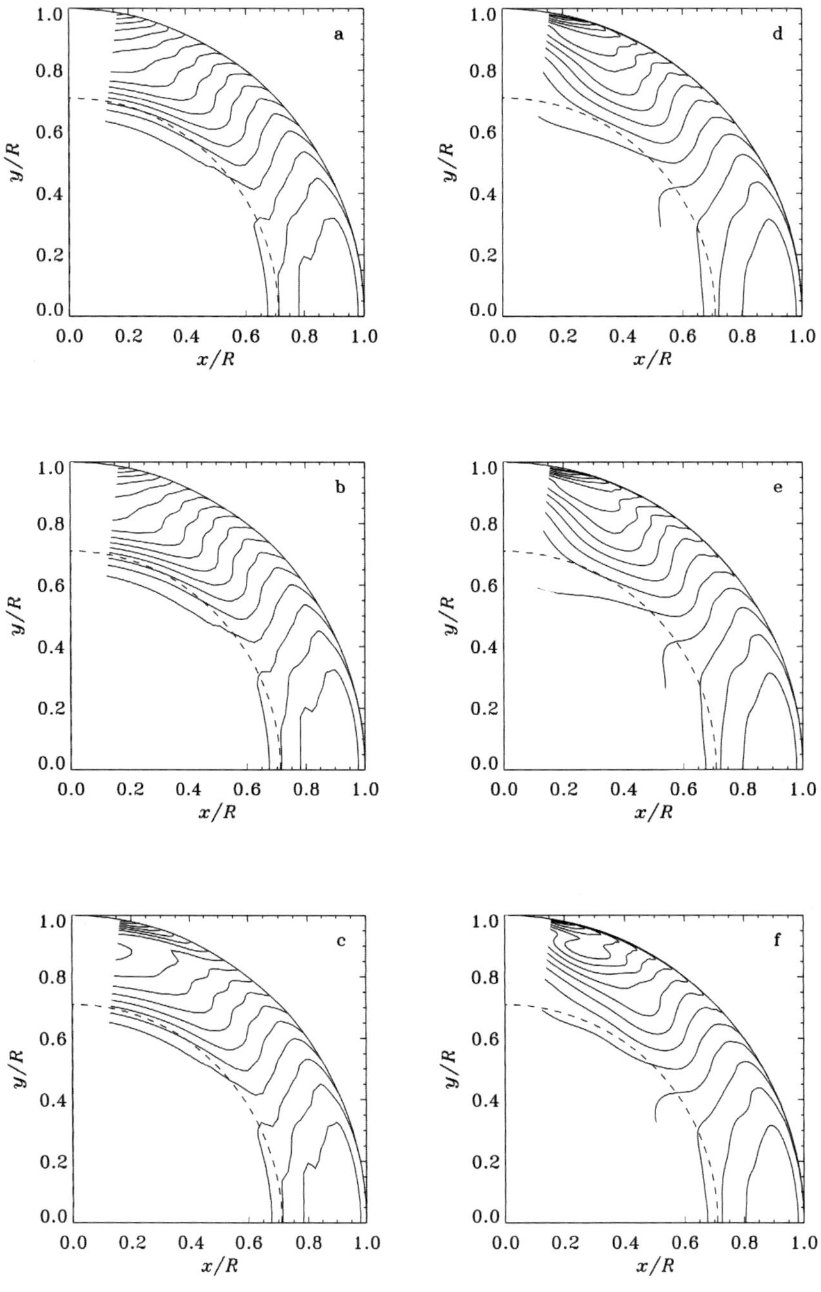

Figure 2. Contour plots of inferred rotation profiles for RLS (a–c) and OLA (d–f) inversions of HMI (top row), MDI FD (middle row) and MDI Vw (bottom row) splittings. The contours are at 10 nHz intervals and are not plotted below $0.5R_{\odot}$ or above 75° latitude. The dashed line represents the base of the convection zone.

During the overlap period between HMI and MDI, it is important to cross-check that the data from the two instruments are consistent, so that, for example, monitoring of temporal variations in the internal rotation can continue uninterrupted. Here we present extremely preliminary results from a 36-day sequence of observations starting May 7.

2. Data and Analysis

The data consist of Ritzwoller-Laveley rotational splitting coefficients [2] up to 35th order, derived using an improved version of the peak fitting code that has historically been used for the MDI data [3]. For MDI, we use two sets of splittings, one [FD] based on full-disk observations as obtained in "Dynamics" mode and one [Vw, or vector-weighted] based on the "structure" mode in which MDI has usually observed, where the images are more tightly apodized, convolved with a Gaussian vector and subsampled by a factor of five. It has recently been found [4] that certain features, such as the high-latitude jet [5] seen in inversions of the vector-weighted splittings, are absent in inversions of full-disk data. As these features are also absent in, for example, inversions of GONG data, they are believed to be artifacts [6].

Figure 1 shows the distribution of fitted modes in the $l - \nu$ plane.

The inversions were carried out using the same inversion codes — 2-dimensional regularized least squares [RLS] and subtractive optimally localized average [OLA] — and regularization parameters that have been used in our earlier work for coefficients based on 72-day sets of MDI observations [7–9]. Particularly in the OLA case, these parameters may not be optimal for the shorter data set.

3. Results

In Figure 2 we show contour plots of the inferred rotation profiles from the three data sets and the two inversion methods. Figure 3 shows superimposed radial cuts at selected latitudes. The results for HMI and MDI-FD are extremely similar, with only a small difference at high latitudes. These results differ somewhat at high latitudes from those for MDI-Vw, which show traces of the high-latitude jet and the "bump" in the coefficients for frequencies around 3.5 mHz.

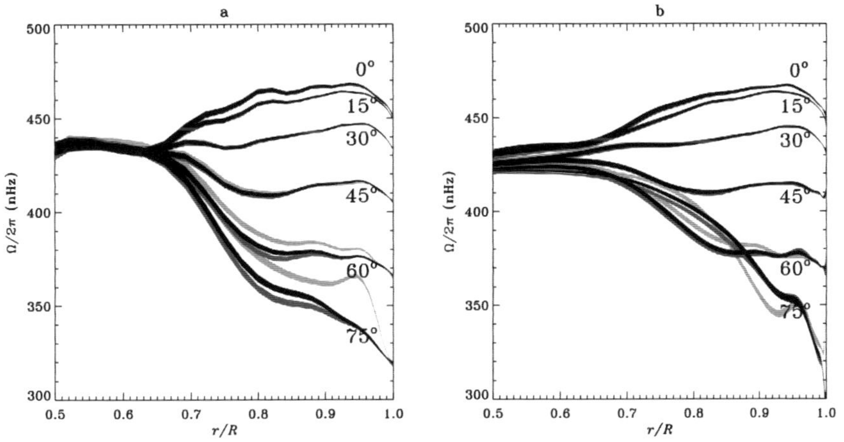

Figure 3. Comparison of inferred rotation profiles at selected latitudes as a function of fractional radius, for RLS (a) and OLA (b) inversions of HMI (black), MDI FD (dark gray) and MDI Vw (light gray) splitting coefficients.

4. Discussion

The preliminary global rotation inversion results for HMI are in good agreement with those from MDI full-disk data, apart from a slight discrepancy at high latitudes deep in the convection zone. The noise levels are also very similar. These results give us confidence that HMI and MDI observations can be combined for long-term monitoring of phenomena like the so-called torsional oscillation. However, because we have full-disk data only for the MDI Dynamics observations, covering about twenty per cent of the lifetime of MDI, it is important that the issues with analysis of the structure-mode data be resolved, and efforts to this end are underway. Until this issue is resolved, some extra care will be needed when combining MDI and HMI data.

Acknowledgments

SOHO is a project of international collaboration between ESA and NASA. HMI data courtesy of SDO (NASA) and the HMI consortium. The National Center for Atmospheric Research is sponsored by the National Science Foundation. We also thank the conference organizers for making it possible for these late-breaking results to be displayed in Aix.

References

[1] Scherrer P H, Bogart R S, Bush R I, Hoeksema J T, Kosovichev A G, Schou J, Rosenberg W, Springer L, Tarbell T D, Title A, Wolfson C J, Zayer I and MDI Engineering Team 1995 *Solar Phys.* **162** 129–88

[2] Ritzwoller M H and Lavely E M 1991 *Astrophys. J.* **369** 557–66

[3] Schou J 1992 *Ph.D. Thesis, Aarhus University*

[4] Larson T and Schou J 2009 *Astronomical Society of the Pacific Conference Series* (*Astronomical Society of the Pacific Conference Series* vol 416) eds M Dikpati, T Arentoft, I González Hernández, C Lindsey, & F Hill pp 311-4

[5] Howe R, Antia H, Basu S, Christensen-Dalsgaard J, Korzennik S G, Schou J and Thompson M J 1998 *Structure and Dynamics of the Interior of the Sun and Sun-like Stars SOHO 6/GONG 98 Workshop Abstract, June 1-4, 1998, Boston, Massachusetts* ed Korzennik S pp 803–8

[6] Schou J, Howe R, Basu S, Christensen-Dalsgaard J, Corbard T, Hill F, Komm R, Larsen R M, Rabello-Soares M C and Thompson M J 2002 *Astrophys. J.* **567** 1234–49

[7] Schou J, Antia H M, Basu S, Bogart R S, Bush R I, Chitre S M, Christensen-Dalsgaard J, di Mauro M P, Dziembowski W A, Eff-Darwich A, Gough D O, Haber D A, Hoeksema J T, Howe R, Korzennik S G, Kosovichev A G, Larsen R M, Pijpers F P, Scherrer P H, Sekii T, Tarbell T D, Title A M, Thompson M J and Toomre J 1998 *Astrophys. J.* **505** 390–417

[8] Howe R, Christensen-Dalsgaard J, Hill F, Komm R W, Larsen R M, Schou J, Thompson M J and Toomre J 2000 *Astrophys. J.* **533** L163–6

[9] Howe R, Rempel M, Christensen-Dalsgaard J, Hill F, Komm R, Larsen R M, Schou J and Thompson M J 2006 *Astrophys. J.* **649** 1155–68

HMI global helioseismology data analysis pipeline

Tim Larson, Jesper Schou

Hansen Experimental Physics Lab (HEPL), Stanford University, Stanford, CA

E-mail: `tplarson@sun.stanford.edu`

Abstract. The HMI global helioseismology data analysis pipeline is based largely on the MDI medium-l program. All of the modules that ran in the SOI Science Support Center have been ported for use in the SDO Joint Science Operations Center (JSOC) and given greater functionality. Many errors and approximations which are present in the standard MDI pipeline have been corrected and improvements have been added. Scripts have been written to automate the submission of compute jobs to our local cluster; it is now possible to go from dopplergrams to mode parameters with the push of a button. JSOC dataseries have been created to hold all intermediate data products, timeseries, window functions, and mode parameters. Here we discuss the operation of the pipeline, the structure of the data it generates, and access to the same.

1. Overview

The pipeline begins with dopplergrams as input. These are apodized in fractional solar radius and remapped to a regular grid in longitude and sin(latitude). Each image is then decomposed into its spherical harmonic components of degree l and azimuthal order m, although it must be noted that since we cannot see the entire surface of the Sun, the modes cannot be perfectly separated by this operation. A critical component of the pipeline is thus the generation of a leakage matrix which quantifies how much each mode leaks into its neighbors. The next step is to construct timeseries of the complex spherical harmonic amplitudes. These can be chunked in l and adjusted in length to suit our I/O and memory restrictions, but they will eventually be retiled into longer timeseries containing only a single l each. At this point gaps and discontinuities in the timeseries are identified. Next the timeseries are detrended by polynomial subtraction and gaps are filled using an autoregressive algorithm. Then a Fourier transform is performed, and the peaks in the transforms themselves are fit using a maximum likelihood minimization, which also takes into account the leakage matrix. The result is the peak frequency, amplitude, and linewidth for each l and radial order n the code was able to fit. Also fit for are the a-coefficients, which specify a polynomial giving the m dependence of the peak frequency. Modes with the same l and n are assumed to have the same amplitude and linewidth for all m. Once we have these mode parameters they can be inverted for sound speed and differential rotation.

2. Improvements

One advantage of this pipeline is that the same software can also process the MDI data that has been ingested into JSOC. In this case we are able to correct in the remapping for change in plate scale, p-angle error, cubic distortion from optics, and alleged CCD tilt. In the HMI data, these type of corrections have already been applied to the dopplergrams, but the remapping still benifits from a correction to the Carrington inclination and an improved algorithm for interpolation. The gapfilling implemented increases the duty cycle and improves the results at high frequencies relative to the standard MDI gapfilling. The peakbagging can take into account distortion of eigenfunctions by differential rotation, the horizontal component of oscillations, and asymmetric line profiles. In this last case another parameter is fitted for each l and n and assumed to be the same for all m. For the effect of these various corrections on mode parameters, see our previous work in [1].

3. Data

We have the option of outputting all intermediate data products, but the pipeline will normally only produce data corresponding to timeseries. The keywords descibing the time period that a particular data record refers to are given in table 1. T_STEP will be a constant in any given dataseries, and equal to the CADENCE keyword of the input dopplergrams. Therefore a time period can be uniquely specified by T_START and NDT, and its length will be NDT*T_STEP. T_START can be specified as a date string or as an offset from the MDI epoch of 1993.01.01_TAI. When a data record additionally refers to a range of spherical harmonic degrees, these must be additionally specified in order to select a unique record. When a data record contains only a single l then we will have LMIN=LMAX, so these two are usually redundant.

Table 1. Keywords.

Keyword	Type	Definition
T_START	time	the beginning of the time interval a record corresponds to.
NDT	int	number of time points represented.
T_STEP	float	length of a time step in seconds, usually a constant.
T_STOP	time	the beginning of the following timeseries, or T_START + NDT * T_STEP.
T_OBS	time	the midpoint of a timeseries as given by (T_START + T_STOP) / 2.
LMIN	int	minimum spherical harmonic degree represented.
LMAX	int	maximum spherical harmonic degree represented.

4. Processing plan

Our current plan is to run the first part of the pipeline once per day, producing one day timeseries with $l = 0$ to $l = l_{max}$. Currently $l_{max} = 1800$, and these timeseries shall be archived. Every 36 or 72 days the one day timeseries will be retiled into longer timeseries with one l each. These are detrended and gapfilled and the resulting timeseries shall also be archived. Peakbagging and whatever other algorithms are in the pipeline will be run, and Fourier transforms and power spectra will be created as needed. The resulting mode parameters shall be archived, and finally will be inverted. All of these data products, as well as window functions, shall be available for download. The source code and scripts that create them are also available online.

5. Other codes and future work

The pipeline described above processes spherical harmonic degress in the range $l = 0$-300. Higher than this individual modes blend into ridges and other algorithms must be employed. One such algorithm that we have implemented in the pipeline is the WMLTP method of Reiter and Rhodes, which can fit m-averaged power spectra to above $l = 1000$ [2]. Another high-l algorithm pending integration with JSOC is the ridge-fitting method of Korzennik and Rabello-Soares [3]. More generally, any code that processes timeseries, power spectra, or Fourier transforms and that takes input from fits files can easily be incorporated into the global pipeline.

Another module under development shall simulate MDI medium-l data, which is convolved with a gaussian and subsampled, from HMI dopplergrams, as well as perform arbitrary binning. This has the dual advantage of reducing I/O and memory usage and also of giving HMI better continuity with MDI. We also plan to add corrections for difference in light travel time from different points on the disk and geometric errors related to the height of formation of spectral lines.

Acknowledgments
This work has been supported by NASA contract NAS5-02139.

References

[1] Larson, T. & Schou, J. 2008, *J. Phys.: Conf. Series*, **118**, 012083
[2] Reiter, J. *et al.* 2010, in preparation
[3] Rabello-Soares, M. C., Korzennik, S. G. & Schou, J. 2008, *Solar Phys.*, **251**, 197

HMI time-distance pipeline: an overview and data products

J. Zhao[1], S. Couvidat[1], R. S. Bogart[1], T. L. Duvall, Jr.[2], A. G. Kosovichev[1], J. G. Beck[1] and A. C. Birch[3]

[1] W.W.Hansen Experimental Physics Laboratory, Stanford University, Stanford, CA 94305-4085, USA

[2] Laboratory for Astronomy and Solar Physics, NASA Goddard Space Flight Center, Greenbelt, MD 20771, USA

[3] North West Research Associates, CoRA Division, 3380 Mitchell Lane, Boulder, CO 80301, USA

E-mail: junwei@sun.stanford.edu

Abstract. The Helioseismic and Magnetic Imager onboard Solar Dynamics Observatory provides uninterrupted high-resolution observations of solar oscillations over the entire disk. This gives a unique opportunity for mapping subsurface flows and wave-speed structures and investigating their role in the Sun's dynamics and magnetic activity on various scales by methods of local helioseismology. A data analysis pipeline for the time-distance helioseismology analysis has been developed and implemented at the SDO Joint Science Operation Center (JSOC) at Stanford. It provides near-real time processing of the helioseismology data. We provide an overview of this pipeline, including the data flow procedures, measurement and inversion codes, and our data products.

1. Introduction

The Helioseismic and Magnetic Imager onboard *Solar Dynamics Observatory* (*SDO*/HMI, [1]) observes the solar full-disk intensity, Doppler velocity, and vector magnetic field of the photosphere with high spatial resolution and high temporal cadence. Similar to the Michelson Doppler Imager (MDI, [2]), an instrument onboard a previous space mission *Solar and Heliospheric Observatory* (*SOHO*), the HMI Dopplergrams are primarily used for helioseismology analysis to investigate the interior structure and dynamics of the Sun. Helioseismology data analysis pipelines are planned for near real-time analyses of the HMI observations in order to provide the analysis results to the helioseismology and solar physics communities. The time-distance analysis pipeline is one of the pipelines for local helioseismology studies. It is designed for routine derivations of nearly full-disk subsurface wave-speed perturbations and horizontal flow fields every 8 hours, as well as synoptic flow maps for every Carrington rotation. It can also be used to analyze specific target areas and time periods.

2. Acoustic travel time measurement

SDO/HMI continuously observes the full-disk Sun, providing Doppler velocity, continuum intensity, line-depth, line-width, and magnetic field maps with a 45-sec cadence, and also vector magnetic field measurements with a cadence of 12 minutes. Each full-disk image has

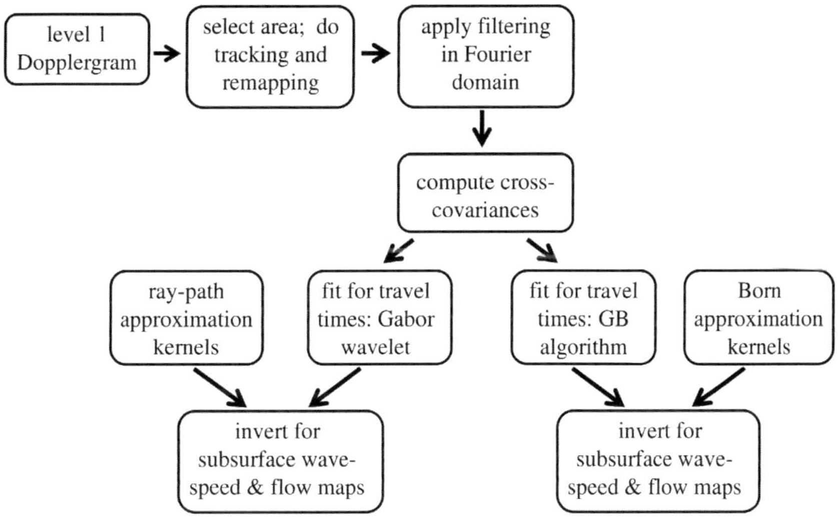

Figure 1. Flow chart for the HMI time-distance helioseismology data analysis pipeline.

4096×4096 pixels with a spatial resolution of 0.504 arcsec pixel^{-1} (i.e., approximately, 0.03 heliographic degree pixel^{-1} at the solar disk center). The Doppler observations are primarily used for helioseismology studies.

As illustrated in Figure 1, the primary input for the pipeline is Dopplergrams, although in principle, the HMI intensitygrams and line-depth data can also be analyzed in the same manner. Users of the pipeline can select specific areas for analysis, preferably within 60° from the solar disk center. In practice, the users provide the Carrington longitude and latitude of the center of the selected area, and the middle time of the selected time period, then the pipeline code selects an area of roughly $30° \times 30°$ centered at the given coordinate, and for a time interval of 8 hours with the given time as the middle point. The data for this selected area and the time period are then tracked to remove the solar rotation, and remapped to Postel's projected coordinates (Postel's projection is also known as azimuthal equidistant projection) relative to the given area center. Normally, the tracked area is 512×512 pixels with a spatial sampling of 0.06° pixel^{-1}; and the temporal sampling is kept concordant with the observational cadence.

Each tracked and remapped Dopplergram datacube is filtered in the 3D Fourier domain. Solar convection and f-mode oscillation signals are removed first, and then phase-speed filtering is applied following the procedures prescribed by [3]. For the travel-time measurements, for each central point, we select 11 annuli with various radii and widths chosen from our past experiences with MDI analyses. All the phase-speed filtering parameters, including the central phase-speed, the filter width, and the corresponding inner and outer annulus radii, are shown in Table 1. The filter is a Gaussian function of phase speed. After the filtering, the data are transformed back to the space-time domain for cross-covariance computations for fittings to get acoustic travel times.

3. Subsurface wave-speed perturbation and flow field inversions

As illustrated in Figure 1, the acoustic travel times are derived by two different fitting methods: the Gabor wavelet function and the GB algorithm. Then, to infer the subsurface wave-speed perturbations and flow velocities, the Gabor wavelet fitted acoustic travel times are inverted using the ray-path approximation sensitivity kernels, and the GB algorithm fitted times are inverted using the Born approximation sensitivity kernels. Born approximation kernels are calculated based upon the GB definition of acoustic travel times. In this analysis pipeline, we

Table 1. Phase-speed filtering parameters used for the selected travel distances (annulus ranges).

annulus No.	annulus range (heliographic degree)	phase speed (μHz/ℓ)	FWHM (μHz/ℓ)
1	0.54 – 0.78	3.40	1.0
2	0.78 – 1.02	4.00	1.0
3	1.08 – 1.32	4.90	1.25
4	1.44 – 1.80	6.592	2.149
5	1.92 – 2.40	8.342	1.351
6	2.40 – 2.88	9.288	1.183
7	3.12 – 3.84	10.822	1.895
8	4.08 – 4.80	12.792	2.046
9	5.04 – 6.00	14.852	2.075
10	6.24 – 7.68	17.002	2.223
11	7.68 – 9.12	19.133	2.039

employ the MCD inversion method [4] with a horizontal regularization [3].

For both the wave-speed and flow field inversions, and for both the ray-path and Born approximation kernels based inversions, we select a total of 11 inversion depths as follows: 0 – 1, 1 – 3, 3 – 5, 5 – 7, 7 – 10, 10 – 13, 13 – 17, 17 – 21, 21 – 26, 26 – 30, and 30 – 35 Mm. There are a total of 11 depth intervals. The inversion results provide the wave-speed perturbations and flow velocities averaged in these layers. Due to the lack of acoustic wave coverage in the deep interior, the reliability of inversion results decreases with the depth. Thus, only inversion results shallower than 20 Mm are given in the pipeline output.

In recent years, several studies have been carried out to validate the time-distance measurements and inversions. To validate the derived subsurface flow fields, [5] and [6] have analyzed realistic solar convection simulations and found satisfactory inversion results for shallow depths covered by the simulations. Validations of the wave-speed perturbation inversions based on numerical simulations with preset structures have also been performed, and a publication is under preparation. Meanwhile, numerical simulations for magnetic sunspot structures with flows are also under development. Validations of the time-distance helioseismology techniques will be carried out as well using these simulations.

4. Data products

The time-distance data analysis pipeline is used for routine production of nearly real-time full-disk (actually, nearly full-disk covering $120^\circ \times 120^\circ$ area on the solar disk) wave-speed perturbation and flow field maps every 8 hours. These maps are then used to construct the corresponding synoptic maps for each Carrington rotation. The pipeline can also be used for specific target areas, such as active regions. For the routine production of the full-disk wave-speed and flow maps, for each day of HMI observations, we select three 8-hour periods of 00:00 – 07:59UT, 08:00 – 15:59UT, and 16:00 – 23:59UT. For each analysis period, we select 25 regions, with the central locations at 0°, $\pm 24^\circ$, and $\pm 48^\circ$ in both longitude and latitude, where the longitude is relative to the central meridian at the middle time of the selected period.

Figure 2 shows a sample of selected areas on the solar disk. The total number of areas is 25, 5 rows and 5 columns. Because of the Postel's projection, the boundaries of these areas are often not parallel to the latitude or longitude lines. It is also evident that many areas overlap, and some areas overlap two times and some overlap four times. The travel times and inversion results are averaged in these overlapped areas. For each full-disk map and each synoptic map,

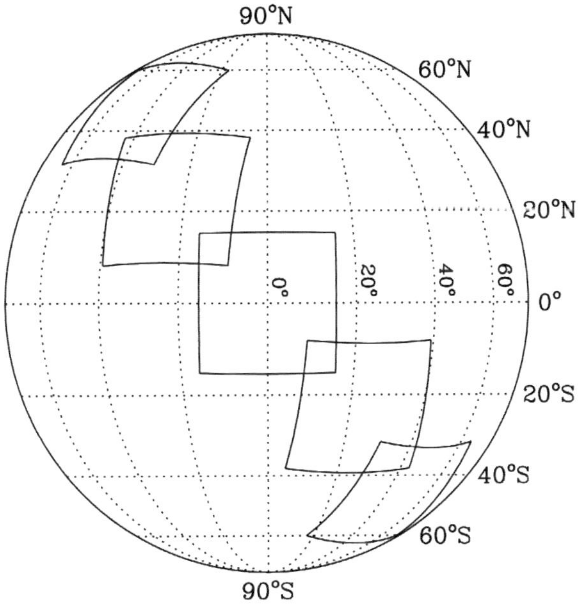

Figure 2. Schematic plot showing how areas are selected for a routine calculations of the full-disk wave-speed and flow maps. Not all of the 25 selected areas are showed in this plot.

the east-west velocity (v_x), the north-south velocity (v_y), and wave-speed perturbation (c_s) in each depth layer are derived with a horizontal spatial sampling of 0.12° pixel^{-1}. For each of 25 areas, the inversion results are first obtained in the Postel's projected coordinates, and then converted into the longitude – latitude coordinates.

As already mentioned, the pipeline can also be run for specific target areas and specific time intervals. Users are required to provide the Carrington coordinate of the center of the target area, and the middle time of the time interval.

5. Summary

We have developed a time-distance data analysis pipeline for *SDO*/HMI Doppler observations. This pipeline performs acoustic travel time measurements based on two different fitting methods, and conducts inversions based on two different inversion kernels calculated in the ray-path and Born approximations. The pipeline gives nearly real-time routine products of full-disk wave-speed perturbations and flow field maps in the range of depth 0 – 20 Mm every 8 hours, and provides the corresponding synoptic wave-speed perturbation and flow field maps for each Carrington rotation. In addition to these routine production, the pipeline can also be used for analysis of specific target areas for specific time intervals.

References

[1] Schou J, et al. 2010 *Solar Phys.* submitted
[2] Scherrer P H, et al. 1995 *Solar Phys.* **162** 129
[3] Couvidat S, Gizon L, Birch A C, Larsen, R M, Kosovichev, A G 2005 *Astrophy. J. Supple. Ser.* **158** 217
[4] Jacobsen B H, Møller I, Jensen, J M, Efforsø, F 1999 *Phys. Chem. of the Earth A - Solid Earth and Geodesy* **24** 15
[5] Georgobiani D, Zhao J, Kosovichev A G, Bensen D, Stein R F, Nordlund Å. 2007 *Astrophys. J.* **657** 1157
[6] Zhao J, Georgobiani D, Kosovichev A G, Benson D, Stein R F, Nordlund Å. 2007 *Astrophys. J.* **659**, 848

Dynamos in Stellar Convection Zones: of Wreaths and Cycles

Benjamin Brown

Dept. of Astronomy and Center for Magnetic Self Organization in Laboratory and Astrophysical Plasmas, University of Wisconsin, Madison, WI 53706, USA

E-mail: bpbrown@astro.wisc.edu

Abstract. We live near a magnetic star whose cycles of activity are driven by dynamo action beneath the surface. In the solar convection zone, rotation couples with plasma motions to build highly organized magnetic fields that erupt at the surface and undergo relatively regular cycles of polarity reversal. Despite our proximity to the Sun, the nature of its dynamo remains elusive, but observations of other solar-type stars show that surface magnetism is a nearly ubiquitous feature. In recent time, numerical simulations of convection and dynamo action have taken tremendous strides forward. Global-scale organization and cyclic magnetism are being achieved by several groups in distinctly different solar and stellar simulations. Here I will talk about advances on the numerical front including wreath-building dynamos that may occupy stellar convection zones. I will discuss the interplay between the new simulations, various classes of mean-field models, and current and upcoming solar and stellar observations.

1. Introduction

The 22-year solar activity cycle stands out as one of the most remarkable and enigmatic examples of magnetic self-organization in nature. The magnetism we see at the surface as sunspots likely originates in the solar convection zone, where turbulent plasma motions couple with rotation and magnetic fields to drive strong dynamo action. Magnetism is a ubiquitous feature of solar-type stars and there are well known scaling relationships between the amount of surface magnetism and stellar properties, such as rotation rate and stellar-type [e.g., 1]. Other late-type stars undergo similar magnetic cycles, with periods ranging from several years to several decades [2, 3]. As in the case of the solar cycle, these must arise through hydromagnetic dynamos operating in their convective envelopes. However, the observational landscape is complex, with few well-established trends to constrain dynamo models [e.g., 4, 5].

Indeed, explaining the origin of globally organized fields and cyclic behavior in the Sun has inspired and challenged astrophysical dynamo theory for over a century and continues to do so. The tremendous growth of computational resources, coupled with the insights afforded by helioseismology about the solar internal structure and differential rotation, has lead to an explosion of dynamo modeling efforts. These range from sophisticated 2D mean-field models that incorporate data assimilation techniques to fully 3D simulations that can capture non-linear dynamics of solar convection and self-consistently establish solar-like differential rotation profiles. Both classes of models are being applied to stars other than the Sun, sampling different spectral types and exploring how rotation affects stellar convection and dynamo action.

Here I will briefly review the state of dynamos for the Sun and sun-like stars that rotate somewhat faster, as the Sun did in its youth. These studies are suggesting new modes of global-scale dynamo action and are raising exciting questions about the nature of the solar dynamo.

2. Mean-field models of stellar dynamos

Simulations of the solar dynamo and solar convection are generally broken into two classes: the two dimensional (2D) mean-field models and the three-dimensional (3D) simulations. By going to 2D, typically in radius and latitude, the computations can be made much more tractable and high resolution simulations can be pursued for very long intervals of time. It is not unusual for such models to simulate several tens of solar cycles, and reproducing the entire observational record of sunspots is quite feasible [e.g., 6].

The trouble lies in the treatment of turbulence and non-linear correlations. Dynamo action generally requires correlations between the non-axisymmetric, fluctuating velocity and magnetic fields. In mean-field models these non-axisymmetric flows are not directly simulated and instead must be captured through some assumed model. This is a very difficult and outstanding problem in turbulence theory. Generally, an attempt is made to model the coherent effect of the fluctuations in terms of the global-scale (mean) fields, and this is often embodied as an "α-effect," though many variations have been explored. Cyclonic convection coupling with rotation is often associated with the α-effect and thus α is thought typically to depend on the kinetic helicity of the convection, but the α-effect remains very difficult to constrain observationally. Comparable difficulties underlie descriptions of the turbulent processes that transport angular momentum to self-consistently establish the observed solar differential rotation. This mean internal rotation profile, however, can be measured in detail throughout the solar convection zone, and the observed profile is directly incorporated in the mean-field models [e.g., 7].

The α-effect is most important for the regeneration of poloidal (north-south) magnetic field; mean toroidal fields can be generated from a mean poloidal field by the shear of differential rotation in what is generally called an "Ω-effect." These $\alpha\Omega$ dynamos and their variants have become central to the language used to discuss solar and stellar dynamos. Some live in the convection zone alone while others rely on the interface at the base of the convection zone, the tachocline, to generate cyclic reversals of global-scale polarity.

Many recent solar dynamo models have also emphasized the meridional circulation as a potentially important factor in promoting cyclic magnetic activity [e.g., 6, 8–14]. In these *Flux-Transport* models, the equatorward migration of emerging bipolar active regions over the course of the solar cycle is attributed to the equatorward advection of toroidal flux in the lower convection zone by the mean circulation. Many Flux-Transport models are also *Babcock-Leighton* models whereby the principle source of mean poloidal field generation is the buoyant emergence and subsequent dispersal of fibril toroidal flux concentrations, often modeled as a non-local α-effect [15]. Although the physical origin of the Babcock-Leighton mechanism is distinct from the turbulent α-effect underlying distributed and interface dynamo models, the fields generated are still helical in nature and this helicity ultimately arises from the rotation of the star.

The review by Charbonneau [15] is an excellent place to read in more detail about these models, while [16] delves into the details of the turbulence models themselves. Comparisons between many of the codes used in different mean-field models have been undertaken by [17]. Mean-field models still do not unambiguously reproduce the solar cycle, despite heroic efforts leading up to the current cycle [e.g., 6]. To be clear, these models are much further along than the 3D models, which are beginning to regularly produce cyclic solutions only in recent time [e.g., 18–21]. Mean-field models can be run for many solar cycles and are an invaluable tool for exploring the extensive parameter space of solar and stellar convection, and several explorations have been made of global-scale circulations in solar-like stars [e.g., 22–26]. However, many of the important underlying variables are difficult to constrain (e.g., the dependence of α on radius

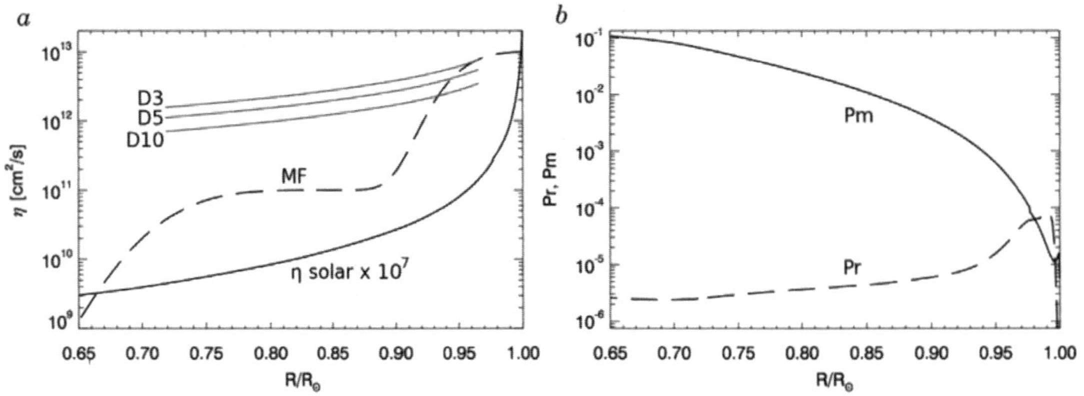

Figure 1. (a) Radial profiles of magnetic diffusivity η used in various models. Shown is the double-step profiles used in the Babcock-Leighton models of [7] and similar to that used in [6, 27] (MF,- - - -). Three profiles from 3D MHD ASH dynamos are shown for comparison; these correspond to cases D3, D5 and D10 (grey lines). The molecular diffusivity for a hydrogen plasma at solar conditions is also shown, multiplied by 10^7 for display purposes. (b) The Prandtl number $\mathrm{Pr} = \nu/\kappa$ (dashed) and magnetic Prandtl number $\mathrm{Pm} = \nu/\eta$ (solid) for a hydrogen plasma at solar conditions. Simulations use values of order unity.

and latitude in the Sun; how α varies with stellar properties such as rotation rate and mass; the applicability of α-effects to modeling the turbulent induction observed in 3D models; etc.). We turn now to a discussion of fully non-linear 3D convection driven dynamos, which are beginning to provide the opportunity to better constrain these unknown quantities.

3. The gap between simulations and the Sun

Numerical studies of the solar dynamo have a rich history, with the the first 3D magnetohydrodynamic (MHD) convective global-scale solar dynamo simulations attaining cyclic behavior in the early studies of Gilman [28]. Those Boussinesq simulations were quickly joined by fully non-linear global-scale anelastic simulations, which captured the stratified nature of the solar convection zone as well [29, 30]. Computational resources have grown at a tremendous rate, and with them the complexities of the models studied. Modern simulations typically have higher resolutions and evolve for longer intervals of time; in one simulation we will examine later (case D5), this represents roughly a factor of a million more computation than was possible in early studies [28]. This is is in surprisingly good agreement with Moore's law doubling over the almost thirty year interval separating these simulations, but it helps clarify the huge gap remaining between solar convection and the highest resolution simulations: another century of growth might provide the resources to directly simulate solar convection on global-scales.

Stellar convection spans a vast range of spatial and temporal scales which remain well beyond the grasp of direct numerical simulation. Models of stellar dynamos must make various tradeoffs, either building up from the diffusive scales or building down from the global-scales. These are respectively called *local* or *global* simulations; the latter will be our focus here. The highest resolution modern 3D simulations, running on massively parallel supercomputers, capture roughly 1000^3 total points and typically can evolve for some $10^6 - 10^7$ timesteps. In the Sun, the smallest scales of motion are set by diffusion and are likely of order 1 mm [e.g., review 31] while the largest scales or motion are comparable to the solar radius (700 Mm), with a total spectral range of almost 10^{12} in each of three dimensions. Temporal separations are similar, with fast granulation on the surface overturning on roughly five minute timescales while the deep structure of the Sun evolves over a span of gigayears.

Clearly, stellar dynamo studies must drastically simplify the physics of the stellar interior. Molecular values of the magnetic diffusivity η for a hydrogen plasma under the conditions of the the solar convection zone range from roughly 10^2–10^6 cm^2/s as one moves from the tachocline to the near surface regions, while the molecular viscosity ν is of order 10 cm^2/s in the solar convection zone [e.g., 32, 33]. These diffusivities are vastly smaller than the values used in either mean-field models or 3D MHD simulations, and we illustrate this in Figure 1.

In contrast to the solar values, simulations typically employ values of η and ν that are of order 10^{12} cm^2/s; this large value is more similar to simple estimates of turbulent diffusion associated with granulation at the surface where $\nu_t \sim V_t L_t \sim 10^{11}$ cm^2/s given $V_t \sim 1$ km/s and $L_t \sim 1$ Mm [e.g., 31]. Shown in Figure 1*a* are radial profiles of η for both mean-field models and 3D MHD dynamo simulations with the ASH code. Mean-field models often match to a turbulent diffusion consistent with supergranulation at the surface and then taper to a lower value at mid-convection zone (here 10^{11} cm^2/s) [e.g., 6, 7, 27]. Below the tachocline, η_{MF} is tapered further, sometimes approaching the molecular values. In 3D models the choice is often made to scale η with the background density. Here a scaling of $\rho^{-0.5}$ is used, though in other studies the exponent can scale from 0 to -1 [e.g., 34]. The relative mixing from diffusion of vorticity (by viscosity ν), temperature (κ) and magnetism (η) is given by the Prandtl number and magnetic Prandtl number, which are shown in Figure 1*b*. Molecular ratios for a hydrogen solar plasma are tiny, with Pm ranging from 10^{-1} to 10^{-5} while Pr is of order 10^{-5}. Turbulent values are likely of order unity but are not well constrained under solar conditions; simulations typically take Pm and Pr to be near unity as resolving large separations in diffusive scales requires very high resolutions.

4. Convection driven dynamos: the Sun

Despite this daunting separation in parameter space, modern models are making tremendous strides in understanding the non-linear couplings between convection and rotation that build the solar differential rotation. It is now possible to study relatively high Reynolds number convection (fluctuating Re of order a few hundred) in stratified convection zones capturing density contrasts exceeding 100 (e.g., more than 5 density scale heights) between the base of the convection zone and the near-surface layers. The anelastic spherical harmonic (ASH) code has been a very useful tool in global-scale studies [34–36]; tremendous progress has been made in models of photospheric convection as well, but those will not be the focus here [e.g., 37].

Simulations of solar convection self-consistently produce solar-like profiles of differential rotation, with prograde equators, retrograde polar regions and a monotonic decrease of angular velocity with latitude. This profile is achieved partly through the redistribution of angular momentum by turbulent Reynolds stresses in the convection [38–40]. The differential rotation profile is in what is called a "thermal-wind balance", with an accompanying latitudinal gradient of temperature. This is an effect that is well known in the geophysical community, entering the vorticity evolution equation as a baroclinic term. In solar convection, thermal-wind balance leads to more conical profiles of angular velocity Ω [38, 40]. Simulations suggest that in the Sun the accompanying temperature perturbations at the solar surface may be of order 1-10K. Perturbations of this size at the base of the convection zone, consistent with the geostrophic balances likely achieved in the tachocline, can serve to tilt the contours of constant Ω until very good agreement is attained between simulations and helioseismic measurements [41, 42].

Solar dynamo simulations generally produce complex magnetic topologies, with more than 95% of the magnetic energy in the fluctuating (non-axisymmetric) field components [34]. Mean fields are complex, with multipolar structure and transient toroidal ribbons and sheets. Polarity reversals of the dipole component occur but they are irregular in time. The presence of an overshoot region and a tachocline of rotational shear promotes mean-field generation, producing persistent bands of toroidal flux antisymmetric about the equator while strengthening and

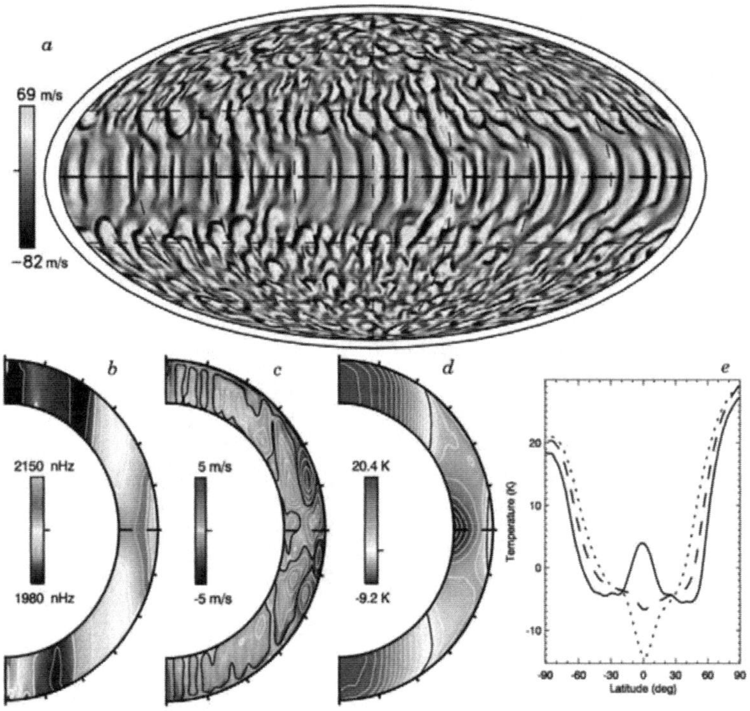

Figure 2. Convection and global-scale flows in case D5 [18]. (*a*) Convective patterns of radial velocity v_r in global Mollweide projection at 0.95 $R_\odot$ with upflows light and downflows dark. Poles are at top and bottom and thin line denotes stellar surface. (*b*) Differential rotation shown with longitudinally averaged angular velocity Ω. The rotation is solar-like, with fast (prograde) equator. (*c*) Meridional circulations with color indicating amplitude and sense of circulation (red counter-clockwise; blue clockwise) and mass flux streamlines overlaid. Compared to solar simulations, the circulations here are broken into several weaker cells in both radius and latitude. (*d*) Profile of mean temperature fluctuations. This profile, with hot poles and cool mid-latitudes, represents the thermal wind balance achieved with the differential rotation. (*e*) Latitude cuts of temperature at fixed radius, sampling top (——), middle (– – –) and bottom (⋯⋯) of convection zone. At the surface, the temperature contrast in latitude can reach 30K. Profiles shown in *b*–*e* have been averaged in time over a 225 day interval.

stabilizing the dipole moment [43–45]. These simulations exhibit notable self-organization through the turbulent pumping of magnetic flux into the tachocline, amplification by rotational shear, and selective shear-induced dissipation of small-scale magnetic fluctuations. Modern simulations build organized magnetic fields and attain cyclic behavior, sometimes with a tachocline playing an important role [19, 20], and sometimes in the convection zone of the Sun alone [21, 46]. We now turn to the special class of wreath-building convection zone dynamos.

5. Convection driven dynamos: rapidly rotating suns

When stars like the Sun are younger they rotate much more rapidly. These stars are observed to have strong surface magnetic activity and are thought to have very active dynamos in their convection zones. Rotational constraints are stronger in more rapidly rotating systems and this can lead to greater correlations as convective structures align with the rotation axis.

Patterns of convection in a simulation of a young, rapidly rotating sun are shown in Figure 2*a* near the stellar surface. Convection fills the domain and near the equator is strongly north-south

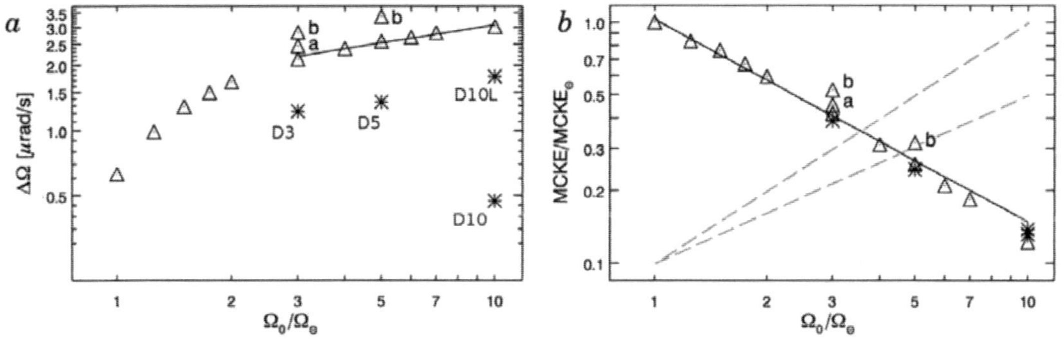

Figure 3. Global-scale flows and stellar rotation [47]. (*a*) Angular velocity shear of differential rotation $\Delta\Omega$ in latitude near the stellar surface shown as a function of rotation rate Ω_0 relative to the solar rotation rate $\Omega_\odot$. Hydrodynamic cases are shown with diamonds while dynamos are labeled and shown with asterisks. $\Delta\Omega$ grows with more rapid rotation in hydrodynamic cases. Cases labeled a and b sample more turbulent states. (*b*) Kinetic energy of the meridional circulations MCKE, normalized to that energy in the simulation at the solar rate. MCKE decreases with more rapid rotation; a powerlaw of $\Omega_0^{-0.9}$ is shown for reference. Grey dashed lines indicate scalings typically used in mean-field models for more rapidly rotating suns (e.g., [48–50], and see [51] for models that follow the $\Omega_0^{-0.9}$ scaling).

aligned. Correlations in these "bannana-cells" transport angular momentum and build the profile of differential rotation shown in Figure 2*b*; the equator is fast, the poles are slow and the angular velocity contrast is larger than in the Sun. The meridional circulations in contrast are weak and multi-celled in both radius and latitude (Fig. 2*c*). Accompanying the angular velocity profile is a large latitudinal gradient of temperature, shown Figures 2*d*, *e*. Near the surface, there can be 30K contrasts between the hotter poles and cooler mid-latitudes.

The angular velocity contrast of the differential rotation $\Delta\Omega$ and the kinetic energy contained in the meridional circulations (MCKE) is shown for many rapidly rotating suns in Figure 3. Generally, we find that $\Delta\Omega$ grows with more rapid rotation, while MCKE drops strongly [47]. The decrease of MCKE is a surprise and may hold important implications for flux-transport dynamos in the mean-field framework [51]. The growth of $\Delta\Omega$ is roughly in agreement with observations of surface differential rotation in other stars, though substantial disagreement remains between different groups of observers [e.g., 52, 53]. In hydrodynamic cases (triangles) this shear continues to grow with faster rotation; in dynamo cases it may begin to saturate as Lorentz forces become important and react back on the differential rotation (asterisks). In all cases, the growth of $\Delta\Omega$ with rotation rate Ω_0 is accompanied by a growing latitudinal temperature contrast. The temperature contrast grows from a few K at the solar rate to several hundred K at the fastest rotation rates.

The magnetic fields created in dynamo simulations of rapidly rotating suns are organized on global-scales into banded wreath-like structures. These are shown for a dynamo at three times the solar rotation rate (case D3) in Figure 4*a* [54, 55]. Two such wreaths are visible in the equatorial region, spanning the depth of the convection zone and latitudes from roughly $\pm 30°$. The dominant component of the magnetism is the longitudinal field B_ϕ, and the two wreaths have opposite polarities (red, positive; blue, negative). An even more rapidly rotating dynamo (case D5) is shown in Figure 4*b*. Now the wreaths fill the convection zone and the polar caps. These wreaths show significant time-variation and undergo quasi-regular polarity reversals [18].

The time history of these two cases is shown in Figures 5*a*, *b* (case D3 and D5 respectively). Here the mean (axisymmetric) longitudinal magnetic field $\langle B_\phi \rangle$ is shown at mid-convection zone over an interval of about 4000 days; both cases have a full history of roughly 20,000 days.

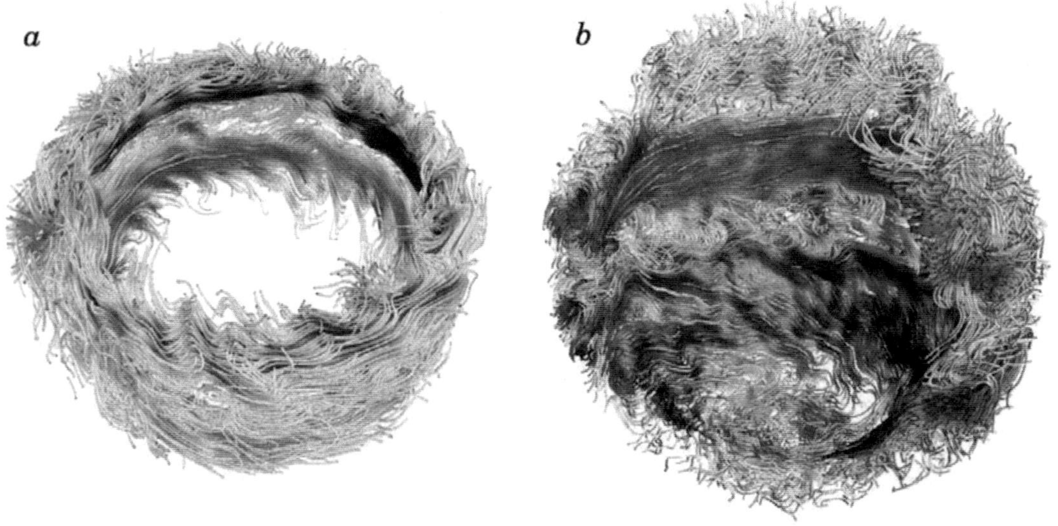

Figure 4. Magnetic wreaths in stellar convection zones. (*a*) Persistent wreaths in case D3. Two wreaths of opposite polarity (red, positive; blue, negative) form above and below the equator. These magnetic structures coexist with the turbulent convection and retain their identity for more than 20,000 simulated days. (*b*) Magnetic wreaths in cyclic case D5. In this simulation the wreaths undergo reversals of polarity on roughly a 1500-day timescale. Relic wreaths from the previous cycle are visible in the polar caps. This snapshot is at same instant as Figure 2*a*.

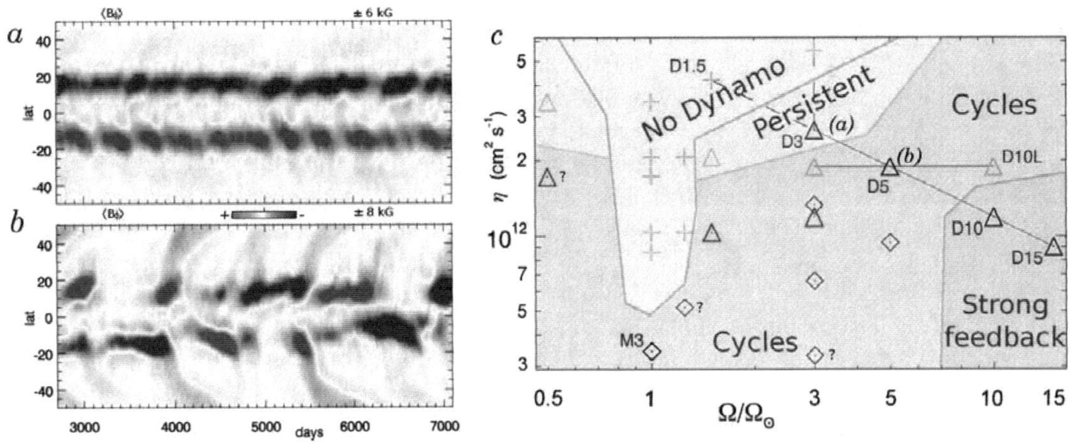

Figure 5. Family of wreath-building dynamo solutions. (*a*) Time-latitude plot of mean (axisymmetric) longitudinal magnetic field B_ϕ at mid-convection zone in persistent case D3 [55]. (*b*) Cyclic case D5 shown for same span of time [18]. Three reversals are visible here, occurring on roughly 1500 day periods. (*c*) Primary control parameters for simulations, with magnetic diffusivity η and rotation rate Ω_0 sampling 0.5–15$\Omega_\odot$. Very approximate dynamo regimes are shown. Time dependence emerges at higher magnetic Reynolds number (lower η) and many dynamos undergo repeated reversals of global-scale polarity. Cases with question marks show significant time-variation but have not been computed for long enough to definitively establish cyclic behavior. At the highest rotation rates the Lorentz force can substantially reduce the differential rotation, but dynamo action is still achieved.

In case D3 we generally find little time variation in the axisymmetric magnetic fields associated with the wreaths; small variations are visible on a roughly 500 day timescale, but the two wreaths retain their polarities for the entire interval, which is significantly longer than the convective overturn time (roughly 10–30 days), the rotation period (9.3 days), or the ohmic diffusion time (about 1300 days at mid-convection zone). We refer to the dynamos in this regime as persistent wreath-builders. Case D5 (Fig. 5*b*) is very different. Here global-scale polarity reversals occur roughly every 1500 days. Three such reversals are shown here. The ohmic diffusion time in this simulation is about 1800 days, while the rotation period is 5.6 days.

These two simulations are part of a much larger family of wreath-building dynamos, which are summarized in Figure 5*c*. Shown here are 26 simulations at rotation rates ranging from $0.5\ \Omega_\odot$ to $15\ \Omega_\odot$. Wreath-building dynamos are achieved in most simulations (17), though a smaller number do not successfully regenerate their mean poloidal fields (9, indicated with crosses). At individual rotation rates (e.g., $3\Omega_\odot$), further simulations explore the effects of lower magnetic diffusivity η and hence higher magnetic Reynolds numbers. Some of these follow a path where the magnetic Prandtl number Pm is fixed at 0.5 (triangles) while others sample up to Pm=4 (diamonds). Near the onset of dynamo action the wreaths are similar to those found in case D3 and persist for long intervals with little variation in time. At higher magnetic Reynolds numbers (lower η and higher Ω_0) we find many simulations that show cyclic reversals of global-scale magnetism (as in case D5). It is difficult to determine what sets the cycle period in these dynamos: cycles appear to become shorter as η decreases, opposite to what might be expected if the ohmic time determined the cycle period. The dependence of cycle period on Ω_0 is less certain. It is striking that coherent magnetic structures can arise at all in the midst of turbulent convection. We find the combination of global-scale spatial organization and cyclic behavior fascinating, as these appear to be the first self-consistent 3D convective stellar dynamos to achieve such behavior in the bulk of the convection zone rather than relying on a stable tachocline of shear.

6. Where we now stand

This is an exciting time in solar and stellar dynamo theory. Stellar dynamo models have progressed tremendously in the past decade. Mean-field models are reaching a point where credible predictions of upcoming solar cycles can be attempted. Meanwhile, several different 3D simulations using distinctly different codes and assumptions have achieved global-scale magnetism and cyclic behavior in simulations of the solar dynamo [e.g., 19–21, 46, 56]. Global-scale organization and cyclic reversals are being found even in simulations without tachoclines. A major challenge now is to understand why such cycles occur. Significant progress can be made on this problem by translating the results of 3D dynamo models into the language of mean-field theory, measuring difficult to constrain quantities such as α and the turbulent electromotive force (emf) that builds the mean poloidal fields. First attempts at diagnosing these quantities are underway [e.g., 18, 55, 57] but now 3D and mean-field modellers must partner to better understand the cyclic dynamo simulations [e.g., 51, 58]. Such efforts will refine the mean-field models but will also yield crucial insights into the operation of the dynamos within the highly turbulent and time-dependent 3D simulations.

It is also crucial that further observational constraints be provided for such 3D models; here is a brief and biased wish list. In the Sun, some sense of the deep meridional circulations would greatly enhance our confidence in modeling results. In particular, it would be useful to determine whether these circulations are multi-cellular in latitude and radius, or whether one large cell extends from the surface to the tachocline. Simulations generally find that thermal-wind balances arise along with the differential rotation. The amplitude of thermal perturbations at the surface is likely to be quite small, of order 1-10K, but the detection (or non-detection) of such a latitudinal gradient would be very useful. Estimates are beginning to be made based on simulations [e.g., 40, 42, 59–61] and attempts have been made to observe this profile in the Sun [e.g., 62, and references therein]. Any estimate of the properties of giant cell convection in the Sun would be tremendously useful as well. Direct detection of these structures would of course be ideal, but much could be learned from indirect observations as well [e.g., 63].

Convective dynamo models in 3D are being applied to other stars, both similar to and different from the Sun. For the solar-type stars, observations of the surface differential rotation $\Delta\Omega$, its variation with stellar mass and rotation rate, and its temporal variation would greatly constrain these models. Additionally, it is crucial to know how basic dynamo properties, such as the amount of surface magnetism and the cycle period, scale with differential rotation $\Delta\Omega$ rather than overall rotation rate Ω_0. The thermal wind balances achieved in more rapidly rotating solar-type stars likely lead to latitudinal temperature contrasts of several hundred K; in more massive and luminous F-type stars, these contrasts may be as large as several thousand K (K. Augustson, private communication). These signatures may be observable.

The research on wreath-building dynamos in rapidly rotating suns has been done in collaboration with Matthew K. Browning, Allan Sacha Brun, Mark S. Miesch, Nicholas J. Nelson and Juri Toomre. I owe them a great debt of gratitude. I also thank Ellen Zweibel and Cary Forest for inspiring the discussion of η and plasma transport. Funding for this research is provided in part through NSF Astronomy and Astrophysics postdoctoral fellowship AST 09-02004. CMSO is supported by NSF grant PHY 08-21899. The simulations were carried out with NSF PACI support of NICS, PSC, SDSC, and TACC.

References

[1] Pizzolato N, Maggio A, Micela G, Sciortino S and Ventura P 2003 *A&A* **397** 147–157

[2] Baliunas S L et al. 1995 *ApJ* **438** 269–287

[3] Oláh K et al. 2009 *A&A* **501** 703–713 (*Preprint* 0904.1747)

[4] Rempel M 2008 *Journal of Physics Conf. Series* **118** 012032.1–012032.10

[5] Lanza A F 2010 *IAU Symposium*, vol 264, ed A G Kosovichev et al. pp 120–129

[6] Dikpati M and Gilman P A 2006 *ApJ* **649** 498–514

[7] Muñoz-Jaramillo A, Nandy D and Martens P C H 2009 *ApJ* **698** 461–478 (*Preprint* 0811.3441)

[8] Wang Y and Sheeley Jr N R 1991 *ApJ* **375** 761–770

[9] Dikpati M and Charbonneau P 1999 *ApJ* **518** 508–520

[10] Küker M, Rüdiger G and Schultz M 2001 *A&A* **374** 301–308

[11] Nandy D and Choudhuri A R 2001 *ApJ* **551** 576–585 (*Preprint* arXiv:astro-ph/0107466)

[12] Rempel M 2006 *ApJ* **647** 662–675 (*Preprint* arXiv:astro-ph/0604446)

[13] Jouve L and Brun A S 2007 *A&A* **474** 239–250 (*Preprint* 0712.3200)

[14] Yeates A R, Nandy D and Mackay D H 2008 *ApJ* **673** 544–556 (*Preprint* 0709.1046)

[15] Charbonneau P 2010 *Living Reviews in Solar Physics* **7** 3:1–91

[16] Brandenburg A and Subramanian K 2005 *Phys. Rep.* **417** 1–4 (*Preprint* arXiv:astro-ph/0405052)

[17] Jouve L et al. 2008 *A&A* **483** 949–960

[18] Brown B P, Miesch M S, Browning M K, Brun A S and Toomre J 2010 *ApJ* submitted

[19] Ghizaru M, Charbonneau P and Smolarkiewicz P K 2010 *ApJ* **715** L133–L137

[20] Käpylä P J, Korpi M J, Brandenburg A, Mitra D and Tavakol R 2010 *Astr. Nachr.* **331** 73–81

[21] Mitra D, Tavakol R, Käpylä P J and Brandenburg A 2010 *ApJ* **719** L1–L4 (*Preprint* 0901.2364)

[22] Rüdiger G, von Rekowski B, Donahue R A and Baliunas S L 1998 *ApJ* **494** 691–699

[23] Küker M and Stix M 2001 *A&A* **366** 668–675

[24] Küker M and Rüdiger G 2005 *A&A* **433** 1023–1030 (*Preprint* arXiv:astro-ph/0409246)

[25] Küker M and Rüdiger G 2005 *Astron. Nachr.* **326** 265–268 (*Preprint* `arXiv:astro-ph/0504411`)
[26] Küker M and Rüdiger G 2008 *Journal of Physics Conf. Series* **118** 012029.1–012029.8
[27] Dikpati M, Gilman P A and MacGregor K B 2006 *ApJ* **638** 564–575
[28] Gilman P A 1983 *ApJS* **53** 243–268
[29] Glatzmaier G A 1984 *Journal of Computational Physics* **55** 461–484
[30] Glatzmaier G A 1985 *ApJ* **291** 300–307
[31] Rieutord M and Rincon F 2010 *Living Reviews in Solar Physics* **7** 2:1–82 (*Preprint* `1005.5376`)
[32] Braginskii S I 1965 *Reviews of Plasma Physics* **1** 205–311
[33] Rieutord M 2008 *Comptes Rendus Physique* **9** 757–765
[34] Brun A S, Miesch M S and Toomre J 2004 *ApJ* **614** 1073–1098
[35] Clune T L, Elliott J R, Glatzmaier G A, Miesch M S and Toomre J 1999 *Parallel Comp.* **25** 361–380
[36] Miesch M S, Elliott J R, Toomre J, Clune T L, Glatzmaier G A and Gilman P A 2000 *ApJ* **532** 593–615
[37] Nordlund Å, Stein R F and Asplund M 2009 *Living Reviews in Solar Physics* **6** 2:1–116
[38] Brun A S and Toomre J 2002 *ApJ* **570** 865–885 (*Preprint* `arXiv:astro-ph/0206196`)
[39] Miesch M S 2005 *Living Reviews in Solar Physics* **2** 1:1–139
[40] Miesch M S, Brun A S, DeRosa M L and Toomre J 2008 *ApJ* **673** 557–575
[41] Rempel M 2005 *ApJ* **622** 1320–1332 (*Preprint* `arXiv:astro-ph/0604451`)
[42] Miesch M S, Brun A S and Toomre J 2006 *ApJ* **641** 618–625
[43] Browning M K, Miesch M S, Brun A S and Toomre J 2006 *ApJ* **648** L157–L160 (*Preprint* `arXiv:astro-ph/0609153`)
[44] Browning M K, Brun A S, Miesch M S and Toomre J 2007 *Astr. Nachr.* **328** 1100–1103
[45] Miesch M S, Browning M K, Brun A S, Toomre J and Brown B P 2009 *Astro. Soc. of the Pacific Conf. Series*, vol 416, ed M Dikpati et al. pp 443–452 (*Preprint* `0811.3032`)
[46] Miesch M S, Brown B P, Browning M K, Brun A S and Toomre J 2010 *Preprint* `1009.6184`
[47] Brown B P, Browning M K, Brun A S, Miesch M S and Toomre J 2008 *ApJ* **689** 1354–1372 (*Preprint* `0808.1716`)
[48] Charbonneau P and Saar S H 2001 *Magnetic Fields Across the Hertzsprung-Russell Diagram* (*Astro. Soc. of the Pacific Conf. Series* vol 248) ed Mathys G et al. pp 189–198
[49] Dikpati M, Saar S H, Brummell N and Charbonneau P 2001 *Magnetic Fields Across the Hertzsprung-Russell Diagram* (*Astro. Soc. of the Pacific Conf. Series* vol 248) ed Mathys G et al. pp 235–238
[50] Nandy D 2004 *Sol. Phys.* **224** 161–169
[51] Jouve L, Brown B P and Brun A S 2010 *A&A* **509** A32:1–11 (*Preprint* `0911.1947`)
[52] Donahue R A, Saar S H and Baliunas S L 1996 *ApJ* **466** 384–391
[53] Barnes J R, Cameron A C, Donati J F, James D J, Marsden S C and Petit P 2005 *MNRAS* **357** L1–L5 (*Preprint* `arXiv:astro-ph/0410575`)
[54] Brown B P, Browning M K, Brun A S, Miesch M S, Nelson N J and Toomre J 2007 *American Institute of Physics Conf. Series* vol 948 pp 271–278
[55] Brown B P, Browning M K, Brun A S, Miesch M S and Toomre J 2010 *ApJ* **711** 424–438
[56] Nelson N J, Brown B P, Browning M K, Brun A S, Miesch M S and Toomre J 2010 *Preprint* `1010.6073`
[57] Racine E, Charbonneau P, Ghizaru M, Bouchat A and Smolarkiewicz P K 2010 *ApJ* submitted
[58] Miesch M S 2008 *Journal of Physics Conf. Series* **118** 012031.1–012031.10
[59] Balbus S A, Bonart J, Latter H N and Weiss N O 2009 *MNRAS* **400** 176–182 (*Preprint* `0907.5075`)
[60] Balbus S A and Latter H N 2010 *MNRAS* **407** 2565–2574 (*Preprint* `1005.5100`)
[61] Brun A S, Antia H M and Chitre S M 2010 *A&A* **510** A33:1–7 (*Preprint* `0910.4954`)
[62] Rast M P, Ortiz A and Meisner R W 2008 *ApJ* **673** 1209–1217 (*Preprint* `0710.3121`)
[63] Hanasoge S M, Duvall T L and DeRosa M L 2010 *ApJ* **712** L98–L102 (*Preprint* `1001.4508`)

A search for coherent structures in subsurface flows

R Komm, R Howe and F Hill

National Solar Observatory, Tucson, AZ 85719, USA

E-mail: `rkomm@nso.edu`

Abstract. We search for coherent patterns in horizontal subsurface flows obtained from Global Oscillation Network Group (GONG) Dopplergram data using ring-diagram analysis. The existence of north-south aligned downflow patterns near the equator has been predicted by numerical models of the solar convection zone. We analyze time series of daily flow measurements near the solar equator focusing on the vertical velocity component and the derivative of the zonal flow in the east-west direction. To reduce the influence of surface magnetic activity, we analyze observations during the minimum phase of the solar cycle. We find coherent equatorial structures that persist for several days in the zonal velocity derivative and the vertical velocity component and are not associated with surface magnetic activity. We use a cross-correlation analysis to measure the strength and rotation rate of these coherent patterns. Our results are consistent with other studies that have observed north-south aligned patterns in supergranulation.

1. Introduction

Global numerical models of the solar convection zone show coherent downflow patterns aligned in the north-south direction in the equatorial region [1, 2]. Such downflow lanes, associated with giant cells, represent the largest and longest lived features of the convection zone in these models. These patterns are detectable in the radial velocity, but they are most noticeable in the derivative of the zonal flow in the east-west direction [2]. While the north-south aligned patterns might be obscured near the surface by granular and supergranular motions, they should be observable with local helioseismic techniques in deeper layers.

We search for such patterns in horizontal subsurface flows obtained from Global Oscillation Network Group (GONG) Dopplergram data using ring-diagram analysis. To reduce the influence of magnetic features, we study subsurface flows during the recent solar cycle minimum. Since the resolution of the standard ring-diagram analysis is comparable to the size of active regions and thus much larger than the narrow downflow lanes, we cannot resolve these lanes directly. Instead we look for persistent patterns near the equator that rotate across the solar disk.

2. Data and Analysis

We determine the horizontal components of solar subsurface flows with a ring-diagram analysis using the dense-pack technique [3]. The full-disk Doppler images are divided into 189 overlapping regions with centers spaced by 7.5° ranging from $\pm 52.5^\circ$ in latitude and central meridian distance (CMD). Each region is apodized with a circular function reducing the effective diameter to 15° before calculating three-dimensional power spectra. The data are analyzed in ring-diagram "days" of 1664 minutes. Each of these regions is tracked throughout the sequence of images using the appropriate surface rate. For each dense-pack day, we derive maps of horizontal velocities at 189 locations in latitude and CMD for 16 depths from 0.6 to 16 Mm.

We calculate daily flow maps from GONG Dopplergram data for a period with minimum solar magnetic activity (Carrington rotation 2071 – 2079; 2008 June 9 – 2009 Feb. 9) that includes the recent solar cycle minimum. For each daily flow map, we calculate the vertical velocity component using mass conservation [4] and the derivative of the zonal flow in the east-west direction or zonal velocity derivative for short. Then, we extract an equatorial strip from each daily flow map by averaging over all values centered within $\pm 22.5^{\circ}$ latitude. We limit this strip to $\pm 45^{\circ}$ CMD to avoid artifacts due to the increased errors near the limb. Large-scale trends with CMD have been subtracted [5]. To increase the signal-to-noise ratio, we average in depth and calculate the values for three depth ranges. Even though we study the epoch of the extended solar minimum, active regions or remnants of regions are occasionally present, as shown in Figure 1.

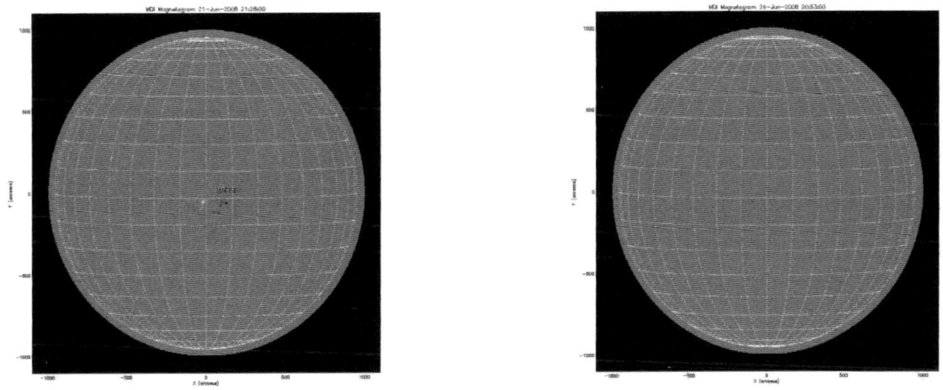

Figure 1. MDI magnetograms of two days during Carrington rotation 2071 showing active region 10999 on 2008 June 21 (left) and no obvious magnetic activity on 2008 June 26 (right).

3. Results

Figure 2 shows, as example, the unsigned magnetic flux averaged over equatorial strips as a function of CMD for all 1664-minute days of Carrington rotation 2071. Except for two small active regions, the unsigned flux is rather flat during most days. The disk passage of these two regions is visible in the zonal velocity derivative as a series of troughs moving across the disk, which implies converging flows or downflows, as expected for active regions. The most obvious pattern that is not related to an active region rotates across the disk from day 13 to 17 crossing disk center on day 15 which is 26 June 2008 (see Figure 1). The corresponding unsigned magnetic flux shows only a slight bump that is small compared to the flux values of the two regions. The flow pattern is noticeable at all depth ranges and is clearly stronger than expected from the unsigned flux variation.

Figure 3 shows the same as Figure 2 for the nine rotations analyzed in this study. Even during this epoch of minimum activity, there are many remnants of activity that rotate across the solar disk. The zonal velocity derivative shows patterns related to magnetic activity but also others that rotate across the disk without a corresponding pattern in the unsigned flux.

To quantify the strength of these features and to determine their rotation rate, we correlate the time series of the central meridian (0° CMD) with the time series at other CMD locations. Figure 4 shows a strong correlation signal in the unsigned flux from -45° CMD and a lag of -3 days to $+45^{\circ}$ CMD and a lag of $+3$ days. These locations of strong correlation align with the

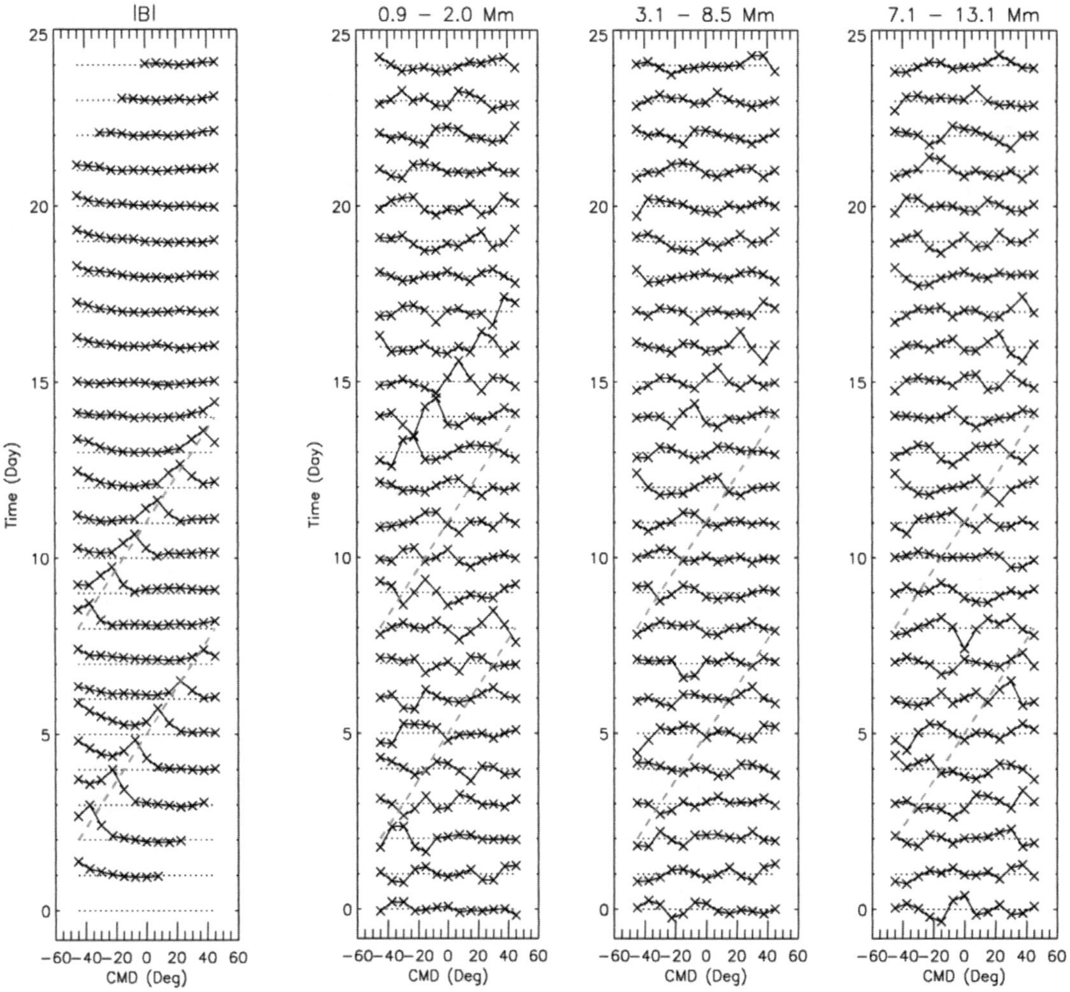

Figure 2. The unsigned magnetic flux and the zonal velocity derivative for three depth ranges (0.9 – 2.0, 3.1 – 8.5 and 7.1 – 13.1 Mm) for each ring-diagram day of CR 2071 averaged over ±22.5° latitude as a function of CMD (shown from left to right). Two small active regions (10998 and 10999) are noticeable rotating across the disk (dashed lines).

average rotation rate of 15 degree CMD per ring-diagram day, which corresponds to a rotation rate of 27.7 days. The zonal velocity derivative shows a similar but weaker pattern. To reduce the influence of magnetic activity, we eliminate 25% of all data points with the highest magnetic activity values. The corresponding correlation signal is greatly reduced for the unsigned flux but only slightly reduced for the zonal velocity derivative. For comparison, we include the correlation of the corresponding error of the zonal velocity derivative. The error is expected to track the presence of magnetic activity and indeed shows a greatly reduced correlation signal when locations with the highest activity level are excluded.

Figure 5 shows the cross-correlation values at 15°-CMD intervals from −30° to +30° CMD as a function of lag time for the three depth ranges. The locations of relative high activity have been excluded from the analysis. The corresponding correlation maxima are shifting from near −2 to +2 days in lag time tracking the average rotation rate. This rotation-rate pattern is noticeable at all three depths and in both flow parameters. For the ±15°-CMD correlations,

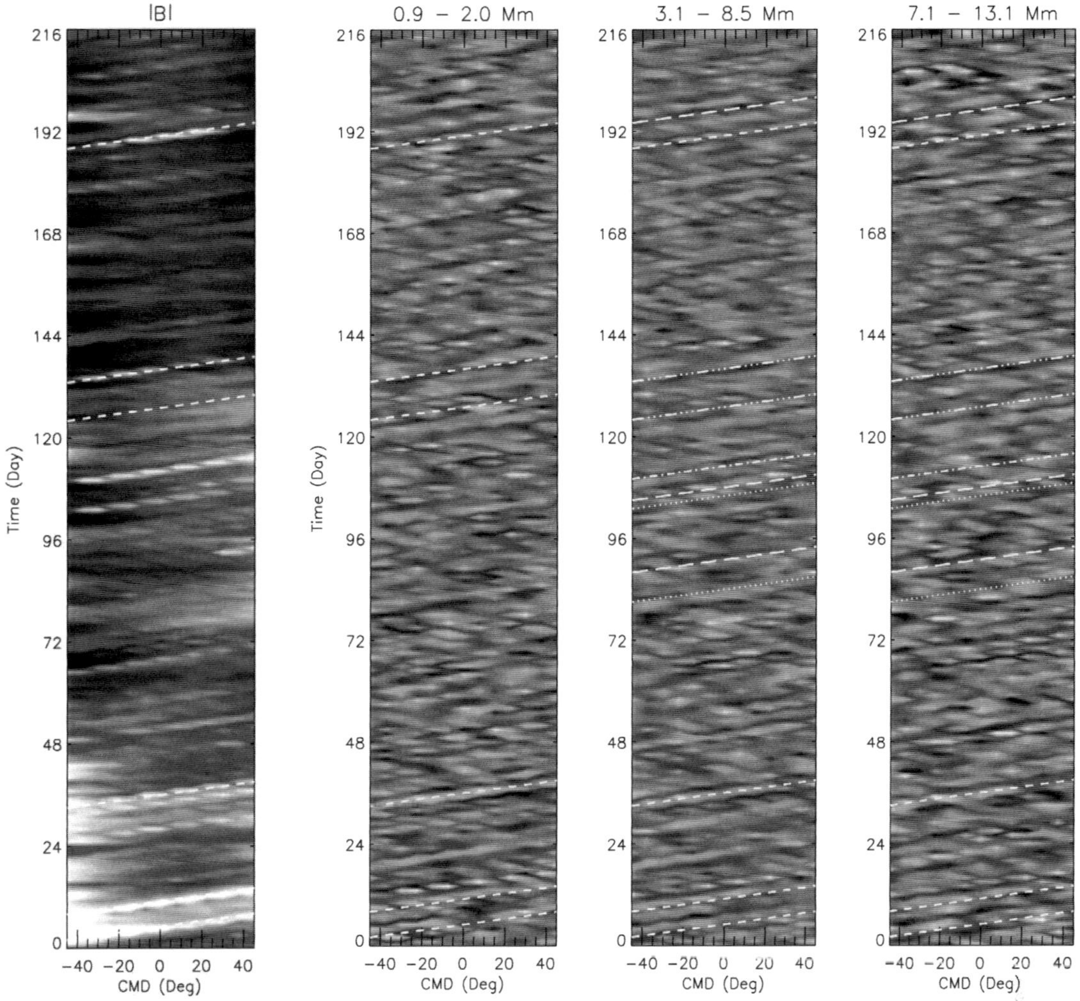

Figure 3. The unsigned magnetic flux and the zonal velocity derivative for three depth ranges (0.9–2.0, 3.1–8.5 and 7.1–13.1 Mm) for each day of CR 2071–2079 averaged over $\pm 22.5^{o}$ latitude as a function of CMD (shown from left to right). Dashed lines indicate active regions.

the peak values at ± 1 days are greatest in the near-surface layer and smallest in the deeper layer. The correlation values of the vertical velocity measurements are slightly smaller than the values of the zonal velocity derivative. The largest difference in amplitude between the two flow parameters occurs for the $\pm 30^{o}$-CMD correlations with maxima at ± 2 days.

We have repeated the analysis after randomizing the time series. The correlation values of the randomized data sets are much smaller than the correlation values of the original data. In addition, their maxima are not organized in a sequence from -2 to $+2$ days lag time as is the case for the original data.

4. Discussion

We find equatorial structures that persist for several days in the values of the zonal velocity derivative and the vertical velocity component. We are not aware of any artifact that might introduce such a coherent pattern and it seems unlikely that they are just a random phenomenon.

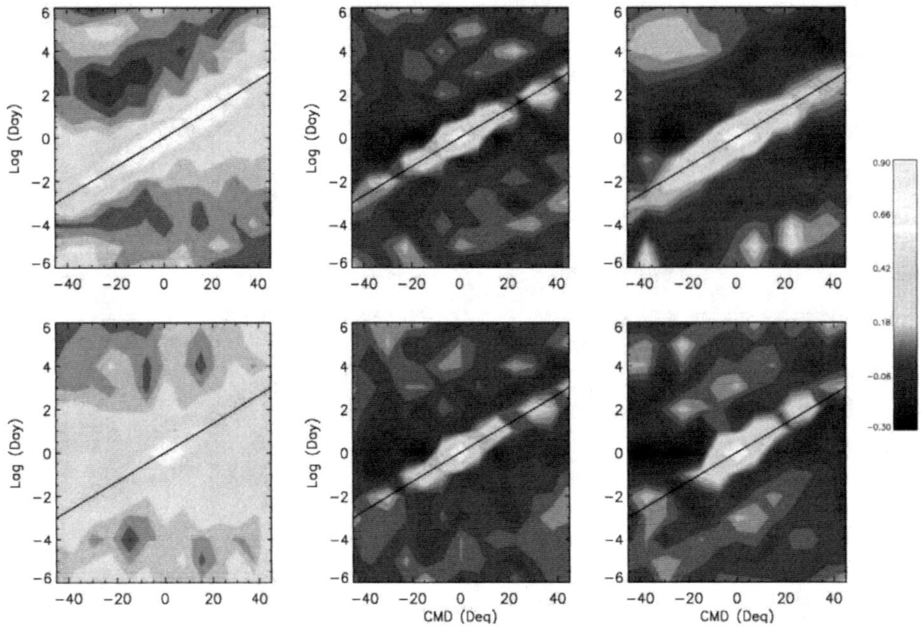

Figure 4. The cross-correlation of unsigned magnetic flux, the zonal velocity derivative (3.1 – 8.5 Mm), and its error (left to right) of the central-meridian time series with the time series at different CMD values. The correlation values are shown as a function of CMD and lag time for the complete data set (top) and after excluding 25% of all locations with the highest unsigned flux (bottom). The solid line indicates the average rotation rate of 27.7 days

Our results are consistent with other studies that have observed north-south aligned patterns in supergranulation [6, 7]. The observed pattern seems to be strongest in the near-surface layers, which might suggest that this is indeed a supergranular phenomenon. However, measurements at deeper layers are noisier and have larger error bars. Therefore, the decrease in correlation with increasing depth does not necessarily imply that the coherent structures are a near-surface phenomenon. In addition, it has been suggested that the north-south aligned supergranular patterns are a consequence of large convective structures [6]. The convection zone models [2] and the supergranulation measurements [6] suggest that our results are indications of large-scale convective structures.

We have limited our analysis to subsurface flows during solar cycle minimum, since these flows are sensitive to the presence of magnetic activity. However, the results have shown that weak magnetic patterns that rotate across the solar disk can be present even during an epoch of minimum activity. To further reduce the influence of magnetic activity, we have excluded locations of relative high magnetic activity from the minimum epoch. Still, some systematic flow variations coincide with very small variations of surface magnetic activity, which are too small to produce the observed flow pattern. This raises the question whether in some cases the bulk of magnetic activity is submerged. We plan to look more closely at the magnetic information and to identify the locations of remnants of active regions. This will allow us to further reduce the influence of magnetic activity in the search for coherent convective structures.

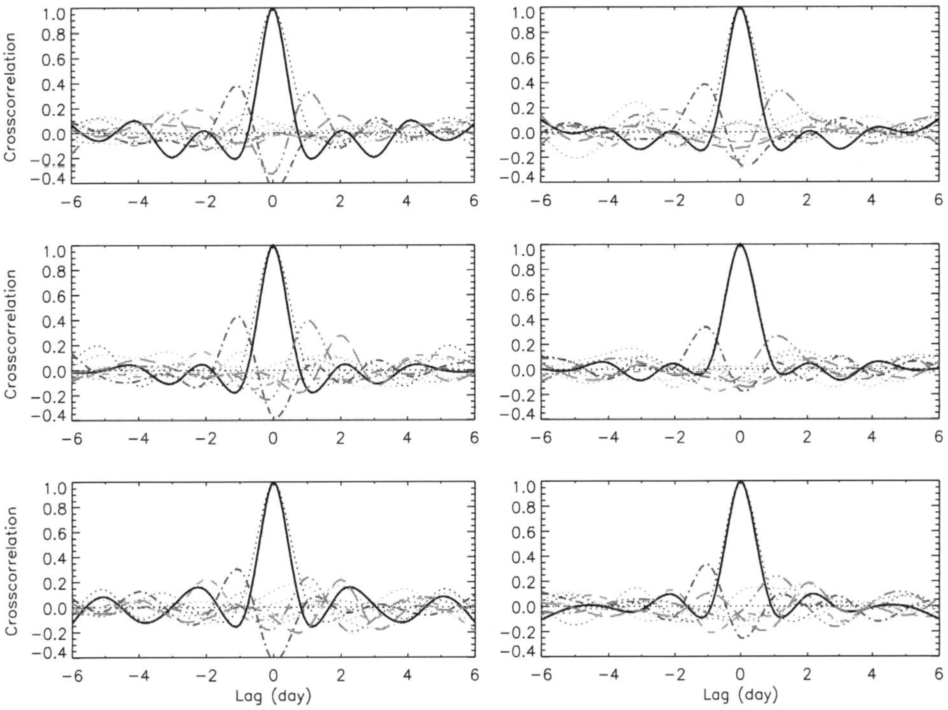

Figure 5. The cross-correlation of the zonal velocity derivative (left) and the vertical velocity (right) at three depth ranges (top: 0.9 – 2.0, middle: 3.1 – 8.5; bottom: 7.1 – 13.1 Mm) in steps of 15° CMD (dashed: −30°; dot-dashed: −15°; solid: 0°; dot-dot-dashed: 15°; long dashed: 30°). The lag time is in units of days; the correlation values are spline-interpolated for smoothness. Dotted lines indicate results of randomized data sets

Acknowledgments

This work utilizes data obtained by the Global Oscillation Network Group (GONG) program, managed by the National Solar Observatory, which is operated by the Association of Universities for Research in Astronomy (AURA), Inc. under a cooperative agreement with the National Science Foundation. The data were acquired by instruments operated by the Big Bear Solar Observatory, High Altitude Observatory, Learmonth Solar Observatory, Udaipur Solar Observatory, Instituto de Astrofísica de Canarias, and Cerro Tololo Interamerican Observatory. SOHO is a mission of international cooperation between ESA and NASA. This work was supported by NASA grant NNG08EI54I to the National Solar Observatory.

References

[1] Miesch MS 2005 *Living Reviews in Solar Physics,* **2** 1 (http://solarphysics.livingreviews.org/Articles/lrsp-2005-1/)

[2] Miesch MS, Brun AS, DeRosa ML and Toomre J 2008 *ApJ* **673** 557–75

[3] Haber DA, Hindman BW, Toomre J, Bogart RS, Larsen, RM and Hill F 2002 *ApJ* **570** 855–64

[4] Komm RW, Corbard T, Durney BR, González Hernández I, Hill F, Howe R and Toner C 2004 *ApJ* **605** 554–67

[5] Komm R, Howe R and Hill F 2009 *Solar Phys.* **258** 13–30

[6] Lisle JP, Rast MP and Toomre J 2004 *ApJ* **608** 1167–74

[7] Nagashima K, Zhao J, Kosovichev A and Sekii T 2010 *AAS Meeting* **216** #400.02

What do global p-modes tell us about banana cells?

Piyali Chatterjee

NORDITA, AlbaNova University Center, Roslagstullsbacken 23, SE 10691 Stockholm, Sweden

E-mail: `piyalic@nordita.org`

Abstract. We have calculated the effects of giant convection cells also know as sectoral rolls or banana cells, on p-mode splitting coefficients. We use the technique of quasi-degenerate perturbation theory formulated by Lavely & Ritzwoller in order to estimate the frequency shifts. A possible way of detecting giant cells is to look for even splitting coefficients of 'nearly degenerate' modes in the observational data since these modes have the largest shifts. We find that banana cells having an azimuthal wave number of 16 and maximum vertical velocity of 180 m s^{-1} cannot be ruled out from GONG data for even splitting coefficients.

1. Introduction

The power spectra of solar convective velocities show distinct peaks representing granules and supergranules but no distinct features at wavenumbers representative of mesogranules or giant cells (Wang 1989; Chou et al. 1991; Straus and Bonaccini 1997; Hathaway et al. 2000). Numerical simulations of solar convection routinely show the existence of mesogranules and giant cells (Miesh et al. 2008, Käpylä et al. 2010). Of particular interest is the existence of sectoral rolls or 'banana cells' having maximum radial velocities of 200-300 ms^{-1}. The giant convective cells have always been elusive to observations at the solar surface. In the past there have been studies which have failed to detect giant cell motions (LaBonte, Howard and Gilman 1981; Snodgrass and Howard 1984; Chiang, Petro and Fonkal 1987) as well as those hinting at their existence (Hathaway et al. 1996; Simon and Strous 1997). With the availability of Dopplergrams from SOHO/MDI, it became possible to study such long-lived and large scale features more reliably. Beck, Duvall and Scherrer (1998) were able to detect giant cells at the solar surface with large aspect ratio (~ 4) and velocities ~ 10 m s^{-1} using the MDI data. Hathaway et al. (2000) used spherical harmonic spectra from full disk measurements to detect long-lived power at $l \leq 64$.

Are banana cells observed regularly in direct numerical simulations (DNS), real or are they an artifact of insufficient resolution? Is it possible for techniques of global helioseismology to throw any light on the existence of giant cells? Traditionally, only degenerate perturbation theory (DPT) has been used to calculate the effect of rotation on the p-modes. But the first order contribution from poloidal flows (of which giant cells are a particular case) calculated using the degenerate perturbation theory vanishes giving rise to the need to use the quasi degenerate perturbation theory (QDPT) which in contrast to DPT couples modes having slightly different unperturbed frequencies. Quasi degenerate perturbation theory was applied to calculate shifts due to flows and asphericity in solar acoustic frequencies by Lavely and Ritzwoller (1992). Roth and Stix (1999, 2003) used QDPT to calculate the effect of giant cells on p-modes and claimed that giant cells could be found by modeling the asymmetries and line broadening in the solar power spectrum. They claimed that finite line width of the multiplets would limit the detection of the frequency splittings to vertical velocity amplitude of 100 m s^{-1} or larger. In Chatterjee and Antia (2009, CA09 now onwards), we used a different approach to examine if the use of quasi degenerate perturbation theory introduces significant corrections in the frequency shifts over that obtained from degenerate perturbation theory. We found the effect of rotation on the odd coefficients is negligible and hence using the degenerate theory is sufficient. Additionally, CA09 also calculated the effect of N-S asymmetric component of rotation, single cell meridional circulation, giant cells and banana cells on p-modes. We also showed that for higher angular degree of the poloidal flows, the 'nearly degenerate' modes would have very large frequency shifts (splitting coefficients). It is these splitting coefficients we can hope to detect in observational data.

In this paper we present results for banana cells but with an angular degree twice that used in CA09. In Sect. 2, we present a brief description of the method for completeness. In Sect. 3 we present the results of our calculations and draw conclusions in Sect. 4.

2. Perturbation of p- & f-modes by giant cells

Any perturbation calculation first requires definition of a base state with unperturbed eigenfrequencies as well as eigenvectors denoted ω_k and $\mathbf{s_k}$ respectively, which we shall call non-rotating spherically symmetric (NRSS) state. In this work we have used a standard solar model with the OPAL equation of state (Rogers and Nayfonov 2002) and OPAL opacities (Iglesias and Rogers 1996) and use the formulation due to Canuto and Mazitelli (1991) to calculate the convective flux. The equations of motion for a mode k with eigenfrequency ω_k for a NRSS model and a model perturbed by addition of differential rotation and/or large scale flow can be respectively represented by,

$$\mathcal{L}_0\mathbf{s_k} = -\rho_0\omega_k^2\mathbf{s_k}, \tag{1}$$

$$\mathcal{L}_0\mathbf{s'_k} + \mathcal{L}_1\mathbf{s'_k} = -\rho_0{\omega'_k}^2\mathbf{s'_k}, \tag{2}$$

where ω'_k is the perturbed frequency and $\mathbf{s'_k} = \sum_{k'\in K} a_{k'}\mathbf{s_{k'}}$ is the perturbed eigenvector. Taking scalar product with $\mathbf{s_j}$ in equation (2) and using the notation $\mathcal{H}_{jk'} = -\int \mathbf{s_j}^{\dagger}\mathcal{L}_1\mathbf{s_{k'}}dV$ and the definition $\mathcal{L}_1\mathbf{s_k} = -2i\omega_{\rm ref}\rho_0(\mathbf{v}.\nabla)\mathbf{s_k}$, we obtain the matrix eigenvalue equation,

$$\sum_{k'\in K}\left\{\mathcal{H}_{jk'} + \delta_{k'j}(\omega_{k'}^2 - \omega_{\rm ref}^2)\right\}a_{k'} = ({\omega'_k}^2 - \omega_{\rm ref}^2)a_j, \tag{3}$$

with eigenvalue $\lambda = ({\omega'_k}^2 - \omega_{\rm ref}^2)$ and eigenvector $X_j = \{a_j\}$. Here $\omega_{\rm ref}$ is a reference frequency which approximates ω'_k. In this work we use $\omega_{\rm ref} = \omega_k$, the frequency of the mode being perturbed. For details on calculation of the matrix elements, the reader is referred to Sect. 3.2 of CA 09. Essentially we use the Wigner-Eckart theorem (equation 5.4.1 of Edmonds 1960) which states that the general matrix element of any tensor perturbation operator can be expanded in terms of Wigner $3j$ symbols whose coefficients of expansion are independent of azimuthal order m and m'.

An important thing to remember is that for two modes with frequencies ω_1 and ω_2, the frequency shift is given by,

$$\delta\nu = \frac{\omega'_2 - \omega_2}{2\pi} \sim \frac{\mathcal{H}_{12}^2}{4\pi\omega_2\Delta} \tag{4}$$

where $\mathcal{H}_{12}$ is the coupling matrix between the two modes and $\Delta = ({\omega_2}^2 - \omega_1^2)$. Traditionally in helioseismology, the frequency shift as a function of m is described in terms of splitting coefficients for all modes that are detected. These coefficients are defined by (e.g., Ritzwoller and Lavely 1991)

$$\omega_{nlm} = \omega_{nl} + \sum_{q=0} a_q^{(nl)}\mathcal{P}_q^l(m), \tag{5}$$

where ω_{nl} is the mean frequency of the multiplet, $\mathcal{P}_q^l(m)$ are the orthogonal polynomials of degree q and $a_q^{(nl)}$'s are the so called splitting coefficients. A multiplet with frequency ω_{nl} is $2l + 1$-fold degenerate in absence of rotation, magnetic field, poloidal flows and asphericity. Following earlier works (Lavely and Ritzwoller 1992; Roth, Howe and Komm 2002) we express the velocity field in terms of spherical harmonics. For completeness, we give the expression here again.

$$\begin{aligned}\mathbf{v}(r,\theta,\phi) &= \mathrm{Re}[u_s^t(r)Y_s^t(\theta,\phi)\hat{\mathbf{r}} + v_s^t(r)\nabla_h Y_s^t(\theta,\phi) - \\ &\quad w_s^t(r)\hat{\mathbf{r}} \times \nabla_h Y_s^t(\theta,\phi)]. \end{aligned} \tag{6}$$

The quantities u_s^t, v_s^t and w_s^t determine the radial profiles of the flows and ∇_h is the horizontal gradient operator. The Re refers to using only the real part of the spherical harmonics as in,

$$\mathrm{Re}[Y_s^t(\theta,\phi)] = \begin{cases} [Y_s^{-t}(\theta,\phi) + Y_s^t(\theta,\phi)]/2 & \text{if } t \text{ is even,} \\ [Y_s^{-t}(\theta,\phi) - Y_s^t(\theta,\phi)]/2 & \text{if } t \text{ is odd.} \end{cases} \tag{7}$$

The first two terms in equation (6) define the poloidal component of the flow whereas the last term is the toroidal component. By the poloidal component, we imply the meridional and non-zonal toroidal flows (average over ϕ direction is zero) e.g., (i) the meridional circulation which carries mass poleward near the surface and sinks near the poles and (ii) the giant convection cells, respectively. These flows are also called large scale flows to distinguish them from other small scale flows like the turbulent eddies which are of the size smaller than the typical scale of global modes used in helioseismology.

The sectoral rolls or banana cells are characterised by $s = t$. CA09 used $s = t = 8$ for banana cells. Visual inspection of the snapshots of vertical velocity in Fig. 1 of Miesch et al. (2006) as well as Fig. 2c of Käpylä et al. (2010) reveals that banana cells observed in DNS of stellar convection has $s \sim 16, t \sim 16$. In addition Miesch et al. (2008) also put the value of the maximum radial velocity at 200 m s^{-1}. So it makes sense to repeat our calculations for a flow with an angular dependence $Y_{16}^{16}(\theta,\phi)$. In this calculation we have used only the estimate of the magnitude of maximum radial velocities and their angular degree from the DNS. The solar structure we use for frequency shift calculations come from a spherically symmetric non-rotating standard solar model (NRSS model) and the rotation splittings needed are obtained from temporally averaged GONG data. These two quantities are very different for DNS since they have much shallower density stratification and sometimes give a rotation profile which is constant on cylinders rather than cones as for the Sun. A more realistic calculation would be to use the stratification and the rotation rate from the DNS to define the unperturbed state. Nevertheless, a first calculation using a NRSS model and GONG rotation profile can provide valuable insights into the possibility of such poloidal flows.

In presence of only the poloidal flow ($w_s^t = 0$) we can apply the equation of mass conservation $\nabla.(\rho_0 \mathbf{v}) = 0$ to get a relation between $u_s^t(r)$ and $v_s^t(r)$ e.g.,

$$v_s^t(r) = \frac{1}{r}\frac{\partial}{\partial r}\left[\frac{\rho_0 r^2 u_s^t(r)}{s(s+1)}\right]. \tag{8}$$

Here $\rho_0(r)$ is the density in a spherically symmetric solar model. So now it only remains to choose $u_s^t(r)$ appropriately and $v_s^t(r)$ will be determined by equation (8). We choose the radial profile of $u_s^t(r)$ as given by equation (19) of Roth, Howe & Komm (2002):

$$u_s^t(r) = \begin{cases} u_0 \dfrac{4(R_{\mathrm{top}} - r)(r - r_b)}{(R_{\mathrm{top}} - r_b)^2} & \text{if } r_b \le r \le R_{\mathrm{top}}, \\ 0 & \text{otherwise.} \end{cases} \tag{9}$$

Here $r_b = 0.7R_\odot$ and $R_{\rm top} = R_\odot$ define the boundaries of region where the flow is confined. The coupling matrix in Eq. (3) calculated for the flow defined above is proportional to u_0. Hence using Eq. (4), we have the frequency shift, $\delta\nu \propto u_0^2$.

3. Results

As explained in CA09, this calculation is little more involved as a non-zero t allows coupling between different m and m' of the p-modes. According to the selection rules imposed by the Wigner $3j$ symbols, the $s = 16, t = 16$ flow couples the mode (n, l, m) with $(n', l', m \pm 16)$. Hence it becomes important to take into account the effect of rotational splitting on Δ before calculating the effect of these kind of poloidal flows. Thus, the difference of the square of frequencies, Δ is no longer independent of m. The couplings within a multiplet i.e., $(n, l, m) \rightleftarrows (n, l, m \pm 16)$ happen to be zero because of the anti-symmetry of the matrix elements (see Eqs. (16)-(18) in CA09).

In the upper panel of Fig. 1, we show the frequency shift of the mode with $(n, l) = (11, 38)$ due to coupling with $(10, 46)$. From Eq. (4), we have the frequency shift, $\delta\nu \propto 1/\Delta$. The $\delta\nu(m)$ shows a discontinuity at $m = 0$ since Δ for the $(11, 38, m) \rightleftarrows (10, 46, m - 16)$ becomes zero at $m = 0$ as shown in the lower panel. The reader is encouraged to compare this figure with Fig. 7 of CA09 where they plot similar curves for coupling of the modes $(18, 61, m) \rightleftarrows (17, 69, m \pm 8)$. It is these discontinuities which give rise to a large value of splitting coefficients, a_q. Also notice the asymmetry of $\delta\nu(m)$ about $m = 0$ in Fig. 1. This means that not only the even coefficients, a_{2q} but also the odd coefficients, a_{2q+1} are non-zero. Usual inversion procedure for rotation neglects giant cells and assumes that the odd splitting coefficients arise only from rotation. If there are additional contributions to these coefficients, rotation inversions may not give correct results. In CA09, we performed a 1.5d rotation inversion on odd coefficients a_1 and a_3 to detect any discernible feature in the inverted profile. However, since the magnitude of the features in the inverted profile are smaller than the inversion errors we concluded that giant cells with $u_0 \leq 100$ m s^{-1} do not have much effect on the rotation inversion. The splitting coefficients a_1, a_2 and a_3 as a function of lower turning point radius and normalised by corresponding errors in the observational data from GONG are shown in Fig. 2. One can easily compare these coefficients for $s = 8, t = 8$ and $s = 16, t = 16$ flows. While the $Y_8^8(\theta, \phi)$ flow with an amplitude of $u_0 = 100$ m s^{-1} already produces a maximum a_2 which is twice the observational error, a Y_{16}^{16} flow with an amplitude $u_0 = 300$ m s^{-1} produces a maximum a_2 which is half the observational error. A similar thing may be said about the coefficients a_1 and a_3 as well. However, we have not performed a 1.5 D inversion for the $s = 16, t = 16$ case. We do not see a clear signal of 'nearly degenerate' modes in the observational data. Hence, like in CA09, in order put an upper limit on the amplitude of the flows we compare the a_2 values from theory with observational errors to calculate the confidence level of the upper limit. For the $Y_{16}^{16}(\theta, \phi)$, the confidence limit estimated by maximum value of a_2/σ_2 is less than 0.6 for a flow with an amplitude $u_0 = 300$m s^{-1} for several GONG data sets. In terms of the maximum radial velocity, $u_r^{\rm max} \sim 180$m s^{-1}. Thus banana cells with such amplitudes cannot be ruled out using the GONG data sets used. Another interesting observation from Fig. 2 is the large values of a_q at $r_t \sim 0.7R_\odot$ for both $t = 8$ and 16. We have checked that this is not due to $r_b = 0.7R_\odot$, but is an intrinsic property of the solar model used. The rotation profile used to calculate the rotational splittings for $t \neq 0$ giant cell flows may also be responsible. Nevertheless, since we also see this behaviour for $s = 8, t = 0$ flow (see Fig. 5 of CA09), we may conclude that the modes which have their turning points near the base of the convection zone are coupled strongly due to such poloidal flows.

4. Conclusions

We have calculated the effects of giant cells with an azimuthal degree 16 on p-mode splitting coefficients. We use the technique outlined in Chatterjee & Antia (2009) to perform this

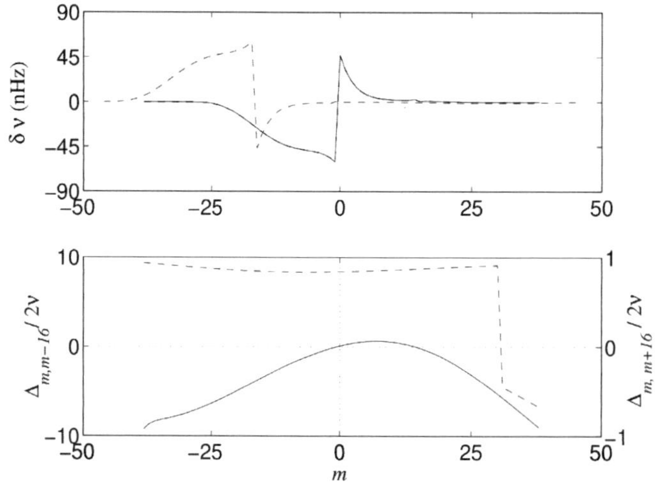

Figure 1. Upper panel: The *solid line* gives the frequency shift because of interaction between modes $(11, 38, m)$ having lower turning radius $r_t = 0.69R_\odot$ and $(10, 46, m \pm 16)$ having $r_t = 0.74R_\odot$, due to the flow with an angular dependence $Y_{16}^{16}(\theta, \phi)$. The frequency shift for the mode $(10, 46)$ is given by the *dashed line*. Lower panel: $\Delta_{m,m-16}/2\nu$ (*solid line*) for the coupling $(11, 38, m) \rightleftarrows (10, 46, m - 16)$; and $\Delta_{m,m+16}/2\nu$ (*dashed line*) for the coupling $(11, 38, m) \rightleftarrows (10, 46, m + 16)$. These values are in μHz.

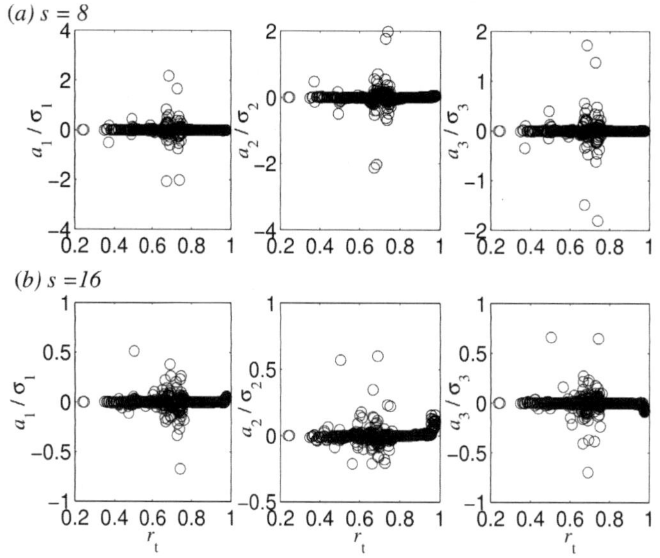

Figure 2. (a) a_1/σ_1, a_2/σ_2 and a_3/σ_2 as a function of lower turning point radius r_t for the $Y_8^8(\theta, \phi)$ kind of flow with $u_0 = 100$ m s^{-1}. (b) Same as above but for a flow varying as $Y_{16}^{16}(\theta, \phi)$ with $u_0 = 300$ m s^{-1}. The σ_1, σ_2 and σ_3 are the errors in the corresponding observational splitting coefficients a_1, a_2 and a_3 from the GONG data set centered about November 2002.

calculation. We find that from observations we cannot rule out the existence of giant cells with an angular variation $Y_{16}^{16}(\theta,\phi)$ and an amplitude of $u_0 = 300$ m s^{-1} since the splitting coefficients are less than the observational errors. It is important to remember that $a_q \propto u_0^2$. In other words we can say that a flow varying as $Y_8^8(\theta,\phi)$ in the Sun can be ruled out with a confidence level estimated by the maximum value of the ratio $a_2/\sigma_2 \sim 2$ which is 30 times more than the confidence level for a flow varying as $Y_{16}^{16}(\theta,\phi)$ with the same amplitude $u_0 = 100$ m s^{-1}. Speaking in terms of maximum vertical velocity, u_r, according to Eq. (6), a $u_0 = 100$ m s^{-1} for a $s = 8, t = 8$ flow corresponds to $u_r^{\max} = 50$ m s^{-1} whereas for a $s = 16, t = 16$ flow implies a $u_r^{\max} = 60$ m s^{-1}.

We do not find any evidence from global helioseismology which may point at the appearance of banana cells with an azimuthal degree $t = 16$ in the DNS to being just an artifact of insufficient grid resolution. But we believe that in the DNS the velocity spectra would peak at the wavenumbers of the banana cells since they are so visually conspicuous in contrast to observations where there is no clear peak at giant cell scales. This may be because we do not resolve very many scales smaller than the banana cells in the DNS. But it is important to remember that we have used very regular analytical expressions for the banana cells. In the Sun, the cells will not only be irregular but also have finite lifetimes. The GONG data sets are averaged over 108 days and this may be somewhat longer than the lifetimes of giant cells and hence the signal may be averaged out. We can look at shorter data sets, but in that case the errors would be larger.

Acknowledgements
We thank the organisers of the GONG 2010 SOHO-24 meeting in Aix-en-Provence for an excellent meeting. This work utilizes data obtained by the Global Oscillation Network Group (GONG) project, managed by the National Solar Observatory, which is operated by AURA, Inc. under a cooperative agreement with the National Science Foundation. The data were acquired by instruments operated by the Big Bear Solar Observatory, High Altitude Observatory, Learmonth Solar Observatory, Udaipur Solar Observatory, Instituto de Astrofisico de Canarias, and Cerro Tololo Inter-American Observatory.

[1] Beck, J.G., Duvall, T.L. Jr. and Scherrer, P.H. 1998 *Nature* **394** 653
[2] Canuto, V.M. and Mazzitelli, I. 1991 ApJ **370** 295
[3] Chatterjee, P. and Antia, H.M. 2009 ApJ **707** 208
[4] Chiang, W.-H. , Petro, L.D. and Foukal, P.V. 1987 *Solar Phys.* **110** 129
[5] Chou, D.-Y., LaBonte, B.J., Braun, D. C. and Duvall, T.L., Jr. 1991 ApJ **372** 314
[6] Edmonds, A.R. 1960 *Angular momentum in quantum mechanics* (Princeton University Press)
[7] Hathaway, D.H., Gilman, P.A., Harvey, J.W., Hill, F., Howard, R.F., Jones, H.P., Kasher, J.C., Leibacher, J. W., Pintar, J.A. and Simon, G.W. 1996 *Science* **272** 1306
[8] Hathaway, D.H., Beck, J.G., Bogart, R.S., Bachmann, K.T., Khatri, G., Petitto, J.M., Han, S. and Raymond, J. 2000 *Solar Phys.* **193** 299
[9] Iglesias, C.A. and Rogers, F.J. 1996 ApJ **464** 943
[10] Käpylä, P.J., Korpi, M., Guerrero, G., Brandenburg, A. and Chatterjee, P. 2010 *Preprint* (arXiv:1010.1250)
[11] LaBonte, B.J., Howard, R.F. and Gilman, P.A. 1981 ApJ **250** 796
[12] Lavely, E.M. and Ritzwoller, M.H. 1992, *Phil. Trans. Roy. Soc. Lon. A.* **339** 431
[13] Miesch, M.S., Brun, A.S. and Toomre, J. 2006 ApJ **641** 618
[14] Miesch, M.S., Brun, A.S., DeRosa, M.L. and Toomre, J. 2008 ApJ **673** 557
[15] Ritzwoller, M.H. and Lavely, E. M. 1991 ApJ **369** 557
[16] Rogers, F.J. and Nayfonov, A. 2002 ApJ **576** 1064
[17] Roth, M., Howe, R. and Komm, R. 2002 A & A **396** 243
[18] Simon, G.W. and Strous, L.H. 1997 *Bull. Am. Astron. Soc.* **29** 1402
[19] Snodgrass, H.B. and Howard, R. 1984 ApJ **284** 848
[20] Straus, T. and Bonaccini, D. 1997 A& A **324** 704
[21] Wang, H. 1989 *Solar Phys.* **123** 21

The rotation rate and its evolution derived from improved mode fitting and inversion methodology

S G Korzennik[1] **and A Eff-Darwich**[2]

[1] Harvard-Smithsonian Center for Astrophysics, Cambridge, MA, USA

[2] Instituto de Astrofísica de Canarias La Laguna, Tenerife, Spain

E-mail: `skorzennik@cfa.harvard.edu, adarwich@iac.es`

Abstract. We present inferences of the internal solar rotation rate and its evolution during solar cycle 23. A full solar cycle of MDI observations have been analyzed using an improved fitting methodology and using time series of various lengths, up to a single 4,608 day long epoch (64 times 72 days or 12.6 yr). We used time series of spherical harmonic coefficients computed by the MDI group, including those resulting from using their improved spatial decomposition. This decomposition includes our best estimate of the image plate scale and of the MDI instrumental image distortion. The leakage matrix used in the fitting includes the effect of the distortion of the eigenfunctions by the solar differential rotation, while the undistorted leakage matrix was itself carefully reviewed and independently recomputed. Rotation inversions were carried out for all available mode sets, fitted for that epoch, including the MDI and GONG "pipe-line" values. The improved inversion method uses an iterative methodology based on a least-squares regularization, but with an optimal model grid determined by the actual information in the input set. This method also allows us to use an optimized irregular grid, with a variable number of latitudes at different depths.

1. Introduction

Since the deployment of the Global Oscillation Network Group (GONG) instruments and the start of operations of the Michelson Doppler Imager (MDI), soon after the launch of the SOHO spacecraft, we have accumulated some 13 years worth of data. Nevertheless, *one* still uses 20 and 18 year old fitting methodologies, respectively, to derive the solar p-mode characteristics. The GONG *pipe-line* frequencies are produced, using the method described in [4], by fitting *overlapping* 108-day long segments, not using any leakage matrix information and fitting a symmetric profile to individually but independently fitted peaks. Some 147 *GONG months* (each 36 days long) were available, spanning 1995.06.29 to 2009.09.07.

The MDI *pipe-line* frequencies are produced using the methodology described in [5], that includes information about the leakage matrix but still fits a symmetric profile, and parametrizes the frequency splitting with a polynomial expansion in m. Some 67 non-overlapping epochs (each 72-days long) were available, spanning 1996.05.01 to 2009.12.07.

More recently, that approach was improved, see [6]. The spatial decomposition was reprocessed, having gained from our insight based on high-degree mode characterization [3]. Specifically, the actual plate scale was used and our best model of the instrumental image distortion was included. Moreover the modeling of the leakage matrix was improved by including the distortion of the eigenvalues by differential rotation, as introduced by [8], and by allowing, or not, for an asymmetric peak profile (57 & 62 epochs were done so far respectively).

1.1. *An improved fitting methodology*

Motivated by the embarrassing aging of these methodologies, while prompted by the availability of long time-series combined with high performance computing, we have developed a new fitting methodology. The salient characteristics of this approach are: (1) the simultaneous fit, for all m, of individual modes, combined with a sanity rejection to avoid fitting non-significant peaks; (2) using an *optimal* multi-tapered spectral estimator; (3) fit an asymmetric profile; (4) include our best estimate of the leakage matrix (including distortion by differential rotation); and (5) use time-series of varying lengths to optimize the trade-off between SNR and temporal resolution; see [1, 2] for additional details. This method has been applied to some degree to GONG data, but mostly to MDI times series. At first, time-series of *standard* spherical harmonic coefficients were analyzed using 2088-day long & 728-, 364-, 182-day long, overlapping, epochs, spanning 1996.05.01 to 2002.01.17. More recently, *improved* time-series of spherical harmonic – *i.e.*, resulting from the new spatial decomposition mentioned earlier – combined with the inclusion of the effect of the distortion by differential rotation on the leakage have been fitted for 64×72, 32×72, 16×72-day long, overlapping, time-series, covering the 1996.05.01 to 2009.07.16 time span.

1.2. *Potential issue & inverse theory*

The archetypal inverse problem can be cast as follow:

$$y_i = \int K_i \, x(p) \, dp \tag{1}$$

where the set of observables, y_i, combined with known kernels, K_i, is used to derive the underlying model, $x(p)$. Since such inverse problems are singular, they require some form of regularization to lift this singularity (like a smoothness constraint). This produces an *estimate* of the solution, $\hat{x}$, that can be related to the *actual* solution through resolution kernels, R, namely:

$$\hat{x} = x \otimes R \tag{2}$$

These resolution kernels, that optimally would be Dirac distributions, depend on the input set, *i.e.* $\{i\}$: the extent of the available observables.

In practice, for inversion of the solar internal rotation, the inverse problem is

$$\delta\nu_{n,\ell,m} = \iint K_{n,\ell,m}(r,\theta) \, \Omega(r,\theta) \, dr \, d\theta \tag{3}$$

where $\delta\nu_{n,\ell,m}$ are the rotational splittings, (r,θ) the radius and co-latitude, $K_{n,\ell,m}$ the known rotational kernels and $\Omega(r,\theta)$ the solar rotation rate. The inversion input set is defined by the available modes (*i.e.*, $\{n,\ell,m\}$ or the $\{n,\ell,a_i\}$ coefficients[1]), whose temporal changes will affect the resolution kernels, R. In order to decouple this and be assured that the changes seen in Ω can be attributed to changes in the sun, we chose to invert a *constant* input set. Such a constraint leads to mode attrition, since the exact same mode set is not fitted for each epoch.

The optimal latitudinal resolution can only be obtained by letting the inversion do it, and is thus achieved when using rotational splittings resulting from fitting individual frequencies rather than a polynomial expansion. While the GONG pipeline produces such individual frequencies, our *constant* input set constraint produces an unacceptable mode attrition. Figure 1 illustrates

[1] These are the coefficients used to characterize the frequency dependance on m as a polynomial expansion.

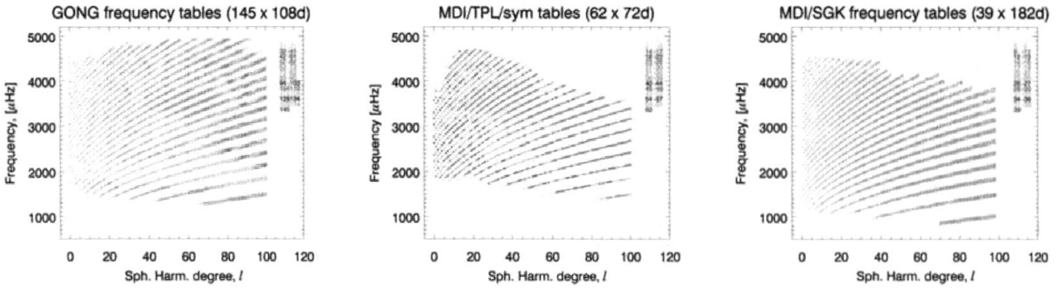

Figure 1. Mode attrition for GONG pipe-line (left) MDI improved pipe-line (middle) and MDI with our fitting (right) frequencies. The colors indicate the occurence of that mode.

this for the GONG and MDI pipe-lines, as well as the attrition when fitting 182-day long time series with our methodology. Also worth pointing out is that fitting using a polynomial expansion ignores the effect of mode visibility. This visibility –the observed amplitude of a mode– has a strong m/ℓ dependence, resulting in a similar dependence of the SNR, since the background noise is nearly independent of m. Using the polynomial expansion coefficients –during or after the fitting– effectively ignores the fact that some peaks are not significant, and is bound to inject *a priori* bias into the resulting splittings.

1.3. Other problems

Since individual modes are *rarely* resolved, inaccuracies in the leakage matrix will result in biased estimates. The closest (in frequency) spatial leak is only $\Delta\nu = \Delta\nu_{\delta m=2,\delta\ell=0}$ away. It is thus resolved if $\Delta\nu \gg \Gamma$, since $\Delta\nu \simeq 2 \times \frac{\Omega}{2\pi} \simeq 0.8\,\mu$Hz.

Error in the plate scale, image distortions and image orientation errors not accounted for will produce inaccurate leakage matrices. Fortunately, the new MDI spherical harmonic coefficients account for plate scale and image distortion. The distortion by differential rotation is, for intermediate degrees, a 1 to 6% effect, but $B_0 = B_0(t)$ is a 3 to 15% effect, while other geometric variations were found to be negligible.

Using a very long time series gives us the unique opportunity to check the leakage matrix, since some of the low order modes are resolved. We found that, despite developing an independent leakage matrix computation, there remains a clear but unaccounted for mismatch for the f-mode – implying a potential bias in those frequency estimates (see animations in the supplementary material for Korzennik et al, (this paper).)

1.4. Inversion method

The inversion methodology used for the profiles presented here is a modified regularized least-squares method, described in [7]. Its main unique features are an iterative approach that leads to an *optimal* model grid, based on input set, while using a *non-uniform* model grid, shown in Fig. 2.

2. Results

The mean solar rotation, derived from 12.6 years of MDI observations (the concatenation of 64×72-day long epochs) is shown in Fig. 3 as well as its associated uncertainty. Note how the unprecedented precision of the obtained rotation splittings, combined with an optimized RLS, allows us to infer a significant solution closer to the core and to the rotational axis than ever

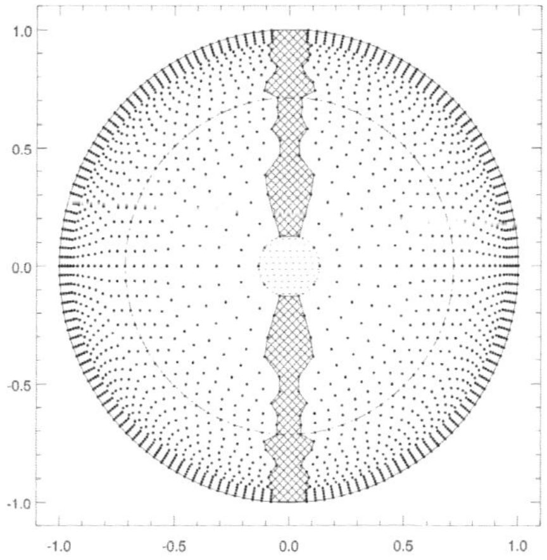

Figure 2. The non-uniform model grid used for the inversion shown in Fig 3. The hatched region is where a localized solution could not be inferred.

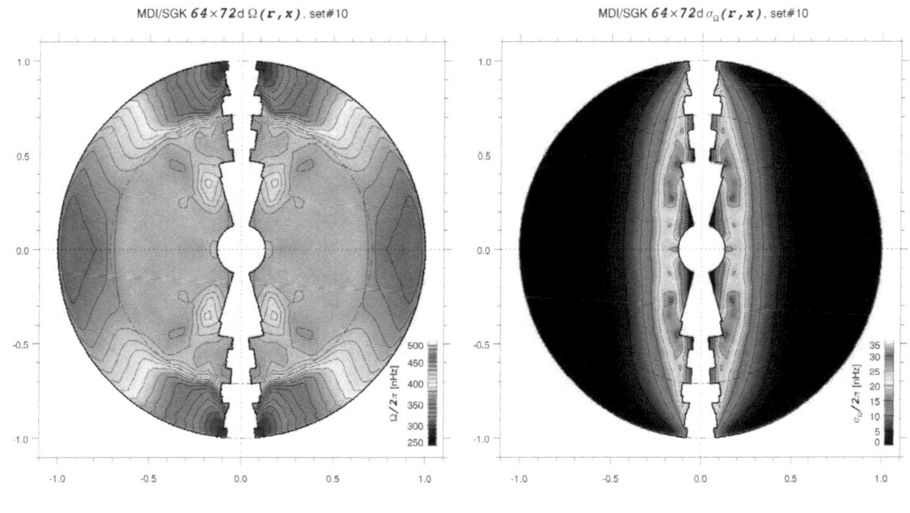

Figure 3. Rotation rate, as a function of depth and latitudes, inferred from 12.6 yr of MDI observations (left) and its associated uncertainty (right).

before. Note how the uncertainty increases as one gets closer to the rotation axis, and thus the dip around $(0.4, 63^{o})$ is barely a one-sigma feature.

Figure 4 shows the rotation rate changes at the surface, using MDI and GONG splittings. The signature of the torsional oscillations is clearly visible for all cases, although the latitudinal and temporal resolution change with the data set used.

Figure 5 presents the rotation rate changes at two depths for a selection of data sets. The latitudinal-averaged change and its Fourier transform are shown, showing inconsistencies between data sets except at the surface (see also the animations in the supplementary material for Korzennik et al, (this paper).)

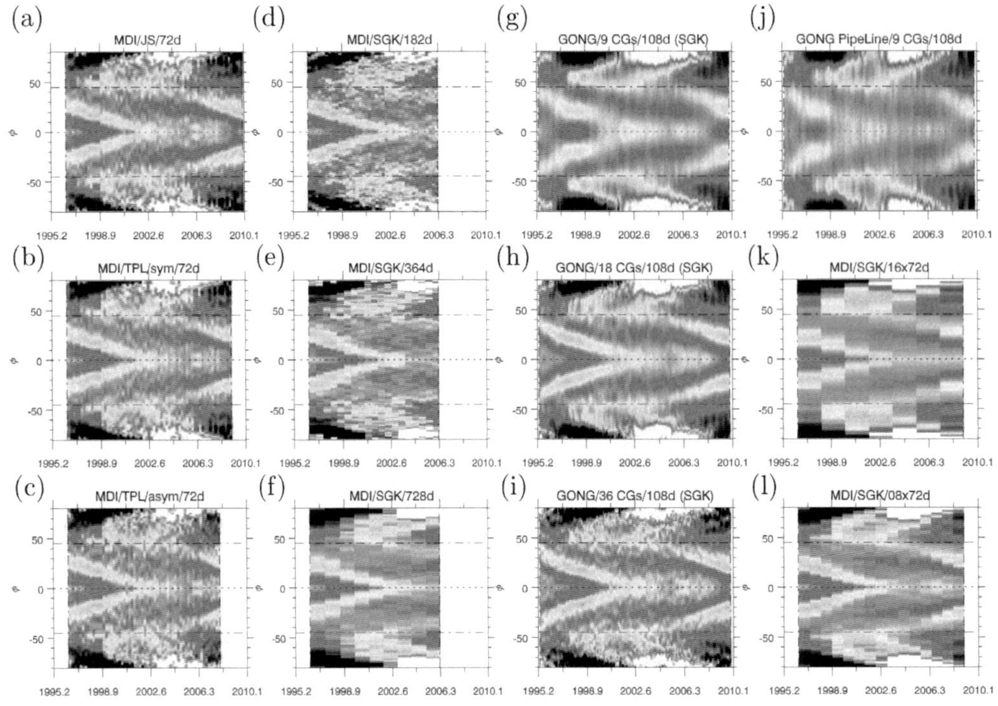

Figure 4. Changes in the rotation rate, as a function of time and latitudes and at the surface, inferred from the available data sets. (a) MDI standard pipe-line, (b) MDI improved pipe-line symmetric fit, (c) MDI improved pipe-line asymmetric fit, (d) MDI our fit to 182-day long series, (e) MDI our fit to 364-day long series, (f) MDI our fit to 728-day long series, (g) GONG pipe-line with our polynomial fit, 9 coefficients only, (h) GONG pipe-line with our polynomial fit, 18 coefficients, (i) GONG pipe-line with our polynomial fit, 36 coefficients, (j) GONG pipe-line with their polynomial fit, 9 coefficients, (k) MDI our fit to 1152-day long series (16×72), and (l) 576-day long series (8×72).

3. Conclusions

Despite some 13 years of data, mode fitting remains an *issue*. The standard pipe-lines produce unacceptable mode attrition, while the leakage matrix does not yet appear to be accurate, as seen in the f-mode. This might be due to our ignorance of the MDI instrumental point spread function, or some residual error in our estimate of the horizontal component.

We have been able to derive what is likely to be the best estimate of the solar mean rotation rate, by fitting a very long time-series. It allows us to push it closer to the core and to the rotation axis. There is a barley significant dip at $(0.4, 63^{\circ})$, that is predominantly seen during the rising branch of the cycle. Inferring the rotation changes is *easy* at the surface, but remains challenging down to the base of the convection zone and difficult below it.

Acknowledgments

The Solar Oscillations Investigation-Michelson Doppler Imager project on SOHO is supported by NASA grant NAS5-3077 at Stanford University. SOHO is a project of international cooperation between ESA and NASA. This work utilizes data obtained by the Global Oscillation Network Group (GONG) Program, managed by the National Solar Observatory, which is operated by

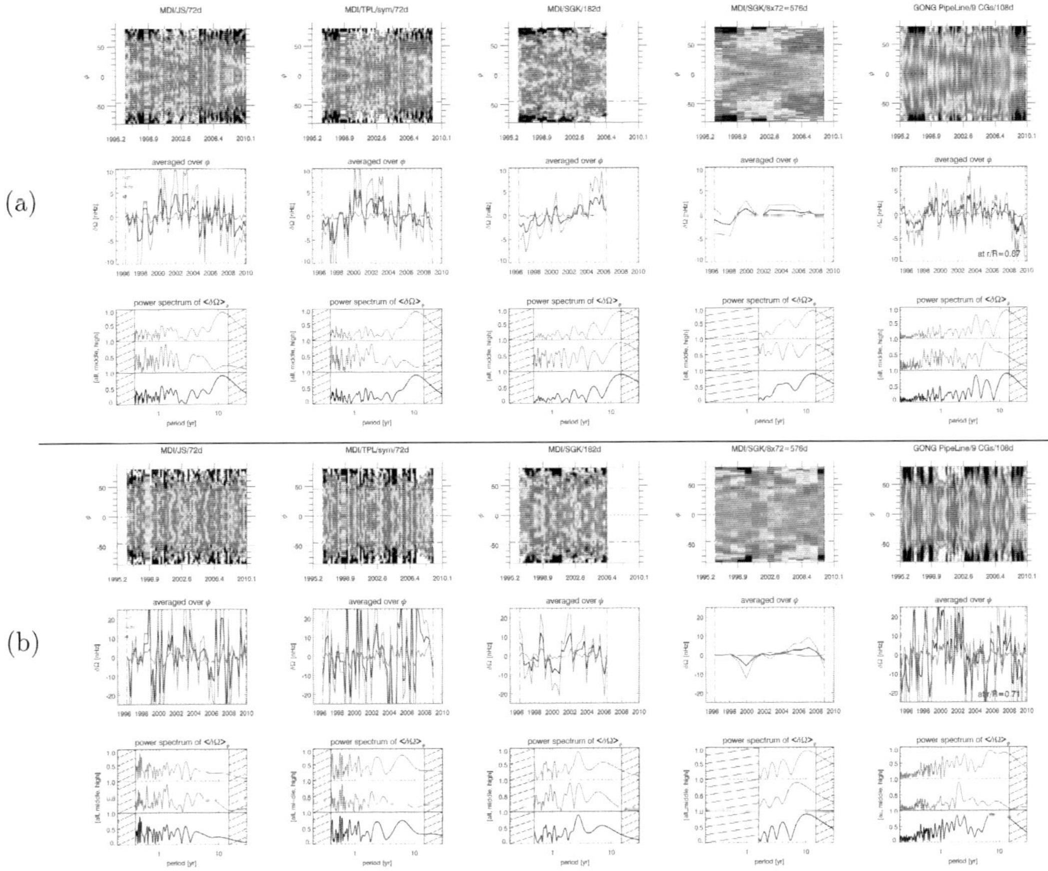

Figure 5. Changes in the rotation rate, for selected data sets (columns), at two depths: (a) $r/R = 0.87$, (b) $r/R = 0.71$, as well as the change averaged over latitude and their power spectra (all, middle and high latitudes are shown in black, red, and, green respectfully). The data sets used (left to right) are the MDI standard pipe-line, the MDI improved pipe-line asymmetric fit, our fit to MDI using 182-day long series, our fit to improved MDI 576-day (8x72) long series, and the GONG pipe-line with their polynomial fit, 9 coefficients. All the figures, including this one and one similar for $r/R = 1.$ are available in the supplementary material for Korzennik et al, (this paper).

AURA, Inc. under a cooperative agreement with the National Science Foundation. This work was in part funded by NASA grant NNX09AB15G, the Spanish grant AYA2004-04462 and the Canarian grant SolSubC200801000243.

References

[1] Korzennik, S. G. 2005, ApJ, 626, 585
[2] Korzennik, S. G. 2008, Journal of Physics Conference Series, 118, 012082
[3] Korzennik, S. G., Rabello-Soares, M. C., & Schou, J. 2004, ApJ, 602, 481
[4] Anderson, E. R., Duvall, T. L., Jr., & Jefferies, S. M. 1990, ApJ, 364, 699
[5] Schou, J. 1992, Ph.D. Thesis.
[6] Larson, T. P., & Schou, J. 2008, Journal of Physics Conference Series, 118, 012083
[7] Eff-Darwich, A., Korzennik, S. G., Jiménez-Reyes, S. J., & García, R. A. 2008, ApJ, 679, 1636
[8] Woodard, M. F. 1989, ApJ, 347, 1176

Assessing the Deep Interior Dynamics and Magnetism of A-type Stars

Nicholas A Featherstone[1], Matthew K Browning[2], Allan Sacha Brun[3] & Juri Toomre[1]

[1]JILA and Department of Astrophysical and Planetary Sciences, University of Colorado, Boulder, CO 80309-0440

[2]Canadian Institute for Theoretical Astrophysics, University of Toronto, Toronto, ON M5S3H8, Canada

[3]DSM/IRFU/SAp, CEA-Saclay & UMR AIM, CEA-CNRS-Université Paris 7, 91191 Gif-sur-Yvette, France

E-mail: `feathern@solarz.colorado.edu`

Abstract. A-type stars have both a shallow near-surface zone of fast convection that can excite acoustic modes and a deep zone of core convection whose properties may be studied through asteroseismology. Many A stars also exhibit large magnetic spots as they rotate. We have explored the properties of core convection in rotating A-type stars and their ability to build strong magnetic fields. These 3-D simulations using the ASH code may serve to inform asteroseismic deductions of interior rotation and magnetism that are now becoming feasible. Our models encompass the inner 30% by radius of a 2 solar mass A-type star, capturing both the convective core and some of the overlying radiative envelope. Convection can drive a column of strong retrograde differential rotation and yield a core prolate in shape. When dynamo action is admitted, the convection is able to generate strong magnetic fields largely in equipartition with the dynamics. Introducing a modest external field (which may be of primordial origin) into the radiative envelope can substantially alter the turbulent dynamics of the convective core, yielding magnetic fields of remarkable super-equipartition strength. The turbulent convection involves a complex assembly of helical rolls that link distant portions of the core and stretch and advect magnetic field into broad swathes of strong toroidal field. These simulations reveal that supercomputing is providing a perspective of the deep dynamics that may become testable with asteroseismology for these stars.

1. Observed Character of Magnetic A-type Stars

The peculiar A stars (Ap) have continued to engender much interest since their initial discovery in the late 19th century [12]. Broadly speaking, these stars exhibit strong and variable spectral lines (relative to solar values) in Si and certain rare earth metals (e.g. Sr and Hg). Of the Si and Sr-Cr-Eu pecliarity classes, most, if not all, possess equally variable and unusually strong magnetic fields [e.g., 13]. Typically, the magnetic Ap stars possess field strengths of a few hundred Gauss, but field strengths ranging from a lower threshold of ~300 G up to 20,000 G have been observed [1].

When variable, these fields appear to change at the stellar rotation rate [e.g. 8; 13], suggesting that the magnetic fields are "frozen in" to the radiative envelope of the star. Since a typical Ohmic decay time is much longer than the lifetime of an A star, these magnetic fields are likely to be remnants of the primordial magnetic field which threaded the molecular cloud whose collapse ultimately formed the star [7]. Such primordial fields likely evolve over long-time scales, ultimately relaxing into a stable, twisted field configuration [2]. A dynamo operating within the core of the A star might also influence the surface magnetic fields. That the cores of these stars likely harbor magnetic dynamos has been suggested by Krause & Oetken [10]. More recently, 3-D numerical simulations have demonstrated that nuclear burning cores of A-type stars can drive vigorous convection capable of generating equipartition strength magnetic fields [3; 4].

Understanding the strong surface fields of these stars in terms of a core-dynamo faces some difficulties. The circulations and magnetic buoyancy effects (on equipartition strength fields) are likely to be too weak for the transmission of interior field to the surface [e.g., 11]. Nevertheless, if the surface fields are indicative of a global-scale magnetic field with roots in the core, then the primordial field must extert some influence on the core dynamo, and vice versa. We explore this possibility by modeling a core dynamo with a large-scale field of primordial origin threading the radiative envelope.

2. Modeling A-type Star Core Dynamos

We model convection and dynamo action in a main-sequence A-type star of $2M_\odot$ rotating at four times the solar rate (with a rotation period of seven days) using the 3-D anelastic spherical harmonic (ASH) code. ASH solves the three-dimensional MHD equations in a rotating, spherical frame under the anelastic approximation, thereby filtering out sound waves which would otherwise severely limit the time steps [e.g., 5]. The anelastic approximation is particularly appropriate in the deeper interiors of A-type stars where the fluid motions are distinctly subsonic and thermodynamic variables are small compared to their mean, horizontally averaged values at a given depth.

Our computational domain extends from the inner 0.02 to 0.3 of the star by radius, the inner 0.15 of which is convectively unstable, with an overlying stable, radiative envelope comprising the remainder of the domain. The computational domain is decomposed into 82 Chebyshev polynomials in the radial and horizontally into spherical harmonics up to degree l of 170. ASH simulations explicitly follow the largest scales of motion, while employing a sub-grid scale (SGS) treatment of the unresolved motions using eddy diffusivities. For this study, our diffusivities scale as the inverse square root of the mean density. We adopt a Prandtl number of 0.25 and a magnetic Prandtl number of 5 throughout the domain. Based on the average rms velocity at midcore and using the core radius as our length scale, the Reynolds number following equilibration is $\sim$136 and the magnetic Reynolds number is $\sim$680. Further details of this model may be found in Featherstone et al. [9].

We have initialized our fossil field system by augmenting the magnetic fields from a mature A-star dynamo simulation (case C4m) of Brun et al. [4]. Their dynamo was started using a well equilibrated hydrodynamic simulation from Browning et al. [3] by adding a small dipole seed field to this system. Persistent dynamo action realized in that simulation yielded equipartion magnetic energies with respect to the convective flows. The magnetic energy was largely comprised of fluctuating (non-axisymmetric) fields, with the mean (axisymmetric) poloidal and toroidal fields constituting only a small fraction ($\sim$ 5%) of the magnetic energy.

Arguments concerning the geometry most likely to support a primordial field against decay have long suggested that such fields require a poloidal and toroidal component if they are to remain stable over the lifetime of a star [e.g., 14]. Recent numerical simulations have shown that such fields may relax into a twisted toroidal shape with a strong dipolar component to the field [2].

We have adopted such a twisted field geometry for our model fossil field, taking B_ϕ to be symmetric in longitude with a Gaussian cross-section of amplitude 30 kG, centered along the equator at $r_0 = 0.17R_{star}$ with a halfwidth of $0.2r_0$. A poloidal field component consistent

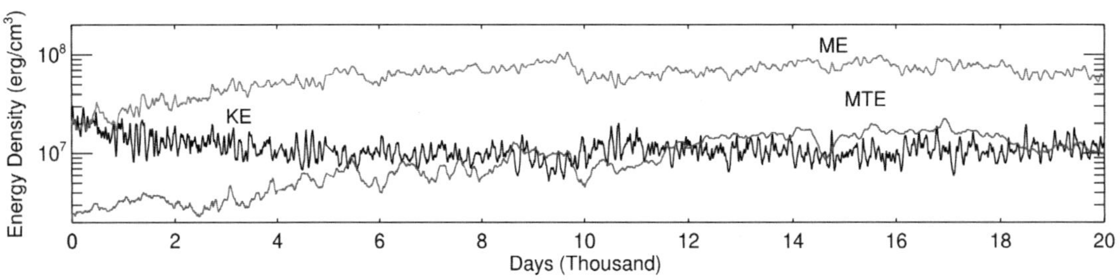

Figure 1. Evolution of volume-averaged energy densities following the imposition of an external mixed magnetic field spanning 20,000 days (or about 2,900 rotation periods). Magnetic energy ME (red) has grown in strength to become about ten-fold greater than the kinetic energy KE (black). The mean toroidal field magnetic energy MTE (blue) has also increased prominently in strength, attaining roughly equipartition levels with KE at ~9,000 days, and becoming slightly super-equipartition at ~13,000 days.

with a current threading through the center of our magnetic torus was then added to give some twist to the field, with the total magnetic energy associated with this field constituting a 10% increase to the magnetic energy of the overall system. The strength of this field was adjusted so that the ratio of energy in the poloidal field to that in the toroidal field was 1:9; this ratio was suggested by the calculations of Braithewaite & Spruit [2] as being the likely one for a stable twisted torus residing in the radiative zone of an A-type star. The twist, however, is not crucial to the discussion that follows. We have explored other fossil field configurations involving purely poloidal fields with a spherical harmonic degree ℓ up to 8, and found similar long-term behavior of the core-fossil field system in all cases.

3. Nature of Super-Equipartition Hydromagnetic State

One of the more striking effects arising from the inclusion of a fossil field in a core dynamo model is readily apparent in the temporal evolution of the system's (globally averaged) energy densities. The time histories of the kinetic energy density (KE), magnetic energy density (ME), and the energy associated with the mean (axisymmetric) toroidal fields (MTE) are presented in Figure 1. The evolution of these quantities is shown over the course of 20,000 days, approximately three magnetic diffusion times across the core. ME and KE vary on the convective turn over timescale (100-200 days), and MTE evolves more gradually on timescales of 500-1000 days.

The inclusion of a fossil magnetic field in our system has led to a notable departure from the energy balance achieved in [4]. A fivefold increase in ME develops over approximately one magnetic diffusion time ($\sim$ 7000 days) following the imposition of the external field. This rise in magnetic energy is accompanied by a halving in the KE of the system, yielding a ratio of ME/KE $\sim$ 10 that is maintained (with some variation) throughout the remainder of the simulation. The core dynamo also shows a propensity for building strong mean fields in the presence of a fossil field. Energies associated with the mean toroidal fields attain levels roughly 20% of the total magnetic energy, growing steadily in time alongside ME, reaching equipartition levels at about 9,000 days and eventually attaining slightly super-equipartition levels of its own. Such a change in the energy balance is remarkable given that the imposed fossil field constituted but only a small perturbation to the magnetic energy of the system.

3.1. Structure of Flows and Fields

The bulk of the ME change seen in Figure 1 is associated with non-axisymmetric magnetic fields largely contained within the core. Typical magnetic field strengths have transitioned

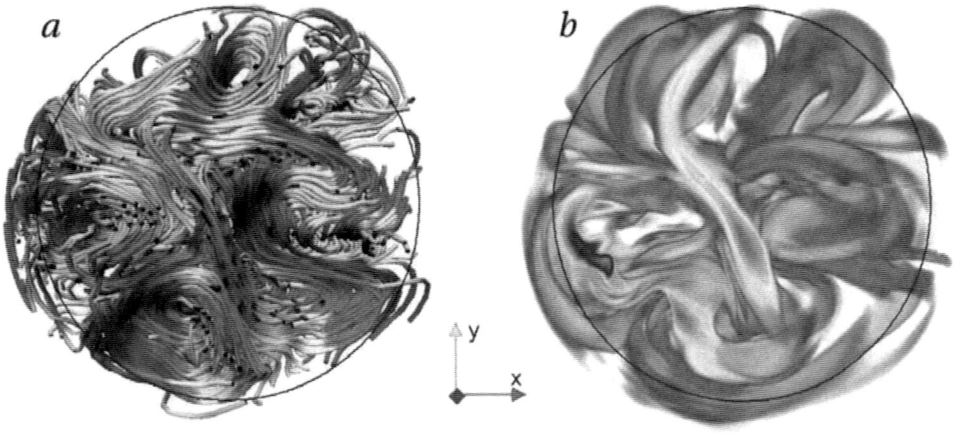

Figure 2. Sampling the evolving (*a*) flow streamlines and (*b*) accompanying magnetic energy density close to the equatorial plane at day 9,000. Violet tones indicate positive motions in the *y*-direction, and yellow tones negative motions. Regions of strong ME are shown in yellow/blue tones. Convective motions freely cross the rotation axis, stretching and advecting magnetic field as they do so.

from ~67 kG in the absence of a fossil field (based on rms values at mid-core) to ~80 kG, and typical flow speeds are ~20 m s^{-1} (vs. ~29 m s^{-1} initially). Magnetic fields and flows in this super-equipartition state are topologically different from those in the initial state as well. Magnetic fields are larger in scale on average and substantial mean fields have developed that are characterized by a coherent dipolar component and strong (100 kG) toroidal bands encircling the equator in the outer core and lower radiative zone. Along with the strength of the convective motions, mean flows have diminished. Differential rotation of the convective core, once characterized by a slow retrograde column at mid-core aligned with the rotation axis, is virtually non-existent in the super-equipartition field state.

Core convection in the presence of such strong magnetic fields is now characterized by four to six evolving cylindrical rolls encircling the rotation axis and aligned with it. A cross section of these rolls in the equatorial plane is displayed in Figure 2*a*. The force balance in these rolls is largely magneto-geostrophic in nature, with pressure gradients balancing Coriolis and Lorentz forces. Where these cylindrical rolls intersect the spherical edge of the convective core, the stable stratification of the radiative zone works to brake the radial component of their flow, resulting in a "tilting" of the otherwise largely horizontal motions. Such behavior is similar to that found in rotating convection studies of flows in spherical geometries with rigid boundaries [cf. 6] and is visible in detailed rendering of the streamlines in Figure 3*b*. Pressure gradients between the equatorial plane and the poles set up by this tilting drives flows along the axes of these rolls, imbuing neighboring rolls with opposite senses of helicity. Poleward axial flows from one roll thus link to the equatorward axial flows of their neighbors. Moreover, owing to the diminished differential rotation, downflows from one set of rolls freely cross the rotation axis undeflected, connecting distant regions of the core which would otherwise remain relatively isolated from one another.

Magnetic fields generated by these convective motions exhibit a similar globally-connected topology. Magnetic energy in the equatorial plane, accompanying the streamlines of Figure 2*a*, is rendered in Figure 2*b* showing that the cylindrical roll-like convection leaves its imprint in the magnetic structures. These roll-like structures influence the magnetic topology over much

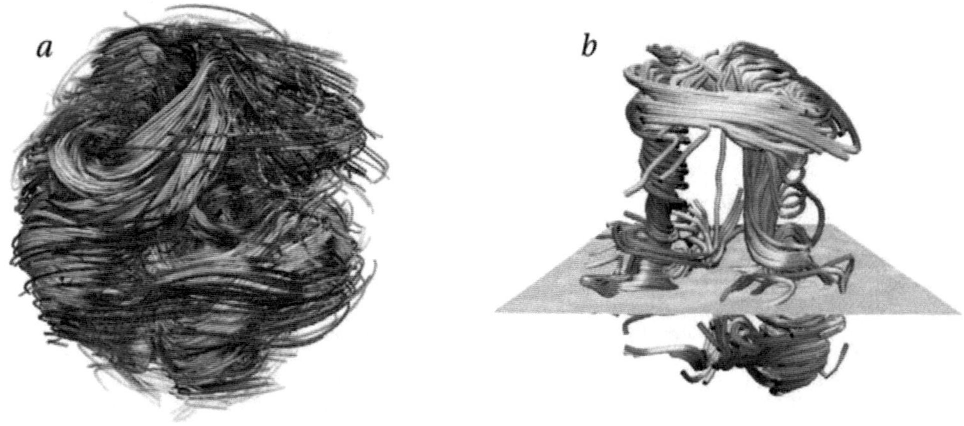

Figure 3. Convective and magnetic stuctures (with rotation axis vertical) near day 15,000. (*a*) Magnetic fields lines colored by their magnetic energy density as realized in the convection zone and region of overshooting; orange/green tones indicate high values, and blue/violet tones low values. (*b*) Interior view of columnar convection in the core of the A-star visualized near day 15,000 using instantaneous streamlines. Streamlines are colored by velocity component along the rotation axis v_z. Blue (yellow) tones indicate northward (southward) motion; equatorial plane indicated by light blue. Many rolls in the A-star interior possess axial flows that freely cross the equatorial plane as seen here. These axial flows tend to link neighboring rolls near the edge of the convective core leading to the global magnetic field topology evident in (*a*).

of the core as can be seen in field line renderings of the full core (Figure 3*a*). Broad flux ropes of strong ($\sim$ 300 kG) field extend from pole to pole encircling the periphery of the cylindrical rolls. Upflows tend to advect these rolls into the region of convective overshooting just outside the core boundary. There, large swathes of field left behind by overshooting convection diffuse away slowly, ultimately joining with other bundles of magnetic flux to create the strong axisymmetric fields observed in that region.

3.2. Diminishing Lorentz Force Feedbacks

One might anticipate that strong Lorentz forces would accompany such high magnetic field strengths, ultimately suppressing the very convective motions that spur their genesis. The super-equipartition state achieved here, in which ME/KE$\sim$ 10 is thus truly remarkable and raises the question of how a system might build and maintain such strong magnetic fields, The generation of magnetic energy is closely aligned with the large-scale topology of the flow patterns. Near the rotation axis, the cylindrical rolls closely abut one another creating a region of strong shearing. A coherent bundle of streamlines is also visible in Figure 2*a* extending in the y-direction, linking a distant portions of the core across the rotation axis, and possessing a strong magnetic signature as seen in Figure 2*b*. Magnetic structures advected across the core in this fashion pass through region of strong shearing flows near the rotation axis and serve as the seed for further field generation. A more detailed analysis of the energy generation has been carried out in Featherstone et al. [9], finding that magnetic energy generation is strongest near the rotation axis and in the region of overshooting where motions deposit and shear magnetic field.

As the core convection generates stronger magnetic fields, the flows and magnetic fields tend to arrange themselves in such a way as to minimize the Lorentz forces. Regions of strong flow and strong field tend to exist separately from one another such that flows tend to persist most strongly in regions where Lorentz forces are already minimal [see 9] . In regions where strong magnetic fields coexist with substantial flows, streamlines and magnetic field lines tend to be nearly parallel. Minimizing the Lorentz force depends on the alignment of the current density **j** with the magnetic field **B**. The current density **j**, however, depends on $\mathbf{v} \times \mathbf{B}$ through Ohm's law, with velocity vector **v**. Minimizing the Lorentz force thus typically implies some alignment of **v** and **B**. A sense of such aligment can be gathered from Figure 2 where the largest structures in magnetic energy seem to run parallel to streamlines. A time averaged sampling of the angle between **B** and **v** throughout the core shows that typically this angle is smaller in the super-equipartition state than in the progenitor solution.

4. Conclusions and Perspectives

Our work suggests that the presence of a primordial field can subtly influence the dynamics within the core, yielding a stronger core dynamo characterized by global-scale magnetic fields and flow configurations. These results are quite different from those obtained earlier in the absence of a primordial field. We have at best scratched the surface of these systems in terms of rotation rates, diffusivities, and other parameters. Of particular interest is whether the strong mean fields generated near the edge of the convective core might become buoyant, ultimately contributing to the large star spots seen at the surface of the Ap and Bp stars. There may be a wide array of dynamo states accessible to these cores of these stars with peculiarities in their flows and fields beyond those enumerated here. The (nearly) non-existent differential rotation of the core in this study is unlikely to yield any discernible asteroseismic signature at the surface of such a star, and measuring the sub-surface nature of the magnetic fields in Ap stars is fraught with difficulties of its own. Nevertheless, further simulation of the A, and more massive O and B, stars may well yield unexpected surprises in terms of the differential rotation of the core and the nature of the dynamo. These may in turn suggest new routes for making seismic inferences about the deepest interiors of these massive stars.

5. Acknowledgements

The authors gratefully acknowledge many helpful conversations with Mark Miesch, Benjamin Brown, and Kyle Augustson. This research is supported by NASA through Heliophysics Theory Program grants NNG05G124G and NNX08AI57G, with additional support for Featherstone through NASA GSRP program by award NNX07AP34H. Browning was supported by NSF Astronomy and Astrophysics postdoctoral fellowship AST 05-02413, and by a Jeffery L. Bishop fellowship at CITA. Brun was partly supported by the Programme National Soleil-Terre of CNRS/INSU (France), and by the STARS2 grant from the European Research Council. The simulations were carried out with NSF PACI support of PSC, NCSA and TACC, and by NASA HEC support at Project Columbia.

[1] Aurière M et al. 2007 *A&A* **475** 1053
[2] Braithwaite J & Spruit H C 2004 *Nature* **431** 819
[3] Browning M K, Brun A S & Toomre J 2004 *ApJ* **601** 512
[4] Brun A S , Browning M K & Toomre, J 2005 *ApJ* **629** 461
[5] Brun A S, Miesch M S & Toomre J 2004 *ApJ* **614** 1073
[6] Busse F H 2002 *Phys. Fluids* **14** 1301
[7] Cowling T G 1945 *MNRAS* **105** 166
[8] Deutch A J 1956 *PASP* **68** 92
[9] Featherstone N A, Browning M K, Brun A S & Toomre J 2009 *ApJ* **705** 1000
[10] Krause F & Oetken L 1976 *Physics of Ap Stars* (Vienna: Univ. Wien)
[11] MacDonald J & Mullan, D J 2004 *MNRAS* **348** 702
[12] Maury A C 1897 *Harvard Ann.* **28** 96
[13] Mestel L 1999 *Stellar Magnetism* (Oxford: Clarendon Press)
[14] Prendergast K H 1956 *ApJ* **123** 498
[15] Preston G W 1971 *PASP* **83** 571

An Introduction to Wave-Trapping in Supergranulation

W. Allen[1], **F. Hill**[1]

E-mail: wallen@nso.edu

Abstract. This paper is an introduction to modelling waves trapped in a supergranular cell. The supergranular cell is generalized to the form of a hexagon with a cylinder inscribed within its boundaries. A cylindrical wave equation is implemented and solved and we account for the edges of the hexagon through boundary conditions. Plots are created of the solution and will serve as a test as to whether the model reflects actual wave conditions inside a single supergranular cell.

1. Introduction

Over the years many aspects of supergranulation regarding the dimensions, origins, and attributes associated with the convection of flows in relation to local magnetic fields have remained a mystery. Supergranules are large scale convective polygonal structures on the Suns surface. Certain characteristics of supergranules that have been observed or estimated include having a maximum mean diameter ranging from 25 to 85Mm, an estimated maximum depth of 10 Mm, a mean lifetime of about 24 hours, and root mean squared horizontal and vertical flow velocities ranging from .3 to .4 km/s and .1 to .2 km/s respectively. There are an estimated 2500 supergranules covering the Suns surface at any one time [1]. Below the Sun's surface the flow of plasma constituents convect in large part due to temperature and pressure gradients. As the flows rise to the surface much of this flow advects across the surface of the Sun out to supergranule boundaries. More recently it has been speculated that there are not only flows inside supergranular cells, but the fluid inside cells could feature oscillatory wave properties. In this paper we attempt to model these waves using a four-dimensional wave equation featuring independent variables $r, \theta, z,$ and t for a single supergranule cell. We first introduce a single template for the supergranular cell. The template for our model takes the form of a hexagon and this shape was chosen since it is sufficiently complex yet not so complex that the geometry would be too difficult to model. The model is obtained from a four-dimensional wave equation written in terms of independent variables $r, \theta, z,$ and t. The solution to the wave equation is found by separation of variables. Various plots of the model will be shown and these plots will serve as a diagnostic test of how the model relates to observations. In the future we hope to compare the oscillation attributes found from the model to actual observations using the HMI (Helioseismic Magnetic Imager) instrument on SDO. The conclusion will summarize our findings and outline goals for future work.

---

[1]National Solar Observatory, 950 N. Cherry Av. Tucson AZ. 85719

2. The Model

The template of the cell used to trap waves for the model supergranule is shown below in Figure 1.

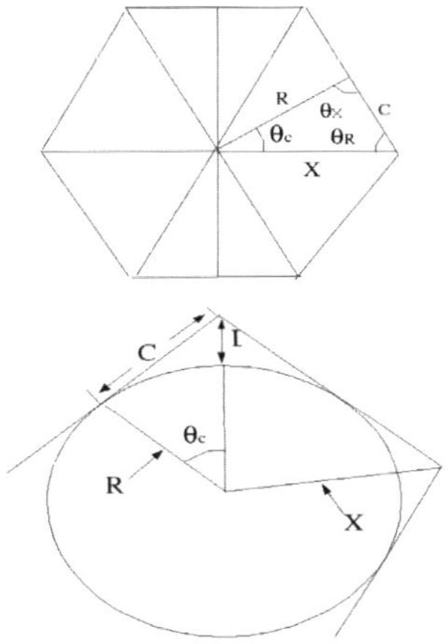

Figure 1. The hexagonal template used for wave-trapping. Boundaries are treated as perfectly reflecting and waves generated originate from the centre of the template.

The template shown above features an inscribed circle within the boundaries of the hexagon. The circle is actually a cylinder and therefore there is a depth associated with the model which extends down to 10Mm. We can use the geometry of the template to write the radial component as,

$$R = \frac{X \sin\theta_R}{\sin\theta_X} = \frac{X \sin\theta_R}{\sin(180° - \theta_R - \theta_C)}, \tag{1}$$

To account for the edges of the hexagon we employ the boundary conditions which yield the eigenvalues:

$$\lambda_{nm} = (\frac{\mu_{nm}}{R}) = \frac{\mu_{nm} \sin(180^\circ - \theta_R - \theta_C)}{X \sin\theta_R} \quad \text{and} \; \beta_{nmp}^2 = \lambda_{nm}^2 - (\frac{p\pi}{l})^2$$

where β_{nmp} combines both horizontal and vertical boundary conditions, μ_{nm} and p are parameters that arise through the implementation of the boundary conditions, and X and l are based upon the horizontal and vertical dimensions of the model supergranule respectively. Cylindrical geometry gives the four dimensional wave equation as,

$$a^2(\frac{\partial^2 U}{\partial r^2} + \frac{1}{r}\frac{\partial U}{\partial r} + \frac{\partial^2 U}{\partial z^2} + \frac{1}{r^2}\frac{\partial^2 U}{\partial \theta_C^2}) = \frac{\partial^2 U}{\partial t^2} \tag{2}$$

The wave equation found in equation (2) will be decomposed through separation of variables into four eigenfunctions and the solution is $U(r,\theta,z,t) = R'(r)M(\theta)Z(z)T(t)$. Written explicitly, the solution becomes,

$$U(r,\theta,z,t) = \sum_{n=0}^{\infty}\sum_{m=1}^{\infty}\sum_{p=0}^{\infty}\{[(A_{nmp}\cos(n\theta) + A^{*}_{nmp}\sin(n\theta))\cos(a\beta_{nmp}t)$$
$$+ (B_{nmp}\cos(n\theta) + B^{*}_{nmp}\sin(n\theta))\sin(a\beta_{nmp}t)]\sin(\beta_{nmp}z)\}J_n(\lambda_{nm}r) \quad (3)$$

where $A_{nmp}, A^{*}_{nmp}, B_{nmp}$, and B^{*}_{nmp} are coefficients, a is a parameter having units of velocity, $J_n(\lambda_{nm}r)$ is a Bessel's function of the first kind, and $\theta = \theta_c$. It is worth mentioning that equation (3) was obtained through the following steps. First determine the boundary conditions from the template in Figure 1. Second use separation of variables to decouple the wave equation into two and later into four eigenfunctions. Boundary conditions yield the eigenvalues λ_{nm}, β_{nmp} and each are used in the construction of the general solution. Finally note that the radial component is modelled so that X =15Mm and time is sampled at t =35 h.

3. Tests On The Model

The solution to the wave equation given in equation (3) can be used in plotting routines to determine if the model solution reflects some of the actual wave properties and fluid flow characteristics found in actual supergranulation. We show three plots found from the solution given in equation (3). The first are a contour plot and polar plot shown in Figure 2. Figure 3 illustrates a plot of the wave attributes of the model.

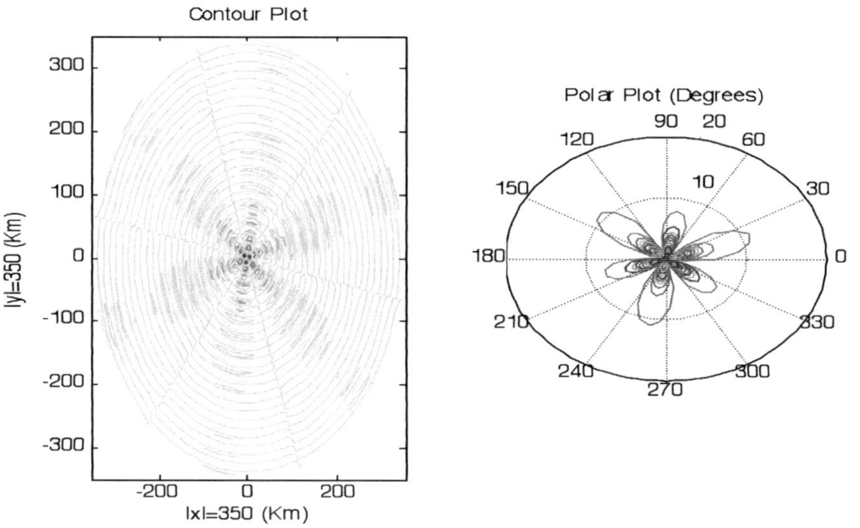

Figure 2. A contour plot shown (left) and a polar plot (right). The contour plot illustrates waves trapped within the template given in Figure 1. Notice that the contour plot is partitioned into six sectors. These sectors show how the waves travel as they are generated from the centre and are reflected by each side due to boundary conditions. The colour scheme shows that the amplitude of the waves are greater near the centre. The polar plot shows various wave modes as generated with an intrinsic time step and the direction of propagation is shown as a function of angle.

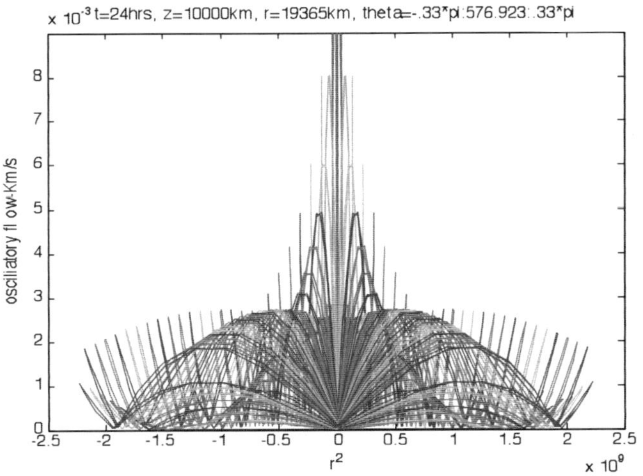

Figure 3. The three-dimensional plot above represents the oscillatory flow as a function of r^2. *As seen in this plot the oscillations diverge from the centre and the magnitude of the oscillatory flow found at the centre is greater than that found at the boundaries.*

The left panel in Figure 2 shows the oscillations as wedges between the contours that branch out from the centre of the plot. The waves emerge and reflect as modelled in an isotropic medium from the centre out to each side of the hexagon where they are seen to reflect off of the boundaries used for the template. On the right of Figure 2 the polar plot shows the distribution of the amplitude of the oscillations with angle for a given time step. Physically the polar plot may also be interpreted through the fact that, in each direction, there is a net dissipative effect that each wave front causes on the medium it propagates in with a larger and larger time step. Finally Figure 3 shows the oscillation pattern as oscillatory flow versus R^2. The modes are hard to discern because the plot is actually in three dimensions as seen in a two dimensional layout. Finally in Figure 3 note that as one approaches the portion of the plot at $R^2 = 0$ the model blows up at that value.

4. Conclusion

This paper illustrates that a cylindrical wave equation constrained by boundary conditions can be used to model waves trapped within a hexagonal template. Waves are generated at the centre of the template and the model shows that the waves diverge from the centre with higher amplitude at that position and subside at the boundaries. This behaviour is in good agreement with the actual fluid flow found in supergranules modelled with 10 Mm depth and 30 Mm diameter. Future work will involve using dynamic boundary conditions so that the shape of the regular polygon template changes with each time step. We also will extend our wave equation to include the magnetic field, pressure gradients as well as gravitational stratification. Finally we hope to couple the theoretical aspects of the model with observation. This will be done by using the Fourier transform of the solution compared with ring diagram analysis generated with the Helioseismic Magnetic Imager from Dopplergrams. The comparison of the model and observations will help fine-tune the independent variables of the model and may enable us to obtain a better estimate for the depth of supergranulation.

References

[1] Bhatnagar A and Livingston W 2005 *Fundamentals Of Solar Astronomy* vol.6 (World Scientific)
[2] Gizon L, Duvall T L Jr., Schou, J 2003a , *Nature* 43 421

Modeling the Near-Surface Shear Layer: Diffusion Schemes Studied With CSS

Kyle Augustson[1], **Mark Rast**[2], **Regner Trampedach**[1], **Juri Toomre**[1]

[1] JILA, [2] LASP, and Dept. of Astrophysical & Planetary Sciences, University of Colorado, Boulder, CO, USA, 80309

E-mail: `kyle.augustson@colorado.edu`

Abstract. As we approach solar convection simulations that seek to model the interaction of small-scale granulation and supergranulation and even larger scales of convection within the near-surface shear layer (NSSL), the treatment of the boundary conditions and minimization of sub-grid scale diffusive processes become increasingly crucial. We here assess changes in the dynamics and the energy flux balance of the flows established in rotating spherical shell segments that capture much of the NSSL with the Curved Spherical Segment (CSS) code using two different diffusion schemes. The CSS code is a new massively parallel modeling tool capable of simulating 3-D compressible MHD convection with a realistic solar stratification in rotating spherical shell segments.

1. Introduction

The solar differential rotation profile exhibits prominent radial shear layers near the top and bottom of the convection zone [1]. The near-surface shear layer (NSSL) occupies the upper 5% of the Sun by radius, whereas the tachocline begins near the base of the convection zone. The dynamics of the NSSL are governed largely by vigorous granular-scale convection that is driven by radiative cooling and large thermodynamic gradients. The collective interaction of these granular-scale flows (average sizes of 1 Mm, lifetimes of 0.2 hr) is a major component in the formation of supergranular (15-35 Mm, 24 hr) and mesogranular (5-10 Mm, 5 hr) scales [2; 3].

Given this wide range of spatial and temporal scales, we currently cannot simultaneously model hundreds to thousands of supergranules, solar granulation, and deeper flows. However, we may still be able to characterize the influence of granulation and deep global flows on the NSSL by coupling CSS with global convection models from below [4] and with surface convection simulations from above [e.g. 3; 5]. Such a coupling, whether it is statistical or direct, requires a careful treatment of diffusion and boundary conditions [6]. For instance, low diffusion is necessary to preserve the spatial structure and advective timescales of the downdrafts flowing into the CSS domain from the surface convection above. Given that our grid is five times coarser than that of typical surface convection simulations, this is no easy task.

We have conducted two numerical simulations in a 20° square patch centered on the equator that encompass most of the NSSL and some of the deep interior, rotating at the solar rate. These simulations explore the effects of turbulent-eddy and slope-limited diffusion schemes with an open lower radial boundary and closed upper boundary; as such they are identical except for the diffusion scheme. The governing equations and numerical approach used in solving them are briefly discussed in §2. The turbulent-eddy and slope-limited diffusion schemes are detailed in §3. The dynamics of the flows established in these simulations are examined in §4.

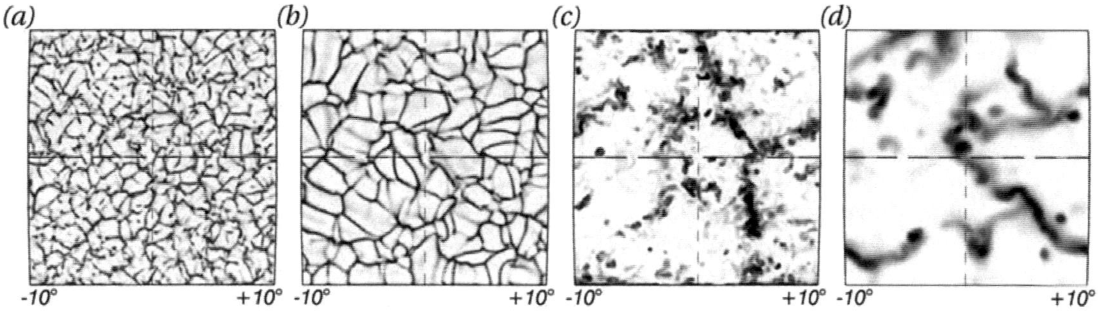

Figure 1. Radial velocities (u_r) at 0.99 $R_\odot$ in (a) the slope-limited diffusion *Case 2*, and (b) the turbulent-eddy diffusion *Case 1*, and at 0.95 $R_\odot$ in (c) *Case 2* and (d) *Case 1*. Dark tones denote downflows and light tones denote upflows. Scaling values are -1.34 km s^{-1}, 0.59 km s^{-1} for (a) and (b) and -0.73 km s^{-1}, 0.24 km s^{-1} for (c) and (d), where positive u_r points outward.

2. Formulating the Problem

The spherical segment domains used in our simulations involve large portions of the Sun's inherent spherical geometry, which is necessary to properly capture the effects of rotation on supergranular scales. We use the Curved Shell Segment (CSS) code to model the 3-D dynamics below the solar photosphere. CSS is a modeling tool that solves the compressible Navier-Stokes equations in rotating spherical segments [6]. To simulate the larger scales of motion that are likely to occur in the solar convection zone, a large-eddy simulation (LES) model is employed. The scales that are not explicitly computed in these simulations are parametrized and included in a sub-grid scale model of turbulent transport and diffusion. The governing equations solved in CSS are:

$$\frac{\partial \rho}{\partial t} + \nabla \cdot (\rho \mathbf{u}) = 0, \tag{1}$$

$$\rho \left[\frac{\partial \mathbf{u}}{\partial t} + (\mathbf{u} \cdot \nabla) \mathbf{u} \right] = -\nabla P - \rho g \hat{\mathbf{r}} + \nabla \cdot \bar{\mathcal{D}} + 2\rho \mathbf{u} \times \mathbf{\Omega} + \rho \Omega^2 \mathbf{R}, \tag{2}$$

$$\rho T \left[\frac{\partial S}{\partial t} + (\mathbf{u} \cdot \nabla) S \right] = \nabla \cdot \left[\kappa_S \nabla S' + \kappa_0 \frac{\partial \bar{S}}{\partial r} \hat{\mathbf{r}} + \kappa_r \nabla T \right] + \Phi, \tag{3}$$

$$\bar{\mathcal{D}}_{ij} = \mu \left[\bar{e}_{ij} - \frac{1}{3} \nabla \cdot \mathbf{u} \bar{\delta}_{ij} \right]. \tag{4}$$

The symbols ρ, $\mathbf{u}$, P, T, S, $\bar{\mathcal{D}}$, $\bar{e}$, $\bar{\delta}$ are the density, velocity, pressure, temperature, specific entropy, viscous stress tensor, stress tensor, and identity tensor respectively; Φ is the viscous heating term; the entropy perturbations are $S' = S - \langle S \rangle$; the equation of state is $P = \rho^\gamma e^{S/C_V}$; $\mathbf{\Omega}$ is the angular velocity of the rotating frame; μ and κ_S are the turbulent-eddy diffusion of momentum and entropy, and κ_r is the radiative thermal diffusion; g is the local acceleration due to gravity, where μ, κ_S, κ_r, κ_0, and g are functions of radius only.

The governing equations (Equations 1-3) are evolved on a uniform spatial mesh. Temporal discretization is accomplished using an explicit fourth-order accurate Runge-Kutta time-stepping scheme. The spatial derivatives are computed using a modified sixth-order compact finite difference scheme. A 3-D domain decomposition divides the full spatial mesh into sub-domains. The boundary information necessary to compute spatial derivatives is passed between nearest-neighbor sub-domains using MPI, while computations are shared among the master and slave cores within a supercomputer node using OpenMP.

Currently there are two diffusion schemes implemented in CSS, a turbulent-eddy diffusion (TED) (Equation 3-4) and the slope-limited diffusion (SLD) described in (§3). In the TED scheme, the momentum and entropy diffusivities (μ and κ_S) are calculated based upon the desired Rayleigh number at the upper boundary with the constraint that the Prandtl number and the dynamic viscosity be constant throughout the domain. There are no directly comparable diffusive parameters in the SLD scheme as the eddy diffusion terms have been dropped. One may still control the level of diffusion, however, with specific choices of which slope-limiter is employed and what characteristic velocity is used at cell interfaces. Figure 1 depicts snapshots of typical flows at two radii from the two diffusion schemes.

3. Slope-Limited Diffusion In CSS

Slope-limited diffusion has many possible formulations, and we have chosen to use a scheme similar to that found in [5]. SLD is based upon a piecewise linear reconstruction within a finite volume (cell) centered on each gridpoint of the solution at each time step. The linear reconstruction leads to a solution that has discontinuities at the cell edges. SLD essentially acts to minimize these discontinuities so that the numerical scheme remains stable. While this model of diffusion is not as physically motivated as the eddy diffusion model, it holds many computational advantages. Indeed, with SLD in CSS, a higher level of turbulence is achieved for a given resolution and there is no need to evaluate second-order derivatives, which decreases memory usage by a factor of three and the execution time per time step by a factor of two.

The slope of the linear reconstruction of the solution is given by a ratio of the downwind and upwind cell-center differences ($r_i = \Delta_l u_i / \Delta_r u_i$). This slope is "limited" by a function $\phi(r_i)$ that belongs to a class of slope limiters that yield total variation diminishing solutions [7]. As in [5], we use a linear combination of two such slope limiters, the minmod and superbee limiters. Using the reconstructed slope, values of the primitive variables are computed at the cell edges as $u_i^e = u_i \pm \frac{1}{2}\phi(r_i)\,\Delta_r u_i$. The diffusive flux at a cell edge is $f_i^e = \frac{1}{2} c_i^e g_i^e \beta_i^e \delta u_i^e$, where g_i^e is a geometric factor and c_i^e is a characteristic velocity at a cell edge, and δu_i^e is the difference of the left and right reconstructed values at a cell-edge. To avoid artificial steepening, a local diffusion coefficient β_i^e is constructed such that when $\delta u_i^e \Delta_e u_i > 0$, $\beta_i^e = (\delta u_i^e / \Delta_e u_i)^2$ and is zero otherwise. This preserves steep gradients and yields a fourth-order upwind diffusion scheme [5].

With the diffusive fluxes at each edge of a given cell as above, the diffusion at the center of a cell is $D_i = \left(f_i^r - f_i^l\right)/\Delta x_i$, where Δx_i is the grid spacing. This is added to the solution after the full Runge-Kutta time step as $u_i = u_i + \Delta t D_i$. Within our simulations, using the true sound speed in the c_i^e proves to be overly diffusive. So, for the computation of the c_i^e, the sound speed throughout the computational domain is fixed to a fraction ($\sim 1/10$) of the surface sound speed.

4. Comparing Two Diffusion Schemes

The simulations encompassing a $20^\circ \times 20^\circ$ domain are relatively low resolution with 128 radial mesh points and 256 points in latitude and longitude. A stellar evolution code is used to establish a realistic initial stratification for the simulations. Since a perfect gas is assumed, the He and H ionization zones cannot currently be simulated. Thus, the impenetrable upper boundary is taken to be 0.995 $R_\odot$ in order to exclude most of the radial extent of these zones. The permeable lower boundary is placed at 0.915 $R_\odot$, yielding a density contrast across the domain of 400. Each case has the lowest diffusivities that allow numerical stability in the TED (*Case 1*) and SLD (*Case 2*) regimes. Using a constant ν with depth in the TED case is rather restrictive in that the diffusion is larger than necessary away from the upper boundary, but for this preliminary analysis it is sufficient to characterize the differences between the two schemes.

Turning to the radial velocity patterns sampled in Figure 1, there is a striking shift toward smaller spatial scales in *Case 2* with considerably less power at large scales. At 0.99 $R_\odot$ the peak of the spatial power spectrum in *Case 1* is around 40 Mm. In *Case 2*, however, there is a broad peak at scales around 20 Mm, more closely matching the supergranular scales of the Sun. As one descends deeper into the simulations (Figures 1c,d), the flows appear significantly more turbulent in *Case 2* than in *Case 1*.

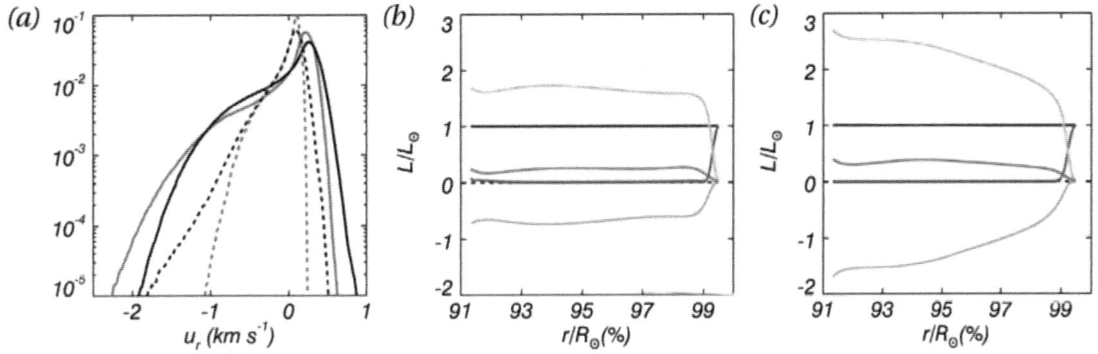

Figure 2. (a) PDFs of radial velocities at 0.99 $R_\odot$ (solid line) and 0.95 $R_\odot$ (dashed), for *Case 2* (black) and *Case 1* (red). Time-averaged radial energy flux balances for *Case 1* (b) and *Case 2* (c), with total flux in purple, radiative flux (blue), enthalpy flux (green), kinetic energy flux (orange), and acoustic flux (red).

Indeed, the downflowing structures of *Case 2* become much more narrow at depth with many more isolated plumes. The greater turbulence also manifests itself in larger extrema of each of the physical variables. For example, the radial velocities at 0.95 $R_\odot$ in *Case 2* have extrema that are twice that of *Case 1* (Figure 2a). In *Case 1*, a precipitous drop in the upflowing radial velocity probability density distribution (PDF) outside of about 0.1 $\mathrm{km\,s^{-1}}$ is indicative of just how uniform the upflows are. In stark contrast, the radial velocity PDF of *Case 2* has twice the dynamic range, emphasizing that the upflows are indeed much more structured than those of *Case 1*. While the radial mass flux PDF has higher velocity wings that contribute only one percent to the overall distribution, the bimodal nature of the temperature perturbation PDF has the effect of significantly amplifying the importance of these outlying values. Thus, the net effect of the larger wings in these PDFs is to shift the mean of the enthalpy flux in *Case 2* to a higher value. The additional velocity in both the horizontal and radial flows greatly alter the kinetic energy distribution, as in *Case 2* where the mean kinetic energy has nearly doubled and correspondingly increased the magnitude of the kinetic energy flux (Figure 2c).

A tentative next step is to have both diffusion schemes active at the same time. This will allow a systematic study on the convergence of solutions as the TED coefficients are lowered to solar values and the SLD takes over as the primary diffusion.

Acknowledgments

The authors thank Neal Hurlburt and Marc DeRosa for most helpful discussions and their contributions to code development and implementation. This work has been supported by NASA grants NNX08AI57G and NNX10AM74H. The computations were carried out on Pleiades at NASA Ames with SMD grant g26133.

References

[1] Thompson M J, Christensen-Dalsgaard J, Miesch M S and Toomre J 2003 *ARAA* **41** 599–643

[2] Rast M P 2003 *ApJ* **597** 1200–1210

[3] Asplund M, Nordlund A and Stein R F 2009 *Living Rev. Solar Phys.* **6**

[4] Miesch M S, Brun A S, De Rosa M L and Toomre J 2008 *ApJ* **673** 557–575

[5] Rempel M, Schüssler M and Knölker M 2009 *ApJ* **691** 640–649

[6] Hurlburt N E, DeRosa M L, Augustson K C and Toomre J 2010 (*ASP Conf. Ser.* vol 440)

[7] LeVeque R J 2002 *Finite Volume Methods for Hyperbolic Problems* (Cambridge Univ. Press)

Characteristics of Solar Meridional Flows

Sarbani Basu[1] **and H. M. Antia**[2]

[1] Department of Astronomy, Yale University, P. O. Box 20801, New Haven CT 06520-8101, U. S. A.

[2] Tata Institute of Fundamental Research, Homi Bhabha Road, Mumbai 400005, India

E-mail: `sarbani.basu@yale.edu`, `antia@tifr.res.in`

Abstract. We have done a ring-diagram analysis of MDI full-disc data to determine the properties of solar meridional flow in the outer 2% of the Sun over the solar cycle 23. The meridional flows show a migrating pattern with higher-velocity flows migrating toward the equator as activity increases. Additionally, we find that the migrating pattern of the meridional flow matches those of the sunspot butterfly diagram and the zonal flows in the shallow layers. A Legendre polynomial decomposition of the meridional flows shows that the latitudinal pattern of the flow was also different during the maximum as compared to that during the two minima. We also find that the dominant component of the meridional flows during solar maxima was much lower than that during the minima of solar cycles 23 and 24.

1. Introduction

Meridional flows from the equator toward the poles have been detected on the solar surface using observations of magnetic features (Komm et al. 1993) as well as direct Doppler measurements (e.g., Hathaway 1996). Splittings of the global modes of solar oscillations are not sensitive to the first order contributions of meridional flows and as a result local helioseismic techniques have been used to study these flows in the solar interior. These techniques so far only give information about the near-surface regions of the Sun. Giles et al. (1997) studied meridional flows using the time-distance technique, while Schou & Kosovichev (1997) and Basu et al. (1998) applied the ring diagram technique to study these flows. They found that the flow velocities in the outer convection zone are similar to those seen at the surface velocities. With the availability of seismic data covering the solar cycle 23, temporal variations of these flows can also be studied.

2. Data and Technique

We use MDI data to study solar meridional flows using ring-diagram analysis (Hill 1988). The technique uses 3D power spectrum (two spatial wave-numbers and the temporal frequency) of solar oscillations. The spectra that we use were obtained along the central meridian for $16^\circ \times 16^\circ$ patches in heliographic longitude and latitudes up to 52.5° in steps of 7.5° in both hemispheres. To reduce noise at each latitude we averaged over a Carrington rotation. We used 15 data sets covering a period from June 1996 to June 2009, with roughly one data set every year. These data cover the entire solar cycle 23 and a very small part of the beginning of cycle 24. The resulting 3D power spectra were fitted to the model of Basu & Antia (1999) to determine the shift in frequencies caused by the horizontal components of solar flows. The meridional flow velocities were obtained by inverting the parameter that describes the shift in frequencies due to

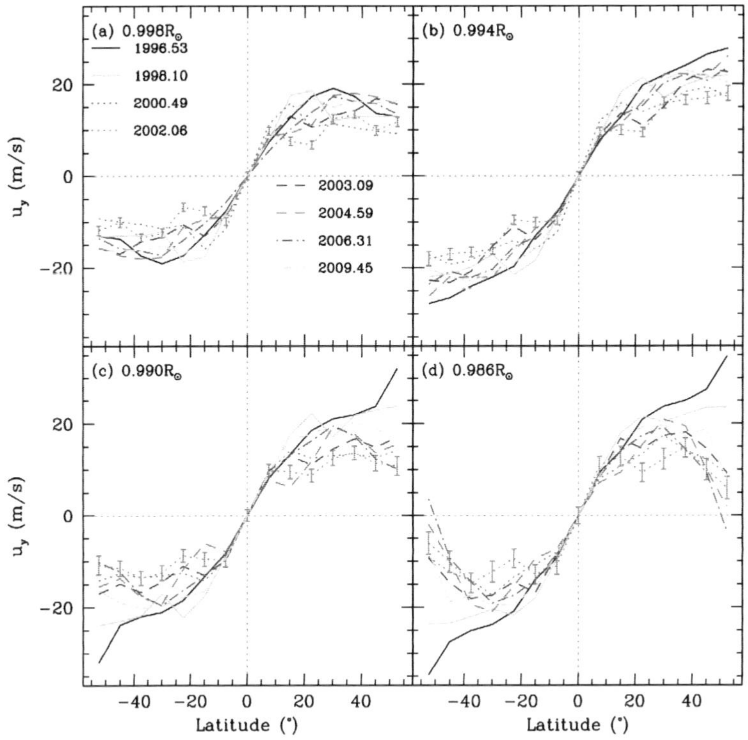

Figure 1. The north-south antisymmetric component of the solar meridional velocity shown as a function of latitude at a few different depths as marked in the panels. These results were obtained using OLA technique for inversion. The results obtained using RLS are similar.

meridional flows. We used both the Optimally Localised Averages (OLA) and the Regularised Least Squares (RLS) techniques to invert for the meridional velocity as a function of depth. The results are robust when two techniques give similar results. The inversions of the flow parameters obtained from the power spectrum of any patch gives the meridional flow velocity as a function of depth in that patch. Inversions of flow parameters of patches that cover different latitudes and times give us the latitude and time dependences of the flow. To determine the details of the latitudinal dependence of the flows at any given time, we decompose the flows in terms of the Legendre polynomials (Hathaway 1996):

$$u_y(r,\theta,t) = -\sum_i C_i(r,t)P_i^1(\cos\theta). \tag{1}$$

Here u_y is the velocity toward the north pole, θ is the angle measured from the north pole. The even values of i correspond to the antisymmetric component of meridional velocity and those are the only ones considered in this work.

3. Results

A sample of the results obtained using the technique described in Section 2 is shown in Figure 1. It should be noted that at some epochs and depths our results show flows across the equator. It is possible that these are artifacts of position-angle errors (Giles et al. 1997) and hence we ignore these in our analysis and concentrate only on the component that is antisymmetric about the solar equator. This is the component that is shown in Figure 1. To show the temporal variations more clearly we show the results at constant depth in Figure 2. This figure shows meridional flow velocity as a function of time and latitude. To make the figure look symmetric we have reversed the sign of meridional velocity in the southern hemisphere. Thus the positive velocity in both hemispheres points toward the respective poles. The results show bands of fast and slow

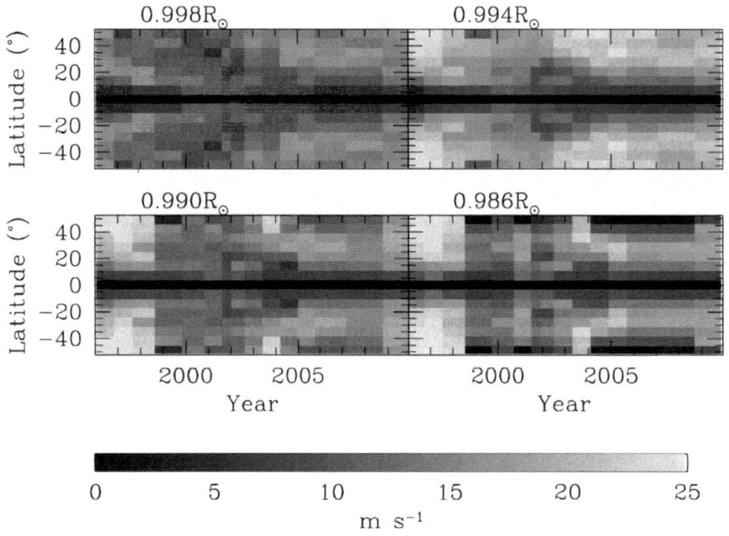

Figure 2. The speed of the north-south antisymmetric component of meridional velocity is shown as a function of latitude and time at a few different depths as marked in the panel. For comparison in one of the panel the position of each sunspot is marked in the time vs latitude plot. In each hemisphere positive velocities imply a flow toward the corresponding pole.

flows migrating toward the equator, very similar to the the zonal flow pattern. The mid-latitude meridional-flow band at a depth of 1.4 Mm coincides with the position of sunspots[1] (i.e. the butterfly diagram) as well as the zonal flow bands at this depth. However, the meridional-flow pattern shows a steeper dependence on depth than the zonal flows and thus the correspondence between the patterns of meridional and zonal flows is lost in deeper layers.

The decomposition of meridional flows in terms of Legendre polynomials shows that the latitudinal behaviour is different at different depths. The behaviour is also different at different phases of solar activity. Although the first even component $C_2(r,t)$ is the dominant component, higher-order components also contribute significantly to the flow pattern. Figure 3, shows the coefficients of various components as a function of time at two different depths. The dominant component, $C_2(r,t)$ is smallest when solar activity is at maximum and is larger during minima. This was also found by Hathaway & Rightmire (2010) from observations at the solar surface. While the $C_2(r,t)$ component shows a clear anti-correlation with solar activity indices, higher-order components show a more complicated behaviour.

Zhao & Kosovichev (2004) and Cameron & Schüssler (2010) have suggested that the temporal variations in the meridional flow can be explained by flows into the active region belts. If this is indeed the case, the higher-order components of the meridional-flow decomposition will be affected to a much larger extent than the C_2 component that the authors calculate. It is not completely clear at this time if this mechanism can explain the depth dependence of the temporal variations in the meridional flows. Furthermore, there are some differences in the meridional flows between the minima of cycles 23 and 24. These cannot be explained by flows into the active latitudes since there were very few, if any, active regions during these times.

4. Summary

Solar meridional flows close to the surface show bands of fast and slow speeds which migrate toward the equator, as noticed earlier by Beck et al. (2002) for the rising phase of cycle 23 and Gonzalez Hernandez et al. (2010) for declining phase of cycle 23. The meridional flow pattern at a depth of 1.4 Mm is very similar to the zonal flow pattern and the fast mid-latitude band coincides with sunspot positions throughout the solar cycle. A decomposition of meridional flows in terms of associated Legendre polynomials show that higher-order components make a

[1] Sunspot positions from http://solarscience.msfc.nasa.gov/greenwch.shtml

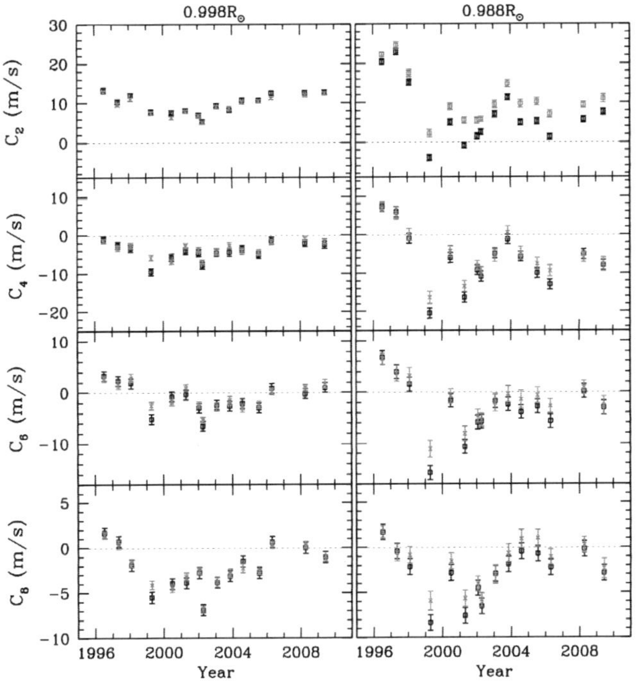

Figure 3. The components of meridional flow in terms of Legendre polynomials are shown as a function of time at two different depths.The black points are results obtained by OLA inversions, the other by RLS inversions.

significant contribution to the flows. The different Legendre components show different time dependences. The dominant component of the flow ($P_2^1(\cos\theta) = 3\cos\theta\sin\theta$) is found to be anti-correlated with global solar activity indices. This is consistent with results of Chou & Dai (2001) and Basu & Antia (2003), as well as with the surface observations of Hathaway & Rightmire (2010).

Acknowledgements

This work utilises data from the Solar Oscillations Investigation/ Michelson Doppler Imager (SOI/MDI) on the Solar and Heliospheric Observatory (SOHO). SOHO is a project of international cooperation between ESA and NASA. SB acknowledges support from NSF grant ATM 0348837 and NASA grant NXX10AE60G.

References

[1] Basu S and Antia H M 1999 *ApJ* **525**, 517
[2] Basu S and Antia H M 2003 *ApJ* **585**, 553
[3] Basu S, Antia H M and Tripathy S 1999 *ApJ* **512**, 458
[4] Beck J G, Gizon L and Duvall T L, Jr. 2002 *ApJ* **575**, L47
[5] Cameron R H and Schüssler M 2010 *ApJ* **720**, 1030
[6] Chou D-Y and Dai D-C 2001 *ApJ* **559**, L175
[7] Giles P M, Duvall T L, Jr., Scherrer P H and Bogart R S 1997 *Nature* **390**, 52
[8] González Hernández I, Howe R, Komm R and Hill F 2010 *ApJ* **713**, L16
[9] Hathaway D H 1996 *ApJ* **460**, 1027
[10] Hathaway D H and Rightmire L 2010 *Science* **327**, 1350
[11] Hill F 1988 *ApJ* **333**, 996
[12] Komm R W, Howard R F and Harvey J W 1993 *Solar Phys.* **147**, 207
[13] Schou J and Bogart R S 1998 *ApJ* **504**, L131
[14] Zhao J and Kosovichev A G 2004 *ApJ* **603**, 776

Zonal Flows Throughout Cycle 23

H. M. Antia[1] **and Sarbani Basu**[2]

[1] Tata Institute of Fundamental Research, Homi Bhabha Road, Mumbai 400005, India
[2] Department of Astronomy, Yale University, P. O. Box 20801, New Haven CT 06520-8101, U. S. A.

E-mail: `antia@tifr.res.in`, `sarbani.basu@yale.edu`

Abstract. We use available GONG and MDI data to study how solar zonal flows evolved over cycle 23 and whether or not there were differences between the minima of cycles 23 and 24. An autocorrelation study shows that as far as the zonal-flows are concerned, cycle 23 lasted for about 11.7 years, considerably shorter than what is seen from magnetic indices. We also find that there were significant differences in the nature of the flows between the minima of cycles 23 and 24.

1. Introduction

With the availability of seismic data over a period exceeding the solar cycle length it is now possible to study not only the temporal variations in the solar rotation rate, but also the cycle to cycle variation in the behaviour of solar rotation. In particular, the current minimum in solar activity is exceptional in terms of its duration and abnormally low activity level and this makes the study of difference in solar dynamics between the two minima interesting. Howe et al. (2009) made the first study of solar rotation between the two minima and concluded that the current minimum is extended because the zonal flow bands are moving more slowly toward the equator during this solar cycle as compared to the last one. With the availability of more data we wish to make a more detailed study of differences and also to determine the 'dynamical' length of solar cycle 23, using seismic data.

2. Data and Technique

We use odd order splitting coefficients from GONG and MDI data for inferring the rotation rate in the solar interior. We use 145 temporally overlapping data sets from GONG, each covering a period of 108 days with a spacing of 36 days between consecutive data sets. These data cover a period from May 1995 to October 2009. We use 69 non-overlapping data sets from MDI, each covering a period of 72 days, starting from May 1996 and ending at April 2010, with a gap around 1998 when the contact with SOHO satellite was lost. We use the 2D Regularised Least Squares (2D RLS) technique as described by Antia et al. (1998) for determining the rotation rate. We then average the results over a solar cycle and subtract this average from the rotation rate at each epoch to get the residual rotation rate $\delta\Omega(r,\theta,t)$, which is the temporally varying component of the rotation rate. Here r is the radial distance, θ the latitude and t is time. In most of the results we show the zonal flow velocity, $\delta v_\phi = \delta\Omega r\cos\theta$. To minimise the effect of systematic errors between the GONG and MDI data (Schou et al. 2002) we calculate the residuals separately for GONG and MDI data.

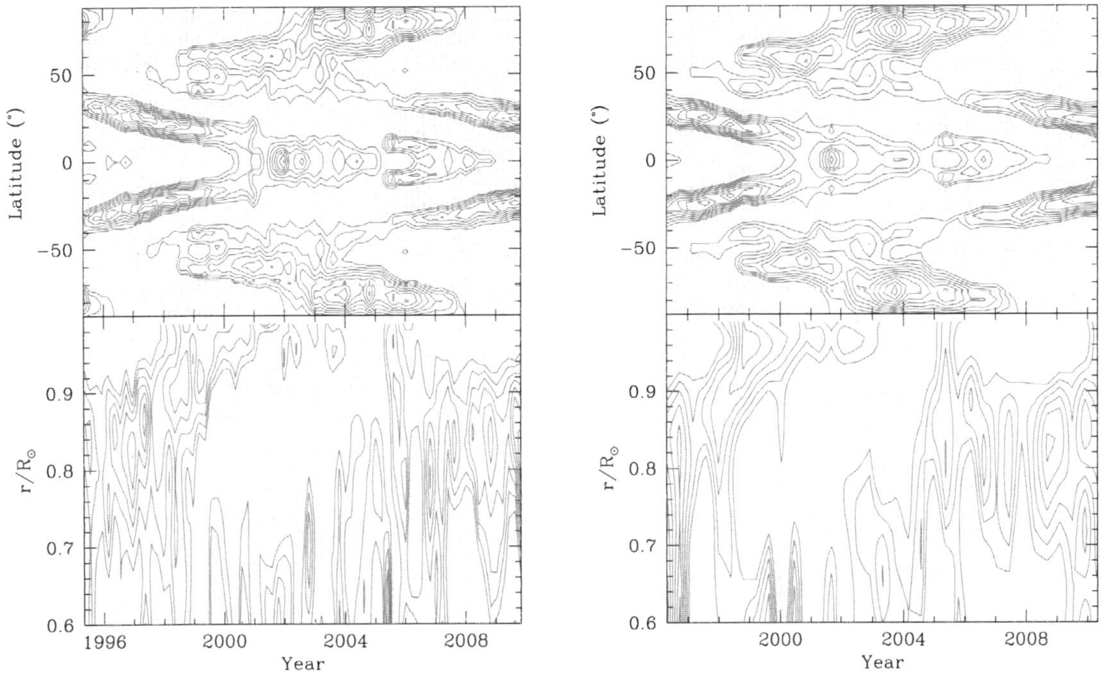

Figure 1. Zonal flow velocities δv_ϕ from GONG (left panel) and MDI (right panel). The top panels show the cuts at $r = 0.98R_\odot$ and the lower panels show the cuts at 15° latitude. The solid red contours represent positive values, while dashed cyan contours show negative values. The zero contour is not shown. The contour interval is 1 m s^{-1}.

To find the dynamical period of the solar cycle we use an autocorrelation analysis with time-shift, using the correlation coefficient

$$C(T,r) = \frac{\sum_{i,j} \delta v_\phi(r,\theta_i,t_j)\delta v_\phi(r,\theta_i,t_j+T)}{\sqrt{\left(\sum_{i,j} \delta v_\phi(r,\theta_i,t_j)^2\right)\left(\sum_{i,j} \delta v_\phi(r,\theta_i,t_j+T)^2\right)}} . \quad (1)$$

Here, θ_i are the latitudes at which the rotation rate is determined by inversions, and t_j are the times when the data are available. The correlation is calculated separately for each value of radius, r. To avoid interpolation we use only those values of T that are integral multiple of the separation between consecutive data sets. The correlation is expected to be maximum when T is equal to the time-period of solar cycle. Thus we can find the dynamical period of solar cycle by finding the maximum of $C(T,r)$. This may give different values at different radii, but in practice the variation is found to be small.

3. Results

We calculate the residual rotation velocity δv_ϕ for GONG and MDI data. The zonal flow pattern and its gradients have been studied by Antia et al. (2008) and in this work we mainly consider the differences in pattern between the last two minima and the length of solar cycle 23. Figure 1 shows two cuts each for GONG and MDI data, one at a radius of $0.98R_\odot$ and the other at a latitude of 15°. There is a good agreement between the GONG and MDI results and the figure shows the well-known bands of faster and slower than average rotation (Howe et al. 2000; Antia & Basu 2000, 2001). At low latitudes these bands move toward the equator, while at high latitudes they migrate polewards. These bands are similar to torsional oscillations seen at the

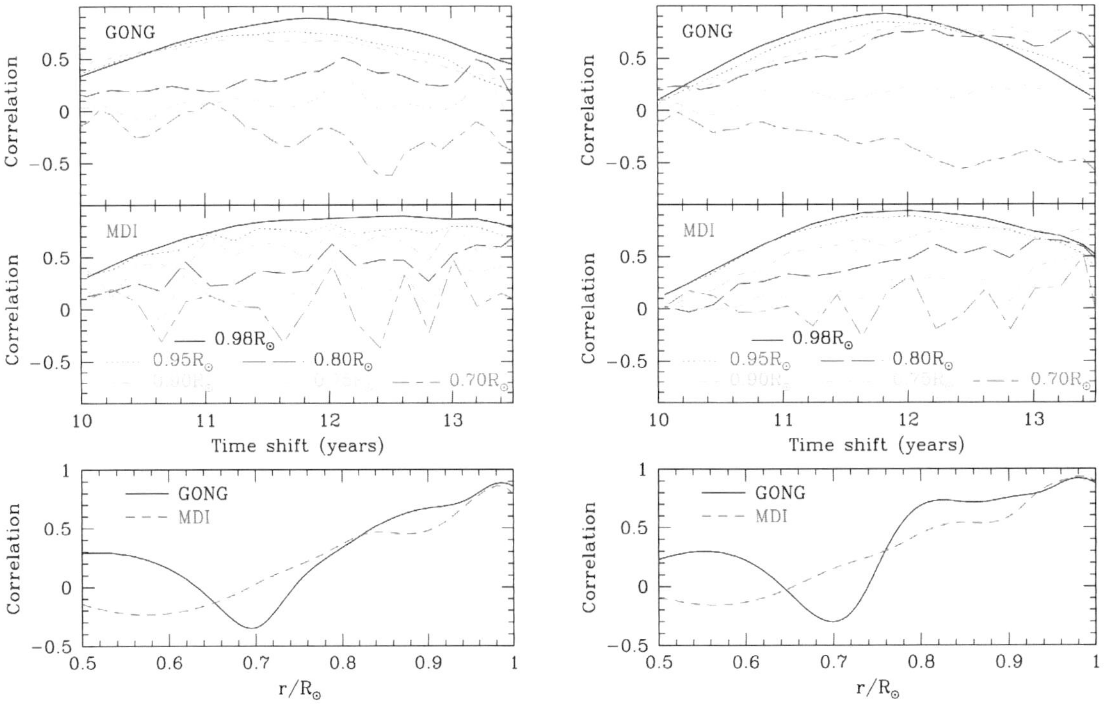

Figure 2. Autocorrelation coefficient (Eq. 1) from GONG (upper panels) and MDI (middle panels) at different depths shown as a function of time-shift. The lowest panel shows the same as a function of depth for T = 11.8 years. The left panels show the results using the entire latitude range, while the right panels show the results in low ($\leq 45°$) latitude region.

solar surface (Howard & LaBonte 1980). The low latitude bands of fast rotation merge during 2000 near the phase of maximum activity, but the equatorial band splits again in 2005 to re-merge around 2006 and the merged fast band ended around end of 2008 after which the rotation rate at equator was slower than average. The end of the fast band occurs around the same time as the minimum of activity as recorded by magnetic indices. This behaviour is different from that during the minimum of cycle 23 and the band was also not well defined during cycle 23 minimum. During cycle 23 the equatorial bands merged around 1996 and ended soon after that close to the time of minimum activity as recorded by magnetic indices. The high latitude band of fast rotation ended just before 2008 about a year before the cycle 24 activity minimum, while during the cycle 23 it ended in 1996 around the same time as minimum activity. Furthermore, as noted by Howe et al. (2009) during the cycle 24 minimum the intermediate latitude fast bands are moving toward the equator at a rate that is slower than what was observed during the previous minimum. This may explain the extended minimum. The equatorial band also has probably extended for longer period during cycle 24 minimum.

We calculated the autocorrelation coefficient (Eq. 1) at various depths and find its maximum value to determine the length of solar cycle 23. The results are shown in Figure 2. The left panel shows the results using entire latitude range, while the right panel shows the results using only low ($\leq 45°$) latitudes. The separation is done because the zonal flow pattern shows different behaviour in the two latitude ranges and is better determined in the low latitude regions. In the near-surface regions the correlation coefficient has a maximum value of close to 1 and hence we can assume that this maximum defines the period of cycle 23. This value is around 11.7 years. The MDI data show a broad peak when all latitudes are included, but when restricted to low latitudes the results are similar to those from GONG data. Thus we find the dynamical period of cycle 23 to be 11.7 years. In deeper layers the correlation coefficient is relatively small and in the tachocline region it shows multiple peaks. It is not clear if this region has any well-defined zonal flow pattern. Apart from the period, the correlation can also be used to estimate the depth to which the zonal flow pattern penetrates inside the Sun. The lowest panels in Figure 2 shows the correlation coefficient for $T = 11.8$ years plotted as a function of radius. We have used 11.8 years instead of 11.7 years as the MDI data are available at an interval of 0.2 years. It is clear that correlation is significant in most of the convection zone and there is a sharp dip near the base of the convection zone. This gives a lower limit on the depth to which the zonal flow pattern penetrates and is consistent with earlier works (Vorontsov et al. 2002; Basu & Antia 2003) that also find that the zonal flow persists throughout the convection zone.

4. Summary

Solar zonal flows show a substantial differences in the flow patterns between the minima of cycles 23 and 24. In addition to the slowing down of the mid-latitude fast band that was reported by Howe et al. (2009), we find that the low-latitude fast band bifurcated in 2005 and merged well before the cycle 24 minimum. This is in contrast to what happened during the cycle 23 minimum. This band is also more extended during cycle 24. The high-latitude fast band was also different. During the cycle 23 minimum it had ended around 1996.5, just at the time of minimum activity, while in the cycle 24 the band ended around 2008.0 about a year before the minimum in activity. Assuming that the zonal flows show a cyclic behaviour we find a dynamical length of cycle 23 to be 11.7 years using autocorrelation analysis, this is substantially shorter than 12.6 year inferred from magnetic indices (Hathaway 2010). Howe et al. (2009) also found a period of around 12 years with a technique similar to the one used here. The correlation also shows that the zonal flow pattern penetrates through most of the convection zone, in agreement with earlier results (Vorontsov et al. 2002; Basu & Antia 2003).

Acknowledgements

This work utilises data obtained by the Global Oscillation Network Group (GONG) project, managed by the National Solar Observatory, which is operated by AURA, Inc. under a cooperative agreement with the National Science Foundation. The data were acquired by instruments operated by the Big Bear Solar Observatory, High Altitude Observatory, Learmonth Solar Observatory, Udaipur Solar Observatory, Instituto de Astrofisico de Canarias, and Cerro Tololo Inter-American Observatory. This work also utilises data from the Solar Oscillations Investigation/ Michelson Doppler Imager (SOI/MDI) on the Solar and Heliospheric Observatory (SOHO). SOHO is a project of international cooperation between ESA and NASA. SB acknowledges support from NSF grant ATM 0348837 and NASA grant NXX10AE60G.

References

[1] Antia H M and Basu S 2000 *ApJ* **541**, 442
[2] Antia H M and Basu S 2001 *ApJ* **559**, L67
[3] Antia H M, Basu S and Chitre S M 1998 *MNRAS* **298**, 543
[4] Antia H M, Basu S and Chitre S M 2008 *ApJ* **681**, 680
[5] Basu S and Antia H M 2003 *ApJ* **585**, 553
[6] Hathaway D H 2010 *Living Rev. Solar Phys.* **7**, 1
[7] Howard R and LaBonte B J 1980 *ApJ* **239**, L33
[8] Howe R, Christensen-Dalsgaard J, Hill F, Komm R W, Larsen R M, Schou J, Thompson M J and Toomre J 2000 *ApJ* **533**, L163
[9] Howe R, Christensen-Dalsgaard J, Hill F, Komm R W, Schou J and Thompson M J 2009 *ApJ* **701**, L87
[10] Schou J, et al. 2002 *ApJ* **567**, 1234
[11] Vorontsov S V, Christensen-Dalsgaard J, Schou J, Strakhov V N and Thompson M J 2002 *Science* **296**, 101

Meridional Flow Observations: Implications for the current Flux Transport Models

Irene González Hernández, Rudolf Komm, Shukur Kholikov, Rachel Howe and Frank Hill

National Solar Observatory, Tucson, Arizona, 857619, USA

E-mail: `irenegh@nso.edu`

Abstract. Meridional circulation has become a key element in the solar dynamo flux transport models. Available helioseismic observations from several instruments, Taiwan Oscillation Network (TON), Global Oscillation Network Group (GONG) and Michelson Doppler Imager (MDI), have made possible a continuous monitoring of the solar meridional flow in the subphotospheric layers for the last solar cycle, including the recent extended minimum. Here we review some of the meridional circulation observations using local helioseismology techniques and relate them to magnetic flux transport models.

1. Introduction

The solar magnetic cycle is generally explained using magneto-hydrodynamic dynamo theory. These models rely on two effects: the Ω effect (stretching of the poloidal field by differential rotation to produce the toroidal component) and the α effect (regeneration of the poloidal field). Solar meridional circulation is a key component of the regeneration of the poloidal field in the current Babcock-Leighton type models [1]. Traditional observations of the meridional circulation relied on the tracking of large or small magnetic elements or direct observation of photospheric Doppler shifts. These techniques are limited to measurements at the solar surface. Local helioseismology techniques have provided the opportunity to infer the meridional circulation profile with depth. In this paper we discuss some of the results obtained by the local helioseismology techniques, such as the temporal variation of the flows with the solar cycle, the extra circulation in the active latitudes and the preliminary inferences of the deep convection zone meridional circulation, and their relation to the flux transport models.

2. Polar Field Reversals

Two transport processes are believed to be involved in the polar field reversals: supergranular diffusion and meridional circulation [2]. While the leading polarity of tilted bipolar active regions cancels by diffusion, the trailing polarity is transported toward the poles canceling and inverting the previous cycle.

The polar field reversal during solar cycle 23 can be clearly seen in Figure 1 around 2001, at maximum activity. Figure 1 also shows that the polar fields during this solar minimum were weaker than the previous one. Wang et al. [2] demonstrated that this variation of the polar fields can be reproduced with a surface transport model that includes variation of the subsurface meridional flow from cycle to cycle. Long-term helioseismic observations from both

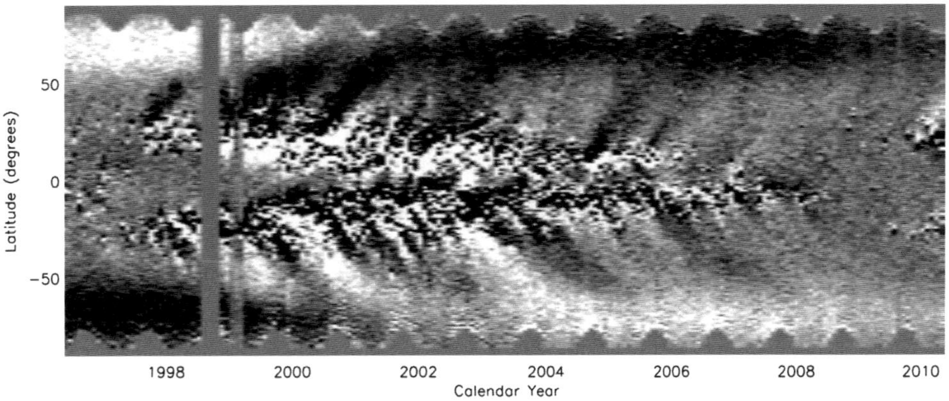

Figure 1. Magnetic field proxy calculated using Michelson Doppler Imager magnetograms obtained during the period 1996-2010 and covering the last solar cycle. The magnetic field is averaged in longitude for each Carrington rotation synoptic map and the sign is preserved. To see the pole reversal, the image is saturated at the active belts.

GONG and MDI provide the opportunity to compare the meridional flows in the previous and the current cycle. Basu and Antia [3], Kholikov et al. [4] and Komm et al. [5] show evidence of the subsurface meridional flow varying between the minimum of solar cycles 23 and 24, with a larger amplitude in this recent extended minimum. Surface observations confirm this behavior [6].

3. Meridional Circulation and Solar Cycle Prediction Models

Meridional circulation is believed to determine the solar dynamo cycle period as well as the Sun's memory about its past magnetic fields [2,7]. Hence, the meridional flow is an essential component of the solar cycle predictions based on flux-transport dynamo models [7]. The assumption used by most modelers is that the meridional flow varies from cycle to cycle. However, it is well established by now that the meridional circulation also varies throughout a particular cycle, with a minimum amplitude at maximum activity and vice-versa [3,6,8,9]. Figure 2 presents the temporal variation of the meridional flow inferred by ring-diagram analysis of GONG data from 2001 to 2010 at 5.8 Mm under the solar surface. Recent work includes helioseismically inferred meridional flow in the subsurface layers that has been temporally averaged as input to their flux-transport model [10], but the variation within the cycle still needs to be considered.

3.1. Extra circulation (jets) of the meridional circulation

Superimposed meridional circulation in the active latitudes was first observed by applying local helioseismology techniques to the Taiwan Oscillation Network data in 2001 [11]. Since then, several studies have investigated the origin of such extra circulation. The recent prolonged solar minimum has provided the opportunity to investigate this behavior of the meridional circulation without contamination from surface activity [9]. In Figure 3, the formation of the bumps at medium latitudes precedes the onset of the activity, confirming the nature of this phenomenon as independent of the surface manifestation of active regions.

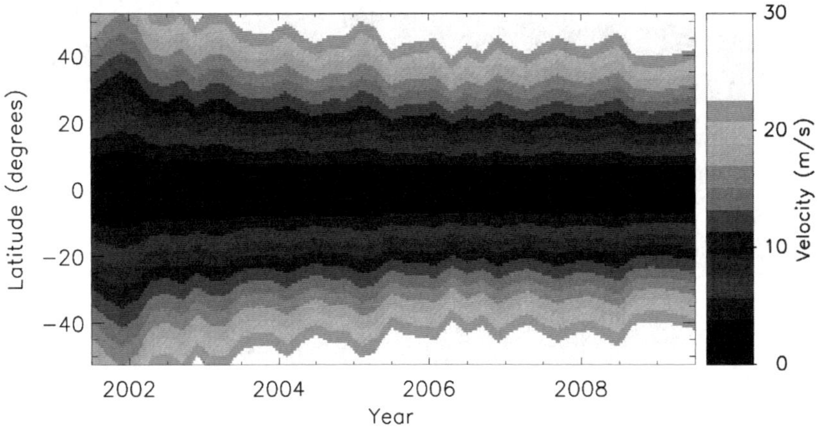

Figure 2. Temporal variation of the fitted polynomial to the meridional circulation observations at a depth of 5.8 Mm obtained from ring-diagram analysis of GONG data. A symmetrical plot averaging both hemispheres is shown. Positive velocities are taken toward each respective pole.

3.2. Meridional Flow deep in the convection zone

The return flow at or near the base of the convection zone is estimated to be of the order of 1-2m/s. To achieve that accuracy with helioseismic observations, time series that span at least one full solar cycle are needed [12]. Time distance analysis of spherical harmonic time series obtained from GONG observations that span from 1995 to 2009 have been used for this purpose [4]. Preliminary results show an increase of time differences at a depth of about $0.77R_{\odot}$. This can be an indication of large perturbations of the meridional flow or some other properties of the deep convective zone. The long series uses for the analysis reduce the uncertainties in the measurements to less than 0.02 seconds.

4. Summary and Future work

Proper determination of the meridional circulation at the surface and the solar interior is key for the current flux transport models. Local helioseismology has contributed substantially to the overall picture, but more work is needed to understand the details. The relationship between the extra circulation of the meridional circulation and the well established torsional oscillation of the zonal flows needs to be investigated. Gizon and Rempel [13] presented a model to account for independent observations of the meridional flow residuals at the surface and at a depth of 60 Mm., in which the flow at these two depths was anticorrelated. The meridional flow profile deep in the convection zone remains elusive. Preliminary results using data from a full solar cycle to infer the meridional flow profile until the base of the convection zone look promising.

The Solar Dynamics Observatory is expected to bring new insights into the meridional circulation, particularly in the subsurface behavior of the flow at higher latitudes (see preliminary results by Chakraborty et al. [14]).

4.1. Acknowledgments

This work utilizes data obtained by the Global Oscillation Network Group (GONG) program, managed by the National Solar Observatory, which is operated by AURA, Inc. under a cooperative agreement with the National Science Foundation. The data were acquired by instruments operated by the Big Bear Solar Observatory, High Altitude Observatory, Learmonth

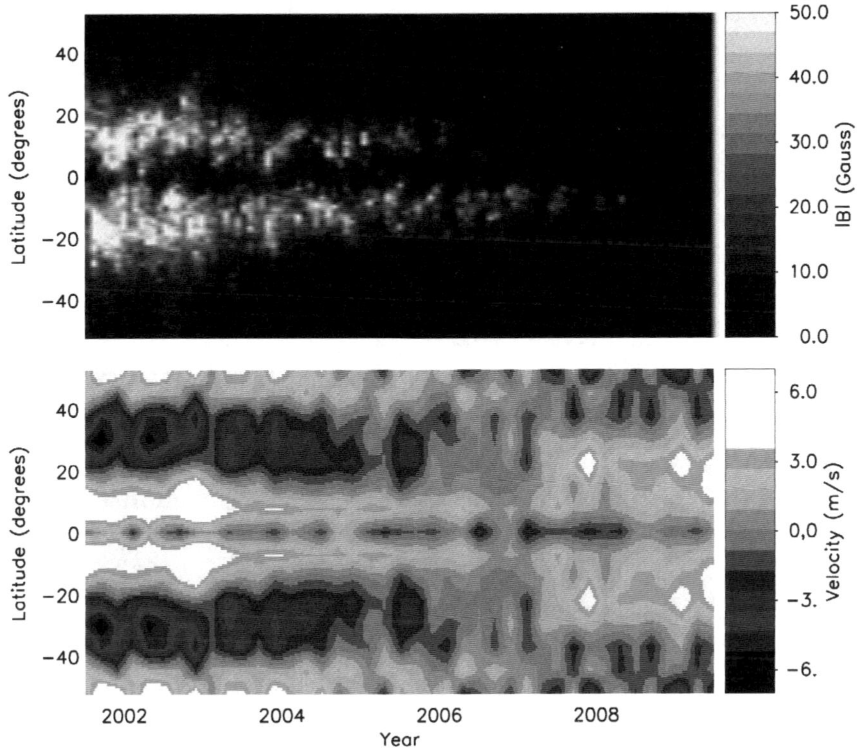

Figure 3. Temporal variation of the meridional circulation residuals at a depth of 5.8 Mm (bottom panel). Positive velocities are directed towards the poles. A symmetrical plot averaging both hemispheres is shown. The top panel presents the location and magnetic strength of the activity during the same period (calculated from MDI synoptic magnetograms).

Solar Observatory, Udaipur Solar Observatory, Instituto de Astrofísica de Canarias, and Cerro Tololo Interamerican Observatory. SoHO is a project of international collaboration between ESA and NASA. This work has been supported by the NASA Living with a Star - Targeted Research and Technology program.

[1] Charbonneau P 2005 *Living Reviews in Solar Physics*
[2] Wang Y-M, Robbrecht E and Sheeley N 2009 *ApJ* **707** 1372
[3] Basu S and Antia H M 2010 *ApJ* **717** 488
[4] Kholikov S, González Hernández, I, Hill F and Leibacher 2010 these proceedings
[5] Komm R, Howe R, Hill F, González Hernández I and Haber D A 2010 these proceedings
[6] Hathaway D H, Rightmire L 2001 *Science* **327** 1350
[7] Dikpati M and Gilman P A 2006 *ApJ* **649** 498
[8] González Hernández I, Kholikov S, Hill F, Howe R and Komm R 2008 *Solar Phys.* **252** 235
[9] González Hernández I, Howe R, Komm R and Hill F 2010 *ApJ* **713** L16
[10] Muñoz-Jaramillo A, Nandy D and Martens P C H 2009 *ApJ* **698** 461
[11] Chou D.-Y. and Dai D.-C. 2001 *ApJ* **559** L175.
[12] Braun D C and Birch A 2008 *ApJ* **689** L161
[13] Gizon L. and Rempel M. 2008 *Solar Phys* **251** 241
[14] Chakraborty S, Rabello-Soares M. C and Bogart R 2010 these proceedings

The torsional oscillation and the new solar cycle

R Howe,[1] F Hill,[1] R Komm,[1] J Christensen-Dalsgaard,[2,3] T P Larson,[4] J Schou,[4] M J Thompson[3,5] and R Ulrich[6]

[1] National Solar Observatory, 950 N. Cherry Ave., Tucson AZ 85719, USA
[2] HEPL, Stanford University, Stanford, CA 94305, USA
[3] High Altitude Observatory, National Center for Atmospheric Research, P.O. Box 3000, Boulder, CO 80307, USA
[4] Department of Physics and Astronomy, Aarhus University, DK-8000 Aarhus C, Denmark
[5] School of Mathematics & Statistics, University of Sheffield, S3 7RH, UK
[6] Department of Physics and Astronomy, University of California, Los Angeles 90095, USA

E-mail: rhowe@noao.edu

Abstract. We present updated observations of the pattern of migrating solar zonal flows known as the torsional oscillation, covering 15 years of helioseismic measurements with GONG and MDI and 30 years of surface Doppler observations from Mount Wilson. We compare the behavior of the flows during the extended solar minimum following Cycle 23 with that in earlier minima. We demonstrate that the timing of the migration of the zonal flow belts may be of some use in predicting the start of the new cycle. We also note that the behavior of the high-latitude part of the pattern currently differs from that seen early in the previous cycle, with the high-latitude poleward-migrating branch still not established.

1. Introduction

The migration of zonal flow bands from mid-latitudes to the equator and poles over the course of the solar cycle, first reported and dubbed the "torsional oscillation" by Howard and LaBonte [1] has been observed for 15 years [2; 3] by the Global Oscillation Network Group [GONG] and Michelson Doppler Imager [MDI] [4]. and for over 30 years in surface Doppler measurements at the Mount Wilson Observatory [5; 6]. In recent years there has been a revival of interest in the relationship between these flows and the solar cycle, due to the extended solar minimum following cycle 23 [7].

2. Data and analysis

The helioseismic data consist of medium-degree rotational splitting coefficients derived from 145 overlapping 108-day intervals of GONG observations and 69 non-overlapping 72-day intervals of MDI observations, covering the period from May 1995 to April 2010. The rotation profile was inferred using 2-dimensional regularized least-squares inversions, and a temporal mean was subtracted from each set of inversions at each location in depth and radius to reveal the temporal variation in the rotation.

The Mount Wilson Doppler data, taken with the Babcock Magnetograph at the 150 foot tower, cover the period 1976 – 2009. The data are sampled at 34 points equidistant in the sine of the latitude, giving a separation of approximately 4° at low latitudes. Each Carrington rotation is divided into 18 pixels, giving a temporal resolution of about 1.5 days. For this study, the data

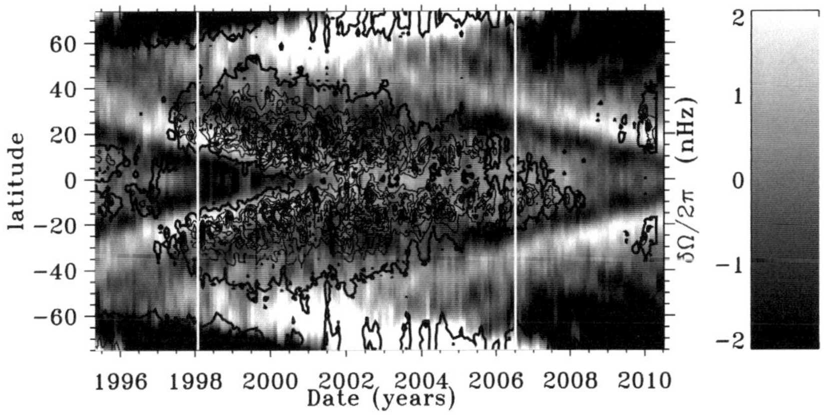

Figure 1. Rotation-rate residuals from RLS inversions at $0.99R_\odot$, for combined GONG and MDI data. The vertical white line at left (right) indicates the time at which the flow profile at latitudes below 45° best matches the most recent (earliest) observations; black contours represent unsigned magnetic field strength from KPVT and SOLIS.

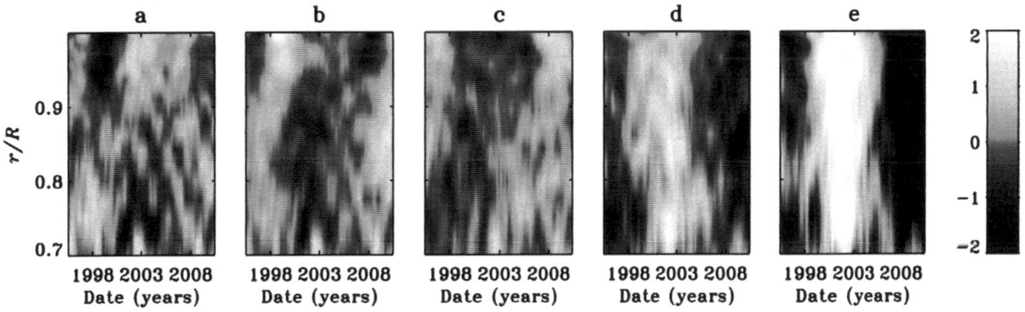

Figure 2. RLS Rotation-rate residuals as a function of depth at latitudes 0° (a) to 60° (e) at 15° intervals. The grayscale is in nHz, and is non-linear (varying most rapidly across the zero contour).

were further averaged in time to give a sampling interval of approximately 2 months, similar to the time period used for the MDI global inversions.

3. Results

3.1. Helioseismic observations

Figure 1 shows the combined MDI and GONG zonal-flow residuals at $0.99R_\odot$ as a function of latitude and time. Clearly, as we reported in Howe *et al.* [7], the time for the flows to migrate from the 1995 position to the current position was longer in the declining phase of Cycle 23 than in the previous cycle, by 1.5 to 2 years. As in the previous cycle, significant magnetic activity started to appear once the flows reached a latitude of about 20°. The other noticeable feature of this plot is that the rotation rate poleward of the faster-rotating belts is still slower than it was in 1995, and there is little sign of the appearance of the new poleward branch, which was already starting to be visible in 1997.

Figure 2 shows the combined RLS rotation-rate residuals as a function of time and depth at selected latitudes. For clarity, the data have been smoothed with a one-year running mean. As in the previous cycle, at lower latitudes we can see hints that for a given latitude the rotation speeds up in the middle of the convection zone a couple of years earlier than at the surface.

3.2. High-Latitude Residuals

Figure 3 shows the GONG and MDI RLS rotation-rate residuals for selected higher latitudes at a depth of $0.99R_\odot$. At 52.5° we can see that the rotation rate was increasing in 1997, whereas in 2009 – 10 it is still trending down or just starting to increase again. At 67.5° the 1997.6 epoch seems to mark a minimum in the rotation rate, and it appears that this may be true in the recent data also. Finally, at higher latitudes the rotation rate was decreasing through 1997 to a minimum in 1999, whereas in the 2009 – 2010 observations it appears to have either reached a (higher) minimum after decreasing during the declining phase of Cycle 23, or to be still decreasing.

The "double peak" seen in the rotation rate at $0.99R_\odot$, 78.8° is curious. Could it reflect different behavior in the northern and southern hemispheres? The new magnetic activity cycle in the southern hemisphere does seem to be more delayed than in the northern one.

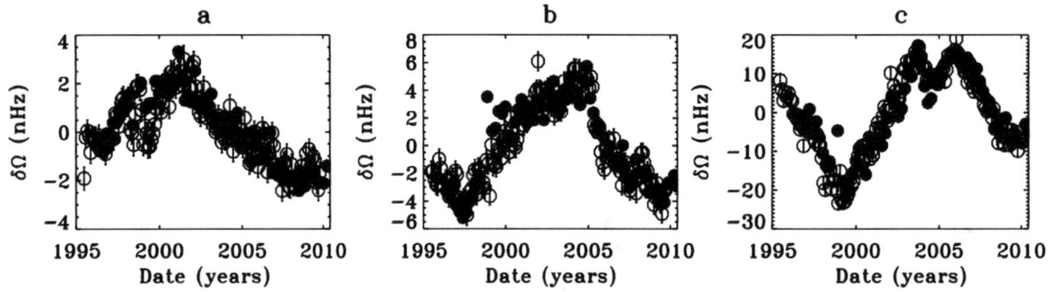

Figure 3. Rotation-rate residuals at latitudes 52.5° (a), 67.5° (b), and 78.8° (c), for GONG (open symbols) and MDI (closed symbols).

3.3. Doppler Observations

For historical context, in Figure 4 we show the Mount Wilson observations since 1975. The data have been symmetrized about the equator. The black contours show unsigned magnetic field strength. For each cycle, we see that the epoch most closely matching the current low-latitude flow pattern — when the equatorial branch of the pattern reaches about 20° —corresponds to a time when significant and widespread activity had begun. In the previous two cycles the first hint of high-latitude zonal flow enhancement is also visible around this epoch, at least in the southern hemisphere, whereas in Cycle 24 it seems to be delayed.

4. Discussion

The torsional oscillation signal may provide some indication of the timing of the solar cycle in advance of the appearance of surface activity. At the current time, though the low-latitude branch for Cycle 24 is well established and the cycle is (finally) underway, the high-latitude branch has yet to appear. It will be interesting to see how the zonal flow pattern evolves with the new cycle, both in continuing observations from GONG and in new data from HMI.

Acknowledgments

This work utilizes data obtained by the Global Oscillation Network Group (GONG) Program, managed by the National Solar Observatory, which is operated by AURA, Inc. under a

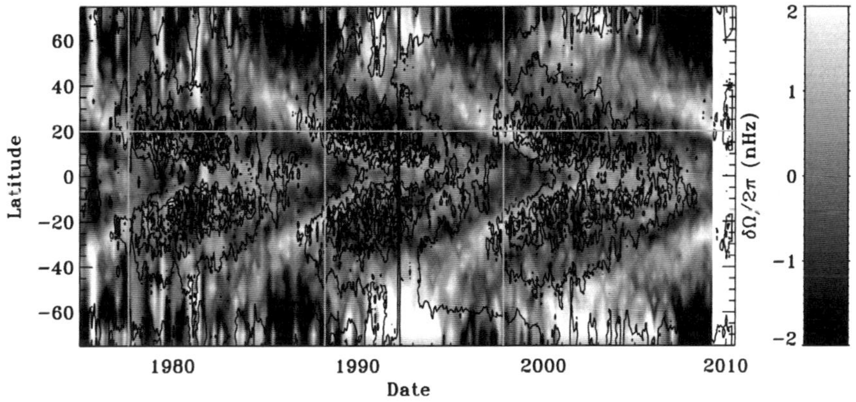

Figure 4. Rotation-rate residuals from Mt. Wilson data. The vertical light lines indicate the epochs where the zonal flow profile below 45° best matches the most recent helioseismic measurements; black contours show the unsigned magnetic field strength The horizontal line at 20° North indicates the latitude of the flow belts at these epochs.

cooperative agreement with the National Science Foundation. The data were acquired by instruments operated by the Big Bear Solar Observatory, High Altitude Observatory, Learmonth Solar Observatory, Udaipur Solar Observatory, Instituto de Astrofísica de Canarias, and Cerro Tololo Interamerican Observatory. The Solar Oscillations Investigation (SOI) involving MDI is supported by NASA grant NNX09AI90G to Stanford University. SOHO is a mission of international cooperation between ESA and NASA. RK was supported by NASA contracts S-92698-F, NAG 5-11703, and NNG05HL41I. The Mt. Wilson observations have been supported over several decades by a series of grants from NASA, NSF and ONR and are currently supported by NSF/ATM. The Mt. Wilson observatory is managed by the Mt. Wilson Institute under agreement with the Observatories of the Carnegie Institution of Washington. NSO/Kitt Peak magnetic data used here are produced cooperatively by NSF/NSO, NASA/GSFC and NOAA/SEL, and SOLIS data by NSF/NSO and NASA/LWS. The National Center for Atmospheric Research is sponsored by the National Science Foundation.

References

[1] Howard R and LaBonte B J 1980 *Astrophys. J.* **239** L33–6

[2] Howe R, Christensen-Dalsgaard J, Hill F, Komm R W, Larsen R M, Schou J, Thompson M J and Toomre J 2000 *Astrophys. J.* **533** L163–6

[3] Howe R, Christensen-Dalsgaard J, Hill F, Komm R, Schou J and Thompson M J 2005 *Astrophys. J.* **634** 1405–15

[4] Scherrer P H, Bogart R S, Bush R I, Hoeksema J T, Kosovichev A G, Schou J, Rosenberg W, Springer L, Tarbell T D, Title A, Wolfson C J, Zayer I and MDI Engineering Team 1995 *Solar Phys.* **162** 129–88

[5] Ulrich R K 2001 *Astrophys. J.* **560** 466–75

[6] Howe R, Komm R, Hill F, Ulrich R, Haber D A, Hindman B W, Schou J and Thompson M J 2006 *Solar Phys.* **235** 1–2

[7] Howe R, Christensen-Dalsgaard J, Hill F, Komm R, Schou J and Thompson M J 2009 *Astrophys. J.* **701** L87–90

Rotation-rate variations at the tachocline: an update

R Howe,[1] R Komm,[1] F Hill,[1] J Christensen-Dalsgaard, [2,3] T P Larson,[4] J Schou,[4] M J Thompson[3,5] and J Toomre[6]

[1] National Solar Observatory, 950 N. Cherry Ave., Tucson AZ 85719, USA

[2] Department of Physics and Astronomy, Aarhus University, DK-8000 Aarhus C, Denmark

[3] High Altitude Observatory, National Center for Atmospheric Research, P.O. Box 3000, Boulder, CO 80307-3000, USA

[4] Hansen Experimental Physics Laboratory, Stanford University, Stanford, CA 94305, USA

[5] School of Mathematics & Statistics, University of Sheffield, S3 7RH, UK

[6] JILA and Department of Astrophysical and Planetary Sciences, University of Colorado, Boulder, CO 80309, USA

E-mail: rhowe@noao.edu

Abstract. After 15 years of GONG and MDI observations of the solar interior rotation, we revisit the issue of variations in the rotation rate near the base of the convection zone. The 1.3-year period seen in the first few years of the observations disappeared after 2000 and has still not returned. On the other hand, the agreement between GONG and MDI observations suggests that variations seen in this region have some solar origin, whether a true rotation-rate change or possibly mere stochastic variation; we present a numerical experiment supporting this contention.

1. Introduction

The region of strong rotational shear at the base of the solar convection zone, known as the tachocline, plays an important role in many models of the dynamo that drives solar activity. The report [1] of apparently periodic variations in the rotation rate at this location, with a period of approximately 1.3 years, therefore attracted considerable attention. Other workers [2] did not reproduce the finding. The periodicity was not seen in observations after the solar maximum [3]. With new cycle 24 finally underway, it seems worthwhile to provide an update on these observations.

2. Data and Analysis

The data consist of 145 sets of medium-degree rotational splittings from overlapping 108-day periods of GONG observations and 69 sets of rotational splitting coefficients from non-overlapping 72-day periods of MDI [4] observations, together covering the period from May 1995 to May 2010. From each data set, two-dimensional rotation profiles were inferred using both regularized least-squares [RLS] and optimally localized average [OLA] inversions, as in our earlier work [1]. A temporal mean at each location was subtracted from each set of inversions to yield the rotation residuals.

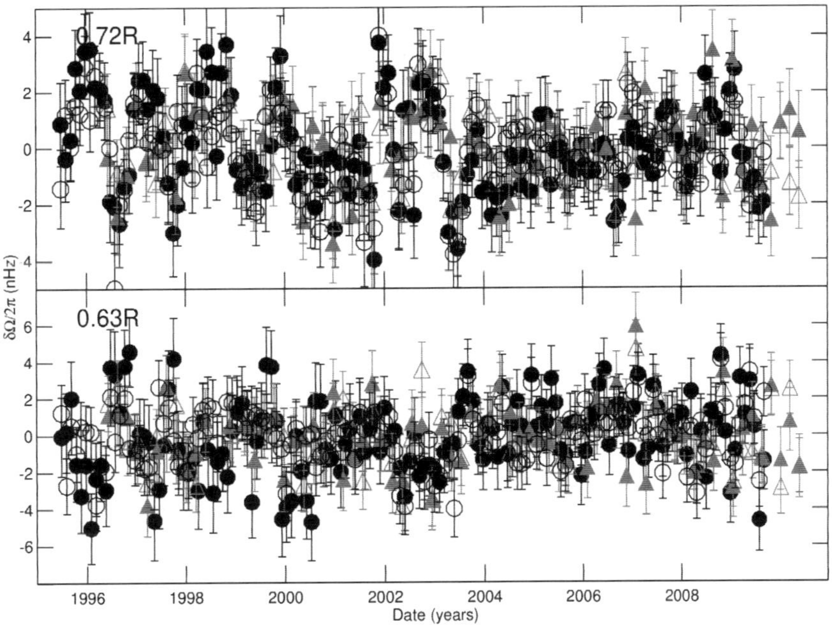

Figure 1. Rotation-rate residuals at the equator at fractional radii 0.72 (top) and 0.63 (bottom), for RLS (filled symbols) and OLA (open symbols) inversion of GONG (circles) and MDI (triangles) data.

3. Results

In Figure 1 we show the rotation-rate residuals at the equator for the two instruments and the two inversion methods, for the depths of $0.72R_{\odot}$ and $0.63R_{\odot}$.

We quantify the power in the residuals as a function of frequency by fitting sine waves. In Figure 2 we show the results of this fitting for different subsets of the observation period, for GONG RLS inversions at $0.72R_{\odot}$ at the equator. Though the 1.3-year signal is quite clear in the first five years of observations, its significance decreases as more data are included, and it is entirely absent when the last 10 years of observations are analyzed alone.

Even though the 1.3-year signal appears only in the early years of the data set, the variation of the $0.72R_{\odot}$ equatorial rotation rate in MDI and GONG looks remarkably similar during the whole period. We quantify this by interpolating both data sets to a common time base and examining the correlation between them, as seen in Figure 3. The correlation coefficients obtained for RLS and OLA inversions are 0.69 and 0.48 respectively. In the next section we will examine the significance of these correlations for the RLS case.

3.1. Numerical simulation

How significant are these correlations? To test this for the RLS case, we generated simulated inversions by integrating artificial rotation profiles over the averaging kernels for the $(0.72R_{\odot}, 0^{\circ})$ location. For 1000 realizations of noise distributed according to the formal errors of the real input data, the noise was propagated through to the synthetic inversion profiles via the inversion coefficients [5]. The noise realizations for GONG and MDI were independent. We then examined the correlation between the synthetic GONG and MDI time series, again interpolated to the common time base. Two cases were considered; one in which the artificial profile contains only

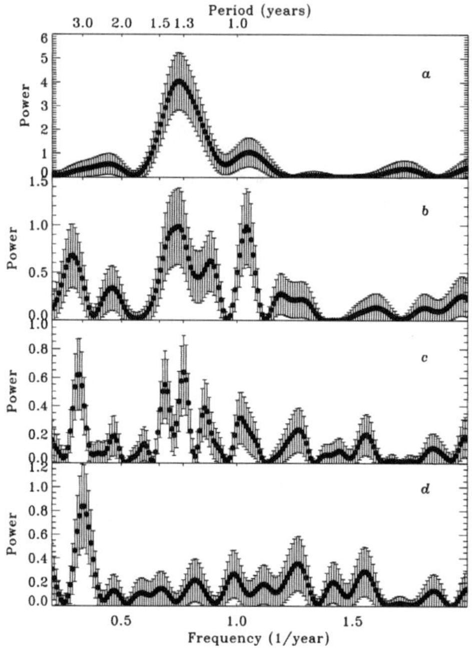

Figure 2. The amplitude of sine-wave fits to the GONG RLS residuals at $0.72R_{\odot}, 0°$ as a function of frequency, for periods 1995 – 2000 (a), 1995 – 2005 (b), 1995 – 2010 (c), and 2000 – 2010 (d).

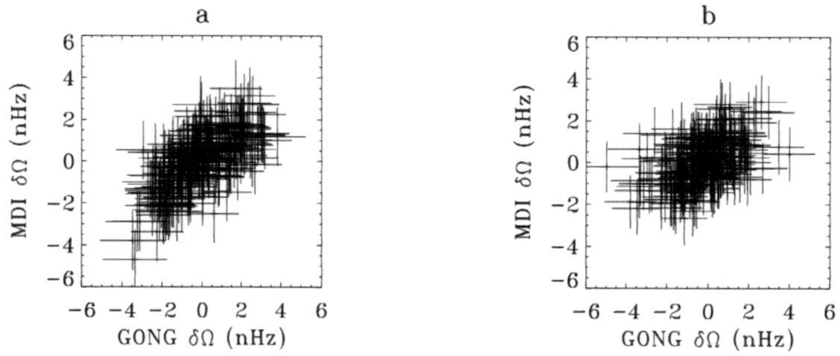

Figure 3. Scatter plots of the interpolated rotation-rate residuals for RLS (a) and OLA (b) inversions of GONG and MDI observations at the $(0.72R_{\odot}, 0°)$ location.

an emulation of the torsional oscillation signal at all latitudes and at radii down to $0.72R_{\odot}$, and one in which a 1.3-year oscillation, coherent over the region $0.7 \leq r/R_{\odot} \leq 0.8$ and latitudes up to 30 deg, was also included. Figure 4 shows how the distribution of correlation coefficients changes as the amplitude of the input signal increases. Note that the amplitude of the observed torsional oscillation signal is about 1 nHz. We would expect the correlation to increase as the signal rises above the noise, but it does not seem to be feasible to reach the observed correlation with the torsional oscillation signal alone at realistic levels; the shorter-period signal (1.3 years in this test) increases the correlation more rapidly. However, this test is not designed to exclude the possibility that the fluctuations in the rotation rate arise from the stochastic nature of the modes.

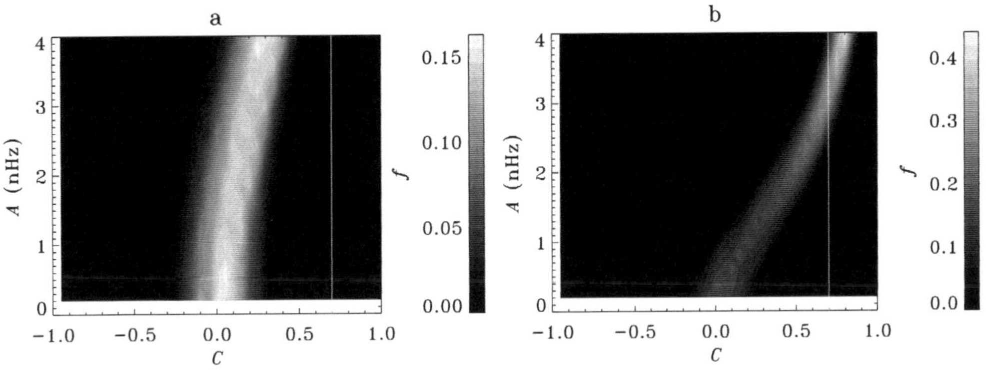

Figure 4. Distribution of MDI – GONG correlation coefficients for RLS rotation residuals at $0.72R_\odot, 0°$, for 1000 realizations of artificial data with only the torsional oscillation signal (a) and with an extra 1.3-year sinusoidal signal (b). The strength of the periodic signal in the artificial data varies along the y axis. The vertical white line in each panel indicates the correlation coefficient from the observations.

4. Discussion

The 1.3-year signal at the base of the convection zone disappeared after 2000 and so far has not returned. The level of correlation between GONG and MDI observations (at least for RLS inversions) does suggest that the variations seen here at least have some solar origin common to the GONG and MDI observations, and is consistent with the presence of a short-period signal at the base of the convection zone. It will be interesting to see what happens in the new cycle, with HMI observations superseding MDI.

Acknowledgments

This work utilizes data obtained by the Global Oscillation Network Group (GONG) Program, managed by the National Solar Observatory, which is operated by AURA, Inc. under a cooperative agreement with the National Science Foundation. The data were acquired by instruments operated by the Big Bear Solar Observatory, High Altitude Observatory, Learmonth Solar Observatory, Udaipur Solar Observatory, Instituto de Astrofísica de Canarias, and Cerro Tololo Interamerican Observatory. The Solar Oscillations Investigation (SOI) involving MDI is supported by NASA grant NNX09AI90G to Stanford University. SOHO is a mission of international cooperation between ESA and NASA. RK was supported by NASA contracts S-92698-F, NAG 5-11703, and NNG05HL41I. The National Center for Atmospheric Research is sponsored by the National Science Foundation.

References

[1] Howe R, Christensen-Dalsgaard J, Hill F, Komm R W, Larsen R M, Schou J, Thompson M J and Toomre J 2000 *Science* **287** 2456–60

[2] Basu S and Antia H M 2003 *Astrophys. J.* **585** 553–65

[3] Howe R, Christensen-Dalsgaard J, Hill F, Komm R, Schou J, Thompson M J and Toomre J 2007 *Advances in Space Research* **40** 915–8

[4] Scherrer P H, Bogart R S, Bush R I, Hoeksema J T, Kosovichev A G, Schou J, Rosenberg W, Springer L, Tarbell T D, Title A, Wolfson C J, Zayer I and MDI Engineering Team 1995 *Solar Phys.* **162** 129-88

[5] Howe R, Rempel M, Christensen-Dalsgaard J, Hill F, Komm R, Larsen R M, Schou J and Thompson M J 2006 *Astrophys. J.* **649** 1155–68

Subsurface structure of the Evershed flows in sunspots

Irina N. Kitiashvili[1], Alexander G. Kosovichev[1], Nagi N. Mansour[2], Alan A. Wray[2]

[1]Stanford University, Stanford, CA 94305, USA
[2]NASA Ames Research Center, Moffett Field, CA 94040, USA

E-mail: `irinasun@stanford.edu`

Abstract. The radial outflows in sunspot penumbrae, known as the Evershed effect, are of significant interest for understanding the dynamics of sunspots. Local helioseismology has not been able to determine the depth of these flows nor their relationship to mass circulation in sunspots. Recent radiative MHD simulations have provided a convincing explanation of the Evershed flow as a natural consequence of magnetoconvection in the strongly inclined magnetic field region of the penumbra. The simulations reproduce many observational features of penumbra dynamics, including the filamentary structure, the high-speed non-stationary "Evershed clouds", and the "sea-serpent" behavior of magnetic field lines. We present the subsurface structure of the Evershed effect, obtained from numerical simulations, and determine the depth of the radial outflows for various magnetic field strengths and inclinations. The simulations predict that Evershed flows are rather shallow and concentrated in the top 0.5 – 1 Mm layer of the convection zone. This prediction can be tested by local helioseismology methods.

1. Introduction

Investigation of the high-speed radial outflows in sunspots ("Evershed flows", [2]) has a century-long history. However, only recently has it become possible to obtain detailed high-resolution observations and to construct close-to-reality numerical models. The improvement of observational capabilities has led to different models of sunspot penumbrae and Evershed flow; however, these models, listed in Table 1, can explain only some observational features. Recent improvements in computational capabilities have led to the development of realistic MHD simulations [3, 10]; these have revealed connections between overturning convection and radial outflows in the inclined magnetic field of sunspot penumbrae.

To explore the idea of overturning convection in a sunspot penumbra, we have performed realistic radiative MHD simulations using the "SolarBox" code [4] and have compared the simulation results with observational features. The code was developed at the NASA Ames Research Center and the Stanford Center for Turbulence Research for numerical simulation of the top layers of the convective zone and lower atmosphere. It solves the radiative MHD equations with real gas effects in a rectangular computational domain. The code is based on a LES (Large-Eddy Simulation) formulation and includes various subgrid-scale turbulence models. It takes into account several physical phenomena: compressible fluid flow in a highly stratified medium, three-dimensional multi-group radiative energy transfer between the fluid elements, a real-gas (tabular) equation of state, ionization and excitation of all abundant species, and magnetic effects.

Table 1. Models of a sunspot penumbra.

Model	Ref.	Model	Ref.
Magnetic field gaps	[11]	Thin magnetic flux tube model	[9]
Siphon model	[7, 12]	Overturning convection	[3, 10]
Convective rolls	[1]		

In the paper, we use the fields from numerical simulations described in [5], and discuss the subsurface structure of magnetoconvection in highly inclined magnetic fields.

2. Realistic simulations of the solar convection in an inclined magnetic field

Construction of a realistic sunspot penumbra model is a complicated problem, in particular because it is a part of a sunspot, where strong plasma inhomogeneities co-exist with magnetic fields of highly variable structure and strength. To simplify the problem, we perform separate numerical simulations of different regions of a penumbra, regions in which the mean magnetic field inclination and strength have different values; these quantities have constant spatial averages in each simulation. In this manner, the results of the simulations for different magnetic field strengths and inclinations can be associated with areas located at different distances from the sunspot umbra. In the simulations that will be discussed, the magnetic field varies from 2000 G to 600 G, and the mean inclination of the field is approximately 85° from the surface normal [5].

Figure 1 shows 3D snapshots of the subsurface structure of magnetoconvection for an initial magnetic field strength of 1200 G at 85° inclination. Distributions of the horizontal magnetic field, the kinetic energy, and the magnitude of vorticity are shown. The superimposed curves show selected magnetic field lines to indicate the topology of the magnetic field.

At the surface, the magnetic field has a filamentary structure represented by elongated areas of strong magnetic field along the direction of the field inclination. In the inclined field, strong horizontal velocities are observed with values up to 2 – 4 km/s and occasionally as high as 6 km/s ("Evershed clouds"). The magnetic filaments are a result of the influence of narrow high-speed

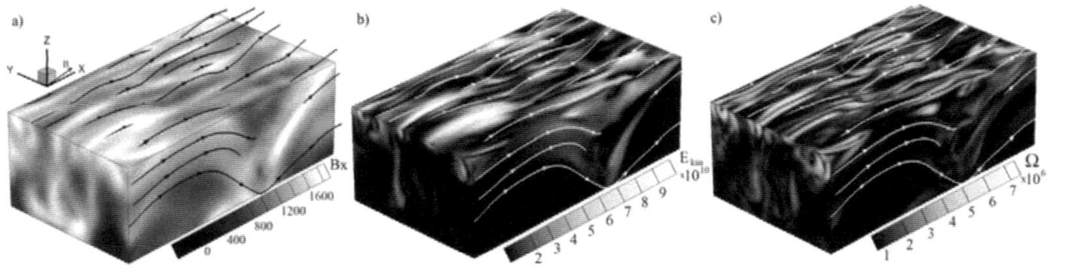

Figure 1. 3D snapshots show distributions of a) the horizontal magnetic field, b) kinetic energy and c) magnitude of vorticity. Curves show the topology of the magnetic field with selected field lines.

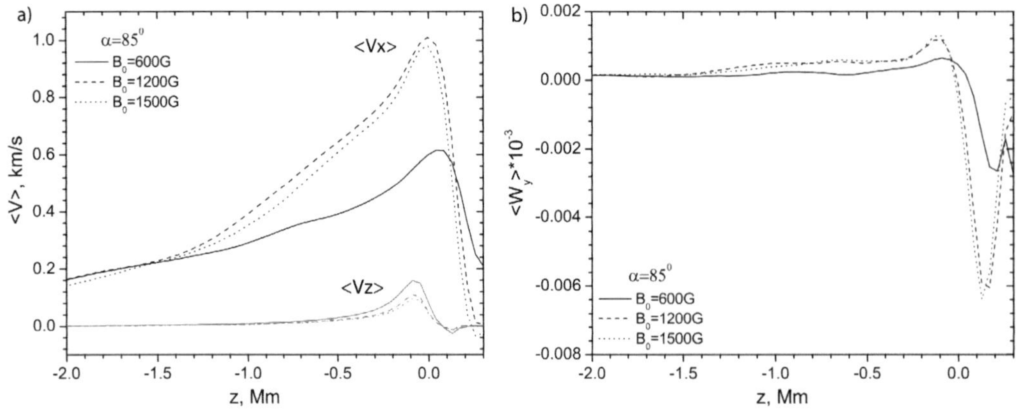

Figure 2. Vertical distribution of the horizontal, <Vx>, and vertical, <Vz>, velocities (panel a) and the y-component of vorticity (panel b) for initial magnetic field strengths of 600, 1200 and 1500G, and inclination $\alpha = 85°$ from vertical.

horizontal flows. These flows push the magnetic field aside and thus cause concentrations of field at the boundaries of the high-speed regions [5]. An additional effect on the field topology is caused by partially frozen magnetic field lines in the plasma. In this situation, upflows of stretched convective cells and downflows between them cause deformation of the magnetic field lines, thus forming thin and wavy magnetic flux tubes (Fig. 1) that are observed as the "sea-serpent" effect [6, 8]. Thus, convective flows in the inclined magnetic field regions of the penumbra form narrow magnetic tubes that demonstrate a complicated oscillatory behavior along the surface in the direction of the field lines and also in the plane perpendicular to the field inclination. The kinetic energy distribution in Figure 1b shows strong local concentrations in the subsurface layers, representing the Evershed flow. The magnitude of vorticity (Fig. 1c) shows concentration of vortical motions along and perpendicular to the field inclination near the surface. The mean horizontal and vertical flows are shown in Figure 2a. The simulations show subsurface vortical motions mostly in the direction perpendicular to the direction of inclination of the mean magnetic field. The magnitude of vorticity increases for stronger magnetic field strengths (Fig. 2b). It is 4 – 5 times smaller in the subsurface layers than at the surface.

Form a physical point of view, the inclination of the magnetic field plays the principal role in creation of subsurface shear flows, the Evershed effect. This inclination breaks the symmetry of the convection and causes deformation of the convective cells that appear as traveling convective waves. By examining the mean Reynolds and Maxwell stresses (Fig. 3), we find that the Maxwell stress is the main source of the mean shear flow – Evershed effect.

3. Conclusion

Radiative MHD simulations of solar magnetoconvection in regions of inclined magnetic field qualitatively and quantitatively describe many observed features of the Evershed effect in sunspots. Realist numerical simulation results indicate that the principal physical mechanism of Evershed flows is related to two effects: 1) the filamentary structurization of convective cells in a strongly inclined magnetic field, and 2) the traveling-wave nature of the magnetoconvection. Our simulations model this phenomenon in realistic solar conditions and reproduce many details of the observations, thus providing a basis for explaining the Evershed effect.

The simulation results provide a synergy of earlier proposed models: propagation of flows in weakly-magnetized gaps in the magnetic field [11], the thin-flux-tube model [9] (Fig. 1), the

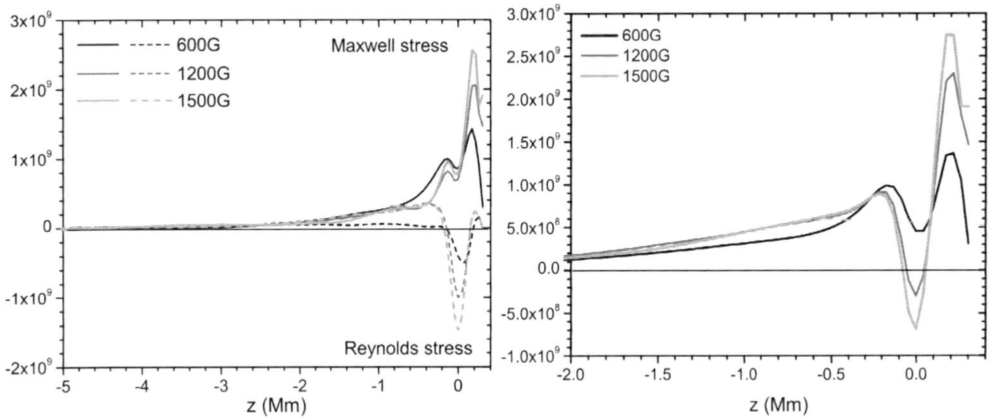

Figure 3. Variations of the Maxwell and Reynolds stresses (left panel), and their sum (right panel) obtained from the simulation results for initial magnetic field strengths of 600, 1200 and 1500 G, inclined at 85° from vertical.

siphon model [7], the overturning convection model [10], and the convective rolls model [1]. In particular, our analysis shows that the oscillatory behavior of the convective rolls in the last model has some similarities with the simulation results. As indicated by this theory, the balance between the Reynolds and Maxwell stresses (Fig. 3) in subsurface layers can be a driving force of the Evershed effect. The horizontal flows originate from convective upflows of hotter plasma, as in ordinary convection, but are channeled by the magnetic field and amplified by the traveling convective waves. The whole process is highly non-linear and stochastic, with high-speed patches appearing randomly, but the simulations also show large-scale organized patterns across the simulation domain. This pattern seem to be associated with traveling waves. The simulations suggest that the Evershed flow is rather shallow, concentrated in the top 0.5 – 1 Mm of the convection zone (Fig. 2a). This prediction may be tested by local helioseismology methods.

References

[1] Busse F H 1987 A new mechanism for the Evershed flow. *Proc. of the Inaugural workshop* 187.

[2] Evershed J 1909 Radial movement in sun-spots. *MNRAS* **69** 454.

[3] Hurlburt N E, Matthews P C and Proctor M R E 1996 Nonlinear compressible convection in oblique magnetic fields. *ApJ* **457** 933.

[4] Jacoutot L, Kosovichev A G, Wray A and Mansour N N 2008 Realistic numerical simulations of solar convection and oscillations in magnetic regions. *ApJ* **684** L51.

[5] Kitiashvili I N, Kosovichev A G, Wray A A and Mansour N N 2009 Traveling waves of magnetoconvection and the origin of the Evershed effect in sunspots. *ApJ* **700** L178.

[6] Kitiashvili I N, Bellot Rubio, L R, Kosovichev A G, Mansour N N Sainz Dalda A and Wray A A 2010 Explanation of the sea-serpent magnetic structure of sunspot penumbrae. *ApJ* **716** L181.

[7] Meyer F and Schmidt H U 1968 The Evershed effect as a wave phenomenon. *Mitteilungen der Astronomischen Gesellschaft Hamburg* **25** 194.

[8] Sainz Dalda A and Bellot Rubio L R 2008 Detection of sea-serpent field lines in sunspot penumbrae. *A&A* **481** L21.

[9] Schlichenmaier R, Jahn K and Schmidt H U 1998 A dynamical model for the penumbral fine structure and the Evershed effect in sunspots. *ApJ* **493** L121.

[10] Scharmer G B, Nordlund Å and Heinemann T 2008 Solar surface magneto-convection. *ApJ* **677** L149.

[11] Severnyi A B 1965 The Nature of solar magnetic fields (The fine structure of the field). *Soviet Astronomy* **9** 171.

[12] Thomas J H 1988 Siphon flows in isolated magnetic flux tubes. *ApJ* **333** 407.

Solar-cycle variation of zonal and meridional flow

R Komm[1], R Howe[1], F Hill[1], I González Hernández[1] and D Haber[2]

[1] National Solar Observatory, Tucson, AZ 85719, USA
[2] JILA, University of Colorado, Boulder, CO , USA

E-mail: `rkomm@nso.edu`

Abstract. We study the variation with the solar cycle of the zonal and meridional flows in the near-surface layers of the solar convection zone. We have analyzed MDI Dynamics-Program data with ring-diagram analysis covering the rising phase of cycle 23, while the analyzed GONG high-resolution data cover the maximum and declining phase of cycle 23. For the zonal flow, the migration with latitude of the flow pattern is apparent in the deeper layers, while for the meridional flow, a migration with latitude is apparent only in the layers close to the surface. The faster-than-average bands of the zonal flow associated with the new cycle are clearly visible. Similarly, a pattern related to the new cycle appears in the residual meridional flow. We also study the flow differences between the hemispheres during the course of the solar cycle. The difference pattern of the meridional flow is slanted in latitude straddling the faster-than-average band of the torsional oscillation pattern in the zonal flow. The difference pattern of the zonal flow, on the other hand, resembles the cycle variation of the meridional flow. In addition, the meridional flow during the minimum of cycle 23/24 appears to be slightly stronger than during the previous minimum of cycle 22/23.

1. Introduction

We study the variation with the solar cycle of the zonal and meridional flows in the near-surface layers of the solar convection zone. We have analyzed Michelson Doppler Imager (MDI) Dynamics-Program data with ring-diagram analysis covering the rising phase of cycle 23, while the analyzed Global Oscillation Network Group (GONG) high-resolution data cover the maximum and declining phase of cycle 23. The combination of GONG and MDI data allows us to study the complete cycle 23.

It is well established that the zonal and meridional flow vary with the solar cycle [1]. Alternating bands of faster- and slower-than-average rotation migrate from mid-latitudes to the equator, a pattern called torsional oscillation [2]. The variation of the meridional flow is of similar amplitude with large (small) values during cycle minimum (maximum) [3, 4]. The ring-diagram analysis allows us to study the flows in each hemisphere, unlike global methods that produce north-south averaged results [2]. In this study, we focus on the onset of solar cycle 24 and the flow differences between the hemispheres during the course of solar cycle 23.

2. Data and Analysis

We determine the horizontal components of solar subsurface flows with a ring-diagram analysis using the dense-pack technique [3]. The full-disk Doppler images are divided into 189 overlapping regions with centers spaced by 7.5° ranging from $\pm 52.5^\circ$ in latitude and central meridian distance (CMD). Each region is apodized with a circular function reducing the effective diameter to

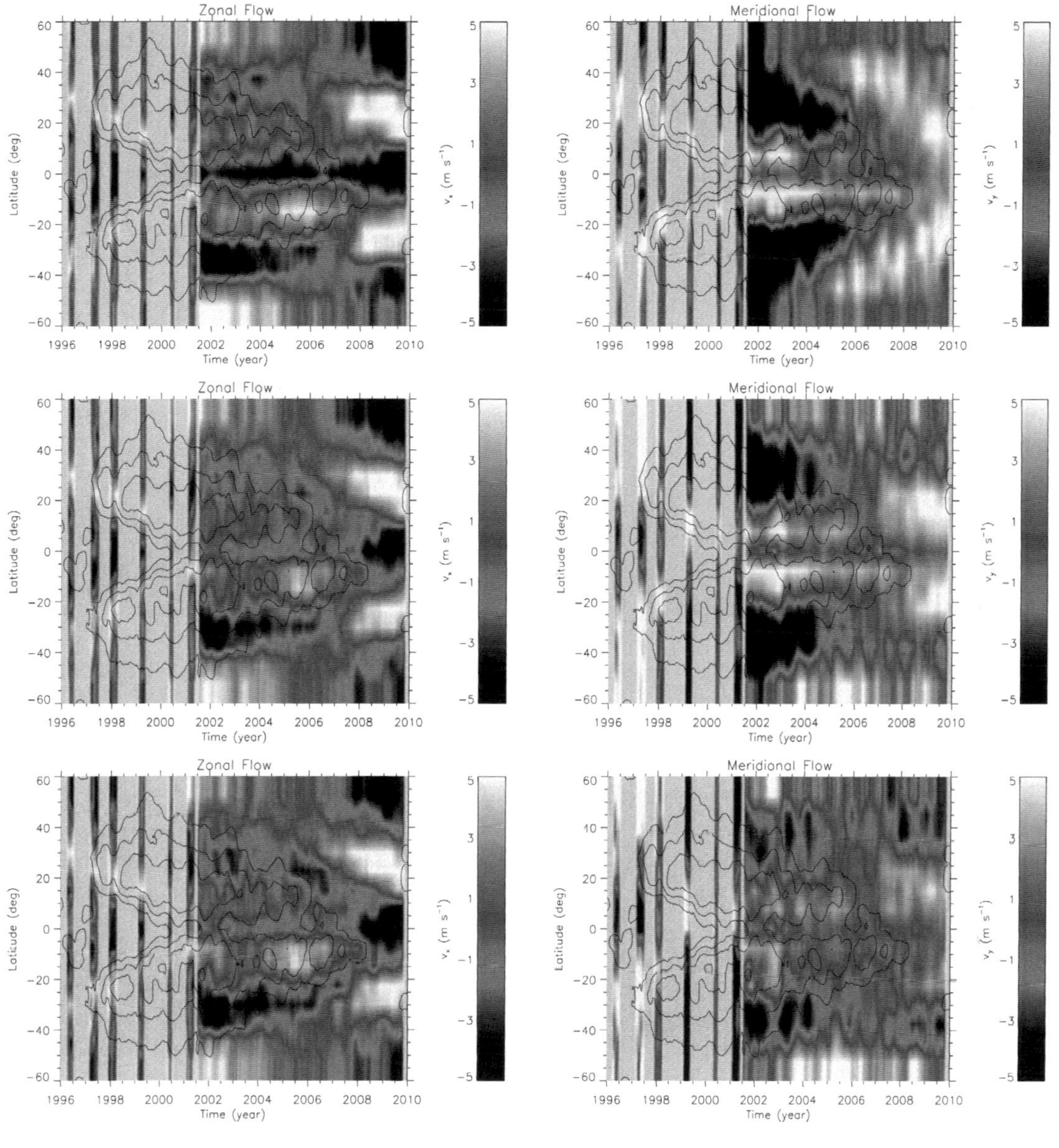

Figure 1. Zonal and meridional flow of solar cycle 23. Rotation-rate (left) and meridional-flow residuals (right) in m/s from SOHO/MDI (before mid-2001) and GONG (after mid-2001) averaged over three different depth ranges (top: 0.9 – 2.0 Mm, middle: 4.4 – 10.2 Mm; bottom: 10.2 – 15.8 Mm). Positive (negative) values indicate faster (slower) zonal flow and poleward (equatorward) meridional flow. Overlaid contours (5, 10, 20, 40 G) show the gross magnetic field strength derived from NSO Kitt Peak and SOLIS synoptic maps.

15° before calculating three-dimensional power spectra. The data are analyzed in "days" of 1664 minutes. Each of these regions is tracked throughout the sequence of images using the appropriate surface rate. For each dense pack day, we derive maps of horizontal velocities at 189 locations in latitude and CMD for 16 depths from 0.6 to 16 Mm.

We then create synoptic flow maps from the daily ones and calculate the average zonal and

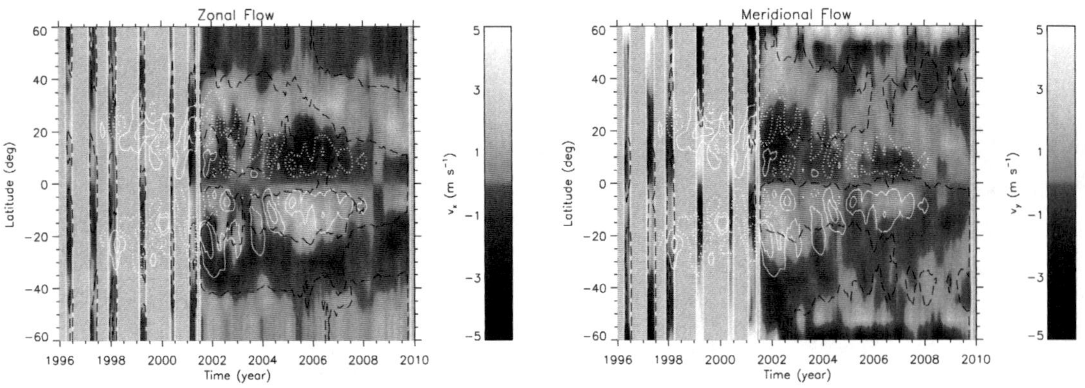

Figure 2. North-south difference of zonal (left) and meridional flow (right) averaged over 4.4 – 10.2 Mm in depth. White contours show the magnetic field difference (solid: 2, 5 G; dotted: −2, −5 G), while the black ones indicate the residual flows shown in Figure 1 (dashed: 0 m/s).

meridional flow for each synoptic map. Here, we use these average flows for each rotation and subtract the mean flow at each latitude. We use 20 Carrington rotations of MDI data covering the rising phase of cycle 23 before mid-2001 and 110 rotations of consecutive GONG data covering cycle 23 after mid-2001 to the beginning of cycle 24.

3. Results

Figure 1 shows the solar-cycle variation of the zonal and meridional flow averaged over three ranges in depth. The zonal flow (left) shows the torsional oscillation pattern for cycle 23 and 24 as bands moving from the mid-latitudes to the equator. The migration with latitude is apparent in the middle and deeper layers that are also probed by global helioseismology [2]. It is less pronounced in the shallow layers, which might be a resolution or observation period issue. The faster-than-average bands of the new cycle are clearly visible and appear before any surface activity of the new cycle is present. The corresponding meridional flow (right) also shows a variation with the solar cycle. However, a migration toward the equator is apparent only in the layers close to the surface. As in the case of the zonal flow, the flow pattern of the new cycle is present before there is new surface activity. The latitudinal pattern of the new cycle appears to be different from the previous one.

Figure 2 shows the north-south difference of zonal and meridional flow averaged over 4.4 – 10.2 Mm (compare with Figure 1 middle row). For the zonal flow (left), the band of positive differences in the south at low latitudes coincides with extra magnetic activity, extending cycle 23. At mid-latitude, the new-cycle branch is stronger in the north implying that more activity will appear in this hemisphere during cycle 24 than in the southern one. The north-south difference of the branch poleward of 40° latitude and that of the new-cycle branch ($20^o - 40^o$) show opposite signs. The difference pattern is aligned with latitude and thus resembles the cycle variation of the meridional flow. For the meridional flow (right), the north-south difference pattern is slanted in latitude straddling the faster-than-average band of the zonal flow.

Figure 3 shows the average meridional flow during the last two minima obtained from MDI data (minimum of cycle 22/23) and GONG data (minimum of cycle 23/24). The amplitudes at mid-latitudes are clearly larger during both minima than during the maximum phase of cycle 23. The maximum of cycle 23 has been observed by both MDI and GONG, which allows us to cross-calibrate the meridional flow measurements. Taking into account that the data are from different instruments, the meridional flow during the minimum of cycle 23/24 has a slightly

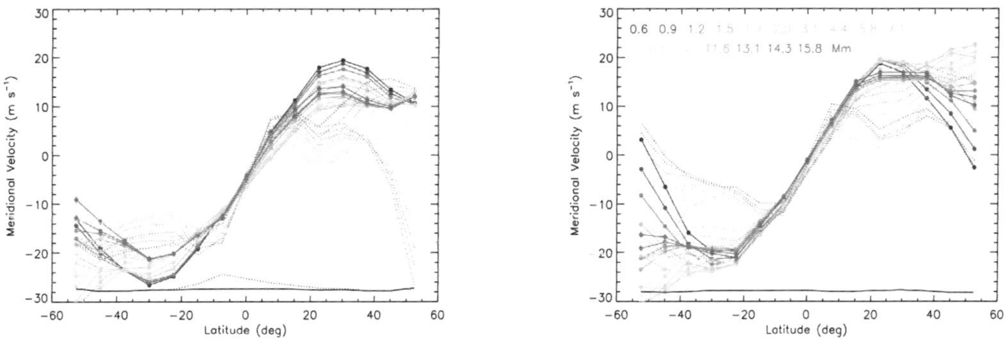

Figure 3. Meridional flow during the minimum of cycle 22/23 (left) averaged over Carrington rotations 1909–1911 from MDI data and the minimum of cycle 23/24 (right) averaged over CR 2076–2080 from GONG data. The dotted lines indicate the meridional flow during the maximum of cycle 23 (CR 1985–1989) from MDI (left) and GONG data (right). Colors indicate depths (blue: shallow; red: deeper). Positive (negative) values imply flows to the north (south).

larger amplitude than the flows during the minimum of cycle 22/23.

4. Discussion

The zonal and the meridional flow clearly vary with the solar cycle. The migration with latitude of the flow pattern is more apparent in the deeper layers than in the shallow ones for the zonal flow, while for the meridional flow it is apparent only in the layers close to the surface. For both large-scale flows, the patterns associated with the new cycle are visible before there is any sign of new-cycle surface activity. For the meridional flow, the latitudinal pattern of the new cycle appears to be different from that of the previous one. We will investigate this more closely as more observations of cycle 24 become available.

We also study the flow differences between the hemispheres during the course of the solar cycle. The difference pattern of the meridional flow is slanted in latitude straddling the faster-than-average band of the torsional oscillation pattern in the zonal flow. The difference pattern of the zonal flow, on the other hand, resembles the cycle variation of the meridional flow. Is this a coincidence or does it indicate a connection between the two flow patterns? We will investigate this relationship in the near future.

Acknowledgments

This work utilizes data obtained by GONG, managed by NSO, which is operated by AURA, under a cooperative agreement with the National Science Foundation. The data were acquired by instruments operated by the Big Bear Solar Observatory, High Altitude Observatory, Learmonth Solar Observatory, Udaipur Solar Observatory, Instituto de Astrofísica de Canarias, and Cerro Tololo Interamerican Observatory. SOHO is a mission of international cooperation between ESA and NASA. This work was supported by NASA grant NNG08EI54I to NSO. Haber's work was partially supported by NASA grants NAS5-02139 and NNX07AH82G.

References

[1] Komm R 2010 *Astron. Nachr.* (in press)
[2] Howe R, Hill F, Komm R, Christensen-Dalsgaard J, Larson, TP, Schou J, Thompson MJ and Ulrich R 2010 *Journal of Physics: Conference Series* (these proceedings)
[3] Haber DA, Hindman BW, Toomre J, Bogart RS, Larsen, RM and Hill F 2002 *ApJ* **570** 855–64
[4] González Hernández I, Kholikov S, Hill F, Howe R, Komm R 2008 *Solar Phys.* **252** 235– 45

Accurate Mapping of the Torsional Oscillations: a Trade-Off Study between Time Resolution and Mode Characterization Precision

A. Eff-Darwich[1,2], S.G. Korzennik[3]

[1] Instituto de Astrofísica de Canarias (IAC), E-38200 La Laguna, Tenerife, Spain
[2] Dept. de Edafología y Geología, Universidad de La Laguna (ULL), E-38206 La Laguna, Tenerife, Spain
[3] Harvard-Smithsonian Center for Astrophysics, 60 Garden St. Cambridge, MA,02138, USA

E-mail: adarwich@iac.es, skorzennik@cfa.harvard.edu

Abstract. One salient result of global helioseismology is the mapping of the so-called torsional oscillations below the solar surface. These subsurface flows are inferred by inverting rotational frequency splitting sets of global modes. These flows extend down to a depth of at least 0.8 R, and are likely associated with the activity cycle of our star. To better understand the mechanisms that drive the solar cycle we need to accurately map these flows, and characterize precisely their penetration depth and their temporal behavior.

We present a study of the spatial (depth and latitude) and temporal variations of the solar rotation rate associated with the torsional oscillation based on state-of-the-art mode fitting of time series of various lengths of MDI observations, namely 1456-, 728-, 364- and 182-day long time series. Such approach allows us to better estimate how much significant information can be extracted from the different time spans and hence trade off time resolution for precision in the inverted profiles resulting from the different mode sets.

1. Introduction

Helioseismic inferences have confirmed that the differential rotation observed at the surface persists throughout the convection zone [9], [6], [2], [4]. The outer radiative zone ($0.3 < r/R_\odot < 0.7$) appears to rotate approximately as a solid body at an almost constant rate ($\approx$ 430 nHz), whereas the innermost core ($0.19 < r/R_\odot < 0.3$) is consistent with that constant rotation, although it might be likely (within error bars) that the core rotates slightly faster than the rest of the radiative region. At the base of the convection zone, a shear layer —known as the tachocline— separates the region of differential rotation throughout the convection zone from the one with rigid rotation in the radiative zone. Finally, there is a subsurface shear layer between the fastest-rotating layer at about $0.95R_\odot$ and the surface. The rotation profile of the sun is not constant: the time-varying component of the rotation is referred to as the torsional oscillations at the surface, while we see hints of variations at the base of the convection zone, both being likely related to the solar activity cycle. However we still need to constrain the rotation profile and fully analyze the nature of the torsional oscillations. Theories about the mechanisms that drive the solar rotation and its variations remain to be tightly constrained by improved helioseismic inversion results. Better inversions mean not only improved inversion methodologies but especially improved rotational splittings.

We have used frequency sets resulting from fitting MDI time-series of various lengths in order to study the effect of the length of the time series on the characterization of the torsional oscillations. Fitting epochs of various lengths leads to averaging these epochs at the power spectrum level. This allows fitting spectra with higher signal to noise ratio and thus extracting the most significant information, as well as leading to values consistent among the different sets. For this purpose, rotational splitting sets were derived from fitting MDI 1456-, 728-, 364- and 182-day long time series [6], [7]. By comparison, the MDI pipeline data are estimated using 72-day long time-series, while the GONG pipeline data are estimated using 108-day long time-series. These standard products were also inverted for the work presented here. An iterative inversion methodology [3] was applied to the data sets to infer the rotation rate in the solar interior.

2. Results and conclusions

All the splitting sets calculated from time series of a given length were inverted and a temporal mean of the rotation rate was computed. The departure of the set of rotation profiles from the temporal mean is plotted as a function of time in order to visualize the torsional oscillations. These torsional oscillations are presented in Fig. 1 at the solar surface as a function of latitude and in Fig. 2 at a latitude of 20 degrees (where the torsional signal is stronger) as a function of depth. While overall the same pattern is observed at the surface, *i.e.* the migration towards the equator of the faster-than-average rotation band, the results are far from identical, in particular at high latitudes (see Fig. 1). One of the reasons for the discrepancy might come from the fact that the MDI- and GONG-pipeline sets contain few low-azimuthal order modes (those more sensitive to high latitudes), since we only consider modes that are fitted at all epochs. By contrast, the data derived from the MDI 1456-, 728-, 364- and 182-day long time series do not suffer such significant mode set attrition.

The reduction of the mode sets for the MDI- and GONG- pipeline data does not only affect low-m modes, but also low-degree modes. As a result, the temporal behavior of the rotation profile in the inner part of the convection zone and the radiative zone is less resolved by the pipeline data. Hence, the torsional oscillations signal seems to dissolve with depth (below $0.8R_\odot$) for the GONG and MDI-pipeline results (see Fig. 2), whereas it remains strong down to approximately the tachocline for the Korzennik's MDI data, in particular for MDI 728- and 364-day long series. The rotation profiles obtained from the inversion of MDI 1456- and 728-day long series show intriguing results below the tachocline: it seems that the time averaging process does not completely remove the oscillatory pattern in the radiative zone that might be seen in the inversions of MDI 182- and 364-day long series; some faster-than-average and slower-than-average rotation bands remain strong below the tachocline. The significance of these results will be tested in future works through the inversion of artificial data sets and spatial resolution analysis of the inversions.

Acknowledgments

The Solar Oscillations Investigation-Michelson Doppler Imager project on SOHO is supported by NASA grant NAS5-3077 at Stanford University. SOHO is a project of international cooperation between ESA and NASA. This work utilizes data obtained by the Global Oscillation Network Group (GONG) Program, managed by the National Solar Observatory, which is operated by AURA, Inc. under a cooperative agreement with the National Science Foundation. This work was in part funded by NASA grant NNX09AB15G, the Spanish grant AYA2004-04462 and the Canarian grant SolSubC200801000243.

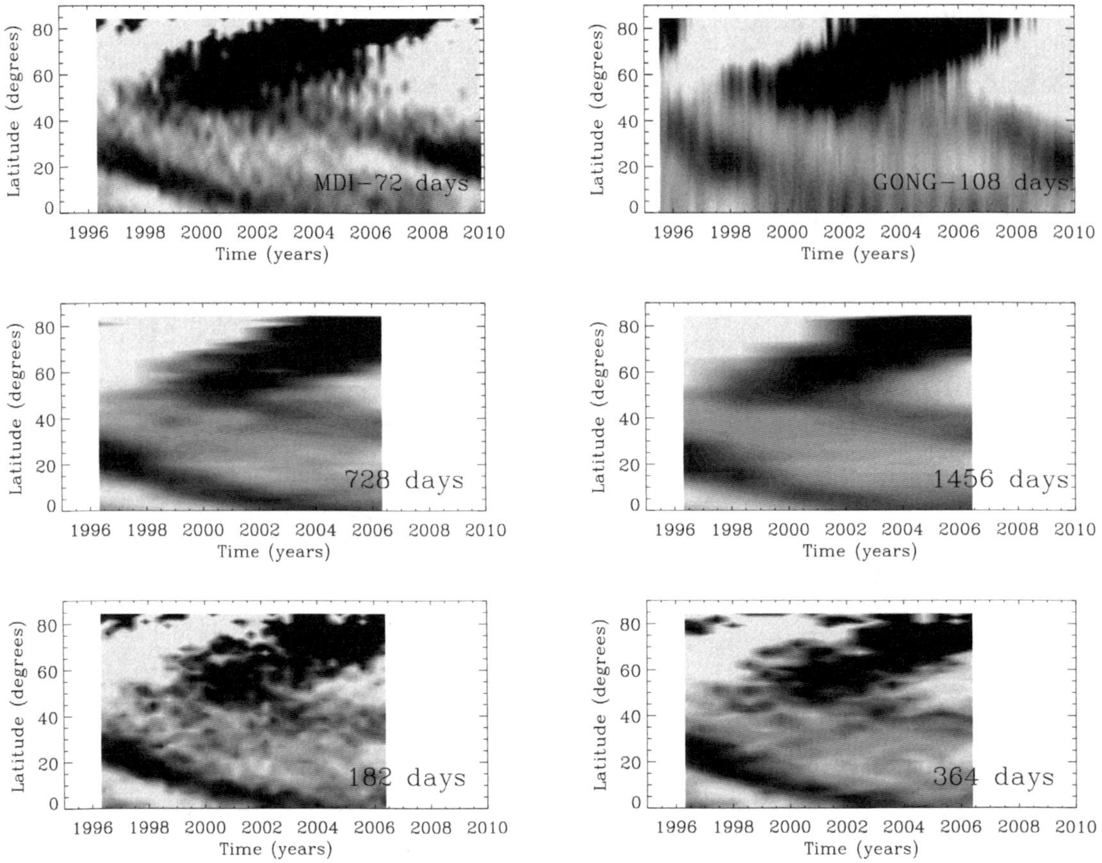

Figure 1. Rotation profile departure from a temporal mean as a function of latitude at $r = R_{\odot}$, inverted from six data sets, namely 108-day long GONG-pipeline, 72-day long MDI-pipeline, 182-day long MDI-Korzennik, 364-day long MDI-Korzennik, 728-day long MDI-Korzennik, and 1456-day long MDI-Korzennik.

References

[1] Anderson, E. R., Duvall, T. L.,Jr., & Jefferies, S. M. 1990, *ApJ*, **364**, 699

[2] Eff-Darwich, A., Korzennik, S. G., Jiménez-Reyes, S. J., & García, R. A. 2008, *ApJ*, **679**, 1636

[3] Eff-Darwich, A., Korzennik, S. G., & García, R. A. 2010, *AN*, in press

[4] Howe, R. 2009, Living Reviews in Solar Physics, **6**, 1

[5] Korzennik, S. G., Rabello-Soares, M. C., & Schou, J. 2004, *ApJ*, **602**, 481

[6] Korzennik, S. G. 2005, *ApJ*, **626**, 585

[7] Korzennik, S. G. 2008, Astronomische Nachrichten, **329**, 453

[8] Schou, J. 1992, Ph.D. Thesis,

[9] Thompson, M. J., Christensen-Dalsgaard, J., Miesch, M. S., & Toomre, J. 2003, Ann. Rev. Astron. Astrophys., **41**, 599

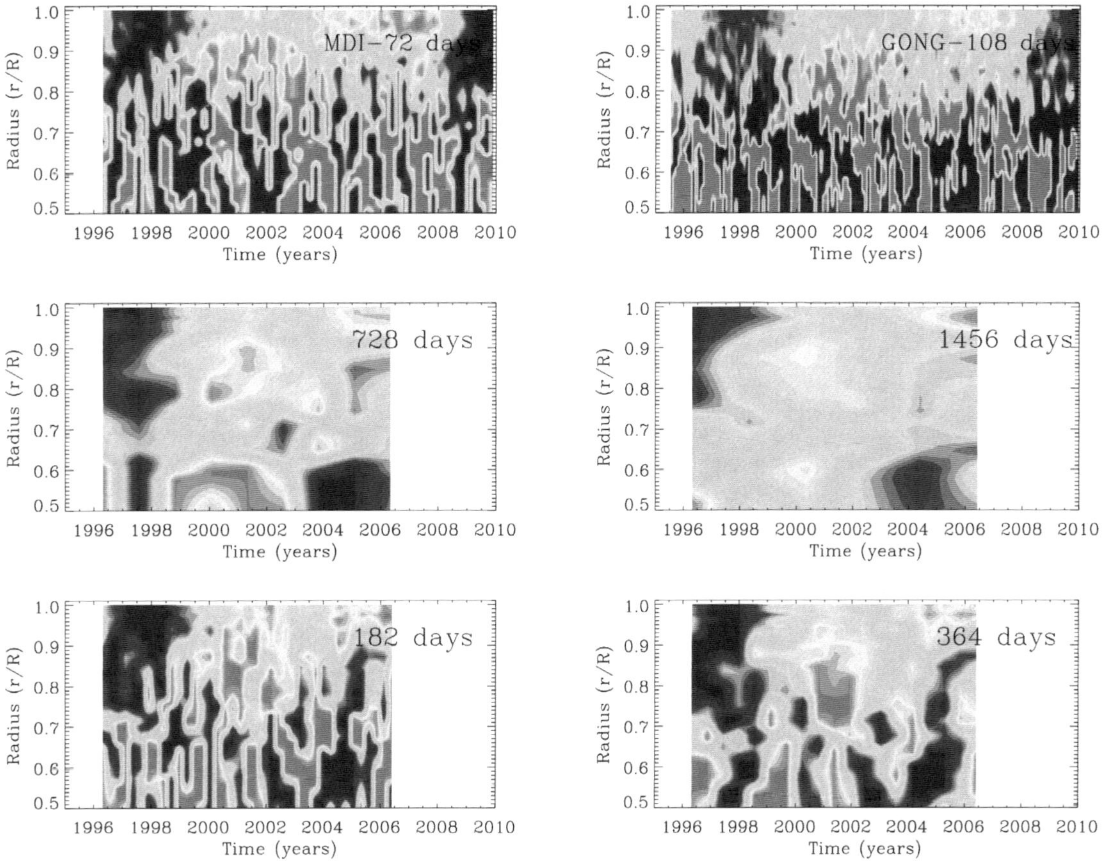

Figure 2. Rotation profile departure from a temporal mean as a function of depth at a latitude of 20 degrees, inverted from six data sets, namely 108-day long GONG-pipeline, 72-day long MDI-pipeline, 182-day long MDI-Korzennik, 364-day long MDI-Korzennik, 728-day long MDI-Korzennik, and 1456-day long MDI-Korzennik.

Observation of oscillation coupling ratios and the meridional flow

A Schad[1,2], M Roth[2] and J Timmer[1,3]

[1] Freiburg Center for Data Analysis and Modeling, University of Freiburg, 79104 Freiburg, Germany

[2] Kiepenheuer-Insitut für Sonnenphysik, 79104 Freiburg, Germany

[3] Freiburg Institute for Advanced Studies, University of Freiburg, 79104 Freiburg, Germany

E-mail: ariane.schad@fdm.uni-freiburg.de

Abstract. Measurements from local helioseismology indicate the existence of a meridional flow in the Sun with a strength up to 15 m/s near the solar surface. The flow profile at depths below ≈ 15 Mm are not accessible. We propose a method using global helioseismic measurements with the prospect to infer the meridional flow profile throughout the solar convection zone and show its performance on simulated data.

1. Introduction

From perturbation analysis it is known that large-scale flows in the solar interior, like convection, giant cells, and differential rotation, lead to couplings between unperturbed eigenfunctions of acoustic modes [1, 2, 3, 4]. These couplings manifest in shifts of the mode eigenfrequencies and distortions of the eigenfunctions. Here we focus on the perturbation of the eigenfunctions due to a meridional flow. In a simulation we show how to use the perturbation signature to infer the meridional flow from global helioseismic observations.

2. The meridional flow and perturbation of solar eigenfunctions

The meridional flow $\mathbf{u}$ is a zonal poloidal flow. It is assumed that the flow is confined between the surface and the bottom of the solar convection zone. In spherical geometry (θ, ϕ) it can be decomposed into radial and horizontal components of different degree s, with corresponding radial flow strengths $u_s(r)$ and $v_s(r)$,

$$\mathbf{u}(r,\theta) = \sum_s \left[u_s(r)\, Y_s^0(\theta,\phi)\, \vec{e}_r + v_s(r)\partial_\theta Y_s^0(\theta,\phi)\, \vec{e}_\theta \right] . \qquad (1)$$

In case of a stationary meridional flow the horizontal flow v_s can be expressed in terms of u_s by means of the mass conservation law.

The coupling of global p-modes due to a meridional flow can be investigated using quasi-degenerate perturbation theory. In algebraic terms the coupling is represented by the supermatrix $\mathbf{Z}$ with respect to the unperturbed mode eigenfunctions $\boldsymbol{\xi}_i^0$ [1].

We regard a single mode $i = (n, l, m)$ and a subset $M_i = \{j_1, ..., j_{N_i}\}$ of modes with quasi-degenerate frequencies coupling with mode i. The corresponding block-matrix $\mathbf{Z}_i$ of $\mathbf{Z}$ is of the form

$$\mathbf{Z_i} = \begin{pmatrix} a_i & H_{ij_1} & H_{ij_2} & \cdots & H_{ij_{N_i}} \\ \bar{H}_{ij_1} & a_{j_1} & 0 & & 0 \\ \bar{H}_{ij_2} & 0 & a_{j2} & & 0 \\ \vdots & & & \ddots & \\ \bar{H}_{ij_{N_i}} & 0 & 0 & & a_{j_{N_i}} \end{pmatrix} . \tag{2}$$

The diagonal elements are determined by the squared unperturbed mode frequency ω_j^2. The off-diagonal matrix elements result from small perturbations due to the flow. From perturbation theory applied to $\mathbf{Z}_i$ the first eigenvector of $\mathbf{Z}_i$ corresponding to mode i can be approximated up to first order by

$$\boldsymbol{\xi}_i = C_{ii}\boldsymbol{\xi}_i^0 + \sum_{j \in M_i} C_{ij}\, \boldsymbol{\xi}_j^0 , \tag{3}$$

with coefficients

$$C_{ii} = 1 \quad \text{and} \quad C_{ij} \;=\; \frac{\bar{H}_{ij}}{\omega_i^2 - \omega_j^2} \quad \text{for } j \in M_i,\, j \neq i . \tag{4}$$

3. Coupling ratios and b-coefficients

The contribution of a perturbed mode i with oscillation amplitude $a_i(t)$ to the observable Doppler velocity field $\mathbf{v}$ is given by

$$\mathbf{v}_i = a_i(t)\Big[C_{ii}\boldsymbol{\xi}_i^0 + \sum_{j \in M_i} C_{ij}\boldsymbol{\xi}_j^0\Big] . \tag{5}$$

Therefore, a projection of the Doppler field onto the spherical harmonic function Y_l^m leads to additional contributions of a mode i to the global solar oscillations of modes in M_i. In the frequency domain we find that the ratios of mode amplitudes $A(\omega)$ at the frequency ω_i between mode i and coupling modes in M_i reflect the coefficients C_{ij} by the relation

$$\frac{A_j(\omega_i)}{A_i(\omega_i)} \propto \frac{C_{ij}}{C_{ii}} \quad \text{for } j \in M_i , \tag{6}$$

and we define the ratios C_{ij}/C_{ii} as coupling ratios.

In Fig. 1 we show an example of simulated coupling ratios for $l = 120$, $n = 5$ in dependence on azimuthal order m for a simulated meridional flow which is described below. The dependence of the matrix elements H_{ij} on azimuthal order m can be expressed in terms of orthogonal polynomials $\mathcal{P}_{ll'}^s$ via the equation

$$H_{nln'l'}(m) \;=\; \sum_s b_{n,l,n',l'}^s \mathcal{P}_{ll'}^s(m) , \tag{7}$$

where we identified the subscript i by its radial order and degree (n, l) and j by (n', l'). The polynomials are given by Wigner-3j-symbols analogously to the definitions in [5]. The b-coefficients are here defined by

$$b_{n,l,n',l'}^s = \int_0^R K_s^{nln'l'}(r) u_s(r)\, dr . \tag{8}$$

They relate the matrix elements to the radial meridional flow strength of corresponding degree s. The integral kernel $K_s^{nln'l'}$ describes the coupling between modes (n, l) and (n', l') due to a meridional flow component of degree s and is given in [4].

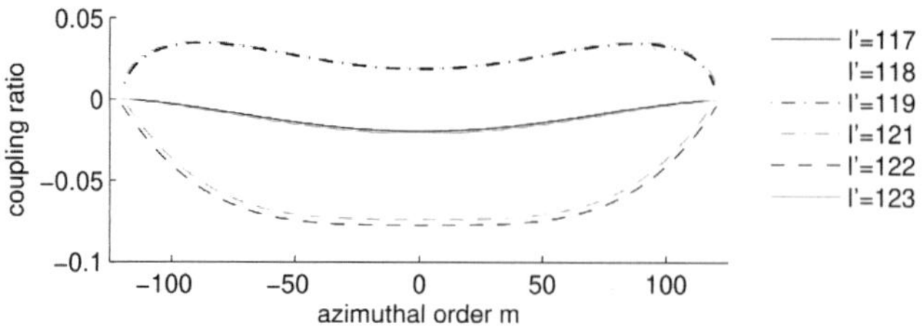

Figure 1. Example of coupling ratios for $l = 120$, $n = 5$ in dependence on azimuthal order m for a simulated meridional flow. The lines belonging to $l' = 119$ and $l' = 121$ overlap as the corresponding coupling ratios are nearly the same.

4. Inference of the meridional flow - a simulation

In a simulation we generated a meridional flow as defined in [4]. The flow consists of components with degree $s = 1, \ldots, 5$ and a return flow at $r = 0.713\,R$. The horizontal flow strength at the surface varies between $0.5 - 56$ m/s. We used Solar Model S [6] to compute the meridional flow kernels and the coupling ratios. We estimated the b-coefficients by fitting polynomials as defined before. The observation error was assumed to be Gaussian with unit variance and independent for all coupling ratios. From application of the angular momentum selection rules and inclusion of restrictions for the observable frequency range we obtain 8048 couplings between p-modes with frequencies of $1.1 - 3.7$ mHz and degree $l = 1 - 299$.

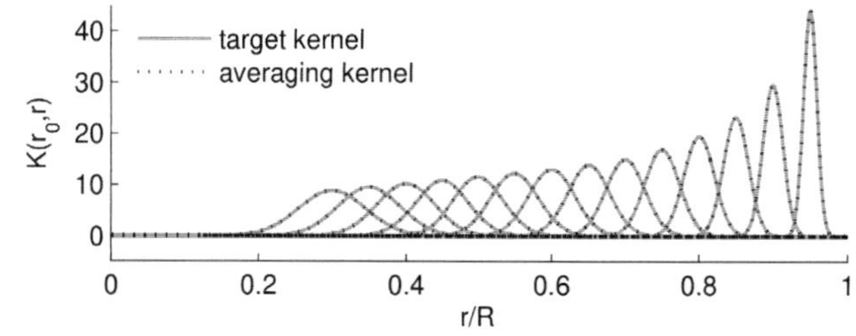

Figure 2. Target kernels and averaging kernels for different depths for $s = 3$.

The meridional flow and the b-coefficients are formally connected by the integral equation of the first kind in Eq. (8). For the problem of inference of the meridional flow we adopted the Subtractive Optimally Localized Averaging (SOLA) technique which is succesfully applied for the inversion of the differential rotation [7]. For the SOLA technique target kernels sensitive to different depths r are defined with a Gaussian shape. Using meridional flow kernels we determine averaging kernels which optimally approximate the target kernels (Fig. 2). This finally yields estimates of the radial flow profiles as shown in Fig. 3. Vertical errorbars for u_s are derived from the error magnification of the SOLA technique. Horizontal errorbars correspond to the radial resolution of the averaging kernels. The horizontal flow profiles are derived from the estimated radial flow profiles using the mass conservation law. The inversion results show a good estimation of the true underlying flow profiles. The vertical errorbars for $s = 3 - 5$ are very small due to the very well agreement of averaging kernels and target kernels. A bias of the estimated flow is observed at $r = 0.7R$ where the simulated flow profile has an imposed artificial discontinuity at the transition region of the convection zone and radiation zone.

5. Discussion

We have derived a method to infer the meridional flow based on the perturbation of the eigenfunctions of global p-modes. In a simulation we have shown that from a determination of the coupling ratios it is possible to infer the meridional flow by means of the SOLA inversion

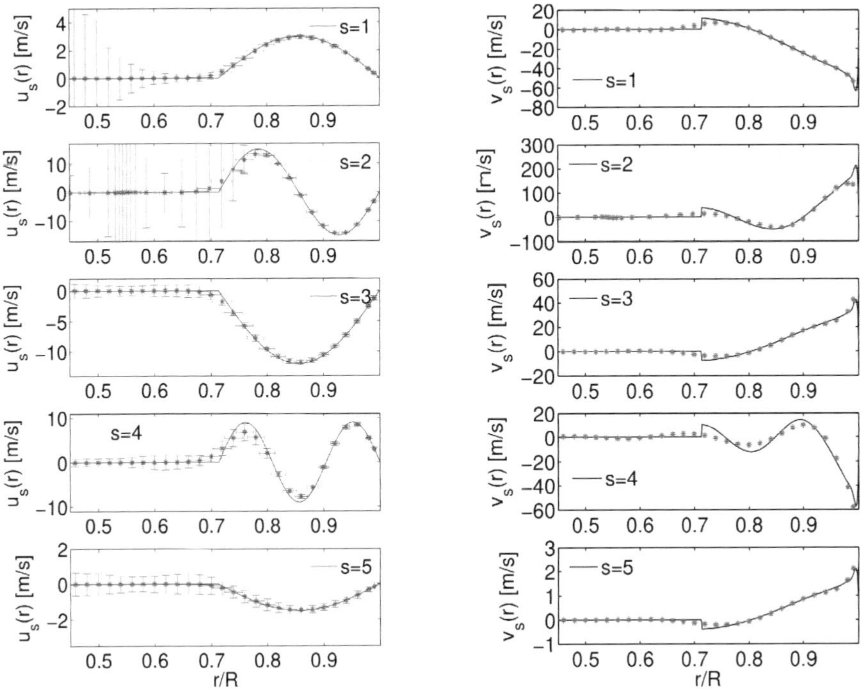

Figure 3. Simulated and estimated profiles of the radial u_s and horizontal v_s flow components in dependence on depth r/R and degree s. Black lines mark the simulated flow profiles. Gray asterisks show the estimated profiles. The vertical errorbars for $s = 3, ..., 5$ are of small size. For reasons of visibility we magnified them by a factor of ten.

technique. In further studies we will investigate the influence of leakage due to incomplete observations of the Doppler velocity field on the determination of coupling ratios from global oscillations. A more extensive treatment of the observation error and its propagation from u_s to v_s throughout the inversion procedure is in work. An application of the method to observed global oscillations will be performed.

Acknowledgments
This work is supported by the Deutsche Forschungsgemeinschaft, Grant No. Ti315/4-2, and by the European Helio- and Asteroseismology Network (HELAS) which was funded as Coordination Action under the European Commission's Framework Programme 6 (FP6).

References

[1] Lavley E M and Ritzwoller H M 1992 *Roy. Soc. Phil. Trans. Ser.* A **339** 431
[2] Roth M and Stix M 1999 *Astron. Astrophys.* **351** 1133
[3] Roth M, Howe R and Komm R 2002 *Astron. Astrophys.* **396** 243
[4] Roth M and Stix M 2008 *Sol. Phys.* **251** 77
[5] Ritzwoller M H and Lavely E M 1991 *Astrophys. J.* **369** 557
[6] Christensen-Dalsgaard J et al 1996 *Science* **272** 1286
[7] Pijpers F P and Thompson M J 1994 *Astron. Astrophys.* **281** 231

Characterizing Convection in Stellar Atmospheres

Joel Tanner, Sarbani Basu, Pierre Demarque & Frank Robinson

Department of Astronomy, Yale University, New Haven CT, U. S. A.

E-mail: `joel.tanner@yale.edu`, `sarbani.basu@yale.edu`, `pierre.demarque@yale.edu`, `frank.robinson@yale.edu`

Abstract. We perform 3D radiative hydrodynamic simulations to study the properties of convection in the superadiabatic layer of stars. The simulations show differences in both the stratification and turbulent quantities for different types of stars. We extract turbulent pressure and eddy sizes, as well as the T-τ relation for different stars and find that they are sensitive to the energy flux and gravity. We also show that contrary to what is usually assumed in the field of stellar atmospheres, the structure and gas dynamics of simulations of turbulent atmospheres cannot be parameterized with $T_{\rm eff}$ and $\log(g)$ alone.

1. Introduction

Large eddy simulations (LES) have been successfully applied to stellar atmospheres to investigate the effect of convection on stellar structure and atmosphere stratification. LES can produce accurate turbulent gas dynamics in regions of efficient and inefficient convection, and treat overshoot self consistently. Statistics from the simulated turbulence can be parameterized and applied to 1D envelope or stellar models. For example, Kupka & Robinson (2007) demonstrate that some effects of convection can be described by averaging the skewness of the vertical velocity.

Some progress has been made in parameterizing the turbulent dynamics in the superadiabatic layer (SAL) in this way. For example, Ludwig *et al.* (1999) and Freytag *et al.* (1999) have undertaken extensive efforts to map the mixing efficiency (or effective α for MLT) using 2D HRD simulations, and constructing 1D envelope models by matching the specific entropy of the models and simulations. Trampedach *et al.* (1999) has also attempted to extract a mixing length parameter from simulations by matching averaged 3D simulation stratifications to 1D envelope models and modifying the turbulent pressure.

Although all attempts to date to parameterize convection have been carried out in the $\log(g)$ and $\log(T_{\rm eff})$ plane, it has been pointed out by other researchers (e.g., Ludwig *et al*, 1999) that these traditional atmospheric parameters are not necessarily suitable for the study of convection in the SAL. In this work, we examine the parameter space over which the turbulent dynamics of convection in the SAL are described.

2. Modelling Stellar Convection

A 3D simulation is characterized by its surface gravity, effective temperature and chemical composition. We get the surface gravity, stellar flux and initial stratification for each simulation from a 1D stellar evolution model. The simulations are evolved until they have thermally relaxed to a new equilibrium state. The simulation domain is located at the top of the convection zone, with the top and bottom of the domain located at approximately 3 and 8 pressure scale heights

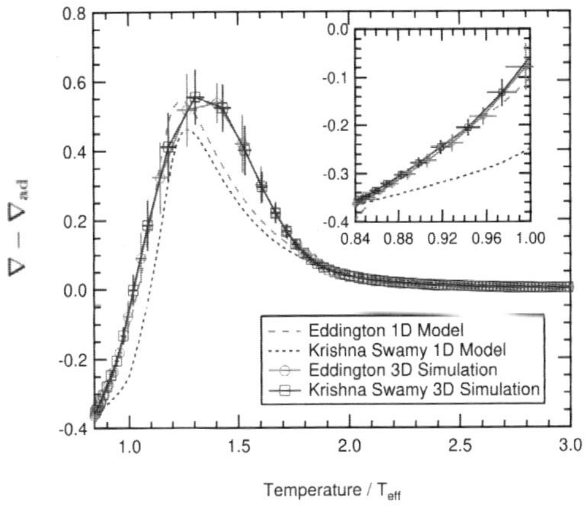

Figure 1. The superadiabaticity of standard solar models (dashed and dotted lines) computed with the Eddington (circles) and Krishna Swamy (squares) T-τ relations. Solid lines show the superadiabaticity from the 3D simulations after thermal relaxation. The simulations are in agreement despite beginning with different stratifications. The inset shows the superadiabaticity above the photosphere.

above and below the photosphere, respectively. The domain is small enough that curvature and radial variation in gravity can be safely ignored. The vertical walls are periodic and the horizontal walls are slip free and impenetrable (closed box). Radiative transfer is treated with the 3D Eddington approximation (Unno & Spiegel, 1966) in the optically thin region. Properties of the simulations are largely insensitive to the boundary conditions. Kupka (2005) has found good agreement between simulations of the solar SAL computed with different codes with varied boundary conditions, radiative transfer models and resolution.

3. Standard Solar Models

We present a comparison of solar simulations to demonstrate that the initial stratification near the surface taken from the 1D models does not determine the relaxed simulated state. The solar simulations begin from two standard solar models with slightly different input physics. One model is constructed with the Eddington T-τ relation, and the other with the semi-empirical Krishna Swamy T-τ (Krishna Swamy, 1966). The mixing length parameter α is adjusted to compensate for the different surface conditions and produce the correct solar radius in both models. The mixing length parameter is 1.83 and 2.14 for the Eddington and Krishna Swamy standard solar models, respectively.

Figure 1 shows the superadiabatic gradient ($\nabla - \nabla_{ad}$) of the two solar models and their corresponding simulations. The two solar models only differ near the surface, and are the same deeper in the stellar interior. Differences can be seen in both the peak value and position of the SAL because of the different T-τ relations and mixing length parameters. The simulations are not affected by the treatment of convection or the atmosphere in the initial 1D model, and produce the same thermally relaxed stratifications through the entire simulated domain.

4. Model Pairs in the Gravity-Temperature Plane

In order to examine the quantities that can adequately characterize the turbulent structure, we prepare a set of stellar models and corresponding simulations at strategic positions in the HR diagram. The initial conditions for our set of simulations is comprised of three pairs of stellar models, with each pair having the same surface gravity and effective temperature but may have other stellar properties that are varied. Figure 2 shows the location of the three model pairs in the $\log(g)$-$\log(T_{\rm eff})$ plane (details are in Table 1).

5. Parameters for Turbulent Atmospheres

To investigate whether the traditional atmospheric parameters of surface gravity and effective temperature can uniquely characterize turbulent atmospheric structure, we compare the

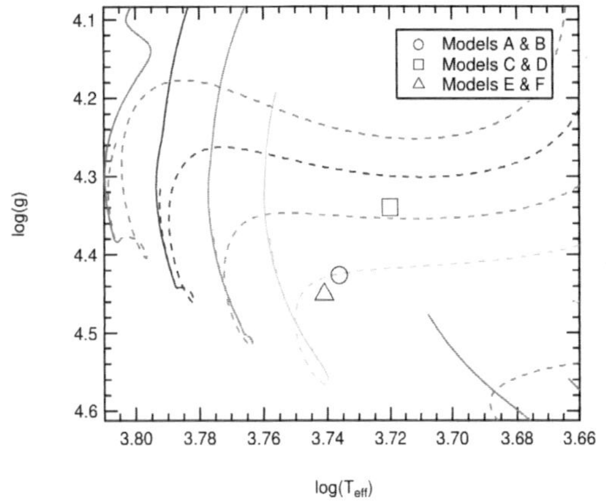

Figure 2. Locations in the $\log(g) - \log(T_{\rm eff})$ plane of three pairs of models and simulations. Each pair comprises two models in different evolutionary stages and corresponding simulations. Model pairs #1 and #2 comprise a pre- and a post-main sequence model, while both models in pair #3 are pre-main sequence. Properties of the simulations and models are compared in Figures 3 and 4.

Models	$\log(g)$	$\log(T_{\rm eff})$	Mass	Radius	α
A & B	4.426	3.736	0.90 & 1.00	0.960 & 1.012	1.8 & 1.8
C & D	4.340	3.720	0.80 & 1.00	1.001 & 1.119	1.8 & 1.2
E & F	4.450	3.741	1.00 & 1.00	0.985 & 0.985	2.1 & 1.5

Table 1. Properties of stellar models used as initial conditions for the three simulation pairs.

superadiabatic excess of the three model and simulation pairs described in section 4.

Fig. 3 compares the model and simulated superadibaticity for the three positions in the $\log(g)$-$\log(T_{\rm eff})$ plane. The first panel shows the structures for models (A & B), which begin with the same structure in the SAL. The relaxed state of the simulations is different than the 1D models, but the simulations are in agreement with each other.

The second panel compares the second model pair (C & D) and their corresponding simulations. Unlike the first model pair, these two models begin with different stratifications below the photosphere, but are both Eddington atmospheres above the photosphere and have the same surface parameters of $\log(g)$ and $\log(T_{\rm eff})$. The simulations produce new stratifications that differ from the 1D models and are not in agreement with each other, despite sharing the same surface parameters. This is contrary to what is predicted in static atmospheres.

Although the second pair demonstrates that $\log(g)$ and $\log(T_{\rm eff})$ do not necessarily uniquely characterize a turbulent atmosphere, the differences in simulation structure could be attributed to the different stellar model parameters of mass and radius. The third model pair (E & F) was constructed with the same mass and radius to remove this ambiguity. The third panel of Fig. 3 compares the SAL of the third pair, and again, the stratification and temperature gradients of the 1D models are identical above the photosphere but differ below, while the 3D simulations have stratifications that differ both above and below the photosphere.

We have demonstrated, by comparing the temperature gradients of the models in the $\log(g)$-$\log(T_{\rm eff})$ plane, that turbulent atmospheres can have different stratifications despite having the same traditional atmospheric surface parameters.

Fig. 4 compares the run of specific entropy in the SAL of the same set of 1D models and corresponding simulations. All model pairs have the same specific entropy above the photosphere, but may have different entropy profiles in the deeper regions which reflect differences in model evolutionary history and interior structure.

The first model pair had the same stratification throughout the simulated domain, and as Fig. 4 shows, the same run of specific entropy. The second and third model pairs both began with

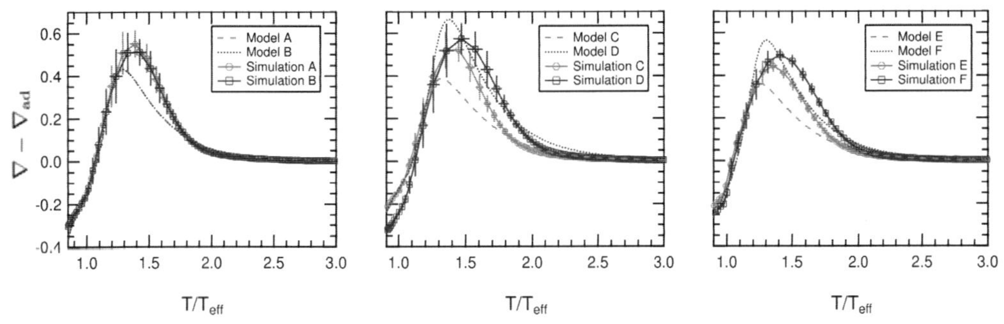

Figure 3. Comparison of the superadiabaticity in 1D models and 3D simulations. Each panel corresponds to a different model pair in the log(g)-log($T_{\rm eff}$) plane. Each model pair is comprised of models at different evolutionary states and their corresponding simulation.

identical structures above the photosphere but resulted in different simulated stratifications. In both cases the run of specific entropy below the photosphere was different in the models and simulations. We suggest that when combined with the specific entropy, the surface gravity and effective temperature can characterize the structure of a turbulent atmosphere.

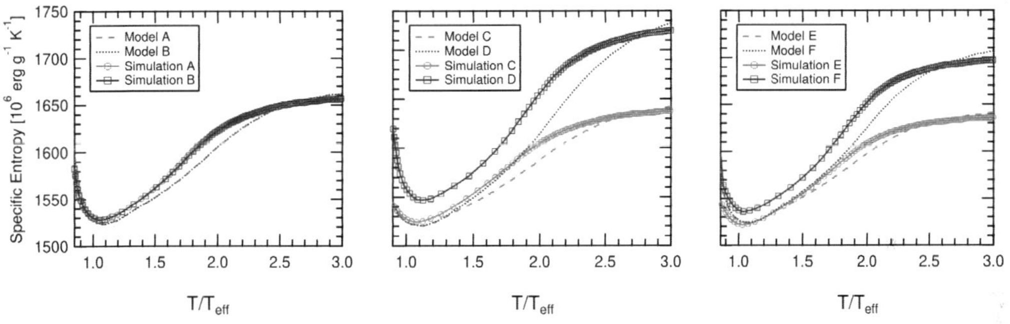

Figure 4. Comparison of the specific entropy in 1D models and 3D simulations. The panels correspond to the same models and simulations as in Fig. 2.

Acknowledgements

This work was partially supported by NASA ATFP grant # NNX09AJ53G to SB and was supported in part by the facilities and staff of the Yale University Faculty of Arts and Sciences High Performance Computing Center. JT is also partially supported by a PGS-D scholarship of the Natural Sciences and Engineering Research Council of Canada.

References

[1] Freytag, B., Ludwig, H.-G. & Steffen, M., 1999, ASPC, 173, 225
[2] Krishna Swamy K.S., 1966, ApJ, 145, 174
[3] Kupka, F., 2005, in Element Stratification in Stars: 40 Years of Atomic Diffusion, Eds., G. Alecian, O. Richard and S. Vauclair, EAS Publ. Ser., 17, 177
[4] Kupka, F. & Robinson, R.J. 2007, MNRAS, 374, 305
[5] Ludwig, H. C., Freytag, B., Steffen, M., Wagenhuber, J., 1995, in Proc. 2nd Liège International Astrophysical Colloquium , eds. A. Noels, D. Fraipont-Caro, M. Gabriel, N. Grevesse, and P. Demarque, p. 213
[6] Trampedach, R., Stein, R.F., Christensen-Dalsgaard, J. & Nordlund, Å, 1999, ASPCS, 173, 233
[7] Unno W., Spiegel E. A., 1966, PASJ, 18, 85

Coupling of solar p-modes of high degree l by joint effects of differential rotation and meridional circulation

S. V. Vorontsov

Astronomy Unit, Queen Mary, University of London, Mile End Road, London E1 4NS, UK
Institute of Physics of the Earth, B.Gruzinskaya 10, Moscow 123810, Russia

E-mail: `S.V.Vorontsov@qmul.ac.uk`

Abstract. Perturbational analysis of the p-mode coupling at high degree l, induced by the solar differential rotation [1] is extended to include, into a unified theoretical description, the effects of global meridional circulation. The predicted observational signatures of the meridional circulation in the global p-mode measurements are briefly discussed.

1. Introduction

Large-scale flows in the solar interior induce the potentially observable signatures in both the oscillation frequencies of global solar p modes, and in their velocity fields (the eigenfunctions). The most pronounced are the effects of the strongest global flow, the differential rotation, which allow its measurements from the frequency splittings. Starting from degree l of about 100, the distortion of the eigenfunctions (the p-mode coupling) by the effects of differential rotation becomes observable in the p-mode measurements [2,3]. The theoretical description of this effect was first suggested by Woodard [4] and later generalized in [1]. The effect grows rapidly with degree l (when frequency separation between the interacting modes of the same p-mode ridge becomes smaller), and needs to be accurately taken into account in any adequate analysis of the observational p-mode power spectra, including frequency measurements (e.g. [3]).

This work represents a direct extension to the analysis reported in [1] to include the effects of a stationary axisymmetric meridional circulation. Apart from a potential possibility of measuring the meridional flow from its effects in global p-mode data, a more immediate motivation of this study is to see what is the degree range where the effects in the observational power spectra can be safely ignored. Technically, this work represents a generalization, by applying the quasi-degenerate normal-mode perturbation technique, of the earlier approach of Woodard [5], which was based on a local analysis of the surface waves.

2. Mode coupling

Solutions to the zero-order equations (no internal flows) are designated with tilde; in operator form,

$$\rho_0 \tilde{\omega}_l^2 \tilde{\mathbf{u}}_l = H_0 \tilde{\mathbf{u}}_l, \quad (1)$$

with the displacement field

$$\tilde{\mathbf{u}}_l = \hat{\mathbf{r}}U(r)Y_{lm}(\theta,\phi) + V(r)\nabla_1 Y_{lm}(\theta,\phi), \tag{2}$$

where ∇_1 is angular part of gradient operator, and time dependence is separated as $\exp(-i\omega t)$. Radial order n and azimuthal order m are dropped from indexing the solutions for shortness, since interaction is limited by modes of the same m and will be considered for modes of the same n only. The unperturbed eigenfunctions are assumed to be normalized as

$$(\tilde{\mathbf{u}}_{l'}, \rho_0 \tilde{\mathbf{u}}_l) = \delta_{l'l}, \tag{3}$$

where scalar product is defined as

$$(\mathbf{u}_1, \mathbf{u}_2) = \int_V \mathbf{u}_1^* \cdot \mathbf{u}_2 \, dv, \tag{4}$$

and the integration is performed over the spherical volume occupied by the Sun.

For configuration with a slow differential rotation and meridional circulation, the equation (1) is replaced with

$$\rho_0 \omega^2 \mathbf{u} = (H_0 + 2\omega\, \delta H)\, \mathbf{u}, \tag{5}$$

where

$$\delta H = -i\rho_0 \, (\mathbf{v} \cdot \nabla) \tag{6}$$

and $\mathbf{v}$ is the sum of the two axisymmetric stationary velocity fields

$$\mathbf{v} = \mathbf{v}_{\mathrm{rot}} + \mathbf{v}_{\mathrm{mer}} \tag{7}$$

which are represented by their vector spherical-harmonic decomposition as

$$\mathbf{v}_{\mathrm{rot}} = -\sum_{s=1,3,\ldots} w_s(r)\hat{\mathbf{r}} \times \nabla_1 Y_{s,0}(\theta,\phi), \tag{8}$$

$$\mathbf{v}_{\mathrm{mer}} = \sum_{s=1,2,\ldots} \left[u_s(r)\hat{\mathbf{r}}Y_{s,0}(\theta,\phi) + v_s(r)\nabla_1 Y_{s,0}(\theta,\phi)\right]. \tag{9}$$

Since the fluid flow is stationary, $\mathbf{v}_{\mathrm{mer}}$ satisfies the mass-conservation equation

$$\nabla \cdot (\rho_0 \mathbf{v}_{\mathrm{mer}}) = 0, \tag{10}$$

with surface boundary conditions $u_s(R) = 0$. The perturbing operator can thus be written as

$$\delta H = \delta H_{\mathrm{rot}} + \delta H_{\mathrm{mer}}. \tag{11}$$

The matrix elements of operator δH_{rot} are evaluated as described in [1], and the matrix elements $(\tilde{\mathbf{u}}_{l'}, \delta H_{\mathrm{rot}} \tilde{\mathbf{u}}_l)$ of the operator of meridional circulation δH_{mer}—as described in the Appendix.

We now restrict the analysis, to make it shorter, to the meridional flows which are symmetric around the equatorial plane (only the components with even s in the equation 9 differ from zero). The selection rules, specified by the matrix elements for the operator of meridional circulation, are then the same as for the differential rotation: only modes with degree l of the same parity can interact.

As in [1], we now implement a quasi-degenerate perturbational analysis, looking for a perturbed solution as

$$\mathbf{u}_l = \sum_{p=0,\pm1,\pm2,\ldots} c_p \tilde{\mathbf{u}}_{l+2p} + \delta \mathbf{u}_l \tag{12}$$

where the expansion coefficients c_p are non necessarily small. These coefficients solve the algebraic system

$$\left(\tilde{\omega}_{l+2p}^2 - \omega_l^2\right) c_p + 2\omega_l \sum_{p'=0,\pm1,\pm2,\ldots} \left(\tilde{\mathbf{u}}_{l+2p}, \delta H \, \tilde{\mathbf{u}}_{l+2p'}\right) c_{p'} = 0, \quad p = 0, \pm1, \pm2, \ldots \tag{13}$$

Assuming the unperturbed eigenfrequencies to be equidistant along the p-mode ridge, we get

$$\omega_l^2 = \tilde{\omega}_l^2 + 2\tilde{\omega}_l \left(\tilde{\mathbf{u}}_l, \delta H_{\mathbf{rot}} \tilde{\mathbf{u}}_l\right), \tag{14}$$

which shows that the meridional circulation does not change the frequency splittings, and

$$2p \, c_p - \sum_{p'=\pm1,\pm2,\ldots} b_{p'} c_{p-p'} = 0, \quad p = 0, \pm1, \pm2, \ldots, \tag{15}$$

which are the recurrence relations between the expansion coefficients c_p, with solutions

$$c_p = \frac{1}{\pi} \int_0^\pi \cos\left\{pt - \sum_{p'=1,2,\ldots} \frac{1}{p'} \mathrm{Re}(b_{p'}) \sin(p't)\right\} \exp\left\{i \sum_{p'=1,2,\ldots} \frac{1}{p'} \mathrm{Im}(b_{p'}) \cos(p't)\right\} dt, \tag{16}$$

where

$$b_{p'} = -\left(\frac{\partial \tilde{\omega}_l}{\partial l}\right)_n^{-1} \left(\tilde{\mathbf{u}}_{l+2p}, \delta H \, \tilde{\mathbf{u}}_{l+2p-2p'}\right) \tag{17}$$

is now complex-valued, with real part governed by the differential rotation, and imaginary part governed by the meridional circulation. Derivation of the solutions (14,16) to the eigenvalue problem (13) follows closely a corresponding derivation described in [1], were the analysis was limited by the effects of differential rotation. The only new feature in the derivation is that the matrix elements of δH are now complex-valued.

When $\mathrm{Im}(b_{p'})$ are all zero (no meridional circulation), the coupling coefficients given by equation (16) are purely real. A slow meridional flow (with $|\mathrm{Im}(b_{p'})| \ll 1$) brings small corrections which are proportional to $\mathrm{v}_{\mathrm{mer}}^2$ in the real part of c_p and to $\mathrm{v}_{\mathrm{mer}}$ in its imaginary part. The effects induced in the observational power spectra are thus proportional to $\mathrm{v}_{\mathrm{mer}}^2$, if we discard small spatial leaks with imaginary amplitudes which may result from some observational or instrumental distortions (e.g. from a CCD tilt).

To evaluate the expected effects of the meridional circulation, we choose zonal (m=0) modes. For these modes, coupling by the meridional flows is stronger than for tesseral modes, and effects of the differential rotation vanish ($\mathrm{Re}(b_{p'}) = 0$ when $m = 0$, see [1]). A simple numerical estimate for a two-cell meridional circulation (one cell in each hemisphere) with maximum surface velocity of 15 m/s shows that the value of $|\mathrm{Im}(c_1)|$ reaches about 0.1 for solar f- and p_1-modes of degree $l = 300$ (which is an upper-degree limit of the MDI "medium-l" measurements). We thus expect that the effects of the meridional circulation are hardly detectable in the observational power spectra at $l < 300$. The effects can well be measurable, however, if analyzed in the correlations between amplitude spectra, as was first suggested by Woodard [5].

Since unperturbed eigenfrequencies are not purely equidistant along the p-mode ridge, the meridional circulation can contribute slightly to the frequency splittings; this contribution has been addressed recently by Roth and Stix [9]. A weak point of the analysis reported in [9] is that when addressing the effects of the meridional flows, the differential rotation has been discarded. It is a distinctive property of the quasi-degenerate perturbation analysis that the result is not necessarily linear (additive) in the perturbation. However, the reported frequency shifts are small, indicating that the effects of the meridional circulation are hardly detectable in the oscillation frequencies.

Acknowledgments
This work was supported by the UK STFC under grant PP/E001459/1. Author thanks an anonymous referee for useful comments.

Appendix: the matrix elements

A complete expression for the matrix elements $(\tilde{\mathbf{u}}_{l'}, \delta H_{\rm mer} \tilde{\mathbf{u}}_l)$ of operator of meridional circulation can be borrowed from Lavely and Ritzwoller [6] (we note that an alternative convention $\exp(i\omega t)$ for time dependence was used in their paper, which is equivalent to changing sign of ω). For modes of high degree l, the result can be significantly simplified with using semiclassical approximation for Wigner's 3-j symbols [7,8]. When $s \ll l$ and $|l' - l| \ll l$, we have

$$\begin{pmatrix} s & l' & l \\ 0 & m & -m \end{pmatrix} \simeq \frac{(-1)^{l'+m}}{(2l)^{1/2}} \left[\frac{(s-l'+l)!}{(s+l'-l)!}\right]^{1/2} P_s^{l'-l}\left(\frac{m}{l}\right), \tag{18}$$

where $P_s^{l'-l}$ is associated Legendre polynomial. For the product of two 3-j symbols, which enter the expression for these matrix elements [6], we get

$$\begin{pmatrix} s & l' & l \\ 0 & 0 & 0 \end{pmatrix} \begin{pmatrix} s & l' & l \\ 0 & m & -m \end{pmatrix} \simeq (-1)^{(s-l'+l)/2+m} \frac{(s-l'+l-1)!!(s+l'-l-1)!!}{2l(s+l'-l)!} P_s^{l'-l}\left(\frac{m}{l}\right) \tag{19}$$

when $s + l' + l$ is even, and zero otherwise.

Assuming radial eigenfunctions of interacting modes (with $|l' - l| \ll l$ and the same radial order n) to be nearly the same, the required matrix elements are reduced to

$$\begin{aligned} (\tilde{\mathbf{u}}_{l'}, \delta H_{\rm mer} \tilde{\mathbf{u}}_l) \quad &\simeq \quad il(l'-l) \sum_{\substack{s=1,2,\ldots \\ s+l'+l=\text{even}}} (-1)^{\frac{s-l'+l}{2}} \left(\frac{2s+1}{4\pi}\right)^{\frac{1}{2}} \frac{(s-l'+l-1)!!(s+l'-l-1)!!}{(s+l'-l)!} \\ &\times \quad P_s^{l'-l}\left(\frac{m}{l}\right) \int_0^R \rho_0 r v_s(r) \left[U^2 + l(l+1)V^2\right] dr. \end{aligned} \tag{20}$$

Note that these matrix elements are purely imaginary, with zero diagonal elements. We also have $(\tilde{\mathbf{u}}_{l'}, \delta H_{\rm mer} \tilde{\mathbf{u}}_l) = -(\tilde{\mathbf{u}}_l, \delta H_{\rm mer} \tilde{\mathbf{u}}'_l)$ since operator δH is self-adjoint; this property is ensured by the relation

$$P_s^{-t}(z) = (-1)^t \frac{(s-t)!}{(s+t)!} P_s^t(z). \tag{21}$$

References

[1] Vorontsov S V 2007 *Mon. Not. R. Astron. Soc* **378** 1499-506
[2] Vorontsov S V and Jefferies S M 2005 *Astrophys. J.* **623** 1202-14
[3] Vorontsov S V, Jefferies S M, Giebink C and Schou J 2009 *Solar-Stellar Dynamos as Revealed by Helio- and Asteroseismology* (Astron. Soc. Pacific Conf. Series vol 416) ed M Dikpati et al pp 301-5
[4] Woodard M F 1989 *Astrophys. J.* **347** 1176-82
[5] Woodard M F 2000 *Solar Phys.* **197** 11-20
[6] Lavely E M and Ritzwoller M H 1992 *Phil. Trans. R. Soc. Lond. A* **339** 431-96
[7] Brussaard P J and Tolhoek H A 1957 *Physica* **23** 955-71
[8] Dahlen F A and Tromp J 1998 *Theoretical Global Seismology* (Princeton: Princeton Univ. Press)
[9] Roth M and Stix M 2008 *Solar Phys.* **251** 77-89

Properties of Supergranulation During the Solar Minima of Cycles 22/23 and 23/24

Peter E. Williams, W. Dean Pesnell
Code 671, NASA Goddard Space Flight Center, Greenbelt, MD, USA
E-mail: peter.williams@nasa.gov

Abstract. The solar minimum at the transition from cycle 23 to 24 was notable for its low level of activity and its extended duration. Among the various fields of study, the evolution of the solar convection zone may provide insight into the causes and consequences of this recent minimum. This study continues previous investigations of the characteristics of solar supergranulation, a convection component strongly linked to the structure of the magnetic field, namely the time-evolution of the global mean of supergranule cell size, determined from spectral analysis of MDI Dopplergrams from the two previous solar minima. Analyses of the global mean of supergranule sizes show a quasi-oscillatory nature to the evolution of this particular supergranule characteristic. Performing similar analyses on realistic, synthetic Doppler images show similar time-dependent characteristics. We conclude that the observed fluctuations are not observational artifacts, and that an underlying trend exists within the evolution of the supergranulation network.

1. Introduction

Supergranulation is a large-scale component of solar convection. Supergranule cells are typically ~35 Mm across and have a lifetime of between 1-2 days. Doppler observations have shown that they have strong divergent flows at the surface (~300 m s^{-1}), around a whole magnitude larger than their accompanying radial flows. The interaction between supergranulation and the solar magnetic field is well seen in Ca II K images of the chromospheric network. While this interaction takes place on local scales, it is of interest to determine the level of interaction between supergranulation and the global field.

Williams & Pesnell [1, 2] have analyzed data from the Michelson Doppler Imager (MDI) [3] taken from the solar minima that occurred in 1996 and 2008, relating to the transition between cycles 22/23 and 23/24, respectively. Physical characteristics were quantified by averaging values over the whole dataset for a given year. While the $1/e$ lifetimes, the time taken for the cross-correlation of a supergranule pattern to drop to $1/e$, are found to be the same for both years, supergranules are slightly larger and their flows weaker in 1996 than in 2008 [2].

The present paper extends this work by initiating a statistical comparison between the datasets to determine the noise characteristics of the average values by calculating some of the characteristics mentioned above for each individual image. This paper focuses on analyzing the fluctuation of the global average of the supergranule size throughout each year in question.

2. Analysis Methods, Results

Fifteen-minute cadence, de-rotated MDI Dopplergrams [4] are processed to remove dominant axi-symmetric flow signals [5], the resulting images comprising contributions from the surface manifestations of convective flows. Each image is remapped to heliographic coordinates and projected onto the spherical harmonics to extract the spectral coefficients in degree, ℓ, and order, m. The coefficients are summed over m to produce a power spectrum with respect to wavenumber, ℓ. The result is a series of spectra that have been averaged to determine a mean value for the supergranule size for a given year [2]. We now analyze them individually to calculate the noise characteristic of that mean value and to study any time-evolution of the global mean of the supergranule size.

Each power spectrum contains a dominant supergranulation feature, the peak of which, ℓ_{peak}, is calculated by fitting the data with a modified Lorentzian, and determining the wavenumber at which the fitting function peaks. This process is repeated for each image resulting in an array of values that is compiled into a time-series. Figure 1 shows these peak wavenumbers versus time in days. The mean peak wavenumber, $\langle \ell_{peak} \rangle$, and standard deviation, σ_{peak}, of the series are calculated; $\langle \ell_{peak} \rangle = 121.7$ and $\sigma_{peak} = 1.5$ for 1996, and $\langle \ell_{peak} \rangle = 124.8$ and $\sigma_{peak} = 2.0$ for 2008. From these peak wavenumbers, the global average for the supergranule sizes are found to be 35.9 Mm and 35.0 Mm for 1996 and 2008, respectively. Both ℓ_{peak} distributions deviate slightly from a Gaussian ($\chi^2 = 5.92$ for 1996 and $\chi^2 = 4.05$ for 2008). The right-hand panel of Figure 1 shows the distribution of values for 1996.

Figure 1 shows fluctuations within the 1996 time-series that have been investigated using Fourier analysis. Firstly, the time-series is smoothed to reduce the noise. Two sets of smoothed data are then produced using two different filters. One set is the result of smoothing with a 1-day (i.e. 96-point) wide boxcar. The other is produced using a 4th degree, zeroth order, 33-point wide Savitzky-Golay [6] smoothing filter. While the boxcar is a traditional, widely-used filter, Savitzky-Golay smoothing retains much of the variation that is lost when using a boxcar (Figure 1). To calculate the frequency distribution of the time-series, the smoothed data are processed to remove their respective means and low-frequency trends, the latter using a 5-day boxcar filter. The datasets (containing around 6000 data points) were padded out to $2^{13} = 8192$ points to removed aliasing prior to frequency distributions being produced by applying a Fast Fourier Transform (FFT) to the datasets.

Figure 2a shows the frequency distributions for both the boxcar and Savitzky-Golay smoothed 1996 data. Both distributions exhibit dominant peaks at around 0.258 cycles per day giving an oscillatory period of around 3.88 days. For 2008, the FFT produced two dominant peaks (Figure 2b), one at 0.211 cycles per day and another at 0.281 cycles per day. These peaks equate to periods of 4.74 and 3.56 days, respectively.

Data simulations, that produce realistic full-disk Dopplergrams from a synthetic spatial power spectrum modeled on spectra observed from MDI data [7], have been used to show that the fluctuations are inherent in the data and not artifacts of the observation process. 20 days of synthetic images are processed to construct a time series that produced frequency signals on the order of what is seen in the MDI data. Although no single dominant peak is seen, two peaks are seen in the same frequency region as those observed from the MDI data, at 0.281 and 0.328 cycles per day, relating to periods of 3.05 and 3.56 days.

3. Discussion and Conclusion

Frequency analysis of the time-series of supergranule sizes from MDI data, to study noise characteristics within the data, suggest that there exists quasi-oscillatory fluctuations of globally-averaged supergranule cell sizes with a period of around 4 days. Although similar signals are seen within the 2008 MDI data and the simulated data, there are no dominant single peaks. Indeed, both of the latter datasets exhibit strong double peaks. However, all the datasets analyzed show

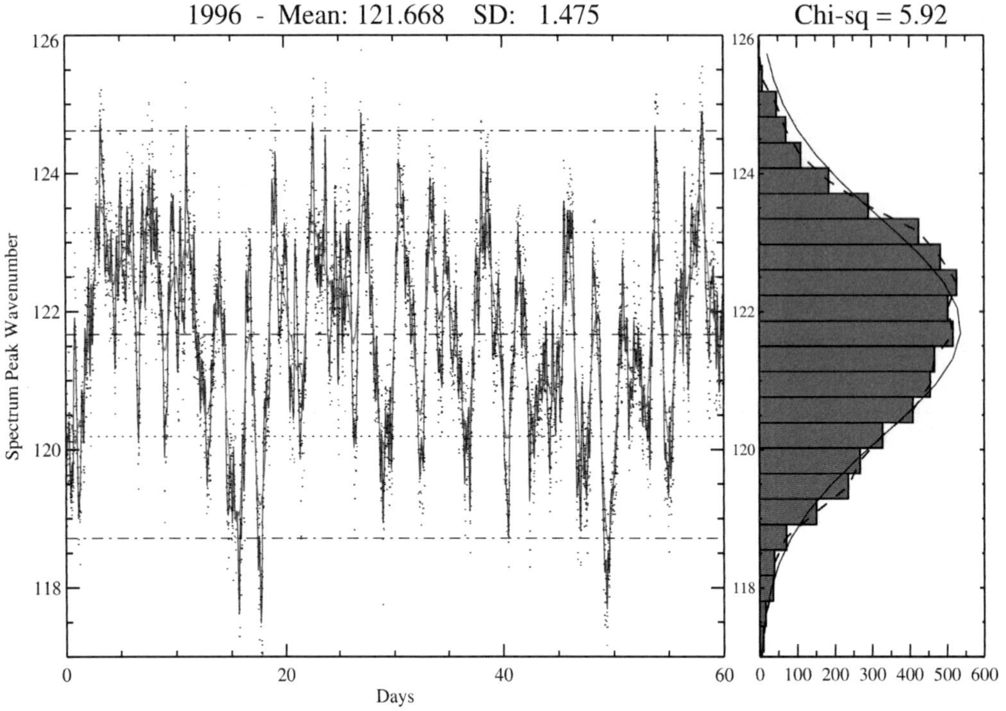

Figure 1. Time-series showing the variation of the peak wavenumber of the fit to the supergranule spectral feature derived from each Doppler image within the 1996 dataset. Each dot represents the peak wavenumber for each spectrum. The red and blue lines are the results of smoothing with a boxcar and a Savitzky-Golay filter, respectively. The time-series values are collected into wavenumber bins producing a histogram that is fitted with a Normal distribution. The histogram shows a slight deviation from a Normal distribution of values.

signals that suggest size fluctuations of between 3-5 days.

We have also constructed Lomb-Scargle periodograms [8] from all the time-series to test the significance of the strong features. The results from these studies do not contradict those from the FFT analyses.

We conclude that correlations exist in time-series of the global average of supergranule sizes, indicating non-random process underlying the evolution of supergranulation. Coherence is observed as peaks within frequency distribution plots that give fluctuation periods on the order of around twice the turnover time of the supergranule pattern. We have also calculated the Hurst exponents [9] for each data-series to be $\sim$0.9, which suggests underlying trends.

Analyses of more datasets, e.g. the FWHM and velocity data [2], MDI data from the intervening years covering the solar cycle between 1996 and 2008, and a longer simulated time-series, should lead to a more rigid conclusion. The data simulations show, however, that the fluctuations are solar in origin and not instrumental effects. Using the observations of these fluctuations, time-dependent properties of these simulations can now be further constrained.

With the Helioseismic Magnetic Imager aboard the Solar Dynamics Observatory now delivering 4096$\times$4096 resolution Dopplergrams every 45 seconds, it will be interesting to see whether these fluctuations appear in this new data.

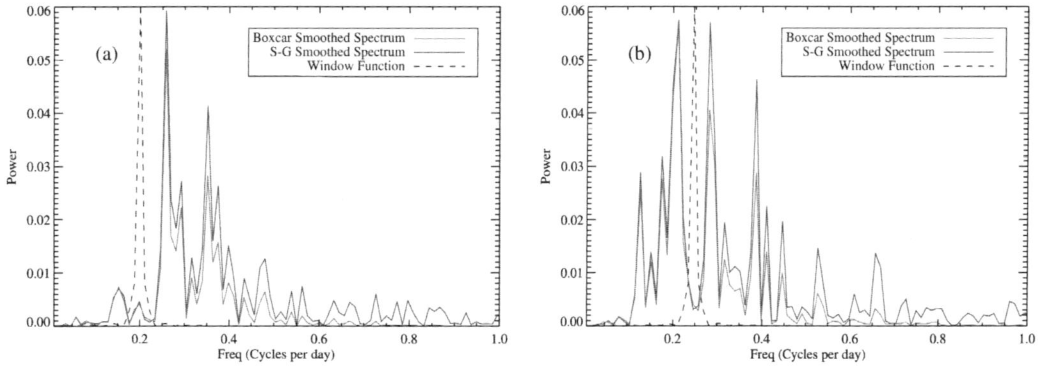

Figure 2. Frequency distributions derived from the (a) 1996 data and (b) 2008 data showing the results of FFTs applied to both of the filtered and reduced time-series, along with the window function.

Acknowledgments

Peter Williams is supported by the Solar Dynamics Observatory via the NASA Postdoctoral Program, managed by Oak Ridge Associated Universities of Oak Ridge, Tennessee. SOHO is a project of international cooperation between ESA and NASA. The authors wish to thank John Beck at Stanford University for producing the de-rotated Doppler images.

References

[1] Williams, P. E., & Pesnell, W. D. 2010a, in *ASP Conf. Series 428, SOHO-23: Understanding a Peculiar Solar Minimum*, ed. S. R. Cranmer, J. T. Hoeksema, & J. L. Kohl (San Francisco: ASP), 127

[2] Williams, P. E., & Pesnell, W. D. 2010b, *Solar Physics*, submitted

[3] Scherrer, P. H. et al. 1995, *Solar Physics*, **162**, 129

[4] Hathaway, D. H. 1988, *Solar Physics*, **117**, 1

[5] Hathaway, D. H. 1992, *Solar Physics*, **137**, 15

[6] Press, W. H., Flannery, B. P., Teukolsky, S. A., Vetterling, W. T. 1995, *Numerical Recipes: The Art of Scientific Computing* (New York: Cambridge University Press)

[7] Hathaway, D. H., Williams, P. E., & Cuntz, M. 2006, *ApJ*, **644**, 598

[8] Horne, J. H., & Baliunas, S. L. 1986, *ApJ*, **302**, 757

[9] Mandelbrot, B. B., & Wallis, J. R. 1969, *Water Resources Research*, **5**, 2, 321

SONG – getting ready for the prototype

F Grundahl[1], J Christensen-Dalsgaard[1], U Gråe Jørgensen[2,3], S Frandsen[1], H Kjeldsen[1] and P Kjærgaard Rasmussen[2]

[1]Department of Physics and Astronomy, Aarhus University, Ny Munkegade, 8000 Aarhus C, DK

[2] Niels Bohr Institute, University of Copenhagen, Juliane Maries Vej 30, 2100 Copenhagen, DK

[3] Centre for Star and Planet Formation, University of Copenhagen, Geological Museum, Øster Voldgade 5-7, 1350 Copenhagen, DK

E-mail: fgj@phys.au.dk

Abstract. The Stellar Observations Network Group, SONG, is a project which aims at building a network of eight identical telescopes distributed geographically around the globe to allow long-term, high-duty-cycle observations of stellar oscillations and to search for exoplanets via the microlensing technique. At each of the network nodes a 1 m telescope with a high-resolution spectrograph and two lucky-imaging cameras is placed. The instruments and telescope, for the prototype node, are currently being built and installation at Observatorio del Teide, Tenerife, Spain is foreseen for early 2011.

1. Introduction

It is well known that observing solar-like oscillations from the ground is a very challenging task. Perhaps the most difficult aspect is organizing the access to observatories spread across the globe, which can provide the long, continuous, time coverage needed for such observations. The Stellar Observations Network Group – SONG – is an initiative to build a global network of 8 identical 1 m telescopes, dedicated to time-domain astronomy within the fields of asteroseismology and exoplanet studies. In order to do this we have obtained funding for designing and building the prototype node for such an observatory. Currently, autumn 2010, we are completing the instruments and the telescope and observatory site is being prepared for "first light" in the spring of 2011. Previous reports on SONG can be found in Grundahl et al. (2007, 2008, 2009).

The scientific goals of SONG are to: 1) study the interiors of stars using asteroseismology and 2) search for, and characterize, extra-solar planets via the radial-velocity and microlensing methods. These two goals put some clear requirements on the instruments needed for the network, and we optimize these such that the scientific goals can be met. Specifically we will study stellar oscillations using radial-velocity observations with a spectrograph designed with the goal of reaching 1 m/s precision for slowly rotating solar-like stars. The targets for asteroseismology will be among the brightest stars in the sky allowing us to use a relatively small telescope. Furthermore the bright stars are also well studied with other techniques, such that their basic parameters: distance, radius luminosity, effective temperature and surface composition are well known. Such detailed information is needed for confronting properly the stellar models with observed oscillation spectra. For the exoplanet observations, the main mode will be imaging observations of microlensing candidates towards the galactic bulge region. This requires high spatial resolution due to the high level of crowding in these regions, and to this end we use the lucky-imaging method (Baldwin et al. 2001). The radial-velocity measurements from the asteroseismic observations can naturally also be used to search for planets orbiting the stars under study.

In addition to these two primary science goals, it is our aim to observe the solar oscillations during daytime. This will be accomplished by installing a window in the dome which allows the telescope to be pointed towards the blue daytime sky and feeding the spectrograph with scattered sunlight, see e.g. Kjeldsen et al. (2008). The purpose of these observations is not only to study the "sun-as-a-star", but also to monitor the performance of the spectrograph, e.g., by comparing with results from ground-based solar networks and space-based observatories.

We note that, once the network is fully developed, it will be possible also to carry out time-critical observations such as photometry and spectroscopy of exoplanet transits, Doppler imaging of active stars, etc.

2. The nodes

We refer to the individual observatories in the network, as nodes. Our aim is to create a network with at least 8 nodes, four in each hemisphere, in order to have a good duty-cycle for any position in the sky. Each node should be placed at already existing observatory sites to avoid building extensive infrastructure.

For a network of 8 telescopes it is not feasible to carry out manned observations; therefore all functions of the observatory should be under computer control. An overall scheduling programme will take care of sending observing requests to the individual nodes which then carry out these when a number of constraints are fulfilled. The instruments must be capable of automatically acquire the desired object in the field of view or on the spectrograph slit, autoguide, and maintain an optimum focus.

Each node will carry two primary instruments: imaging camera(s) at the Nasmyth focus and a high-resolution spectrograph located at a Coudé focus. The Coudé focus is placed in an insulated shipping container next to the telescope pier. A weather station and cloud monitor is attached to the container.

The first node, for the prototype, is located at the Observatorio del Teide on Tenerife in collaboration with Instituto de Astrofísica de Canarias. It will be placed at the site where the STARE telescope (Charbonneau et al. 2000) was located.

2.1. Telescope

The telescope for SONG has an aperture of 1 m and a focal length of 36 m in order to provide an image scale adequate for lucky imaging (see below). It is located on an alt-az mount such that the same design can be used at all nodes. Two Nasmyth instrument ports are available; initially only one will house instruments, but the tertiary mirror is mounted on a rotating platform that allows switching between the two sides in 60 seconds.

The telescope will have a blind pointing of 5 arcseconds (rms), and is equipped with powerful direct-drive motors to allow rapid pointing. It is important to have a high image quality for lucky imaging, so the telescope has a built in Shack-Hartmann wavefront sensor which is used actively to control the thin (5cm) primary mirror.

The dome has a diameter of 5 m and is equipped with ventilation ports, which can be opened during night time, and a cooling unit which operates during the day to minimize heating of the telescope mirror and structure. To facilitate the daytime observations of the Sun, a window will be installed in the dome, such that the telescope can point to the blue sky.

2.2. Spectrograph

The main instrument for the asteroseismic observations is the spectrograph. In order to achieve the high-precision radial velocities needed, an iodine cell will be used as wavelength reference, and we have taken a number of steps in the design to allow us to reach a precision of 1 m/s for the brightest targets. The spectrograph is designed to cover wavelengths between 4800 Å and 6700 Å. Although this is a fairly limited range, it fully covers the region of the iodine absorption spectrum and, importantly, it allows us to use off-the-shelf optical components with very high efficiencies (99%), which helps to maintain a high throughput.

The spectrograph is equipped with 7 slits, allowing resolutions between 60.000 and 180.000 – these are mounted on a motorized stage to allow easy change. For obtaining precise velocities it is highly important that the slit illumination is stable. To this end, we have included a tip/tilt mirror for correcting the input to the slit. Our targets are bright so there is ample light available. Furthermore the Coudé path also includes a facility which continuously monitors the telescope pupil and feeds signals to a mirror on a piezo stage, to ensure a stable pupil location.

The spectrograph pre-slit table is equipped with flat-field and ThAr calibration lamps, as well as the iodine cell. It is important to note that these items can be inserted and removed from the beam as demanded by the application. Thus SONG will also be capable of observing without the iodine cell in a "normal" mode of operation.

The entire spectrograph is mounted on an optical table measuring 900 mm×1200 mm (Fig. 1), which is enclosed in an insulated box. For thermal control, we keep the entire container volume below a fixed temperature and then use heating elements in the insulating box to keep the spectrograph at a few degrees higher temperature. As our detector we use a 2K×2K CCD system from Andor which can read out in less than 5s with a readout noise below 8 electrons, as measured in our laboratory. For time-series applications it is obviously an advantage to be able to read so quickly with low noise. The size of the detector does not cover the full extent of the spectral orders at wavelengths longer than 5350 Å. The focusing camera has been designed to allow this if a larger CCD is used.

All spectra will be stored, flat-fielded and extracted on-site, and subsequently transferred to a central site for further processing (velocity extraction). We are using the REDUCE package by Piskunov & Valenti (2002) for the basic reductions and our own IDL-based code for the velocity measurement.

2.3. Imaging

For the microlensing observations SONG will employ two lucky-imaging (LI) cameras (Andor DU897 models) located at one of the two Nasmyth foci. Baldwin et al. (2001) has illustrated the use of LI. We decided to use two such cameras in order to increase the wavelength coverage, essentially doubling the number of measurements. The wavelength split between the two cameras, using a beamsplitter, is at 6500 Å. The lucky-imaging method should work extremely well on a 1 m telescope at a good site – there is already many reports showing very good results with 2-3 m class telescopes. The ratio between the telescope diameter and typical turbulence scale, D/r_0, is then correspondingly more favourable for the SONG telescope. This implies that a significant improvement in image quality can be obtained, for a large fraction of the available images. To accommodate this, the pixel sampling of the LI cameras is 0.09 arcsecond per pixel (for 2 pixel sampling of the diffraction limit). The field of view is 45 arseconds square – while quite small it is sufficient for our purposes. Since we may wish to observe targets at high zenith distances an atmospheric dispersion corrector is permanently installed in the beam.

In order to obtain the best performance for LI we must be able to focus the telescope very well – for this purpose a small fraction of the light at the Nasmyth focus is reflected to a small camera which will monitor the telescope focus during the observations, and automatically correct any observed drift.

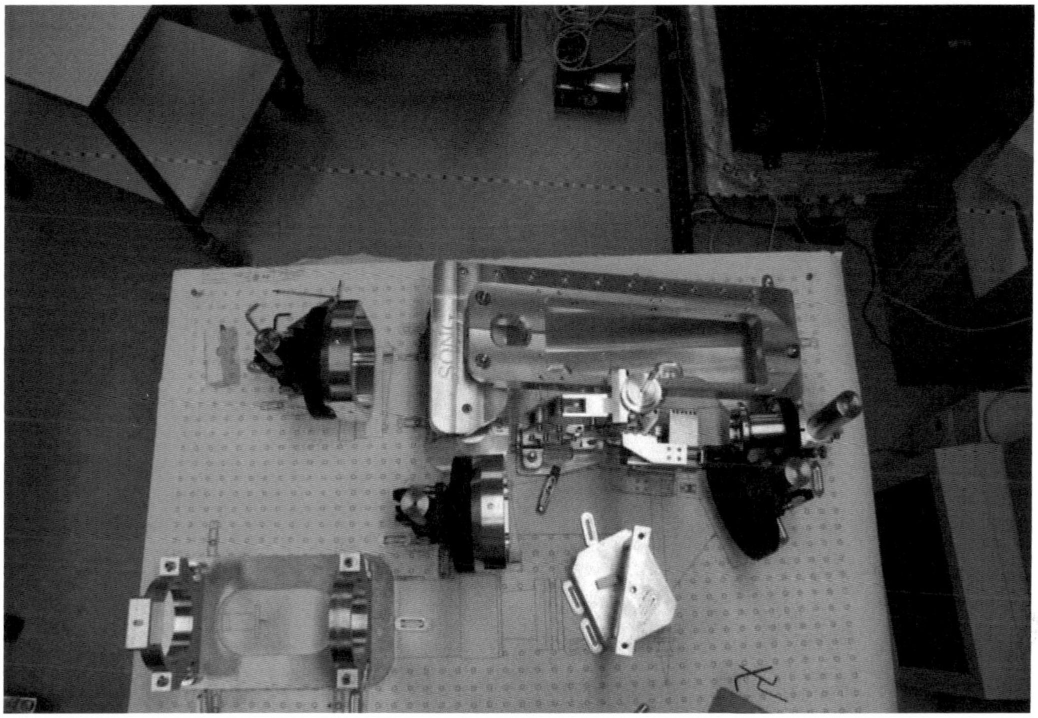

Figure 1. Test assembly of the spectrograph without optics and before anodizing. At the upper right hand corner the mount for the echelle grating can be seen. Mountings for the collimator mirrors, cross-disperser slit and spectroscopic are also present.

Each of the two LI cameras is equipped with a filter wheel with four positions. The two cameras can operate independently, having different frame-rates.

3. Status of the prototype development

There is currently (autumn 2010) a lot of activities ongoing at Aarhus and Copenhagen Universities. The mechanical parts for the instruments are nearly completed, with only minor components missing, and the assembly and integration of the spectrograph is starting. A test container is installed in Aarhus, where we will mount all items and do system integration and testing, before shipping instruments to Tenerife. The software for controlling the prototype, and ultimately the network is being developed, and the setup for automatic execution of observations and copying of data to archives and databases is ready. The site at Observatorio del Teide is being prepared, and installation of the dome support structure and container is scheduled for early 2011, followed by an extensive testing period lasting until the end of 2011.

A single node does not make a network, and we are actively seeking partners and collaborators for the funding and building of more nodes for the network. In China, a group led by Professor Licai Deng (National Astronomical Observatories, Chinese Academy of Sciences) has obtained funding to build a second node in China, to be ready by 2012. Several other groups has expressed strong interest in joining SONG and are actively seeking funds.

Acknowledgments

The authors gratefully acknowledge generous financial support for the SONG project from the Villum Fonden, Carlsbergfondet and The Danish Council for Independent Research | Natural

Sciences

References

Baldwin J E, Tubbs R N, Cox G C, Mackay C D, Wilson R W and Andersen M I 2001 *A&A.* **368** L1-L4

Charbonneau D, Brown, T M, Latham, D W and Mayor M 2000, *ApJ* **529**, L45

Grundahl F, Christensen–Dalsgaard J, Arentoft T, Frandsen S, Kjeldsen H, Jørgensen U G and Kjaergaard P 2008 *Commumications in Asteroseismology.* **157** 273

Grundahl F, Christensen–Dalsgaard J, Kjeldsen H, Jørgensen U G, Arentoft T, Frandsen S and Kjærgaard P 2009 *Solar-Stellar Dynamos as Revealed by Helio- and Asteroseismology: GONG 2008/SOHO 21* 11–15 August 2008, Boulder, Colorado USA, (ASP Conf. Ser. vol 416) ed M Dikpati et al (San Francisco: ASP) p 579

Grundahl F, Kjeldsen H, Christensen–Dalsgaard J, Arentoft T and Frandsen S 2007 *Communications in Asteroseismology.* **150** 300

Kjeldsen H, Bedding T R, Arentoft T, Butler R P, Dall T H, Karoff C, Kiss L L, Tinney C G and Chaplin, W J 2008 *ApJ* **682** 1370

Piskunov N E, Valenti J A 2002 *A&A.* **385** 1095

PLATO : PLAnetary Transits and Oscillations of stars

Claude Catala

Observatoire de Paris, LESIA, 5 place Jules Janssen, Meudon, France

E-mail: `claude.catala@obspm.fr`

Thierry Appourchaux

Institut d'Astrophysique Spatiale, Université Paris-Sud, Orsay, France

E-mail: `thierry.appourchaux@ias.u-psud.fr`

and the PLATO Mission Consortium

Abstract. PLATO is a M-class candidate in the ESA Cosmic Vision program. PLATO's objective is to characterize exoplanets and their host stars in the solar neighbourhood. While it builds on the heritage from CoRoT and *Kepler*, the major breakthrough will come from its strong focus on bright targets ($m_V \leq 11$). The PLATO targets will also include a large number of very bright ($m_V \leq 8$) and nearby stars.

The prime science goals of PLATO are: (i) the detection and characterization of exoplanetary systems of all kinds, including both the planets and their host stars, reaching down to small, terrestrial planets in the habitable zone; (ii) the identification of suitable targets for future, more detailed characterization, including a spectroscopic search for bio-markers in nearby habitable exoplanets.

These ambitious goals will be reached by ultra-high precision, long (few years), uninterrupted photometric monitoring in the visible of very large samples of bright stars, which can only be done from space. The resulting high quality light curves will be used on the one hand to detect planetary transits, as well as to measure their characteristics, and on the other hand to provide a seismic analysis of the host stars of the detected planets, from which precise measurements of their radii, masses, and ages will be derived.

The PLATO space-based data will be complemented by ground-based follow-up observations, in particular very precise radial velocity monitoring, which will be used to confirm the planetary nature of the detected events and to measure the planet masses.

The full set of parameters of exoplanetary systems will thus be measured, including all characteristics of the host stars and the orbits, radii, masses, and ages of the planets, allowing us to derive planet mean densities, and estimate their temperature and radiation environment. Finally, the knowledge of the age of the exoplanetary systems will allow us to put them in an evolutionary perspective.

1. Main science goals

PLATO will address the question of the existence, distribution, evolutionary state, and characteristics of exoplanets in the solar neighbourhood. Answers to these questions are essential to understand how planetary systems, including our own, are formed and evolve, and also as a first and necessary step to understand whether life can exist elsewhere in the Universe, and locate potential sites for life. Since the discovery of the first exoplanet in 1995, this field has seen a remarkable development, with about 400 exoplanets known as of the end of October, 2009. Most of these objects are giant planets in close-in orbits, but continuous progress in the precision of radial velocity observations is now enabling the detection of "Super-earths", with masses just a few times that of the earth.

The field of exoplanet search has been recently boosted by the launch of the CoRoT satellite in Dec. 06 [1], followed by that of *Kepler* in March 09 [2]. The discovery of CoRoT-7b, the very first small telluric, rocky planet with measured radius and mass, and therefore with a known density, has opened up a new era, in which the CoRoT extended mission and *Kepler* will now play a major role.

Both CoRoT and *Kepler* target faint stars, up to $m_V = 15$ and beyond, which makes their ground-based follow-up difficult, in particular in radial velocity monitoring. As a consequence, ground confirmation and mass measurements are restricted to the largest of the CoRoT and *Kepler* planets, which severely impacts the scientific return of these missions. While we can today with CoRoT, and soon *Kepler*, detect the passage of a planet the size of our own world, it is impossible to confirm the presence of any such object found by either spacecraft. Moreover, even in cases where radial velocities can be measured to the required precision to confirm the planetary nature of the detected event and measure the planet-to-star mass ratio, our knowledge of the faint host stars is still too poor to allow us to derive estimates of the planet radii, masses and ages to a sufficient accuracy to significantly constrain their structure and state of evolution.

The goal of PLATO is to alleviate these difficulties by focusing on bright stars, 3 to 4 magnitudes brighter than CoRoT and *Kepler*, and also by including in its target list a large sample of very bright ($m_V \leq 8$) and nearby stars. This will bring three decisive advantages: (i) the groundbased follow-up observations will be highly facilitated, and the required precision will be reached to confirm small, terrestrial planets in the habitable zone and to measure their masses; (ii) the host stars of the detected planets will be studied in detail, *in particular via seismic analysis* using the PLATO data themselves; seismic analysis, i.e. the measurement of stellar oscillations, will be used to probe the internal structure of these stars, and determine their radii, masses, and age in a precise and reliable way; (iii) the detection of exoplanets orbiting very bright and nearby stars will provide the best targets for subsequent detailed follow-up observations, both from space (e.g. JWST) and from the ground (e.g. E-ELT), including in particular spectroscopy of their surfaces and atmospheres, in the search for bio-markers.

The main goal of PLATO is to detect and characterize a significant number of exoplanets through their transits in front of their host stars, and provide measurements of their radii and masses, hence of their mean densities. It will also provide estimates of the ages of the detected exoplanetary systems. The PLATO strategy to reach this objective relies on the detection and measurement of photometric transits, which can be used to derive the ratio R_p/R_* of the planet radius to that of its host star. Follow-up observations from the ground will be used to measure the mass ratio M_p/M_* (actually the mass function). The planet radius and mass can be determined if the radius and mass of its host star are known.

The precision on planet parameters that is needed for constraining its internal composition is of the order of 10% for the mass and 5% for the radius (see Fig. 1). Failure to reach this level of precision results in a degeneracy of planet interior models. Unfortunately, the currently achieved precision is much worse, mainly due to uncertainties on host star parameters. It is clear from Fig. 1 that seismically determined masses and radii of host stars, such as provided by PLATO are required to derive useful constraints on planetary interiors. An example of the spectacular improvement in our estimates of planet parameters when seismic analysis of the host star is performed was recently given by [3] who analyzed *Kepler* asteroseismic data of HAT-7a, a cool dwarf which is harbouring a giant planet.

In addition, seismic analysis of the planet host stars will allow us to determine the ages of exoplanetary systems, and therefore to place exoplanets in an evolutionary context.

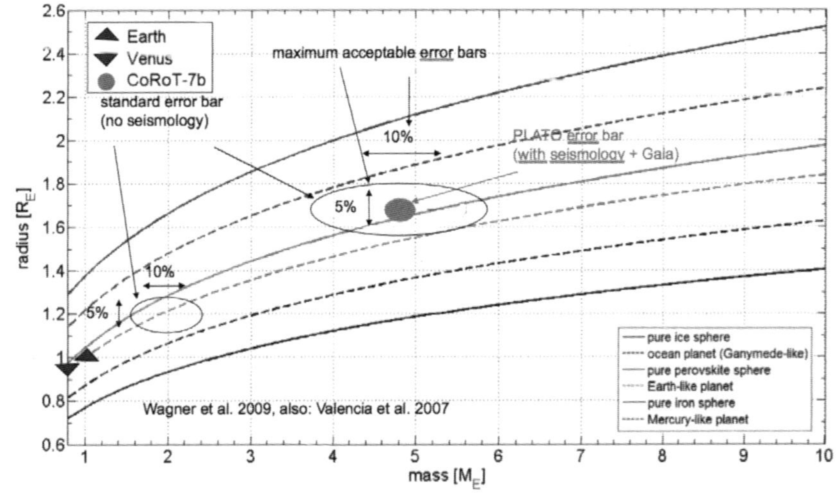

Figure 1. Comparison of planet mass and radius with model predictions. Current precision on planet mass and radius are not decisive for constraining the internal composition of planets, while measurements with PLATO will represent a qualitative jump, and provide stringent constraints on the models

2. Observation strategy

The main PLATO science product will be a very large sample of ultra-high precision stellar light curves in the visible, obtained on very long time intervals (up to 3 years) with very high duty cycle (≥95%). The requirement is to obtain a photometric precision better than 2.7×10^{-5} in 1 hr for more than 20,000 cool dwarfs/subgiants brighter than $m_V \approx 11$. This photometric precision corresponds to about 1 ppm in one month, and is sufficient to detect and measure solar-like oscillations in these stars, detect the transit of planets with the size of the Earth in their photometric lightcurves, and measure the transit characteristics with high precision.

PLATO will also monitor more then 1,000 cool dwarfs brighter than $m_V \approx 8$ for 2 to 3 years. In addition to this main sample, PLATO will obtain lightcurves with a precision better than 8×10^{-5} in 1 hr for 250,000 stars down to $m_V \approx 13$. Finally, the PLATO star sample will also include about 10,000 M dwarfs down to $m_V \approx 16$, observed with a photometric noise level lower than 8×10^{-4} in 1 hr, enabling the detection of transits from small terrestrial planets.

In order to reach these goals, PLATO will observe two successive very wide fields, for 2 to 3 years. These two long monitoring sequences will be followed by a one- or two-year step&stare phase, during which a number of fields will be monitored for several months each. This step&stare phase will bring flexibility to the mission, allowing for instance to survey a very large fraction of the whole sky (up to 50%), as well as to re-visit particularly interesting targets identified during the long monitoring phases.

The spacecraft is intended to be launched in 2018 on a Soyuz-Fregat rocket for injection into an orbit around the L2 Lagrangian point for a nominal lifetime of 6 years, with a possible extension of 2 years, which is compatible with the observation strategy outlined above.

3. Payload

PLATO is presently under study, with the payload being designed by the PLATO Mission Consortium, while the spacecraft is subject to two parallel industrial studies. The payload concept provides at the same time a very wide field-of-view, and a large overall collecting area. Wide fields-of-view are required to obtain large samples of bright stars, while large collecting areas are necessary to reach the desired photometric precision.

The PLATO payload includes a set of 34 identical fully 6-lens refractive cameras, with 120mm pupil and field-of-view of about 37°. Each camera has its own CCD-based focal plane, with four $4{,}510^2 \times 18\mu m$ pixel CCDs. Two out of 34 cameras are devoted to very bright stars with m_V between 4 and 8, and have their CCDs operated in frame transfer mode at a cadence of 2.5 sec (fast cameras), while the remaining 32 cameras observe stars fainter than m_V=8, with their CCDs working in full frame mode at a cadence of 25 sec (normal cameras). The total dynamical range is $4 \leq m_V \leq 16$. The two fast cameras will observe through broad band filters.

The light and centroid curves from each camera are computed on board, then transmitted to the ground, where they are co-added to reach the desired precision.

The 32 normal cameras are grouped in 4 sections of 8 cameras, each section having its line of sight offset from the next one by 35% of the field-of-view. This overlapping line-of-sight arrangement is depicted in Fig. 2 and results in an overall instrument field-of-view of about 50° diameter, or about 2,200 deg^2, each star being observed either by 8, 16, 24, or 32 telescopes, with an effective collecting area of 0.09, 0.18, 0.27 or 0.36 m^2, depending on its position in the field. The surveyed area after two long pointings will thus be 4,400 deg^2, while it will go up to 22,000 deg^2 (i.e. 50% of the whole sky) when a two-year step&stare phase is added during which 4 fields per year are monitored for 3 months each.

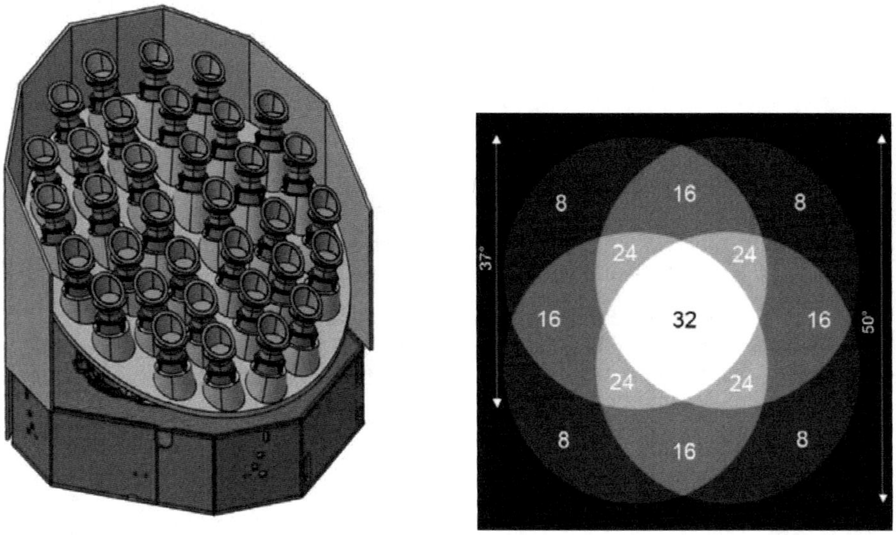

Figure 2. The PLATO payload concept. *Left:* global view; *Right:* overlapping line-of-sight concept; numbers of cameras monitoring the various parts of the field are indicated

4. Performances

A full end-to-end simulator was used to estimate the expected level of noise of the instrument. It takes into account all sources of noise, including photon noise, confusion by neighbouring sources, readout noise, satellite jitter, etc, by including all characteristics of the observed field (star positions and magnitudes, zodiacal light, etc) and of the instrument (optical PSF, detector characteristics, etc).

Fig. 3 presents these results for the region of the field observed by all 32 normal cameras, and shows that photon noise level is approached closely at magnitudes brighter than m_V=11, and that non photonic noise remains well below photon noise at least down to m_V=12. A level of noise of 2.7×10^{-5} in 1 hr is reached down to m_V=10.3, while a noise of 8×10^{-5} in 1 hr is obtained around m_V=12.3. At fainter magnitudes, the noise becomes significantly higher than pure photon noise, due to the contribution of contaminating sources, but remains below 8×10^{-4} in 1 hr down to m_V=15.5. Fig. 3 also shows the results expected from the two fast cameras.

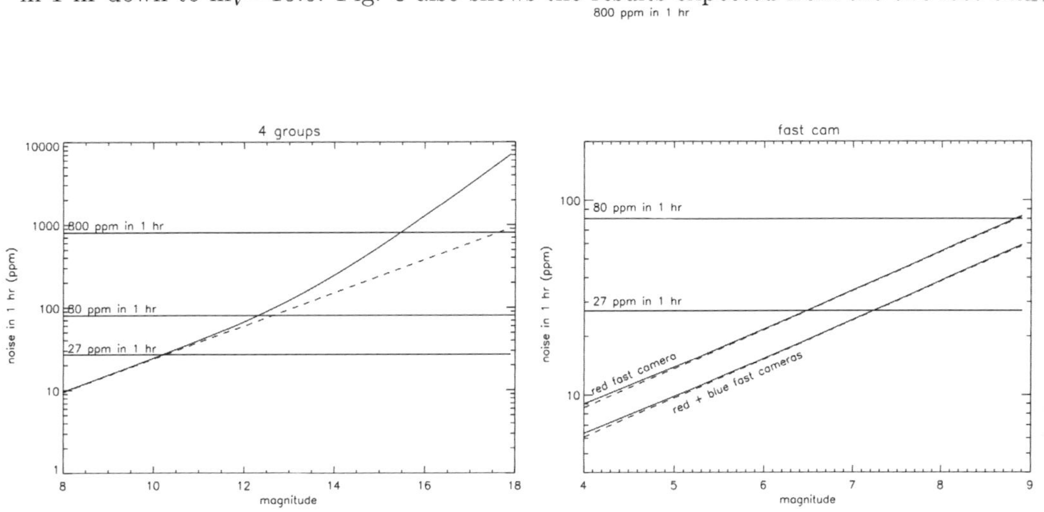

Figure 3. Expected noise level for PLATO instrument. Dashed line: photon noise only. Full line: all sources of noise included. Top: for all 32 normal cameras; Bottom: for the two fast cameras

Table 1 shows the expected numbers of dwarfs/subgiants later than F5, observable with PLATO and *Kepler* at various levels of noise, and down to various magnitudes. The three PLATO concepts are considered in this table, concept C corresponding to the PPLC design presented here. The gain compared to *Kepler* is clearly shown in this table, indicating in particular that PLATO will observe as many stars down to 8th magnitude as *Kepler* down to 11th.

The comparison with *Kepler* is further illustrated by Fig. 4, showing the expected total number of detectable planets as a function of mass and orbital semi-major axis, for both missions. The numbers indicated are those of the expected detectable planets by both missions, while the sizes of the coloured regions show the respective discovery potential of both missions (blue: PLATO and green: *Kepler*). There is no underlying planet formation model, instead we simply assume that each star has one (and one only) planet in the parameter range considered. The planet is considered detectable if it can be seen in transit by the satellite AND confirmed by follow-up radial velocity measurements with a reasonable amount of telescope time.

These results take into account all sources of noise for the radial velocity follow-up (oscillation, granulation, activity level), limiting the required observing time to reasonable values. Future developments in our understanding of the interplay between activity level and induced radial-velocity jitter might help correct for the spurious effect and further improve our characterization ability.

surveyed area	PLATO 4496 deg^2		*Kepler* 100 deg^2	
noise level (ppm in 1hr)	# cool dwarfs & subgiants	mag range	# cool dwarfs & subgiants	mag
27	22,000	8.7 – 10.2	1,300	11.2
80	345,000	11.0 – 12.2	25,000	13.6
magnitude	# stars		# stars	
6	90		0	
8	1,326		30	
9	4,500		100	
10	16,600		370	
11	58,000		1,300	

Table 1. Numbers of targets from PLATO and *Kepler*, at various photometric noise levels and various magnitude limits. All star numbers refer to cool dwarfs and subgiants only. Numbers for PLATO correspond to the two long monitoring phases (2 to 3 years) only.

As can be seen in Fig. 4, we expect a vastly increased number of planet detections with PLATO , in particular in the lower mass range, where *Kepler* is not expected to produce confirmable planets. The main reason is related to the much larger field of view of PLATO , and consequently the brightness of the candidates. The main conclusion is that, although *Kepler* will bring considerable progress, it is unlikely that it can detect unambiguously small planets in far-away orbits, mainly due to the faintness and the limited number of its targets. Only PLATO with its extended surveyed area and its main focus on bright cool dwarfs, will allow us to reach real earth analogue systems and extend the search for exoplanets to small terrestrial planets in the habitable zone of their stars.

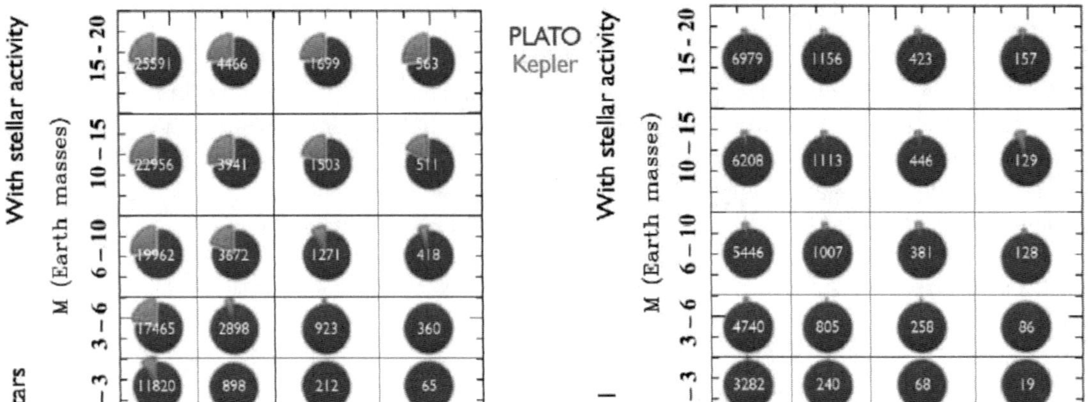

Figure 4. Estimated total numbers of detected transiting planets, which can be confirmed by ground-based radial velocity observations, for PLATO (in the PPLC concept) and *Kepler*. Left: for all stars; right: for stars with mV $\leq$11 only (see text for details).

5. Conclusion

PLATO will represent a real breakthrough in our study of exoplanetary systems, in particular by providing a full census of exoplanets in the solar neighbourhood, down to low-mass, earth-size planets in the habitable zone. Thanks to its capacity to determine in detail all characteristics of the planet host stars, using seismic analysis based on PLATO data themselves, it will allow us to measure the radii, masses and ages of the planets to unprecedented accuracies, and will also identify suitable targets for future more direct characterization facilities. PLATO will reach these ambitious goals by focusing its observations on a large sample of bright stars.

References

[1] Baglin, A., *et al.* 2006, *The CoRoT Mission, Pre-Launch Status*, ESA SP-1306, Eds. M. Fridlund, A. Baglin, J. Lochard, L. Conroy, p. 11-538

[2] Borucki, W.J., *et al.* 2003, The Kepler Mission: Finding the Sizes, Orbits and Frequencies of Earth-size and Larger Extrasolar Planets, *ASP CS*-294, p. 427

[3] Christensen-Dalsgaard, J.; Kjeldsen, H.; Brown, T. M.; Gilliland, R. L.; Arentoft, T.; Frandsen, S.; Quirion, P.-O.; Borucki, W. J.; Koch, D.; Jenkins, J. M., 2010, ApJ 713, L164

The Stellar Imager (SI) – A Mission to Resolve Stellar Surfaces, Interiors, and Magnetic Activity

Jørgen Christensen-Dalsgaard[1], Kenneth G Carpenter[2], Carolus J Schrijver[3], Margarita Karovska[4] , and the SI Team

[1]Department of Physics and Astronomy, Aarhus University, Denmark
[2]Code 667 NASA-GSFC, Greenbelt, MD 20771 USA
[3]LMATC 3251 Hanover St., Bldg. 252, Palo Alto, CA 94304, USA
[4]60 Garden St., Cambridge, MA 02138, USA

Emails: jcd@phys.au.dk, Kenneth.G.Carpenter@nasa.gov, schryver@lmsal.com, karovska@head.cfa.harvard.edu

Abstract. The Stellar Imager (SI) is a space-based, UV/Optical Interferometer (UVOI) designed to enable 0.1 milli-arcsecond (mas) spectral imaging of stellar surfaces and of the Universe in general. It will also probe via asteroseismology flows and structures in stellar interiors. SI will enable the development and testing of a predictive dynamo model for the Sun, by observing patterns of surface activity and imaging of the structure and differential rotation of stellar interiors in a population study of Sun-like stars to determine the dependence of dynamo action on mass, internal structure and flows, and time. SI's science focuses on the role of magnetism in the Universe and will revolutionize our understanding of the formation of planetary systems, of the habitability and climatology of distant planets, and of many magneto-hydrodynamically controlled processes in the Universe. SI is a "Landmark/Discovery Mission" in the 2005 Heliophysics Roadmap, an implementation of the UVOI in the 2006 Astrophysics Strategic Plan, and a NASA Vision Mission ("NASA Space Science Vision Missions" (2008), ed. M. Allen). We present here the science goals of the SI Mission, a mission architecture that could meet those goals, and the technology development needed to enable this mission. Additional information on SI can be found at: `http://hires.gsfc.nasa.gov/si/`.

1. Science goals of the mission

The Stellar Imager (SI) is a NASA Vision Mission (VM), developed by an international team, to study 1) solar and stellar magnetic activity and their impact on space weather, planetary climates, and life and 2) magnetic processes in general and the roles they play in the origin and evolution of its structure and the transport of matter throughout the Universe

SI's primary science goal will be addressed by observing and measuring spatial and temporal stellar surface magnetic activity patterns through ultra-high angular resolution (sub-milliarcsec) UV imaging, and by measuring via disk-resolved asteroseismology the internal structure and flows that produce it, in a sample of stars covering a broad range of masses, radii, and activity levels. These observations will lead to an improved understanding of the underlying dynamo process(es) and thus enable improved forecasting of solar (and stellar) activity on time scales of days to centuries. This, in turn, will facilitate an improved understanding of the impact of stellar magnetic activity on life on earth and on exo-planets found around more distant stars.

1.1. Key Questions

There are a number of important questions that SI needs to address in order to achieve its goal of understanding dynamos and magnetic activity, including:

- what do the internal structure and dynamics of magnetically active stars look like?
- what sets the dynamo strength and pattern in individual stars, from dwarfs to supergiants?
- how can active stars form polar spots?
- what can we expect next from the Sun, on time scales from hours to centuries?
- why do 2 in 3 Sun-like stars show no cycles?
- what causes solar-type "Maunder minima" or "grand maxima"?
- how does stellar activity drive all aspects of "space weather" and affect planetary climates and life around solar-type and evolved stars?
- how do dynamos evolve with time?
- how do dynamos differ in dwarf vs. giant stars?

Only with the answers to such questions will it be possible more fully to constrain theoretical dynamo models and enable true forecasts of future solar and stellar magnetic activity. These questions will be addressed by spatially resolving stellar disks to map evolving atmospheric activity as a tracer of dynamo patterns and by asteroseismic probing (to at least degrees of order 60) of internal stellar structure and flows in stars of various masses, radii, and activity levels. Such a "population study" will provide answers far more rapidly than by continuing our close-up observations of the Sun over many decades as we observe it moving through the multiple and different activity cycles that are needed to obtain a full set of observational constraints - and some of these data would never be obtainable from the Sun alone, as it is only one example of how dynamos operate and magnetic activity is produced.

2. Proposed mission architecture

2.1. Design requirements

SI's science goals require it to have the following capabilities:

- Wavelength coverage: 1200-6600 Å
- access to UV emission lines from Ly-alpha 1216 Å to Mg II 2800 Å for stellar surface imaging
    - Important diagnostics of most abundant elements
    - much higher contrast between magnetic structures and background
    - smaller baselines (UV saves a factor 2-4 vs. optical, active regions 5 times larger)
    - ~10 Å UV pass-bands, to isolate, e.g., C IV (formed at 100,000 K) & Mg II h&k (10,000 K)
- broadband, near-UV or optical (corresponding to temperatures of 3,000-10,000 K) for high temporal resolution, spatially-resolved asteroseismology to resolve internal structure
- angular resolution of 50 micro-arcsec at 1200 Å (120 mas at 2800 Å)
    - resolution of ~1000 pixels over the surface of nearby (~4 pc) dwarf stars and over the surface of the many giant and supergiant stars within ~2 kpc
- energy resolution/spectroscopy possible for detected structures
- a long-term (~10 year) mission to study stellar activity cycles: individual telescopes/hub(s) must be able to be refurbished or replaced

2.2. "Strawman" design

The VM Study developed a baseline mission design that satisfies all of the above requirements. This design is for a space-based, UV-Optical Fizeau Interferometer with 20-30 one-meter primary mirrors, mounted on formation-flying "mirrorsats" distributed over a parabolic virtual surface whose diameter can be varied from 100 m up to as much as 1000 m, depending on the angular size of the target to be observed. The individual mirrors are fabricated as ultra-smooth, UV-quality flats and are actuated to produce the extremely gentle curvature needed to focus light on the beam-combining hub that is located at the prime focus from 1 - 10 km distant. The focal length scales linearly with the diameter of the primary array, i.e., a 100 m diameter array corresponds to a focal length of 1 km and a 1000 m array to a focal length of 10 km. The typical configuration has a 500 m array diameter and 5 km focal length. A one-meter primary mirror size was chosen to ensure that the primary stellar activity targets can be well observed with good signal/noise. The mirrorsats fly in formation with a beam-combining hub in a Lissajous orbit around the Sun-Earth L2 point. The satellites are controlled to mm-micron radial precision relative to the hub and the mirror surfaces to 5 nm radial precision, rather than using optical delay lines inside the hub for fine tuning the optical path lengths. A second hub is strongly recommended to provide critical-path redundancy and major observing efficiency enhancements. The observatory may also include a "reference craft" to perform metrology on the formation, depending on which metrology design option is chosen. The VM Study identified two launch concepts that are quite feasible, assuming 1m diameter primary mirrors, with current vehicles. Depending on the number of hubs to be launched initially, one or two Delta IV launches will suffice to lift the entire observatory to Sun-Earth L2.

Additional details on the architectural concept can be found in [1] and in the complete VM report at `http://hires.gsfc.nasa.gov/si/`.

3. Technology development needed to enable the mission

The major technology challenges to building SI are:

- formation-flying of ~30 spacecraft
    - deployment and initial positioning of elements in large formations
    - real-time correction and control of formation elements
    - staged-control system (km → cm → nm)
    - aspect sensing and control to 10's of micro-arcsec
    - positioning mirror surfaces to 5 nm
    - variable, non-condensing, continuous micro-Newton thrusters
- precision metrology over multi-km baselines
    - 2nm if used alone for path length control (no wavefront sensing)
    - 0.5 microns if hand-off to wavefront sensing & control for nm-level positioning
    - multiple modes to cover wide dynamic range
- wavefront sensing and real-time, autonomous analysis and control
- methodologies for ground-based validation of distributed systems
- additional challenges (perceived as easier than the above)
    - mass-production of "mirrorsat" spacecraft: cost-effective, high-volume fabrication, integration, and test
    - long mission lifetime requirement
    - light-weight UV quality mirrors with km-long radii of curvature, using active deformation of flats
    - larger format (6 K × 6 K) energy resolving detectors with finer energy resolution (R=100)

The major challenges in this list are being attacked via a number of ground-based testbeds [2] to develop and assess precision (to the cm level) formation flying algorithms and closed-loop optical control of tip, tilt, and piston of the individual mirrors in a sparse array, based on feedback from wavefront analysis of the science data stream. The GSFC Fizeau Interferometer Testbed (FIT) is developing closed-loop optical control of a many-element sparse array, with 7-elements in Phase 1, and 18-elements in Phase 2. GSFC, MIT, and MSFC are collaborating on an experiment, the Synthetic Imaging Formation Flying Testbed (SIFFT), utilizing the MIT SPHERES hardware on the MSFC Flat Floor facility to test cm-level formation flying algorithms. The GSFC Formation Flying Testbed (FFTB) is a software simulation facility that has been used to develop deployment of array spacecraft and the multi-stage acquisition of target light from the individual mirrors by the beam-combiner. In addition, there are relevant high precision metrology development efforts at SAO [3] and JPL [4]. The ultimate goal of all these efforts is to demonstrate staged-control methodologies covering over 12 orders of magnitude, from km down to nm scales. We are also studying alternative optical designs for SI to optimize its imaging and spectral energy resolution capabilities [5].

The results from these testbeds and studies will be combined with experience from existing ground-based interferometers, to enable a small, space-based UV/Optical Pathfinder mission, which will use a small number of elements (3-5) with smaller baselines (20-50m) and frequent array reconfigurations (to fill in the Fourier uv-plane and enable high quality imaging) to both accomplish important new science and demonstrate in space the technologies needed for the full-up SI. One or more such Pathfinder missions (others are possible in the IR and X-ray as pathfinders for MAXIM/Black Hole Imager (BHI) and the Sub-millimeter Probe of the Evolution of Cosmic Structure (SPECS) will lay the ground-work for the long-baseline, Strategic "Vision" Missions that will do true high angular resolution interferometric imaging, including SI, Life Finder, and Planet Imager.

Acknowledgments

Contributions are gratefully acknowledged from the wide range of science and technology investigators and collaborators on the Stellar Imager Vision Mission Team (see [1]). This work was supported, in part, by Vision Mission Study grants from NASA HQ to NASA-GSFC and from GSFC to Smithsonian Astrophysical Observatory, Seabrook Engineering, SUNY/Stonybrook, U. Colorado/Boulder, and STScI. Substantial complementary internal institutional support is gratefully acknowledged from all of the participating institutions.

References

[1] Carpenter K G, Schrijver C J, Karovska M and the SI Vision Mission Team 2006 "The Stellar Imager (SI) Vision Mission", in *Advances in Stellar Interferometry*, eds J D Monnier, M Scholler and W C Danchi, *Proceedings of the SPIE*, 6268, pp. 626821

[2] Carpenter K G, Lyon R G, Liu A, Dogoda P, Petrone P, Mozurkewich D, Miller D, Mohan S and Stahl P 2006 "Steps Toward a UV/Optical Interferometer in Space: FIT, SIFFT, FFTB", in *BAAS*, 38, p 1129 (available from `http:/hires.gsfc.nasa.gov/si/`)

[3] Phillips J D and Reasenberg R D 2005 "Tracking Frequency Laser Distance Gauge," *Review of Scientific Instruments*, 76, 064501

[4] Lay O P, Dubovitsky S, Peters R D, Burger J P, Ahn S-W, Steier W H, Fetterman H R and Chang Y 2003 "MSTAR: a submicrometer absolute metrology system", *Opt. Lett.* 28, 890

[5] Mozurkewich D, Carpenter K G and Lyon R G 2007 "Beam Combiner for Stellar Imager and its Application to Full Aperture Imaging", in *UV/Optical/IR Space Telescopes: Innovative Technologies and Concepts III*, *Proceedings of the SPIE*, 6687, 66870K

The Solar Orbiter Mission and its Polarimetric and Helioseismic Imager (SO/PHI)

Achim Gandorfer[1], Sami K Solanki[1], Joachim Woch[1], Valentin Martínez Pillet[2,3], Alberto Álvarez Herrero[3] and Thierry Appourchaux[4]

[1] Max-Planck-Institut für Sonnensystemforschung, Max-Planck-Straße 2, D-37191 Katlenburg-Lindau, Germany

[2] Instituto de Astrofísica de Canarias, C/ Vía Láctea, s/n, E38205 - La Laguna (Tenerife), Spain

[3] Instituto Nacional de Técnica Aeroespacial, E-28850, Torrejón de Ardoz, Madrid, Spain

[4] Institut d'Astrophysique Spatiale, CNRS-Université Paris XI UMR8617, 91405 Orsay Cedex, France.

E-mail: gandorfer@mps.mpg.de

Abstract. We briefly outline the scientific and instrumental aspects of ESA's *Solar Orbiter* mission. Special emphasis is given to the Polarimetric and Helioseismic Imager, the instrument with the highest relevance for helioseismology applications, which will observe gas motions and the vector magnetic field in the photosphere at high spatial and temporal resolution.

1. Introduction

Solar Orbiter will be hopefully selected in 2011 and finally launched in 2017. Then the mission will be Europe's follow-up of the successfull SoHO observatory of ESA and NASA. Like SoHO, *Solar Orbiter* will not be particularly focussed on Helioseismology, but the mission will offer unique opportunities to study surface flows and to probe the solar dynamo. *Solar Orbiter* is more than a pure helioseismology mission: it is an integrated and complete approach to heliophysics in all senses: As an encounter mission it takes unique advantage of its orbit design. Approaching the Sun as close as 0.28 AU, and reaching heliographic latitudes of up to 34°, its suite of instruments will combine remote sensing techniques (typical for observatory like missions) with in-situ analysis of the inner heliosphere.

2. Science Goals

The fundamental science questions, around which *Solar Orbiter* is designed, can be listed as follows:

- How and where do the solar wind plasma and magnetic field originate in the corona?
- How do solar transients drive heliospheric variability?
- How do solar eruptions produce energetic particle radiation that fills the heliosphere?
- How does the solar dynamo work and drive the connections between the Sun and heliosphere?

In order to be able to answer these scientific goals *Solar Orbiter* needs to address the following science targets, namely

- to determine in-situ the properties and dynamics of plasma, fields and particles in the near-Sun heliosphere
- to investigate the detailed structure of the Sun's magnetised atmosphere
- to identify the links between activity on the Sun's surface and the resulting evolution of the corona and inner heliosphere, especially during solar co-rotation passes
- to observe and characterise the Sun's polar regions and equatorial corona from high latitudes.

3. Mission Profile

Solar Orbiter draws its unique capabilities by taking particular advantage of its special orbit characteristics.

3.1. Orbit Design

After separation from the launch vehicle, *Solar Orbiter* will start its three-and-a-half year transfer orbit. Subject to a first Gravity-Assist-Manouvre (GAM) at Venus, and two subsequent GAMs at Earth, the spacecraft will lose orbital energy, which allows *Solar Orbiter* to come close to the Sun. After a second GAM at Venus, *Solar Orbiter* begins its operational phase. From then on its orbit is in a three-to-two resonance with Venus, such that after each third orbit the inclination of the orbital plane with respect to the ecliptical plane can be increased by Venus gravity assists. This particular and unique feature gives *Solar Orbiter* access to the high latitude regions of the Sun.

While the in-situ instrument suite will be operational over the full orbit, the remote sensing instruments will be used in three distinct science phases per orbit, the perihelion passage, and the phases of maximum and minimum solar latitude.

The perihelion passage harbours the unique potential of a corotating vantage point, from which *Solar Orbiter* can follow the evolution of surface structures and solar features not only from close-by, but in addition under practically unchanged geometrical viewing conditions for several days.

This will allow the orbiter to fullfill its prime science goal, to study the magnetic coupling of the different layers of the solar atmosphere from below the surface, through the photosphere, the chromosphere, into the corona and to the inner heliosphere, where the consequences of the remotely observed magnetic surface phenomena will be sensed directly by the in-situ instrumentation package.

3.2. Instrument Suite

The *Solar Orbiter* Instrumentation can be grouped in three major packages, each consisting of several instruments:

- Field Package: Radio and Plasma Wave Analyser and Magnetometer.
- Particle Package: Energetic Particle Detector and Solar Wind Plasma Analyser
- Solar remote sensing instrumentation: Visible-light Imager and Magnetograph, Extreme Ultraviolet Spectrometer, EUV Imager, Coronagraph, and Spectrometer/Telescope for Imaging X-rays, Heliospheric Imager.

The most important mission aspect is the combination of remote observing with in-situ measurements.

The suite of instruments is listed on the official ESA project webpage (http://sci.esa.int/solarorbiter). For the sake of completeness we will repeat it here.

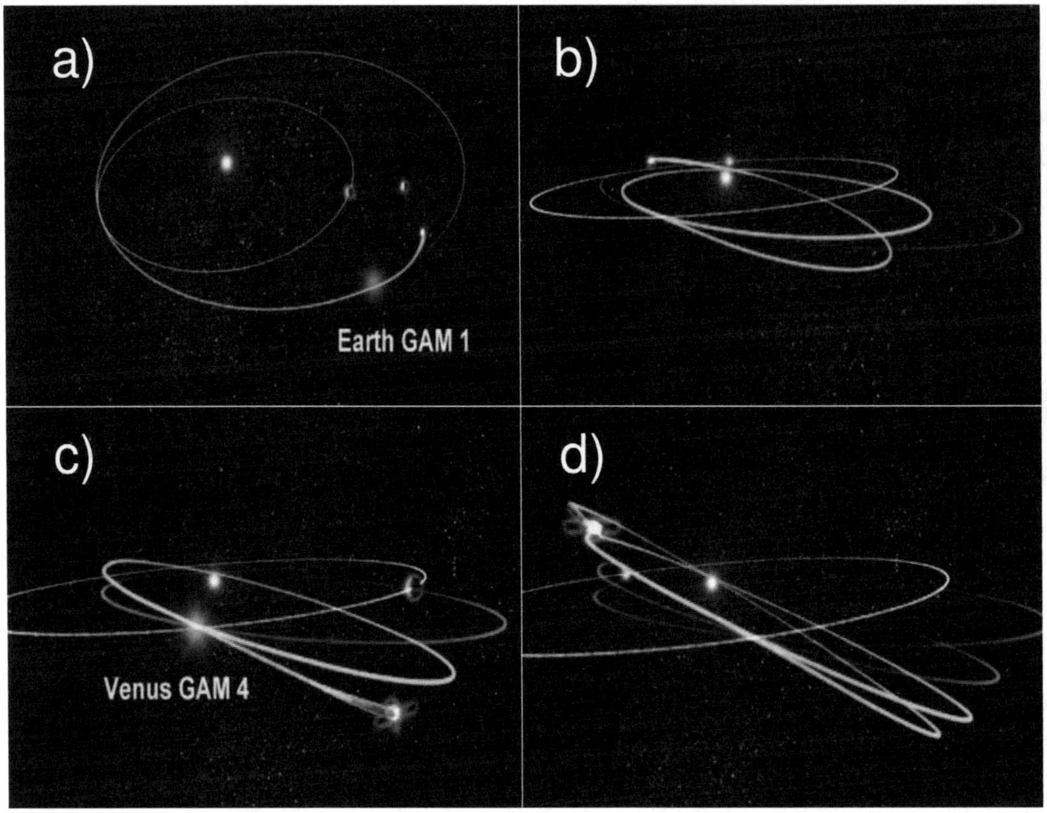

Figure 1. Sketch of different snapshots of Solar Orbiters journey: a) Thanks to a first Gravity Assist Manouvre (GAM) at Earth the spacecraft loses orbital energy and comes closer to the Sun. b) With the third GAM at Venus the operational orbit begins. The orbit is already inclined with respect to the ecliptic plane. c) The spacecraft is in a three-to-two resonance with Venus; every third orbit the inclination can be increased by Venus gravity assist. d) At the end of the operational orbit the inclination will be 34°. The material for these sketches is taken from the Solar Orbiter mission video by EADS/Astrium, which can be found at the ESA mission homepage: http://sci.esa.int/solarorbiter

The in-situ instrumentation package consists of the following instruments:

- *Energetic Particle Detector* (EPD) EPD will measure the properties of suprathermal and energetic particles. Scientific topics to be addressed include the sources, acceleration mechanisms, and transport processes of these particles. Principal Investigator of EPD is Dr. Javier Rodrguez-Pacheco, University of Alcala, Spain.
- *Magnetometer* (MAG) The magnetometer will provide in-situ measurements of the heliospheric magnetic field. This will facilitate detailed studies into the way the Sun's magnetic field links into space and evolves over the solar cycle; how particles are accelerated and propagate around the solar system, including to the Earth; how the corona and solar wind are heated and accelerated. Principal Investigator of MAG is Dr. Tim Horbury, Imperial College London, United Kingdom.
- *Radio and Plasma Waves* (RPW) The RPW experiment is unique amongst the *Solar Orbiter* instruments in that it makes both in-situ and remote sensing measurements. RPW will measure magnetic and electric fields at high time resolution using a number

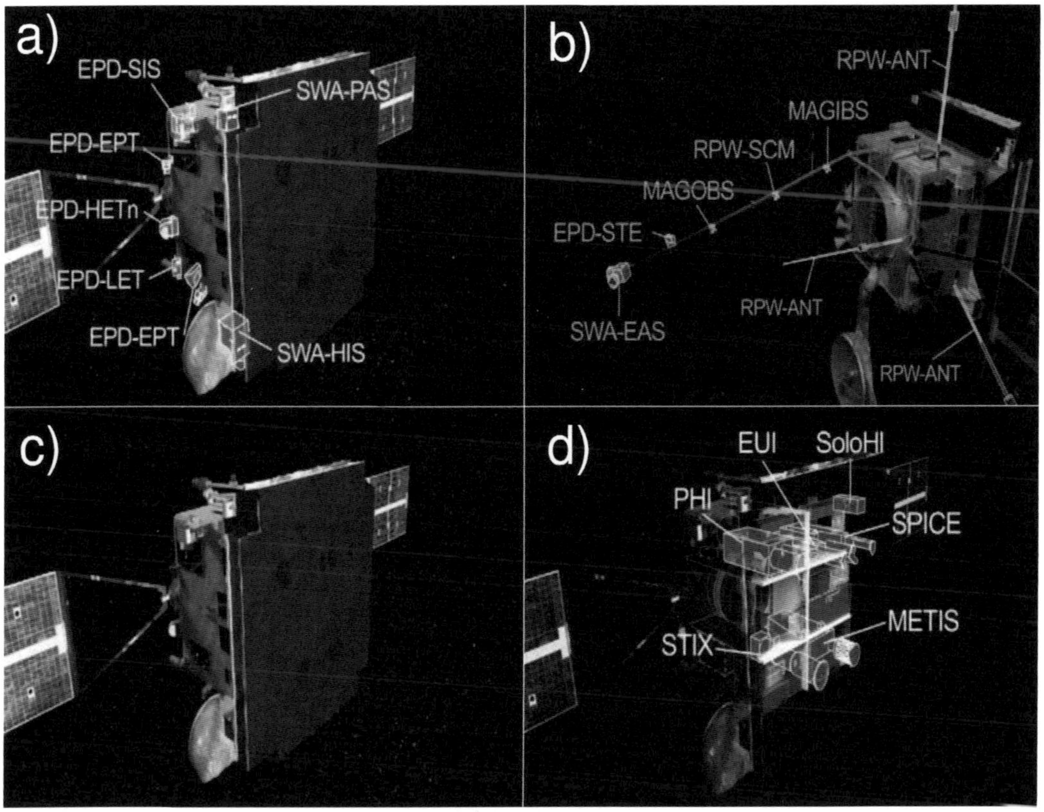

Figure 2. Payload suite of the *Solar Orbiter* spacecraft: a) The in-situ instruments are arranged at the periphery of the spacecraft body. b) Instruments needing clean electromagnetic environment are mounted on dedicated booms on the backside of the spacecraft. c): View of the heat-shield assembly, which protects the spacecraft from the intense solar radiation at perihelion passages. d) Sketch of the remote sensing instrument package, which is arranged behind dedicated feedthroughs in the heat-shield. Optical instrument like PHI are protected by heat-rejecting entrance windows. The material for these sketches is taken from the Solar Orbiter mission video by EADS/Astrium, which can be found at the ESA mission homepage: http://sci.esa.int/solarorbiter

of sensors/antennas, to determine the characteristics of electromagnetic and electrostatic waves in the solar wind. Principal Investigator of RPW is Dr. Milan Maksimovic, LESIA, Observatoire de Paris, France.

- *Solar Wind Plasma Analyser* (SWA) The Solar Wind Plasma Analyser, SWA, consists of a suite of sensors that will measure the density, velocity, and temperature of solar wind ions and electrons, thereby characterising the solar wind between 0.28 and 1.4 AU from the Sun. In addition to determining the bulk properties of the wind, SWA will provide measurements of solar wind ion composition for key elements (e.g. the C, N, O group and Fe, Si or Mg). Principal Investigator of SWA is Dr. Christopher Owen, Mullard Space Science Laboratory, United Kingdom.
- *Suprathermal Ion Spectrograph* (part of EPD) This experiment will measure energetic particles ejected from the Sun. Data will be compared to other solar and interplanetary processes to understand solar system space weather. Understanding the connections

between the Sun and its planets will allow better prediction of the impacts of solar activity on humans, technological systems and even the presence of life itself in the universe. Principal investigator of the Suprathermal Ion Spectrograph, which is funded by NASA, is Dr. Glenn Mason, Applied Physics Laboratory in Columbia, Maryland, USA.

In addition to the in-situ instruments the *Solar Orbiter* instrumentation comprises a suite of remote sensing instruments:

- *Extreme Ultraviolet Imager* (EUI) EUI will provide image sequences of the solar atmospheric layers above the photosphere, thereby providing an indispensable link between the solar surface and outer corona that ultimately shapes the characteristics of the interplanetary medium. EUI will also provide the first-ever images of the Sun from an out-of-ecliptic viewpoint (up to 34° of solar latitude during the extended mission phase). Principal Investigator of EUI is Dr. Pierre Rochus, CSL, Belgium.
- *Coronagraph* (METIS/COR) METIS/COR will simultaneously image the visible and ultraviolet emission of the solar corona and diagnose, with unprecedented temporal coverage and spatial resolution, the structure and dynamics of the full corona in the range from 1.2 to 3.0 (from 1.6 to 4.1) solar radii from Sun centre, at minimum (maximum) perihelion during the nominal mission. This is a region that is crucial in linking the solar atmospheric phenomena to their evolution in the inner heliosphere. Principal Investigator of METIS/COR is Dr. Ester Antonucci, INAF- Astronomical Observatory of Turin, Italy.
- *Polarimetric and Helioseismic Imager* (PHI) The Polarimetric and Helioseismic Imager, PHI, will provide high-resolution and full-disk measurements of the photospheric vector magnetic field and line-of-sight (LOS) velocity as well as the continuum intensity in the visible wavelength range. The LOS velocity maps will have the accuracy and stability to allow detailed helioseismic investigations of the solar interior, in particular of the solar convection zone. Principal Investigator of PHI is Dr. Sami Solanki, Max-Planck-Institut für Sonnensystemforschung, Germany.
- *Heliospheric Imager* (SoloHI) This instrument will provide revolutionary measurements to pinpoint coronal mass ejections or CMEs. To this end the instrument is designed as a low stray light, wide angle, visible camera. Principal Investigator of SoloHI (funded by NASA) is Dr. Russell A. Howard, US Naval Research Laboratory, USA.
- *EUV Spectrometer* (SPICE) This instrument will provide an extreme ultraviolet spectrometer or optical instrument that will measure different wavelengths of light emitted from the sun. Data will advance our understanding of the various dynamics of the sun to better understand the affects on Earth and the solar system. Principal Investigator of SPICE (funded by NASA) is Dr. Don Hassler, Southwest Research Institute, Boulder, USA.
- *X-ray Imager* (STIX) STIX provides imaging spectroscopy of solar thermal and non-thermal X-ray emission. STIX will provide quantitative information on the timing, location, intensity, and spectra of accelerated electrons as well as of high temperature thermal plasmas, mostly associated with flares and/or microflares. Principal Investigator of STIX is Dr. Arnold O. Benz, Institute of Astronomy, ETH Zurich, Switzerland.

Note that the high resolution instruments are all designed to observe the same target region on the solar surface with an identical angular sampling of 0.5 arcsec per pixel. This is of fundamental importance to address the magnetic coupling between the different atmospheric layers, which will be seen using the different instruments.

4. Solar Orbiter Polarimetric and Helioseismic Imager SO/PHI

The instrument, which harbors the greatest potential for helioseismology and the studies of magnetic fields and (sub-)surface flows in the photopheric layers, is the visible light imager and magnetograph, called *Polarimetric and Helioseismic Imager.*

4.1. Science Goals

The *Polarimetric and Helioseismic Imager* PHI onboard *Solar Orbiter* obtains information on gas flows/motions and vector magnetic fields in a two-dimensional field-of-view on the visible solar surface. It will thus probe the deepest layers of the Sun (including the solar interior by helioseismology) of all the instruments on *Solar Orbiter.* Since the magnetic field anchored at the solar surface produces most of the structures and energetic events in the upper solar atmosphere and significantly influences the heliosphere, PHI plays a key role in reaching the science goals of *Solar Orbiter.* Extrapolations of the magnetic field observed by PHI into the Sun's upper atmosphere and heliosphere will provide the information needed for other optical and in-situ instruments to analyse and understand the data recorded by them in a proper physical context.

4.2. Measurement principle

PHI makes use of the Doppler- and Zeeman-effects in a single selected spectral line of neutral iron at 617.3 nm. To retrieve the encoded physical information it must measure two-dimensional intensity maps at six wavelength points within this line, while measuring four polarisation states at each wavelength point.

4.3. Instrument Concept

PHI is a diffraction limited, wavelength tunable, quasi-monochromatic, polarisation sensitive imager.

4.4. Instrument Implementation

PHI consists of two telescopes, which feed one filtergraph and one focal plane array: The High Resolution Telescope (HRT) will provide a restricted FOV of 16.8 arcmin squared and achieve a spatial resolution that, near the closest perihelion pass, will be about 200 km on the Sun.

It is designed as an off-axis Ritchey-Chrétien telescope with a decentered pupil of 140 mm diameter.

The Full Disk Telescope (FDT), with a FOV of 2.1° squared and a pixel size of 730 km (at 0.28 AU), will provide a complete view of the full solar disk during all orbital phases.

The FDT is designed as a refractive telescope. The two telescopes are used sequentially and their selection is made by a feed selection mechanism.

Both telescope apertures are protected from intense solar flux by special heat-rejecting entrance windows, which are part of the heat-shield assembly of the spacecraft. They are purely dielectric broad-band reflectors with a narrow notch in the reflectivity curve around the science wavelength of the instruments. With more than 80% transmittance at the science wavelength, in combination with almost perfect blocking from 200 nm to the far infrared, the heat load into the instruments can be effectively decreased, while preserving the high photometric and polarimetric accuracy of PHI.

The filtergraph unit FG is based on heritage from the Imaging Magnetograph eXperiment (IMaX, Martínez Pillet et al. 2010) onboard the successful Sunrise balloon-borne observatory (Barthol et al. 2010): A $LiNbO_3$ etalon in a telecentric configuration selects a passband of 100mÅ width. Applying a voltage across the crystal allows changing the refractive index of the material, and thus tuning the passband in wavelength across the spectral line. A 3 Å wide prefilter acts as an order sorter for the Fabry-Pérot channel spectrum. The polarimetric analysis

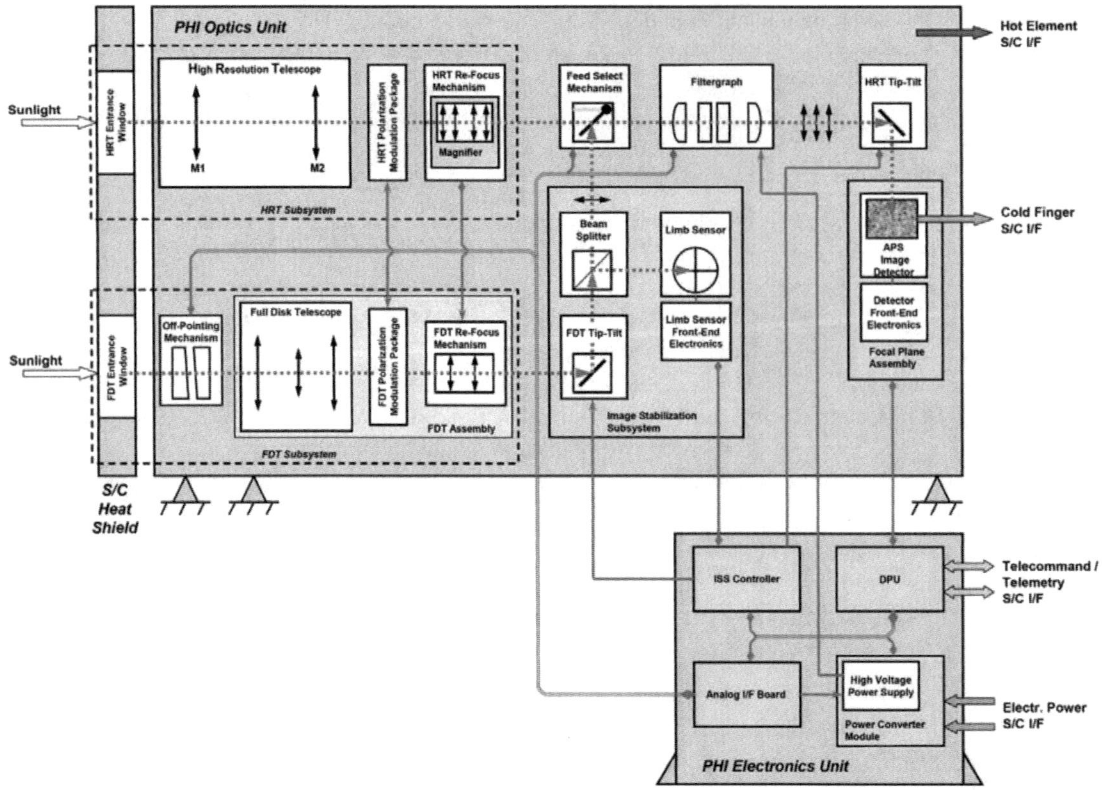

Figure 3. Functional diagram of the PHI instrument (prepared by R. Meller, MPS).

is perfomed by two Polarisation Modulation Packages (PMP) in each of the telescopes. Each PMP consists of two nematic liquid crystal retarders, followed by a linear polariser as analyser. The modulation scheme is the same as the one used in IMaX (Martínez Pillet et al. 2010).

In addition to these instrument units there are a number of other functional systems: An internal image stabilisation system based on a fast steerable mirror greatly reduces residual pointing error by the spacecraft to levels compatible with high resolution polarimetry. The error signal is derived from intensity differences within a full solar image on a quadrant diode. This image is provided also by the FDT via a beam splitter. An off-pointing mechanism in front of the FDT ensures that the full disk image is always centered on the quadrant cell, even when the other instruments, and thus the spacecraft, point to the limb of the Sun.

The focal plane assembly is built around a 2048 by 2048 pixel Active Pixel Sensor (APS), which is especially designed and manufactured for the instrument. It will deliver 10 frames per second which are read out in synchronism with the switching of the polarisation modulators.

The limited telemetry rate and the large amount of scientific information retrieved from the PHI instrument demand a sophisticated on-board data reduction. The measurement technique of PHI, i.e. the determination of the full Stokes vector at several wavelengths, is ideally suited to apply a robust and reliable technique to obtain maps of the physical quantities magnetic field strength and direction, filling factor, line-of-sight velocity and, continuum brightness. A non-linear, least-square, inversion technique is used, numerically solving the radiative transfer equation on board.

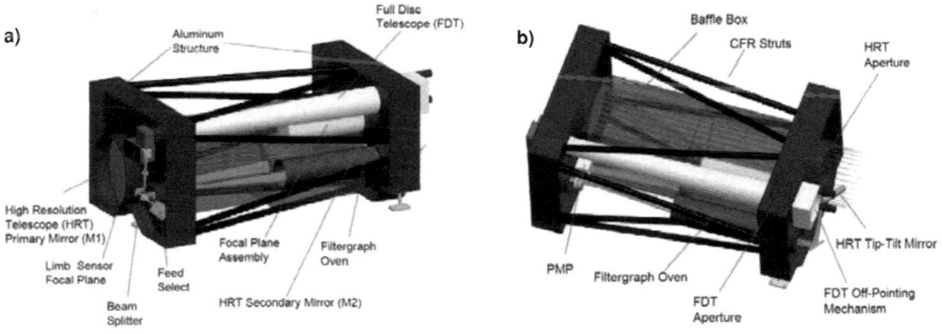

Figure 4. CAD model sketch of the PHI instrument: a) as seen from the backside, b) as seen from the front side of the instrument.

5. Outlook

Solar Orbiter will provide unique opportunities to study the Sun, from its interior to the inner heliosphere. A core instrument is the Polarimetric and Helioseismic Imager PHI, which will allow the vector magnetic field and gas flows/motions to be observed at high spatial resolution, thanks to the close-by observing conditions during the perihelion passages. PHI will for the first time ever provide measurements of the polar magnetic fields, that can not be studied from within the ecliptic plane: Although the polar regions are sometimes visible also from Earth (thanks to the inclination of the solar rotation axis of 7°), a quantitative measurement of the polar field is hindered by the strong angle dependence of the Zeeman effect, which reduces the polarised signal from polar fields to values below the noise level. Only thanks to the unique polar view from an orbit inclined by as much as 34° a reliable estimate on the polar flux and the flux transport to the poles during late phases of the activity cycle, a crucial observation for understanding the solar dynamo, can be obtained.

By co-observations with other instruments on the ground and in space, PHI also promises to probe, for the first time ever, the concept of *stereoscopic helioseismology*, by combining Dopplergrams of the same target region on the solar surface as seen from two different viewing directions. Although this has to be further studied and simulated in detail, *Solar Orbiter* and its Polarimetric and Helioseismic Imager have the potential to establish this new diagnostic technique.

References

[1] http://sci.esa.int/solarorbiter
[2] Barthol P, Gandorfer A, Solanki S, et al. 2010 *Solar Physics* in press
[3] Martinez Pillet V, del Toro Iniesta J C, Álvarez Herrero A, et al. 2010 *Solar Physics* in press

AUTHOR INDEX

AUTHOR INDEX

Institute of Physics
Dirac House, Temple Back
Bristol BS1 6BE UK

proceedings.com